Protein Conformation as an Immunological Signal

Protein Conformation as an Immunological Signal

Edited by

Franco Celada

University of Genoa
Genoa, Italy

Verne N. Schumaker

University of California
Los Angeles, California

and

Eli E. Sercarz

University of California
Los Angeles, California

Plenum Press • New York and London

Library of Congress Cataloging in Publication Data
Main entry under title:

Protein conformation as an immunological signal.

"Proceedings of an EMBO workshop...held October 1–4, 1981, in Portovenere (La Spezia) Italy"—T.p. verso.
  Bibliography: p.
  Includes index.
  1. Antigens—Congresses. 2. Immunoglobulins—Congresses. 3. Complement (Immunology)—Congresses. 4. Lymphocytes—Congresses. 5. Immunochemistry—Congresses. I. Celada, Franco. II. Schumaker, Verne N. III. Sercarz, Eli E. IV. European Molecular Biology Organization. |DNLM: 1. Protein conformation—Congresses. 2. Antigens. Immune response—Congresses. 3. Antigen–antibody reactions—Congresses. 4. Complement—Congresses. QU 55 P96615 1981|
QR186.5.P76 1983   616.7′9   83-13999
ISBN-13: 978-1-4613-3780-5   e-ISBN-13: 978-1-4613-3778-2
DOI: 10.1007/978-1-4613-3778-2

Proceedings of an EMBO workshop on
Protein Conformation as an Immunological Signal,
held October 1–4, 1981,
in Portovenere (La Spezia) Italy

© 1983 Plenum Press, New York
Softcover reprint of the hardcover 1st edition 1983
A Division of Plenum Publishing Corporation
233 Spring Street, New York, N.Y. 10013

PREFACE

This volume is the collection of papers presented during a four day meeting, the EMBO workshop "Protein Conformation as an Immunological Signal" that took place at Portovenere (La Spezia), Italy, October 1-4, 1981.

The motivation that drove us to organize this meeting was the feeling that distinct groups of researchers, active in key areas of modern immunology, sometimes fail to communicate with each other simply because of different traditional affiliations. Yet it is urgent that "molecular" and "cellular" people cooperate more if immunology is to continue the exportation of new concepts to other disciplines. In fact, the deep meaning of molecule-molecule and cell-cell interaction, the generation of signals and their effective transmission which results in elicitation, control or suppression of responses cannot be unraveled without the experts on antibody structure or complement activation sharing their views with the experts on T cell, B cell and macrophage membrane receptors as well as the experts on factors that carry the information released by these cells.

Whether the meeting was scientifically fruitful, the reader can judge after having digested these pages. We, the organizers, are not sure whether the optimal amount of interaction had taken place; especially considering how hard it is to overcome the scientist's catch 22: You have to know something quite well before you get really interested in it. In any event, we are convinced that Portovenere was one of the most successful attempts we have witnessed. And then there was the scenery: the wide and astonishing beauty of the sea breaking against the rocks, the Gothic village with its striped churches, the towering castle that glittered under sudden showers, the madrigals sung in the chapel, against the background of the roar of the waves.

F. Celada
V. N. Schumaker
E. E. Sercarz

v

ACKNOWLEDGMENTS

The organizers would like to express their gratitude to the European Molecular Biology Organization for the substantial contribution to this meeting, as well as to thank Sandoz, Ltd. and Merck and Co. for their valuable contributions. We especially thank three local Italian governmental bodies: the Regione Liguria, the Provincia di La Spezia, and the Commune di Portovenere, for providing generous support as well as the marvellous concert.

The meeting and this monograph would have been literally impossible without the inspired help and organization furnished by Anna Rubartelli and Louisa Di Rosa in Genoa, and Jeff Kissinger and Vicky Godoy in Los Angeles.

CONTENTS

TOPIC VII.   THE RECOGNITION BY T CELLS OF PROTEIN ANTIGENS

TOPIC IX. EFFECTOR MECHANISMS FOR CELL SIGNALING RESULTING
FROM ANTIGEN-ANTIBODY INTERACTIONS

I: INFLUENCE OF ANTIGEN UPON CONFORMATION OF ANTIBODY

INTRODUCTION

A central question of molecular immunology is "How does the binding of antigen to antibody induce the immune response?" In order to explore this question at the molecular level, three models have been devised[1,2] for the generation of the immunologic signal between the Fab and Fc regions of antibodies: 1) the allosteric model; 2) the distortive model; and 3) the associative model. For IgG, using complement activation as a measure of immune response, most of the evidence indicates that association is a necessary step for signal generation. What is not clear, however, is whether association of IgG is the sufficient step for complement activation or whether a conformational change in IgG structure induced by antigen must also accompany the association. For IgM, on the other hand, a conformational change appears to be essential for complement activation, since a single molecule of IgM acquires the ability to bind and activate Cl when it becomes attached to a cell surface through two or more combining sites. IgM is already a pentameric molecule, and this "built-in" association also appears to be a necessary feature for efficient complement activation. Therefore, signal generation in IgM has been described as a mixture of models, associative, but also requiring a conformational change.

There is direct evidence that at least small conformational changes occur when hapten is bound to antibody; however, these changes in structure may be restricted to the readjustment of the variable domain to the better binding conformation, and their biological significance is not clear. In the first paper of Topic I, Israel Pecht reviews the body of evidence from many spectroscopic studies of the kinetics of hapten binding to IgA and IgM to show

1

that binding occurs in two steps with distinct lifetimes, one of
which he suggests is associated with a conformational transition in
protein structure. Dr. Pecht suggests that the common features of
this conformational transition may involve relative motion between
the heavy and light chain domains and, thus, extend throughout the
variable region of the Fab and, perhaps, into the constant domains.

Evidence for more extensive changes in antibody structure upon
hapten binding is presented by Peter Zavodszky and his colleagues
in the second paper, in which results are presented of deuterium
exchange, magnetic resonance, calorimetric and light scattering
measurements on rabbit anti-SIII pneumococcal polysaccharide IgG
antibody before and after interaction with antigen. The results are
interpreted in terms of a dynamic ensemble of rapidly interconverting
conformations, a model for protein structure in solution which must
complement and extend the static, statistically-averaged structure
deduced from protein crystallography. The equilibrium of these
conformational transitions can be affected by a site-filling monovalent ligand, and the authors address the question of whether concerted conformational fluctuations can serve as a signal-transmitting
mechanism over large distances within a macromolecule.

A new approach to the detection of conformational change in
immune complexes is presented in the third paper by Parodi and his
colleagues. They have used monoclonal antibodies in an attempt to
detect new determinants which might appear upon heat-induced association of immunoglobulins. They found significant increases in binding affinity of the monoclonal antibodies when aggregated antigens
were compared to monomeric antigens, but no really "new" determinants.

In summary, the evidence for small conformational changes in
antibody structure upon hapten binding is good, and there is evidence
for much more extensive alterations in dynamic structure. The use
of monoclonal antibodies as a valuable technique for the detection
of conformation changes also is presented. The biological significance of conformational change as a signal generating mechanism has
not been established for IgG. However, conformational changes appear
to be necessary for the activation of complement by IgM.

REFERENCES

1. H. Metzger, Effect of Antigen Binding on the Properties of
    Antibodies, Adv.Immunol. 18:164 (1974).
2. H. Metzger, The Effect of Antigen on Antibodies: Recent Studies,
    Contemp.Topics Molec.Immunol. 7:119 (1978).

ANTIBODY-HAPTEN BINDING KINETICS, CONFORMATIONAL

TRANSITIONS AND DOMAINS INTERACTIONS

Israel Pecht

The Weizmann Institute of Science
Rehovot
Israel

Upon binding antigen, gross changes primarily in the spatial relation of the Fab domains relative to that of the Fc are expected to occur in the antibody structure[1,2]. The question whether conformational changes are caused in the combining site has been of great concern in the context of understanding the mode of effector-function activation mechanism[3-5]. Therefore, the effect induced in antibodies by binding their haptens or even larger parts of their respective antigens, has been examined by a wide range physical and chemical methods[3-5]. The direct structural studies by crystallography failed to detect differences between the free combining sites and their complexes with haptens in the two cases examined to date[6-7]. In contrast several spectroscopic and other physical methods provided clear evidence for structural changes induced by hapten binding[8-10]. In the following the results of the time resolved analysis of antibody-hapten reaction will be described in detail. The kinetic studies of these reactions provided a very detailed insight into the elementary steps of hapten recognition by the antibody combining site. Furthermore, conclusive evidence for a conformational change induced upon hapten binding emerged from these investigations[11-14].

The pioneering studies of the kinetics of antibody-hapten interactions[15-17] were done on normally induced, heterogeneous antibody preparations and were satisfactorily described by the simple, single step association-dissociation equilibrium:

$$Ab + H \rightleftharpoons AbH$$

The first kinetic examination of a homogeneous antibody reacting with its hapten, namely the IgA secreted by the plasmacytoma MOPC 315 and dinitrophenyl haptens has already indicated a more complex sequence of events though it failed to show evidence of a conformational transition[18,19]. A systematic correlation has been carried out for the above immunoglobulin, between the kinetic parameters of association and dissociation and the structure of some fifty odd derivatives of the hapten[20]. This correlation has shown that little variation occurs in the rate of binding of the hapten and that it is best described by a diffusion controlled encounter complex formation which usually undergoes a very fast conversion into the final complex:

Ab + H $\rightleftharpoons$ [AbH] $\rightleftharpoons$ AbH

(For a detailed description and discussion of this problem see[14]).

For the dissociation step of the AbH complex a rather large dependence of the rate on the structure of the hapten was observed. Thus the lifetime of the complex dictates its overall stability. For the case of protein 315 an interesting hypothetic model of the combining site has been proposed from the kinetic data[20]. This structure has later been further examined by building a three dimensional model of it and by detailed nuclear magnetic resonance studies[21,22]. As a crystallographic study of this protein is now in progress, probably, an unambiguous test of the proposed structural features of that combining site will be feasible in the near future.

With hardly any exception, all homogeneous antibody-hapten systems examined, were shown to undergo a conformational transition induced by binding the hapten[11-14,23]. (Even protein 315 which has recently been re-examined using an improved resolution data analysis system was found to exhibit a rather small but clearly resolved second relaxation step representing the above transition[24]). A common mechanism for the reaction of an antibody and its antigen has now been established from the kinetic studies of an increasing number of systems, differing in their specificity and immunoglobulin class[23].

The experimental approach adopted in essentially all our studies has been that of the chemical relaxation temperature jump method[25]. The equilibrium between the antibody and its hapten is perturbed by inducing a fast change in temperature (2 to 4 degrees within a few microseconds or even less depending on the specific method used for heating, cf. reference[25] for further details). In Figure 1 the relaxation of the reactants to the appropriate concentrations dictated by the new temperature can then be monitored through given probes of the system. Most of the reactions discussed were studied using the intrinsic fluorescence of the antibodies or by conjugating the haptens with suitable probes[26]. Characteristic for almost all the systems which we have investigated is the observation of two, usually well resolved relaxation times. The time domains in which these appear, depend on the particular system studied and on the concentration of the reagents. Still, characteristic for the faster of the two relaxation is the time scale of hundreds of microseconds to several dozens of milliseconds, while the slower relaxation may extend into the range of hundreds of milliseconds. The concentration dependence of the two relaxations are also distinct. The fast one shows a linear dependence on the reactants concentration, whereas the slower is either concentration independent or shows a limiting dependence only. Thus, the fast relaxation has already been assigned in preliminary analysis of the data to the association-dissociation step between the antibody and the hapten and the second relaxation to a monomolecular process, involving the antibody molecule or its hapten complex.

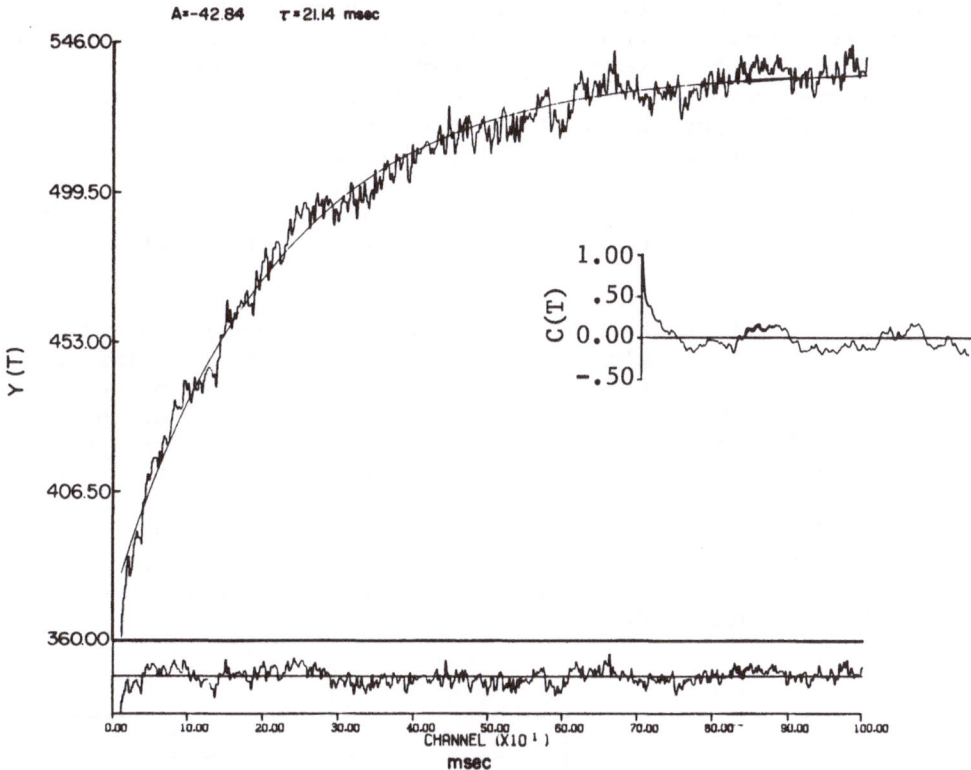

Fig. 1.   A temperature-jump induced chemical-relaxation trace of the conformational transition of hapten-free immunoglobulin. The 1μm solution of the heterologous chain recombinant produced from α chain of protein 460 and λ chain of protein 315 in PBS, was heated by 5.2°C to 7°C. Protein intrinsic fluoresence excited at 280mm was monitored through a cutoff filter above 320nm. The experimental trace is a result of accumulated 3 individual jumps. The data were fitted to a single exponent drawn in as the smooth line. The auto-correlation function is shown in the right top insert and the difference between the experimental data and the fitted curve is given in the lower insert.

The detailed examination of all systems in terms of concentration dependences of both the relaxation times and their amplitudes, has indeed shown that a conformational transition is taking place[11-14,27].

In the following a few illustrative cases will be described. The IgA secreted by MOPC 460 is specific for nitroaromatic haptens and has been the first case where two relaxation times have been observed. In analysing the kinetics of binding to ε-N-(2,4 dinitrophenyl)-L-lysine, several mechanistic schemes have been considered[11]. The two simpler mechanisms which involve a transition occurring only in either the free antibody or in its complex, (Figure 2, Schemes 1 and 2) did not yield satisfactory agreement between the experimental data and the theoretically calculated parameters. Hence in Scheme 3 the more complex scheme involving a conformational change in both, free and hapten-bound antibody has been examined. This scheme was found to provide a good fit with all experimental data. Namely, the combined analysis of the concentration dependence of both relaxation time and amplitudes resulted in a set of parameters describing the kinetic and thermodynamic behavior of the system which was internally consistent and adequately fitting the experimental results[11,28] (Table 1).

To examine the generality of the observed mechanism several other homogeneous immunoglobulins have been examined in a similar fashion. A series of murine IgA molecules, all shown to be specific for galactan was investigated[29]. The binding equilibria of the hapten β-D(1-6)-galactotriose to the homogeneous IgA secreted by plasmacytomas X-24, J-539 and T-601 were investigated by the T-jump chemical relaxation method. All antibody hapten equilibria were found to exhibit two well resolved relaxation times and the analysis of all the results has shown that one and the same mechanism describes satisfactorily the behavior of the three different proteins binding the same hapten[12].

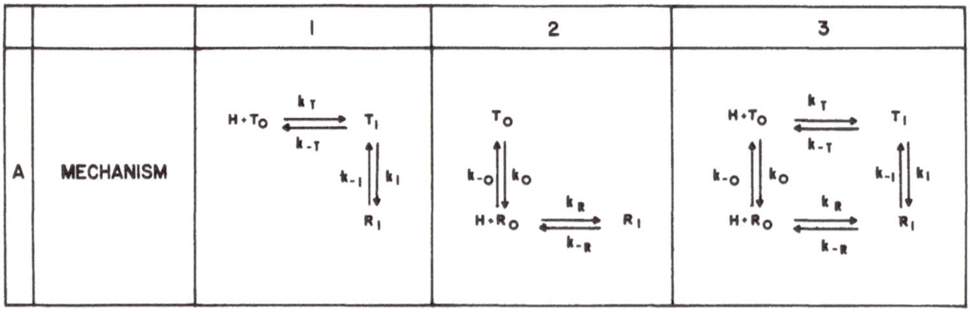

		1	2	3
A	MECHANISM	$H \cdot T_O \underset{k_{-T}}{\overset{k_T}{\rightleftharpoons}} T_I$ $\quad k_{-I} \uparrow\downarrow k_I$ $\quad\quad R_I$	T_O $k_{-O} \uparrow\downarrow k_O$ $H \cdot R_O \underset{k_{-R}}{\overset{k_R}{\rightleftharpoons}} R_I$	$H \cdot T_O \underset{k_{-T}}{\overset{k_T}{\rightleftharpoons}} T_I$ $k_{-O} \uparrow\downarrow k_O \quad k_{-I} \uparrow\downarrow k_I$ $H \cdot R_O \underset{k_{-R}}{\overset{k_R}{\rightleftharpoons}} R_I$

Fig. 2

Kinetic and Thermodynamic Parameters for the Reaction of the Immunoglobulins and Their Hybrids with Haptens[a]

Parameter	X 24[b]	J 539[b]	T 601[c]	H24L539[c]	H24L601[c]	H601L24[c]	M 460[d]	H460L315e
$K(M^{-1})$	1.2×10^5	1.2×10^5	1.0×10^5	2.9×10^5	2.1×10^5	1.2×10^5	6.5×10^4	1.6×10^4
$K_T(M^{-1})$	4.5×10^4	5.0×10^4	2.0×10^4	1.0×10^5	9.0×10^3	2.3×10^4	1.9×10^4	1.1×10^4
$K_R(M^{-1})$	2.7×10^5	2.0×10^5	1.6×10^6	6.3×10^5	5.4×10^5	5.0×10^5	2.2×10^5	2.2×10^4
K_O	0.50	0.93	0.05	0.55	0.60	0.25	0.3	0.81
K_1	3.0	3.7	4.3	3.3	36.5	5.6	3.4	1.1
ΔH(kcal/mol)	-10.6	-11.3	-5.4	-10.2	-6.2	-5.1	-15.8	-10.4
ΔH_T(kcal/mol)	-10.6	-12.8	-6.2	-10.8	-14.6	-5.0	-14.1	-15.1
ΔH_R(kcal/mol)	-10.5	-10.6	-3.6	-9.8	-6.5	-5.5	-15.0	-4.9
ΔH_O(kcal/mol)	0.01	-0.63	-1.75	-0.46	0.91	0.50	0.64	-9.2
ΔH_1(kcal/mol)	0.00	0.53	0.81	0.57	9.00	-0.04	-0.26	0.97
ΔF	0.25	0.20	0.22	0.24	0.16	0.21	-0.5	-0.07
ΔF_T	0.31	0.20	0.26	0.25	0.25	0.26	-0.47	-0.11
ΔF_O	0.41	0.30	0.22	0.44	0.26	0.26	0.02	-0.07
ΔF_1	0.10	0.009	-0.03	0.19	0.007	0.002	-0.01	0.01
$k_T(M^{-1}sec^{-1})$	4.5×10^5	1.2×10^6	1.2×10^6	2.1×10^6	4.7×10^5	1.2×10^6	1.3×10^8	8.8×10^6
$k_{-T}(sec^{-1})$	10.0	24.8	60.9	20.0	52.0	54.7	6.3×10^3	886
$k_O(sec^{-1})$	1.05	1.77	0.78	0.50	2.90	2.36	29	24
$k_{-O}(sec^{-1})$	2.09	1.91	14.20	0.91	4.90	9.37	96	29.6
$k_1(sec^{-1})$	0.30	5.72	11.40	5.36	33.30	0.06	44	66
$k_{-1}(sec^{-1})$	0.10	1.56	2.65	1.64	0.91	0.01	13	40

[a]Enthalpy and fluorescence parameters for the binding of X 24 were obtained by us recently from the data of Vuk-Pavlovic et al. (1978) using more powerful methods of analysis. All data were obtained at 25°C pH 7.4 phosphate buffered saline. We estimated the error for most parameters at ±20%. For ΔH_O, ΔH_1, ΔF_O, and ΔF_1 the estimated error is ±50%. bfrom ref.12, cfrom ref.13, dfrom ref.11, efrom ref.14.

Another saccharide-specific antibody examined was of a different immunoglobulin class, namely IgM. This was the α-(1-3) dextran binding protein secreted by MOPC-104E[30]. The reactions of that antibody with several α-D-(1-3) glucopyranosyl haptens of increasing chain length (all derived from nigeran) also showed two relaxation times which were fitting to the above scheme. The detailed analysis of that system is of particular interest as it may provide an insight into the correlation between the complementarity of hapten and the conformational transition[31-32].

Examination of the specific rates summarized in Table 1 shows the marked difference between those observed for saccharide binders and the nitroaromatic binding antibodies. The slowing down of all rates in the saccharide binders has also been observed with lectins[33] or hapten which are peptides[34,35]. It probably reflects the high flexibility of the ligand which has therefore a large number of conformers from which the one fitting into the site has to be selected. Also the saccharide haptens and their respective sites are hydrated one would assume that it would be time consuming to disrupt and form the multiple bounds required for the formation of the protein-saccharide complex (cf. reference 27 for a detailed discussion of that aspect).

Now the question arises as to whether the observed common mechanism obeyed by all the above described systems is inherent to any combination of immunoglobulins heavy and light chains which maintains the binding specificity for a given antigen. In order to answer that question, the heterologous chain recombinants of the group of galactan binding immunoglobulins[36] have been studied. All those heterologous recombinants examined were found to exhibit a similar kinetic pattern[13]. Again the data fitted the scheme described above for their parental proteins. The results obtained with a group of galactan binders allowed a comparison between them and showed that for all proteins small positive free energy differences separate the two conformational states of the free Ig[13]. Conversely, the energy difference between the two conformers of the hapten bound Ig is in all cases negative. Therefore the equilibrium between the T and R conformers is shifted upon hapten binding making R the predominant species (cf. Figure 2).

An interesting attempt to correlate the binding and kinetic parameters for the hybrids with those of their respective parent protein has been made[13]. This was done in order to resolve a predominant role of the heavy or light chains. Only for the two isomerization rates of the bound hybrids (k_1 and k_{-1}) was correlation observed, showing that the values are very close to those of the parental light chain donor. This result was somewhat unexpected because the light chain of J-539 differs from those of T-601 and X-24 at five and six positions respectively[13]. Recently, hypothetic three-dimensional models were constructed for this group of galactan

binders[37],[38]. Examination of these structures revealed that several
of the more important residue substitutions among these proteins are
at the V_H-V_L contact areas. Furthermore some of the residues that
form the contact between the domains are subjected to a large extent
of variation as they belong to the J and D segments of the immunoglobulins. Thus the residues responsible for the domains' contact
formation are selected at the gene level by the combination of the
J_H and D segments to V_H and of J_L to V_L [39]. This preferential selection of H-L combination is expected to affect not only this association but also their mode of antigen binding and the above described
conformational transition induced by it. Conversely, the formation
of a hybrid which maintains the hapten-binding capacity suggests
that the amino acid substitutions are such that the structural and
functional requirements of the newly formed site are maintained.

A further illustration of the above case has recently been
provided by the preparation and the properties of another heterologous
immunoglobulin chain recombinant[14]. This has been prepared from the
λ chain derived from MOPC-460 and the α chain of MOPC-315. It binds
ϵ-N-(2,4 dinitrophenyl)-L-lysine, which is also the specific hapten
binding to both parental proteins[40]. However the properties of the
hybrid combining site are distinct from those of either parent.
Still the same reaction mechanism which has been found to be common
to all other systems previously studied, has been found to be obeyed
by that hybrid[14].

The prevalence of a common reaction mechanism, being expressed
across the sequence variation and the heterologous recombination of
the chains, constitutes strong evidence for the hapten-induced conformational transition being an inherent property of the tertiary
domain structure of the immunoglobulin molecules. The variation
found in the reaction rates are most probably the reflection of the
modulation effected by the structural variance. A large number of
antibody-hapten systems have been investigated kinetically over the
last decade. The homogeneous ones of these have practically all been
found to exhibit a conformational transition linked to hapten binding. For the few heterogeneous antibody preparations that were examined, there is also evidence for the above-mentioned mechanism
being operative (eg. equine anti-lactose antibodies[41]). The extensively studied group of phosphorylcholine binding homogeneous antibodies (S-107, T-15, H-8) have also been examined kinetically.
Though only a single relaxation is observed, the detailed analysis
of the concentration dependence indicates a more complex mechanism
than the single-step binding process[42].

Two questions arise from the above: 1) What is the detailed
structural expression of such a conformational transition? 2) What
is its functional significance? No clear answer can be given to
either of these two questions at present. From the activation
parameters of the conformational transitions as well as from the

fact that it can be monitored via intrinsic changes of extinction and emission, one is drawn to conclude that at least the whole variable module is involved in it. The recent illustration of hapten binding-induced longitudinal interaction over the whole length of the light chain dimer of protein 315[43] constitutes a very instructive model for the extension of the above transitions into the first constant domain of the Fab. The study of Fd' interaction with L, C_L, and V_L[44] constitutes a clear illustration for the feasability of longitudinal interactions within the Fab. These interactions could be attained through changes in the lateral interactions between the domains[45].

The hapten binding linked conformational transition has now been shown to be the common, generally operative mechanism in hapten binding reactions. Still the functional implications of this mechanism remain to be established. The specific rates and activation parameters observed for the conformational transition of the range of different systems clearly support the occurrence of substantial rearrangement. An interesting illustration of the extent of such changes may be found in the study of the transition caused in trypsinogen upon binding the effector Ile-Val peptide. This peptide was shown to induce the transition of the p-guanidobenzoate zymogen into the trypsin-like structure[46]. This involves the conversion from the partially disordered conformation, characterizing trypsinogen, to a trypsin-like highly ordered structure[47]. The kinetic analysis of the reaction[48] yielded specific rates and activation parameters of very similar values to those observed with antibodies.

A minimal hypothesis accounting for this reaction pattern would be that it reflects the readjustment of the variable domain to the better binding conformation. In view of the limited energetic advantage of this step, one is inclined to prefer an alternative interpretation, namely, that it is involved in controlling effector function activity of the immunoglobulins. Notwithstanding the established requirements for cross-linking the antibodies by the di-or polyvalent antigen[5], the consistently observed structural transition may reflect a step in a finer regulation that has still to be evaluated.

REFERENCES

1. R.C. Valentine and N.M. Green, Electron Microscopy of an Antibody-Hapten Complex, J.Mol.Biol. 27:615 (1967).
2. A. Feinstein and E.A. Munn, Conformation of the Free and Antigen-bound IgM Antibody Molecules, Nature, 224:1307 (1969).
3. R.E. Cathou and K.J. Dorrington, The Conformation, Interaction, and Biological Roles of Immunoglobulin Subunits, in: "Subunits in Biological Systems," S.N. Timasheff and G.D. Fasman, eds., Part C, pp. 91-224. Decker, N.Y. (1975).

4. H. Metzger, Effect of Antigen Binding on the Properties of
    Antibody, Adv.Immunol. 18:169 (1974).
5. H. Metzger, The Effect of Antigen on Antibodies: Recent Studies,
    Contemp.Topics Molec.Immunol. 7:119 (1978).
6. D.M. Segal, E.A. Padlan, G.M. Cohen, S. Rudikoff, M. Potter,
    and D.R. Davies, The Three-Dimensional Structure of a
    Phosphorylcholine-Binding Mouse Immunoglobulin Fab and the
    Nature of the Antigen Binding Site, Proc.Natl.Acad.Sci. U.S.A.
    71:4298 (1974).
7. R.J. Poljak, Correlations Between Three-Dimensional Structure
    and Function of Immunoglobulins, CRC Crit.Rev.Biochem. 4:45
    (1978).
8. D.A. Holowka, A.D. Strasberg, J.W. Kimball, J.W.E. Haber, and
    R.E. Cathou, Changes in Intrinsic Circular Dichroism of Several
    Homogeneous Anti-Type III Pneumococcal Antibodies on Binding of
    a Small Hapten, Proc.Natl.Acad.Sci. U.S.A. 69:3399 (1972).
9. J. Schlessinger, Z.Z. Steinberg, D. Givol, J. Hochman, and I.
    Pecht, Antigen-induced Conformational Changes in Antibodies and
    their Fab Fragments Studied by Circular Polarization of Fluor-
    escence, Proc.Natl.Acad.Sci. U.S.A. 72:2775 (1975).
10. L.A. Tumerman, R.S. Nezlin, and V.D. Zagyansky, Increase of the
    Rotational Relaxation Time of Antibody Molecule After Complex
    Formation with Dansyl-Hapten, FEBS Lett. 19:290 (1972).
11. D. Lancet, and I. Pecht, Kinetic Evidence for Hapten-induced
    Conformational Transition in Immunoglobulin MOPC 460, Proc.
    Natl.Acad.Sci. U.S.A. 73:3549 (1976).
12. S. Vuk-Pavlovic, Y. Blatt, C.P.J. Glaudemans, and D. Lancet,
    Hapten-Linked Conformational Equilibria in Immunoglobulins
    XRPC-24 and J-539 Observed by Chemical Relaxation, Biophys.J.
    24:161 (1978).
13. R. Zidovetzki, Y. Blatt, C.P.J. Glaudemans, B.N. Manjula, and
    I. Pecht, A Common Mechanism of Hapten Binding to Immunoglobulins
    and Their Heterologous Chain Recombinants, Biochemistry 19:2790
    (1980).
14. R. Zidovetzki, Y. Blatt, and I. Pecht, Heterologous Immuno-
    globulin Chain Recombinant Carries a Distinct Site for Dinitro-
    phenyl and Obeys the Common Hapten Binding Mechanism.
    Biochemistry 20:5011 (1981).
15. A. Froese, A. Sehon, and M. Eigen, Kinetic Studies of Protein-
    Dye and Antibody-Hapten Interactions with the Temperature-Jump
    Method, Can.J.Chem. 40:1786 (1962).
16. A. Froese, A. Sehon, and M. Eigen, Kinetics and Equilibrium
    Studies of the Reaction Between Anti-p-nitrophenyl Antibodies
    and a Homologous Hapten. Immunochem. 2:135 (1965).
17. L.A. Day, J.M. Sturtevant, S.J. Singer, The Kinetics of the
    Reactions Between Antibodies to the 2,4-Dinitrophenyl Group and
    Specific Haptens, Ann. N.Y. Acad.Sci. 103:611 (1963).
18. I. Pecht, D. Givol, and M. Sela, Dynamics of Hapten-Antibody
    Interaction. Studies on a Myeloma Protein with Anti-2,4-Di-
    nitrophenyl Specificity, J.Mol.Biol. 68:241 (1972).

19. I. Pecht, D. Haselkorn, and S. Friedman, Kinetic Mapping of Antibody Binding Sites by Chemical Relaxation Spectroscopy, FEBS Lett. 24:331 (1972).

20. D. Haselkorn, S. Friedman, D. Givol, and I. Pecht, Kinetic Mapping of the Antibody Combining Site by Chemical Relaxation Spectrometry, Biochemistry 13:2210 (1974).

21. E.A. Padlan, D.R. Davies, I. Pecht, D. Givol, and C. Wright, Model-building Studies of Antigen-binding Sites: The Hapten Binding Site of MOPC-315. Cold Spring Harbor Symp.Quant.Biol. 41:627 (1976).

22. R.A. Dwek, S. Wain-Hobson, S. Dower, P. Gettins, B. Sutton, and S.J. Perkin, Structure of an Antibody Combining Site by Magnetic Resonance. Nature 266:31 (1977).

23. I. Pecht, Dynamic Aspects of Antibody Function, in:"The Antigens," vol. 6, M. Sela, ed., Academic Press, New York (In Press) (1981).

24. A. Oratore, and I. Pecht, (in preparation).

25. M. Eigen, and L. De Maeyer, Relaxation Methods, Tech.Chem. (N.Y.) Part 2, pp. 63-146 (1973).

26. I. Pecht, Insight into Mode of Antibody Action from Intrinsic and Extrinsic Fluorescent Probes, Ann. N.Y. Acad.Sci. 366:208 (1981).

27. I. Pecht, and D. Lancet, Kinetics of Antibody-Hapten Interactions, In:"Chemical Relaxation in Molecular Biology," I.Pecht and R. Rigler, eds., pp. 307-336, Springer Verlag, Berlin and New York, (1977).

28. D. Lancet, Ph.D. Thesis, The Weizmann Institute of Science, Israel, (1978).

29. B.N. Manjula, C.P.J. Glaudemans, E.B. Mashinshi, and M. Potter, Subunit Interactions in Mouse Myeloma Proteins with Anti-galactan Activity, Proc.Natl.Acad.Sci. U.S.A. 73:932 (1976).

30. G. Schepers, Y. Blatt, K. Himmelspach, and I. Pecht, Binding Site of a Dextran-specific Homogeneous IgM: Thermodynamic and Spectroscopic Mapping by Dansylated Oligosaccharides. Biochemistry 17:2239 (1978).

31. G. Schepers, Dissertation, University of Freibury, W. Germany, (1978).

32. G. Schepers, Y. Blatt, K. Himmelspach, and I. Pecht, (in preparation).

33. R.M. Clegg, F.G. Loontiens, and T.M. Jovin, Binding of 4-Methylumbelliferyl -D-Mannopyranoside to Dimeric Concanavalin A: Fluorescence Temperature-Jump Relaxation Study, Biochemistry 16:167 (1977).

34. I. Pecht, Antibody Combining Sites as a Model for Molecular Recognition, In:"Protein-Ligand Interactions," H. Sund, and G. Blauer, eds., pp. 356-371 De Gruyter, Berlin (1975).

35. S. Geller, Ph.D. Thesis, Albert Einstein College of Medicine, New York (1976).

36. B.N. Manjula, E.B. Mushinshi, and C.P.J. Glaudemans, The Formation of Active Hybrid Immunoglobulins from the Heavy and Light Chains of β-(1,6) d-Galactan Binding Murine Myeloma IgA's S-10 J-539. J.Immunol. 119:867 (1977).

37. R.J. Feldmann, M. Potter, and C.P.J. Glaudemans, A Hypothetical Space-filling Model of the V-Regions of te Galactan-binding Myeloma Immunoglobulin J 539, Mol.Immunol. 18:683 (1981).

38. R.J. Feldmann, personal communication.

39. M. Weigert, L. Gatmartan, E. Loh, J. Schilling, and L. Hood, Rearrangement of Genetic Information May Produce Immunoglobulin Diversity, Nature 276:785 (1978).

40. H.N. Eisen, E.S. Simms, and M. Potter, Mouse Myeloma Proteins with Antihapten Antibody Activity. The Protein Produced by Plasma Cell Tumor MOPC-315, Biochemistry 7:4126 (1968).

41. Y. Blatt, F. Karush, and I. Pecht, unpublished.

42. A. Oratore, K. Zidovetzki and I. Pecht, in preparation.

43. R. Zidovetzki, A. Licht, and I. Pecht, Effect of Interchain Disulfide Bond on Hapten Binding Properties of Light Chain Dimer of Protein 315, Proc.Natl.Acad.Sci. U.S.A. 76:5848 (1979).

44. I. Alexandru, D.I.C. Kells, K.J. Dorrington, and M. Klein, Non-covalent Association of Heavy and Light Chains of Human Immunoglobulin G: Studies Using Light Chain Labelled with a Fluorescent Probe, Mol.Immunol. 17:1351 (1980).

45. E.E. Abola, K.R. Ely, and A.B. Edmundson, Marked Structural Differences of the Mcg Bence-Jones Dimer in Two Crystal Systems, Biochemistry 19:432 (1980).

46. W. Bode, and R. Huber, Induction of the Bovine-Trypsin Transition by Peptides Sequentially Similar to the N-Terminus of Trypsin, FEBS Lett. 68:231 (1976).

47. W. Bode, The Transition of Bovine Trypsinogen to a Trypsin-like State upon Strong Ligand Binding. II. The Binding of the Pancreatic Trypsin Inhibitor and of Isoleucine-Valine and of Sequentially Related Peptides to Trypsinogen and to p-Guanidino-benzoate-trypsinogen, J.Mol.Biol. 127:357 (1979).

48. H.H. Nolte and E. Neumann, Kinetics and Mechanism for the Conformational Transition in p-Guanidinobenzoate Bovine Trypsinogen Induced by the Isoleucine-Valine Dipeptide, Biophys.Chem. 10: 253 (1979).

DYNAMIC ASPECTS OF SIGNAL TRANSFER

IN ANTIBODY MOLECULES

Péter Závodszky, Ferenc Kilár, Judit Török
and György A. Medgyesi

Institute of Enzymology, Biological Research Center,
Hungarian Academy of Sciences, Budapest, PO.Box 7, H-1502
National Institute of Haematology and Blood Transfusion
Budapest, PO.Box 44, H-1502, Hungary

Antibodies are the most well known molecular mediators of the immune response. The multiple functions of antigen recognition and the activation of secondary effector processes are associated with discrete domains within the antibody molecule. A schematic representation of the rabbit IgG* molecule is given in Figure 1. The symmetrical molecule is composed of four polypeptide chains linked by disulphide bridges. Each heavy-chain is folded into four longitudinally arranged globular domains and each light-chain is constructed by two globular domains. Within the domains the polypeptide chain is folded to form two rather distorted anti-parallel beta-pleated sheets, arranged in two parallel planes.

Under controlled conditions proteolytic enzymes tend to cleave immunoglobulins between domains. Some of the proteolytic fragments, the Fv, Fab, Facb, Fc, pFc', are shown in Figure 1. The biological functions of the various fragments are also indicated. The antigen binding-site is located in the Fv variable part, while the binding-site of Clq, the first subcomponent of the first component of the complement, and the cytophilic reactions are associated with the Fc part of IgG.

Antibodies recognizing a virtually unlimited range of antigens can be generated while the elimination of antigen is mediated via a

---

*Abbrevations used in this paper: IgG, immunoglobulin G; Fv, Fab, Facb, Fc, pFc', fragments of IgG; Clq, first subcomponent of the first component of complement; E, internal energy; r.m.s., root mean square; L_2, the light-chain dimer; J, spin-coupling constant; T, absolute temperature.

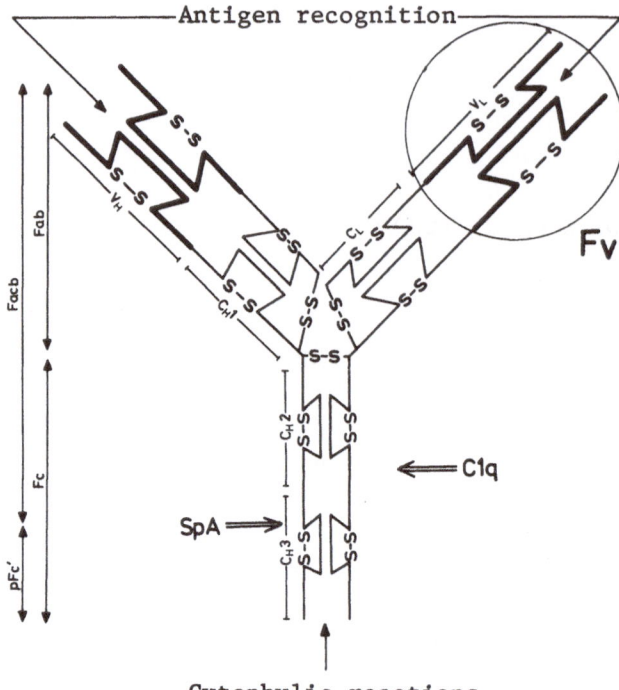

Fig. 1. Schematic structure of IgG molecule. Its proteolytic
         fragments and biological functions of fragments are shown.

small number of effector mechanisms which are triggered by the
formation of antigen-antibody complex. It seems now that the central
molecular event in these effector mechanisms is the recognition of
the Fc portion of the antibody molecule. The mechanism by which
the initial antigen recognition on the variable module is communi-
cated to the Fc region is still a matter of controversy. Henry
Metzger in his recent review[1] discussed allosteric changes, dis-
tortions and aggregations as possible mechanisms as it is diagram-
matically presented in Figure 2. According to the associative model
the clustering of a certain number of antibody molecules is sufficient to activate the complement. The distortive model explains
the effect of antigen binding by inducing distortion in the relative
position of structural domains within the antibody molecule. The
third model assumes allosteric transitions leading to the formation
of secondary binding sites. The associative model is favored in the
recent reviews[1,2] since a large number of attempts to correlate
antigen-induced conformational changes in antibodies with triggering
effector functions led to negative results.

     Studies with paramagnetic probes and on water relaxation carried
out in Oxford by Raymond Dwek's group[3,4,5] did not detect conformational changes in the pFc' part of the molecule around the

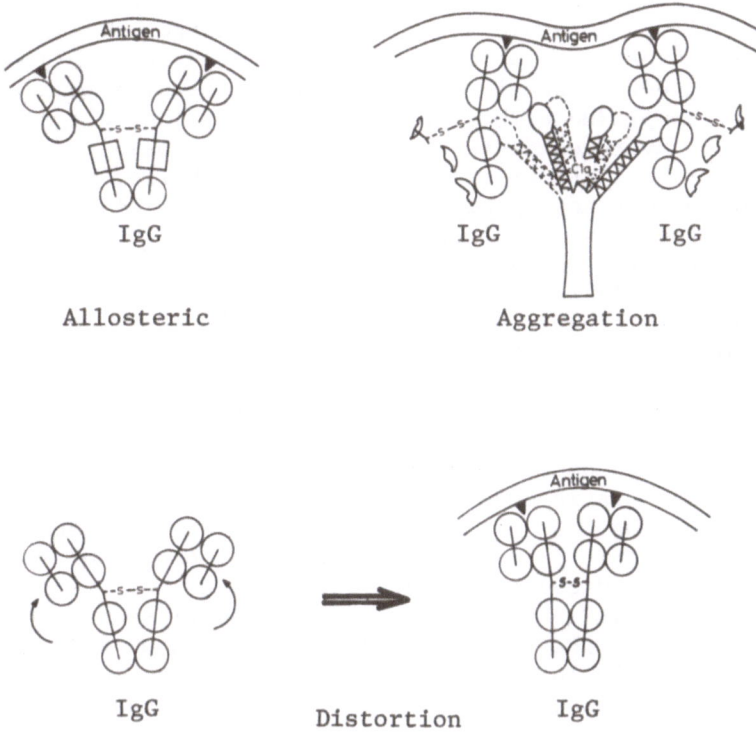

Fig. 2. Possible mechanisms by which antibodies may acquire complement activating capacity upon complex formation with antigen.

gadolinium probe upon hapten binding. No difference in the proteolytic susceptibility of the hinge region was observed upon antigen binding by Jean-Claude Jaton[6].

On the other hand, there are experimental data indicating conformational changes after antigen binding. Changes in tryptophanyl fluorescence[7] or shifts in conformational equilibrium observed by temperature jump chemical relaxation technique in Israel Pecht's laboratory[8] reveal conformational rearrangements upon interaction with haptens.

Apart from some theoretical considerations about the effect of hapten binding on the flexibility of the "elbow bending" by McCammon and Karplus[9] and Huber's[10] comments on the observed disorder in the Fc part of Kol and Zie IgGs, not too much attention has so far been devoted to the intradomain flexibility and motility of the IgG molecules in the interpretation of their concerted multiple function.

We focused on the changes in conformational motility and stability of the antibody molecule to elucidate the effect of antigen
binding from the dynamic point of view. We studied induced homogeneous anti SIII pneumococcal polysaccharide antibody samples without antigen and complexed with site-filling, monovalent oligosaccharide antigen. Hydrogen-deuterium exchange was chosen to detect
changes in the probability of solvent exposure of buried peptide
hydrogens, adiabatic scanning microcalorimetry to follow changes in
conformational stability, high resolution proton NMR to observe
aromatic residues with mobilities independent of the rest of the
protein and small angle X-ray scattering to reveal changes in the
overall electron distribution of the antibody molecule upon antigen
binding. To avoid the disturbing effect of macromolecular interactions all experiments were carried out with solutions of monomeric
antibody molecules saturated with monovalent antigen.

To explain molecular mechanisms and to interpret conformational
changes in proteins static models, generated from X-ray diffraction
data, were used during the last two decades. From the electron
density maps this static picture occurs by expediency rather than
by a reflection of physical reality. Protein molecules are small
objects to apply Newtonian mechanics to their description. From
the size and nature of these "macro"-molecules it is obvious that
they undergo relatively large fluctuations in internal energies.
The mean square variations of these quantities are given by the
following equations:

$$<dE^2> = kT^2mC_v$$

$$<dV^2> = kTVB_T$$

where m is the mass, V is the volume of the molecule, C_v is the heat
capacity at constant volume, B_T is the isothermal bulk compress-
ibility, T is the absolute temperature and k is Boltzmann's constant.
For microscopic particles these fluctuations can be quite significant.
For example for an antibody molecule of molecular weight 150000, rms
fluctuations in internal energy are approximately 320kJ/mole and
volume fluctuations are about 30cm^3/mole. These fluctuations are
significantly bigger than energy changes in the known ligand induced
conformational changes.

The dynamic nature of protein molecules has been recognised by
Linderström-Lang and Schellmann[11] and Straub and Szabolcsi[12]. The
success of X-ray diffraction analysis to determine macromolecular
structure and the widespread use of static models distorted our view
of protein structure and function. The recently developed methods:
high resolution NMR[13,14], hydrogen-deuterium exchange[15,16], fluorescence quenching[17] and time resolved spectroscopy[18] called the
attention again to the dynamic aspects of protein structure.

We used a statistical approach to plan and interpret our experiments. Since the three dimensional structure of proteins is stabilized by weak secondary bonds the energies of which are comparable with the energy of thermal agitation (kT), proteins exist as a dynamic ensemble of rapidly interconverting conformations. The breakage of contacts is a secondary and gaussian stochastic process.

The conformation of an antibody molecule as we detect it by usual physico-chemical methods is the time and number-average of an interconverting heterogeneous population. A more complicated but more complete description of a protein molecule would be the probability distribution of the available reversibly interconverting states. In Figure 3 hypothetical energy distribution functions of a protein in two conformational states are given. Conformational transitions can be interpreted as distortion of the distribution function or as a transition between two distributions.

It was also suggested by Careri[19] that in a highly cooperative structure like a protein molecule the conformational fluctuations are correlated or even concerted, and a complete and correct description of the system involves both its time and spatial structures. These continuous fluctuations lead to segmental opening and closing reactions of the macro-molecule. Usual statistical methods, like ORD, CD, small angle X-ray scattering etc., observe the average conformation and low probability conformations cannot be detected by these methods.

Relaxation methods like hydrogen-deuterium exchange, chemical modification or NMR are more sensitive to detect certain conformations even if they occur at very low probability. The hydrogen-deuterium exchange reports about the probability of solvent exposure of peptide hydrogens. Aase Hvidt[15] suggested two basic mechanisms for the interpretation of hydrogen-deuterium exchange of folded proteins in aqueous solution.

$$N \underset{k_2}{\overset{k_1}{\rightleftharpoons}} I \xrightarrow{k_0} \text{exchange}$$

$$X = n^{-1} \sum_{i=1}^{n} \exp(-k_i t)$$

$$k_i = \frac{k_1 k_0}{k_1 + k_2 + k_0} \quad \text{if } k_1 \gg k_2$$

$$EX_1 \qquad k_i = k_1 \qquad \text{if } k_o \gg k_2$$

$$EX_2 \qquad k_i = \frac{k_1}{k_2} k_o \qquad \text{if } k_2 \gg k_o$$

$$k_o = (10^{-pH} + 10^{pH-6}) \; 10^{0.05(T-298)} \; \text{sec}^{-1}$$

Let us consider one peptide hydrogen of a protein molecule. In the
N conformation it is buried, in the I conformation it is accessible
to the solvent. In the I conformation exchange occurs with a rate
constant of k_o. This chemical exchange rate constant depends on the
pH and the temperature as it is shown. If the motility of the
protein is high ie., the rate constant of refolding (k_2) is much
greater than the rate constant of unfolding (k_1) and than the chemical exchange rate (k_0) the exchange proceeds according to the EX_2
mechanism. This is the case for immunoglobulins in the pH range
between 5 to 9. In this case the data on exchange rates can be
presented as plots of the fraction of unexchange peptide hydrogen
atoms (X) at a given time (t) versus $\log(k_ot)$, where k_o is the pH
and temperature dependent exchange rate constant of solvent-exposed
peptide groups. If no changes in the stability of protein take place
as a function of pH this representation gives the "relaxation spectrum" of the protein, Figure 4. The dotted lines in this plot represent the exchange of hypothetical polypeptides which expose their
peptide hydrogens to the solvent in a cooperative way with equal
probability.

The relaxation spectrum is a distribution function of G difference (G is the free energy) associated with the conformational

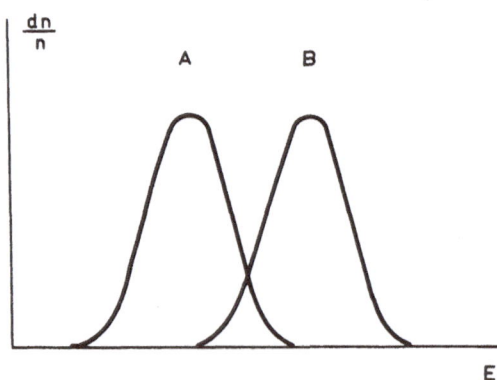

Fig. 3.   Hypothetical probability distribution functions of a
          protein in conformational states A and B.
          E = internal energy

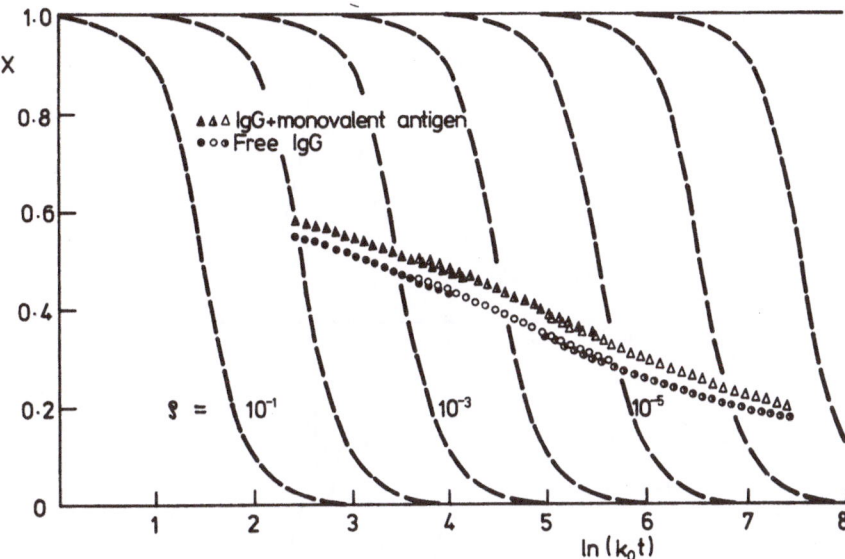

Fig. 4.   Relaxation spectra of free antibody (circles), and anti-
          body-monovalent octasaccharide antigen complex (triangles).
          The measurements were carried out at three pH values:
          5.2 (●,▲), 7.25 (○,▲) and 8.52 (◐,△).   The temperature was
          298 K.   X is the fraction of unexchanged peptide hydrogens,
          k_0 is the chemical exchange rate constant, t is the time
          elapsed from the start of the exchange.   The dashed curves
          represent hypothetical polypeptides which expose their
          peptide hydrogens to the solvent with equal   probability.

transition exposing the peptide hydrogens. This spectrum can be
used as a representation of the conformational stability and of the
conformational motility. According to our comparative hydrogen-
deuterium exchange studies the antibody molecules are among the
most motile proteins of this size.

    This is also reflected in the unusually well resolved 270MHz
proton NMR spectra of the L_2 and Fc fragments, Figure 5. The difference in the resolution, especially in the aromatic region (left
side of the spectrum), is striking if we consider that the molecular
weight of the two fragments is the same. The explanation is that
there is more rotational freedom for the aromatic residues in the
Fc fragment. The substantial segmental flexibility of the Fc
fragment was also detected using two special NMR pulse sequences[20]
which exploit either the multiplet structure spin coupling or the
differential linewidth of resonances. Using Carr-Pulsell A pulse
sequence (Figure 6) the phase of the doublet resonances can be inverted with respect to the singlets, which always appear as positive
peaks. In Figure 7 spectra obtained in this way from pFc' fragment

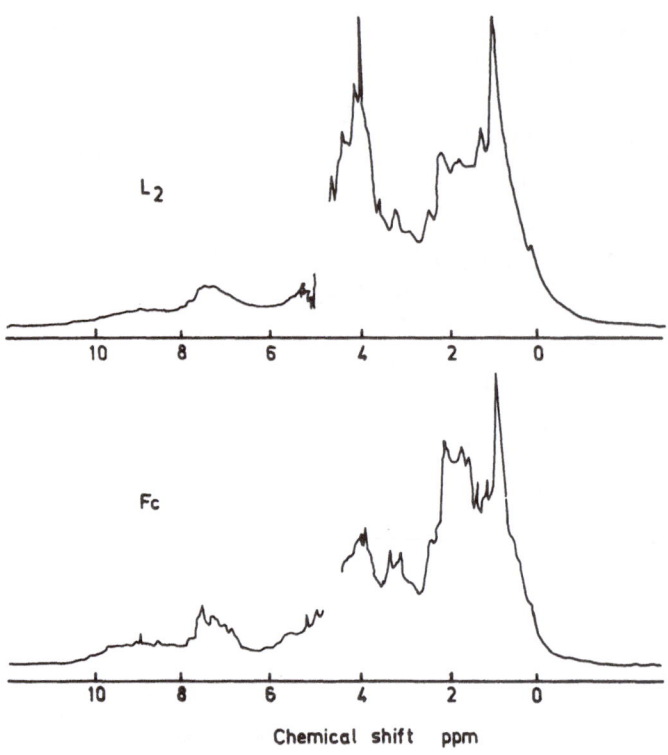

Fig. 5. High resolution ^{1}H NMR spectra of L_2 and Fc fragments.

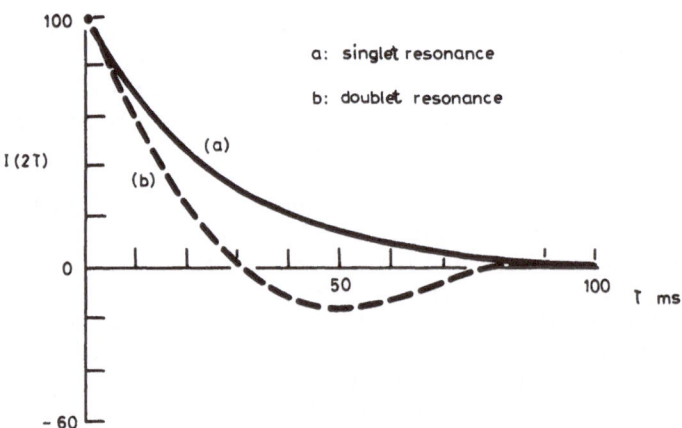

Fig. 6. The intensity of the echo in the Carr-Purcell A pulse
sequence after the 90° pulse. The phase of the singlet
resonance (J=0) always appears as positive peaks, but
the phase of a doublet resonance (J=8.5Hz) will be
inverted after 30ms.

are shown. A number of resonances are observed to invert in the
aromatic region. The pair of doublets at 6.7ppm and 6.96ppm were
assigned to a tyrosine residue from the results of selective de-
coupling experiments. It follows from linewidth calculations that
there has to be motion about the $C - \alpha - C - \beta$ bond

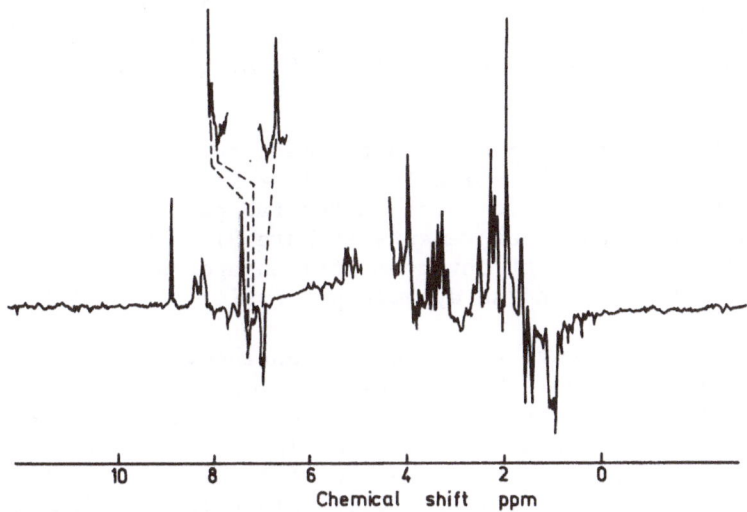

Our microcalorimetric measurements with rabbit IgG and its
fragments[21] revealed longitudinal domain-domain interactions in the
IgG molecule. Heat absorption curves (Figure 8) show biphasic melting for IgG and Fc fragment, while a single transition is observed
for the Fab, Facb and pFc' fragments. The low temperature transition
can be attributed to interactions between the second and third constant domains, the higher temperature transition (349K) corresponds
to the melting of the domains in the IgG.

To characterize further the longitudinal effect we attached
spin label probes to the Fab part of the IgG molecule. Binding

Fig. 7.   Subspectrum for pFc' obtained from a Carr-Purcell A pulse
sequence at 270MHz (T = 313 K, pH 7.2, 0.5mM).

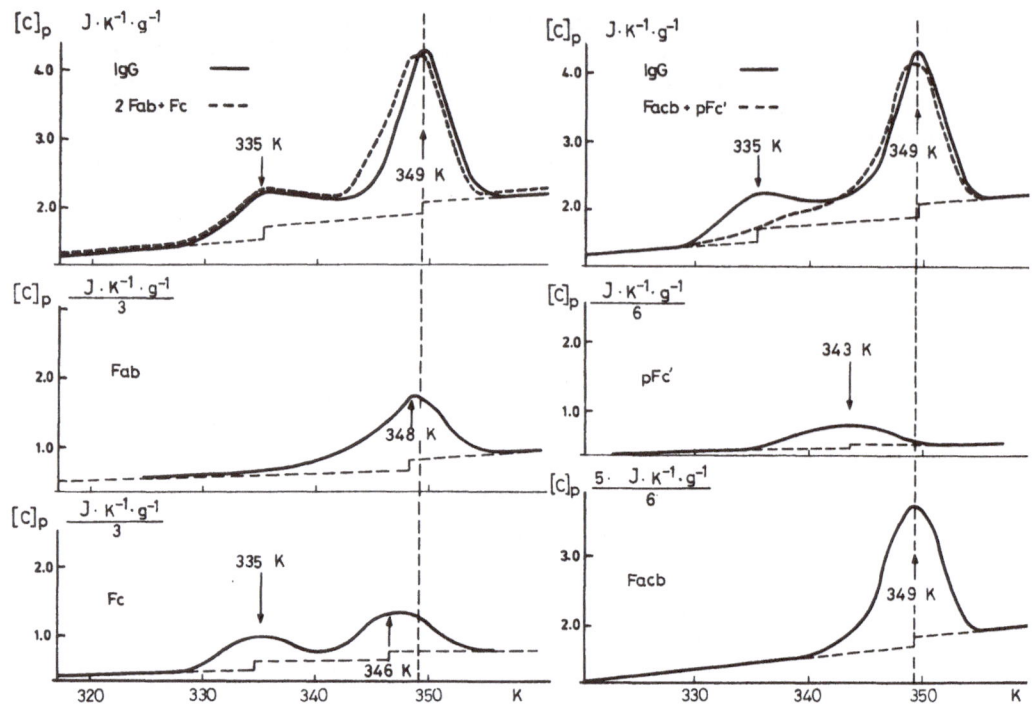

Fig. 8.   Temperature-dependence of the partial heat capacity of
          intact homogeneous anti SIII polysaccharide IgG and its
          proteolytic fragments. The thick dashed curves represent
          the algebraic sum of the heat capacities of the fragments.
          The thin dashed lines show the extrapolated temperature
          dependence of the heat capacity of the samples in various
          conformational states. 0.1 M MES, pH 5.4, conc:0.06-0.15%.

monovalent Staphylococcus protein A fragment to the CH2-CH3 interface on the Fc part of the molecule resulted in an increased rotational freedom of the spin label on the Fab part, as detected by
electron spin resonance measurements (Figure 9). This observation
also supports the view that environmental changes around the Fc
fragment are reflected in the dynamic properties of the Fab part.

     Upon combination with antigens, the antibody molecules exert
new biological functions. Differential sedimentation and small angle
X-ray scattering experiments did not reveal changes in the average
atomic positions shape and volume of the IgG molecule after antigen
binding. The radius of gyration was found to be 4.42 ± 0.015nm for
the unliganded rabbit IgG (Figure 10). This value is significantly
lower than most data reported earlier for various antibodies. Calculations based on various models lead to the conclusion that only
a tight contact between the Fab and Fc parts of the molecule is

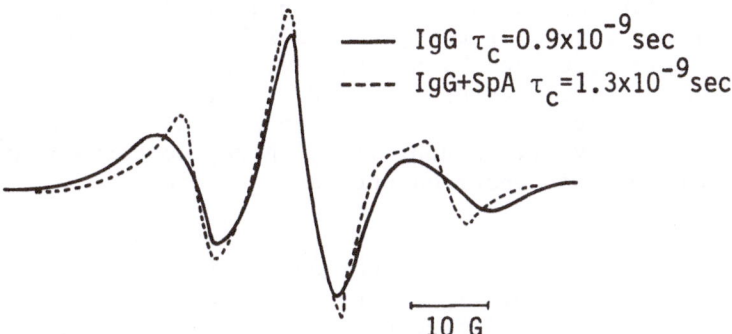

Fig. 9. The effect of monovalent SpA binding on the ESR spectrum
of nitroxyl radical bound to Fab region of IgG.

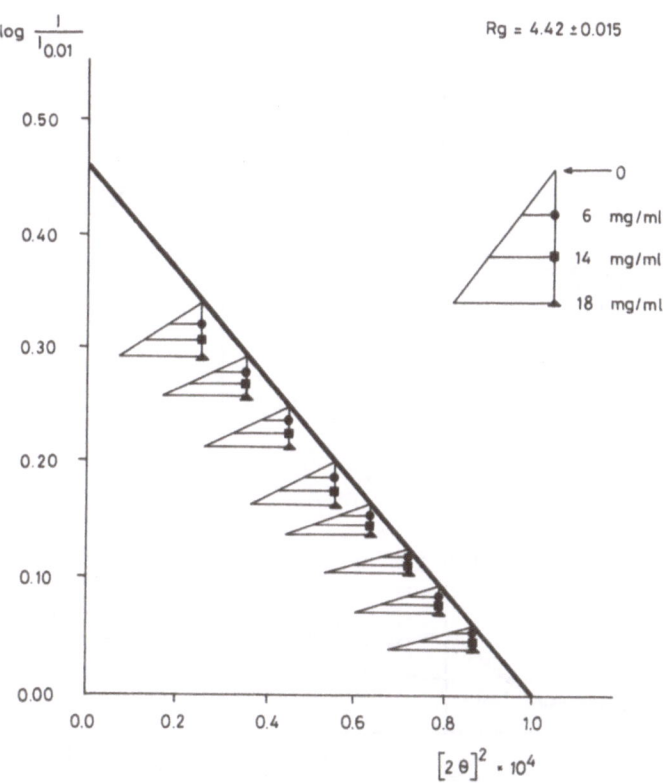

Fig. 10. Determination of the radius of gyration of rabbit
IgG by the Guinier plot. The extrapolation to zero
concentration is seen in the insert. θ is the half
of the scattering angle, I is the intensity of the
scattering.

compatible with the radius of gyration derived from our measurements
by us. Upon addition of antigens a slight increase of the radius
of gyration was observed by a sensitive difference method[22] which
eliminates systematic errors. The 0.05nm increase can be explained
by addition of the mass of the antigen at a distance of 4.7nm from
the centre of gravity, without implying any conformational change.
No volume change was observed upon antigen binding.

At the same time significant shift (+6K) was observed in the
low temperature transition of heat absorption curves (Figure 11),
which reflects a slight tightening of the CH2–CH3 interaction. There
were also significant differences in the hydrogen–deuterium exchange
relaxation spectra (Figure 4). The probability of solvent exposure
of more than half of the peptide hydrogens was increased tenfold,

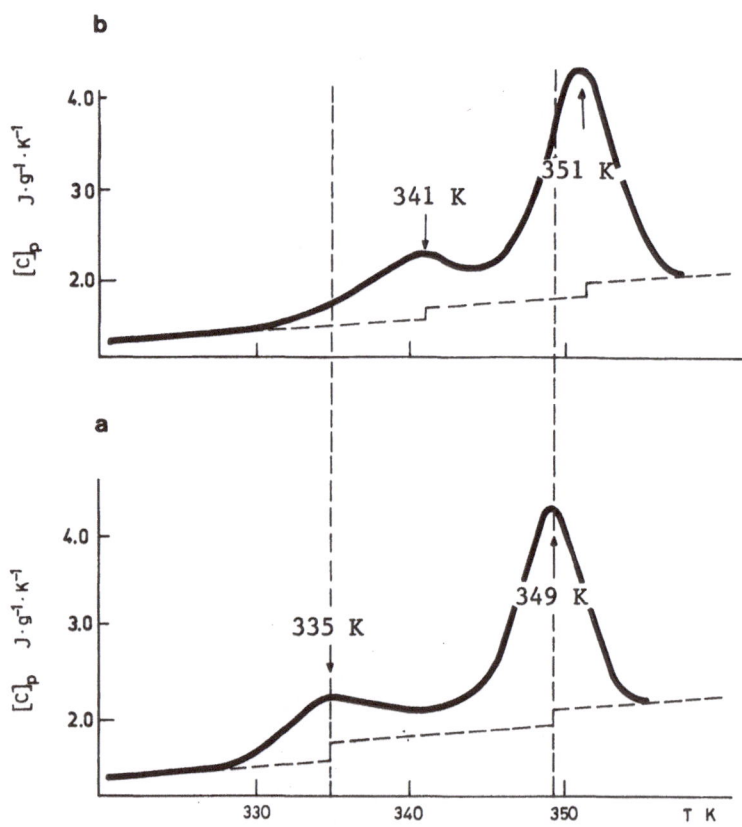

Fig. 11. Temperature dependence of partial heat capacity of free
          antibody (a), and antibody liganded with monovalent
          polysaccharide antigen (b). (0.1 M MES, pH 5.4)
          Heating rate: 1.0 K/min.

which reflects a 4-6kJ/mole increase in the standard free energy of
the reversible conformational transitions exposing the peptide hydrogens to the solvent after antigen binding.

There is an apparent contradiction between these results and
the picture based on experimental data obtained by statistical
methods, reflecting small or no conformational changes. This contradiction can be resolved if we consider the nature of the methods
applied. Relaxation methods like hydrogen-deuterium exchange are
sensitive to shifts in low probability effects as well. In the relaxation spectrum the same change will be observed when we increase
the probability of the "open" conformation from 10^{-7} to 10^{-6}, as
when from 10^{-2} to 10^{-1}. It is obvious that the first effect cannot
to be seen by statistical methods, while the second one might give
a 10 per cent contribution to a statistical parameter.

The same applies to the calorimetric measurements. Since the
energy of intramolecular interactions is a highly sensitive function
of the atomic distances eg. the nonbonded Lennard-Jones 12-6 potential, or the hydrogen bonded 12-10 potential, changes in properties
depending on bonding energies can report about such events. In
contrast, the same events cannot be resolved by techniques based on
the direct measurements of atomic distances (eg. radius of gyration,
sedimentation co-efficient) or by measuring parameters which are
only slightly changing functions of the atomic positions.

Summing up we intend to stress two points:

1. The data presented here support the view that there are longitudinal domain-domain interactions in the IgG molecule and these
longitudinal interactions are able to communicate environmental
effects between distant parts of the molecule.
2. The antibody molecule - like any other protein - is a dynamic
ensemble of rapidly interconverting conformations, and·the equilibrium of these conformational transitions can be affected by the
binding of a site-filling monovalent ligand.

Since protein conformations are specific structures, stabilized
by highly cooperative weak interactions, their fluctuations should
be concerted. Concerted conformational fluctuations may serve as a
possible mechanism of information transfer over large distances
within a macromolecule. "Retuning" some components of the fluctu-
ation spectrum will be reflected in the entire system. The observed
changes in the relaxation spectrum of IgG could explain, in part,
why antigen binding triggers complement fixation. The observed
changes in conformational fluctuations may be parts of the processes
leading to the formation of the complement binding site. Further
essential changes are brought about by the formation of a suitable
density of the same sites in antibody aggregates.

REFERENCES

1. H. Metzger, The effect of antigen on antibodies: recent studies.
    Contemp.Topics Molec.Immunol. 8:119 (1978).
2. D. Beale and A. Feinstein, Structure and function of the constant
    regions of immunoglobulins. Q.Rev.Biophys. 9:135 (1976).
3. S.K. Dower, R.A. Dwek, A.C. McLaughlin, L.E. Mole, E.M. Press,
    and C.A. Sunderland, The binding of lanthanides to non-immune
    rabbit immunoglobulin G and its fragments, Biochem.J. 149:73
    (1975).
4. K.J. Willan, K.H. Wallace, J.C. Jaton, and R.A. Dwek, The use
    of gadolinium as a probe in the Fc region of a homogeneous anti-
    type III pneumococcal polysacharide antibody, Biochem.J. 161:
    205 (1977).
5. D.R. Burton, S. Forsén, G. Karlström, and R.A. Dwek, Proton
    relaxation enhancement /PRE/ in biochemistry. A critical survey.
    Prog. NMR Spectroscopy. 13:2 (1979).
6. J.C. Jaton, J.K. Wright, H. Maeda, A. Schmidt-Kessen, and J.
    Engel, Interactions of monovalent and multivalent oligosaccharide
    ligands with homogeneous anti-polysaccharide antibody and its
    possible effects on the conformation of IgG molecule, in:
    "Immunology 1978", Proceedings of the Fourth European Immunology
    Meeting, J. Gergely, G.A. Medgyesi, and S.R. Hollan, eds.,
    Akadémiai Kiadó, Budapest 277. (1978).
7. J.C. Jaton, H. Huser, D.G. Braun, D. Givol, I. Pecht, and J.
    Schlessinger, Conformational changes induced in a homogeneous
    anti-type III pneumococcal antibody by oligosaccharides of
    increasing size. Biochemistry. 14:5312 (1975).
8. R. Zidovetzki, Y. Blatt, C.P.J. Glaudemans, B.N. Manjula, and
    I. Pecht, A common mechanism of hapten binding to Immunoglobulins
    and their heterologous chain recombinants, Biochemistry. 19:2790
    (1980).
9. J.A. McCammon, and M. Karplus, Internal motions of antibody
    molecules, Nature /Lond./ 268:767 (1977).
10. R. Huber, Conformational flexibility and its functional sig-
    nificance in some protein molecules, TIBS. 12:271 (1979).
11. K.M. Linderström-Lang, and J.A. Schellman, Protein structure
    and enzyme activity, in:"The Enzymes", 2nd Edn., P.D. Boyer,
    H. Lardy, and K. Mybäck, eds., Academic Press, New York 1:443
    (1959).
12. F.B. Straub, and G. Szabolcsi, Molekularnaja Biologija, p.182
    Nauka, Moscow (1964).
13. I.D. Campbell, An NMR view of protein structure, in:"NMR in
    Biology" R.A. Dwek, I.D. Campbell, R.E. Richards, and R.J.P.
    Williams, eds., Academic Press, New York 33 (1977).
14. C.M. Dobson, The structure of lysosyme in solution, in:"NMR
    in Biology", R.A. Dwek, I.D. Campbell, R.E. Richards, and
    R.J.P. Williams, eds., Academic Press, New York 63 (1977).
15. Aa. Hvidt, and S.O. Nielsen, Hydrogen exchange in proteins,
    Adv.Protein Chem. 21:287 (1966).

16. S.Yu. Venyaminov, E. Rajnavölgyi, G.A. Medgyesi, J. Gergely,
     and P. Závodszky, The role of interchain disulphide bridges in
     the conformational stability of human immunoglobulin G1 subclass.
     Hydrogen-deuterium exchange studies, Eur.J.Biochem. 67:81 (1976).
17. A.J. Pesce, C.G. Rosén, and T.L. Pasby, Fluorescence spectroscopy.
     Marcel Dekker, New York 1 (1971).
18. R.H. Austin, K.W. Beeson, L. Eisenstein, H. Frauenfelder, and
     I.C. Gunsalus, Dynamics of ligand binding to myoglobin, Bio-
     chemistry 14:5355 (1975).
19. G. Careri, P. Fasella, E. Gratton, Statistical Time events in
     enzymes: a physical assessment, CRC Crit.Rev.Biochem. 3:141
     (1975).
20. J. Boyd, S.B. Easterbrook-Smith, P. Závodszky, C. Mountford-
     Wright, and R.A. Dwek, Mobility and symmetry in the Fc and pFc'
     fragments as probed by ^{1}H NMR. Mol.Immunol. 16:851 (1979).
21. P. Závodszky, J.C. Jaton, S.Yu.Venyaminov, and G.A. Medgyesi,
     Increase of conformational stability of homogeneous rabbit
     immunoglobulin G after hapten binding, Mol.Immunol. 18:39 (1981).
22. I. Simon, Determination of small alterations in the radius of
     gyration by small-angle X-ray scattering, J.Appl.Cryst. 4:317

COULD AGGREGATION OF HUMAN IMMUNOGLOBULINS

BRING ABOUT APPEARANCE OF NOVEL ANTIGENIC SITES?

B. Parodi, V. Dessi, P. Casali,
R. Strom[1] and F. Celada

Cattedra di Immunologia dell'Università di Genova - XIII
U.S.L. Ospeda le S.Martino, Genova, Italy
[1]Istituto di Chimica Biologica
Università di Roma, Italy

1. INTRODUCTION

Several possibilities have been suggested to explain the effect of antigen binding on such properties of antibody as the fixation of complement and binding to Fc receptor or to Rheumatoid Factor.

There is good evidence that conformational changes may occur in the Fab portion of the antibody molecule following antigen binding[1-6], but it is still not clear whether these changes can be "recognized" by the Fc region and play a role in triggering effector functions of the antibody[7].

On the other hand, polymerization of antibodies by a multi-determinant antigen might be sufficient to enhance the avidity for complement of the Fc fragments of the antibodies[7,8].

The possible existence of aggregation-dependent antigenic sites (ie. determinants present on the aggregated antigen but absent on the monomeric one) was investigated by using specific monoclonal antibodies, raised against aggregated human IgG.

2. MATERIALS AND METHODS

2.1 Antigens

7 S "monomeric" human IgG immunoglobulins (MGG) were prepared from concentrated Ig fraction (Globuman, Berna) by dilution to

31

10mg/ml in PBS (0.15M NaCl, 0.01 M phosphate, pH 7.2) and centrifugation at 1500 x G for 30 min. at 4°C.

Aggregated immunoglobulins (AHG) were obtained by exposure of
a 10mg/ml immunoglobulin solution in PBS buffer to 63°C for 0.5 hours.
By centrifugation in a 40%-10% linear sucrose gradient for 4 hours
at 220.000 x G it was possible to isolate the higher-molecular weight
components (fractions 19 to 37 in Figure 1), the apparent sedimentation coefficient at the center of this peak being, by this procedure, around 40-45S. Analytical ultracentrifugation of this
fraction in the tris buffer pH 7 showed a relatively widespread range
of sedimentation coefficients from 25S to approximately 60S, with a
modal value of 42.8S. Assuming for the aggregates a dyssymmetry
factor f/fo similar to that of monomeric IgG's[9], ie. 1.38 and a
partial specific volume of 0.739, this modal value of 42.8S would
correspond to a molecular weight around 2.2×10^6 dalton, ie. to a
16-mer of the monomeric IgG. If instead, as shown to occur in
chemically-induced rabbit IgG oligomers[10], aggregation results in the
formation of elongated structures, this value of 42.8 would correspond
to a 28-mer, with a molecular weight around 4×10^6 dalton.

2.2 Production and characterization of anti-IgG hybridoma products

Mouse spleen cells sensitized to human pre-formed immune complexes (DNA-anti DNA IgG's or tetanus toxoid-anti tetanus toxoid
IgG's) or to heat aggregated IgG's were fused with P3 x 63Ag myeloma
cells according to Koehler and Milstein[11].

Five different fusions were performed. Clones exhibiting anti-
AHG activity were then isolated in soft agar. As shown in Table 1,
eleven products of these clones were characterized with regard to:

1)   isotype. Products from 8 clones were constituted by IgG_1 sub-
     class, products from three clones displayed combined IgG_1 and
     IgM or IgG_1 and IgG_2 characteristics.

2)   reactivity against Fab or Fc fragment. Products from 7 clones
     and 4 clones specifically reacted with Fc portion and Fab
     portion of HGG, respectively.

3)   reactivity against immunoglobulin classes and subclasses and
     immunoglobulin light chain. Antibodies from 7 clones reacted
     with one, two or three human IgG subclasses, one reacted with
     IgG_1 and IgA, and three bound to K light chain. Human myeloma
     proteins (carrying K light chain except IgA myeloma) and Bence-
     Jones proteins were used as reactants.

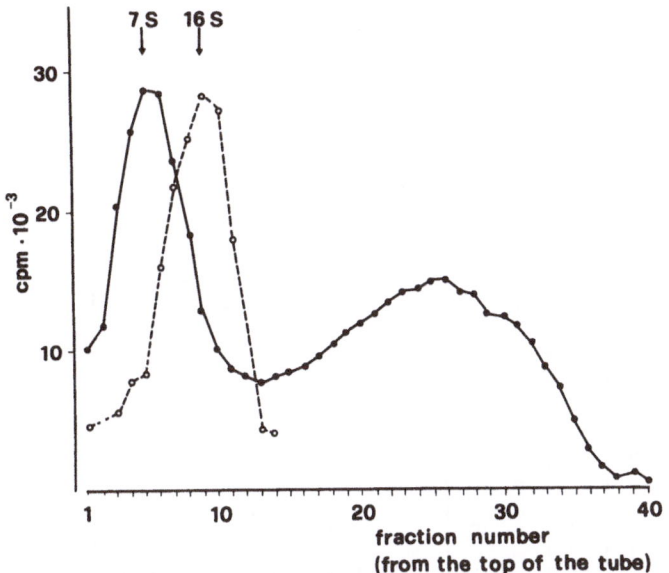

Fig. 1. Sucrose gradient profile of heat-aggregated human
immunoglobulins. The 16S reference peak (dashed profile)
was obtained by addition of a small amount of E.coli
β-galactosidase.

2.3 Procedures for the evaluation of binding

a) Direct binding assay. The ability of the different clone products to bind monovalent or multivalent antigen was assayed by
a fluid phase coprecipitation test: a given amount of monoclonal antibody was allowed to react, in a final volume of 200μl,
with increasing amounts of ^{125}I-labelled antigen (either MGG
or AHC) having a specific radioactivity of 550cpm/μg protein.
Following co-precipitation with rabbit anti-mouse IgG at equivalence and centrifugation at 1500rpm for 20min., the radioactivity of the precipitate was measured.

b) Inhibition of binding of labeled AHG by cold monomeric or
aggregated IgG. This test was carried out in conditions similar
to be binding assay. To a reaction mixture of monoclonal antibody
+ labelled AHG, calibrated to ∿ 80% saturation of the antibody,
increasing amounts of cold MGG and cold AHG were added in parallel
tests. After incubation for 2 hours at room temperature, rabbit
anti-mouse serum was added and the radioactivity of the co-
precipitate measured as in the direct binding assay.

Table 1. Characterization of anti aggregated human IgG monoclonal antibodies

Antigen	Clone	Isotype	Ig fragments		Myeloma proteins					Bence-Jones proteins	
			Fab	Fc	IgG_1	2	3	IgM	IgA	k	λ
AHG	PB 1.1	Y_1,μ	+	+	+						
AHG	PB 1.2	Y_1	+	+	+	+	+				
AHG	PB 1.3	Y_1	+	+	+	+	+				
DNA-αDNA	PB 2.1	Y_1	+		+	+	+	+		+	
DNA-αDNA	PB 3.1	Y_1		+	+	+	+				
DNA-αDNA	PB 3.2	Y_1		+	+	+	+				
TET-αTET	PB 4.1	Y_1,Y_2	+	+	+	+	+				
TET-αTET	PB 5.1	Y_1	+		+	+	+		+		
TET-αTET	PB 5.2	Y_1	+		+	+	+	+		+	
TET-αTET	PB 5.3	Y_1,Y_2		+	+	+	+				
TET-αTET	PB 5.4	Y_1	+		+	+	+	+		+	

2.4 Analysis of binding data

The binding of a given anti-IgG hybridoma product to the immuno-globulin was, as shown in Figure 2, strongly dependent on the aggre-gation state of the antigen. The analysis of the binding data must therefore account for the existence, in the aggregated particle, of several copies of the determinant(s) present in the original monomeric IgG, the whole particle behaving as "bound" under our experimental conditions, when the hybridoma product has combined even to a single determinant.

For a quantitative comparison of the binding to aggregated and non-aggregated antigen, we may assume that, to a first approximation aggregation does not modify the reactivity of the determinant toward the monoclonal antibody, and that the binding of any determinant to any antibody binding site is an independent event, regulated by an intrinsic association equilibrium constant K. The probability P_D that a single determinant and P_A that a whole N-valent aggregate, randomly chosen, be free from interaction with monoclonal antibodies are then[12] such that $P_A = (P_D)^N$. Upon reaction of the antigen with a concentration S of antibody sites, the concentration A_B of antigen particles having a least one bound determinant (apt therefore to be precipitated in the double antibody assay) is expressed (in a double reciprocal plot of $1/A_B$ vs $1/A_F$) as a function of the concentration A_F of completely free antigen particles. It can be shown[13] that, while the intercept q_0 on the ordinate axis is always equal to $1/S_T$, irrespective of the value of N, the slopes m_0 and m_∞, of this binding curve as respectively $1/A_F \to 0$ and $1/A_F \to \infty$ are defined by the ex-pressions:

$$m_0 = \frac{1}{N} \quad \frac{1}{K \cdot S_T} + \frac{N-1}{2} \tag{1}$$

$$m_\infty = \frac{1}{(1 + K \cdot S_T)^N - 1} \tag{2}$$

The intercept q_∞, on the ordinate axis, of the asymptote to the binding curve as $1/A_F \to \infty$ is given by:

$$= \frac{N^2 \cdot K^2 \cdot S_T}{\left[1 + K \cdot ST - \dfrac{1}{(1 + K \cdot S_T)^{N-1}} \right]^2} \tag{3}$$

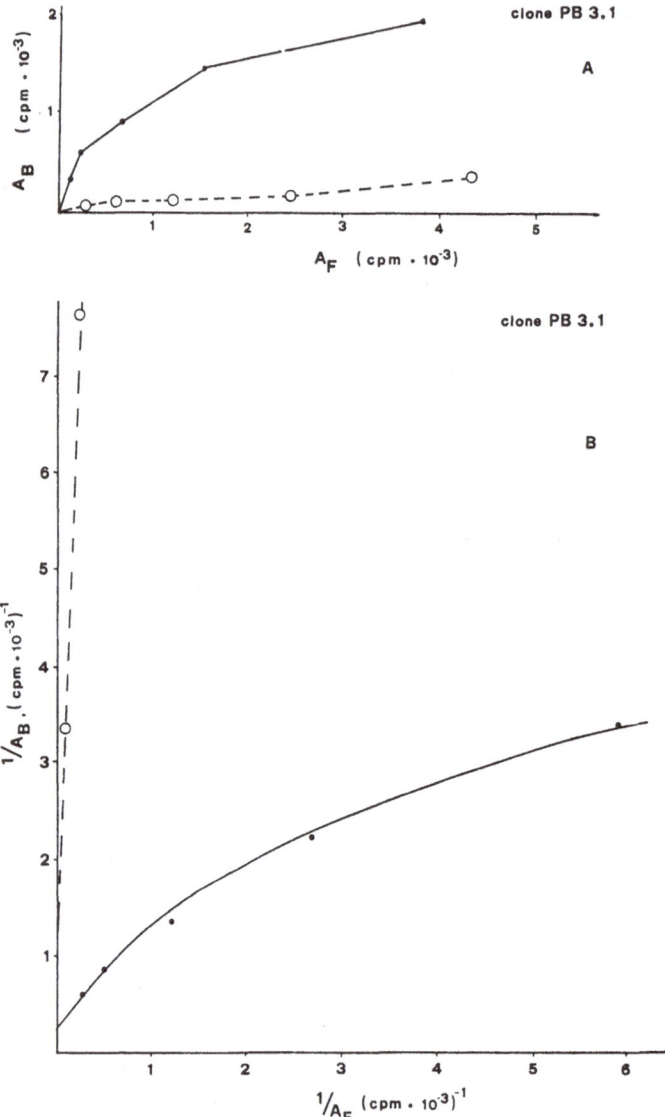

Fig. 2. Binding assay, with MGG (dashed line) and AHG (full line),
         of the monoclonal antibody from clone PB 3.1:

a) A_B vs A_F (direct) plot

b) $1/A_B$ vs $1/A_F$ (double reciprocal) plot

Since our population of aggregates, instead of having a homogeneous
size defined by a single value N in each aggregate, goes from a lower
limit N_{min} to a higher limit N_{max}, with a mean value of <N>,

expressions (1) – (3) need to be modified. It can then be shown that, if we indicate by β the value, for each N and for a given concentration S_T of antibody sites, of

$$\left(\frac{1}{1 + K.S_T} \right)^N$$

and by <β> the weighted mean, for the various N's, of all the values of β, we have:

$$m_O = \frac{1}{<N>} \cdot \left(\frac{1}{K.S_T} + \frac{<N\ (N-1)>}{2<N>} \right) \tag{4}$$

$$m_\infty = \frac{<β>}{1-<β>} \tag{5}$$

$$q_O = \frac{<N> \cdot <N\beta> \cdot K^2 \cdot S_T}{(1+K.S_T)^2 \cdot <β> \cdot (1-<β>)^2} \tag{6}$$

When instead the monoclonal antibody finds on each antigen molecule only one appropriate determinant (as it may be expected to occur when monomeric IgG's are used as antigen), expressions (1) – (3) reduce to:

$$m_O = m_\infty = \frac{1}{K.S_T} \tag{7}$$

$$q_\infty = q_O = \frac{1}{S_T} \tag{8}$$

With monomeric antigen the binding curve in the double reciprocal plot shall therefore be represented by a straight line crossing the ordinate and abscissa respectively at $1/S_T$ and at $-K$. When instead the antigen – notably aggregated IgG's – carries N identical determinants, the binding curve shall have a downward concavity (Figure 3), its asymptotical characteristics being defined by expressions (4) – (6), which reduce to expressions (1) – (3) when the value of N is strictly the same for all antigen molecules.

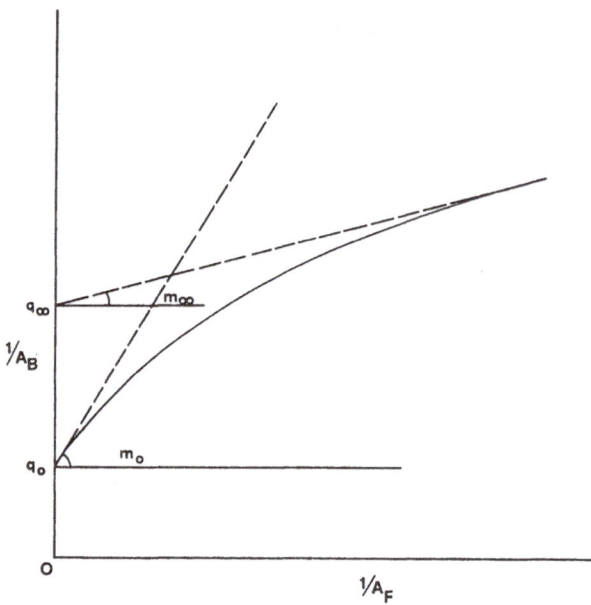

Fig. 3. Graphical derivation of the asymptotical parameters m_0,
m_∞ – characterizing a binding curve of a monoclonal antibody to a multi-determinant antigen.

3. RESULTS AND DISCUSSION

      In the direct binding assay, all eleven clone products exhibited
a higher ability to bind aggregated IgG than monomeric IgG. In
double reciprocal plots of $1/A_B$ vs $1/A_F$, very steep straight lines
were usually obtained with monomeric IgG's, while with aggregated
IgG's the binding curves had a downward concavity (Figures 2b, 4 and
5); the shapes of these curves were however different for the various
clone products. Since the same batch of AHG, with the same value
of <N> and the same distribution around the mean value, were used
in all experiments, such differences suggest that, besides the
purely stochastic binding to the various determinants on the aggregate,
with some monoclonal antibodies the intrinsic affinity for these
determinants may be different from that found toward the monomeric
antigen. The various binding curves were then analysed in terms of
expressions (1) – (3); the use of expression (4) – (7) giving rise
to some problems because of the rather ill-defined size distribution
of the aggregates. In Table 2 are reported the values of some parameters of clones PB 3.2 and PB 4.1, as obtained respectively from
Figures 4 and 5. The estimates of K for monomeric and aggregated
antigen (as derived from the respective values of the slope m_0 and,
in the latter case, also from the values of m_∞) can be compared
(columns 2b, vs 3b or 3c and vs 4b or 4c). It can be seen that,
with the product of clone P. 3.2, similar values of K are obtained

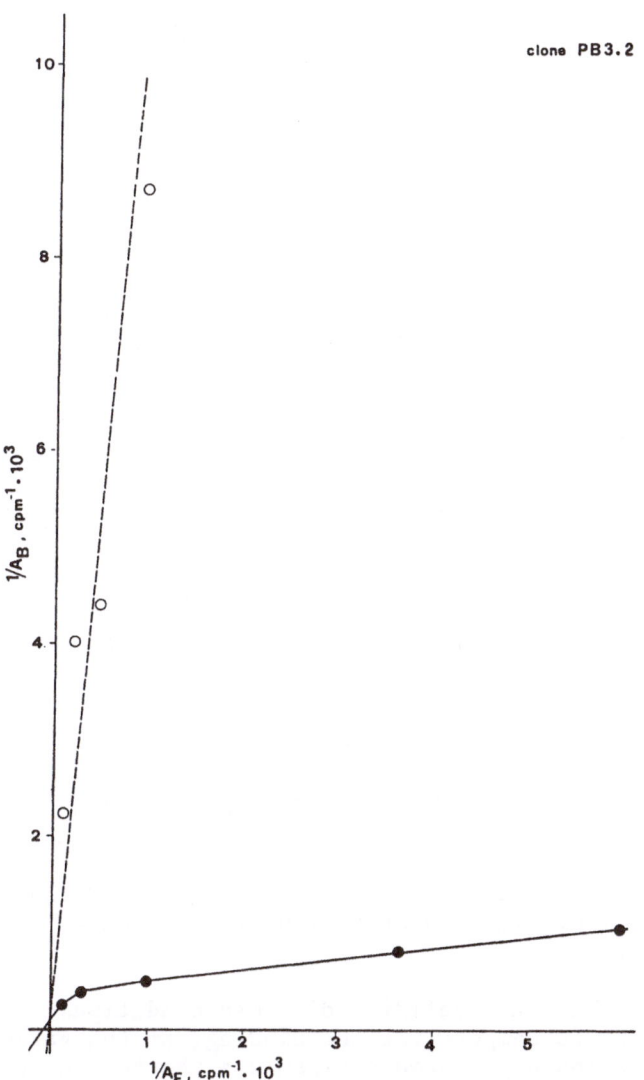

Fig. 4. Double reciprocal plot of the binding to MGG (dashed line)
         and AHG (full line) of the product of clone PB 3.2

for MGG and AHG assuming that the aggregation value of N is 16. With
the product of clone PB 4.1, the best fit is instead with N≈28.
Since, as previously mentioned, the batch of AHG was the same, this
difference seems to pint to the presence, in the product of clone
4.1, of antibody sites having a much higher affinity for the determi-
nant(s) in AHG than in MGG, thus mimicking a larger number of determinants per aggregate. Similar results (not shown in Table 2) could
be obtained utilizing the values of q_∞.

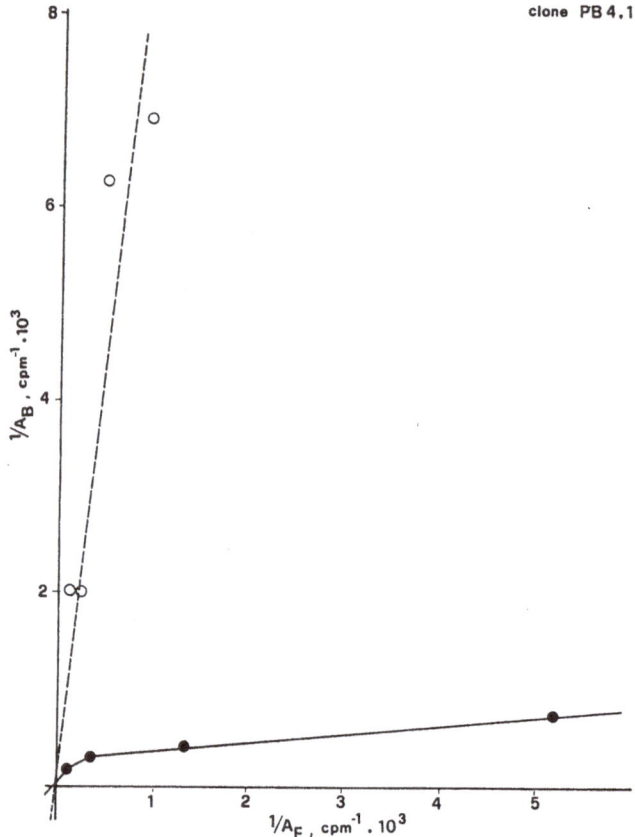

Fig. 5.   Double reciprocal plot of the binding to MGG (dashed line)
          and AHG (full line) of the product of clone 4.1

In order to check the validity of these conditions, the ability
of cold MGG or AHG to compete for the binding, by the various clone
products, of labelled aggregated antigen was tested. As shown in
Figure 6, a comparison of the results obtained with clones PB 3.2
and PB 4.1 appeared to confirm our hypothesis: while with clone
PB 3.2 MGG and AHG had equivalent inhibitory effects, with clone
PB 4.1 aggregated antigen was much more efficient than monomeric
ag (the leveling off at higher concentrations being presumably due
to a saturation effect).

The inhibition assay seems indeed to be more reliable for the
detection of such aggregation-induced potentiation of binding affinity, than the direct evaluation of binding, since it bypasses cumber-
some calculations and avoids asymptotical extrapolations which may
introduce relevant errors. In one of our clones (clone PB 5.1), in

Table 2. Parameters of the binding curves for MGG and AHG of monoclonal antibodies PB 3.2 and PB 4.1, and derived values of S_T and K

	1		2		3			4		
	a	b	a	b	a	b	c	a	b	c
	q_0 (Cpm^{-1})	S_T (μg IgG/test)	m_0 (MGG)	K (MGG)	m_0 (AHG) (mM^{-1})	K(AHG, N=16)	K(AHG, N=28)	m_∞ (AGG)	K(AHG, N=16)	K(AHG, N=28)
PB 3.2	0.12x10^{-3}	15.15	11.2	176.8	1.20	169.2	98.5	0.30	190.0	106.5
PB 4.1	0.08x10^{-3}	22.73	8.75	150.9	0.80	249.1	148.3	0.055	267.6	146.9

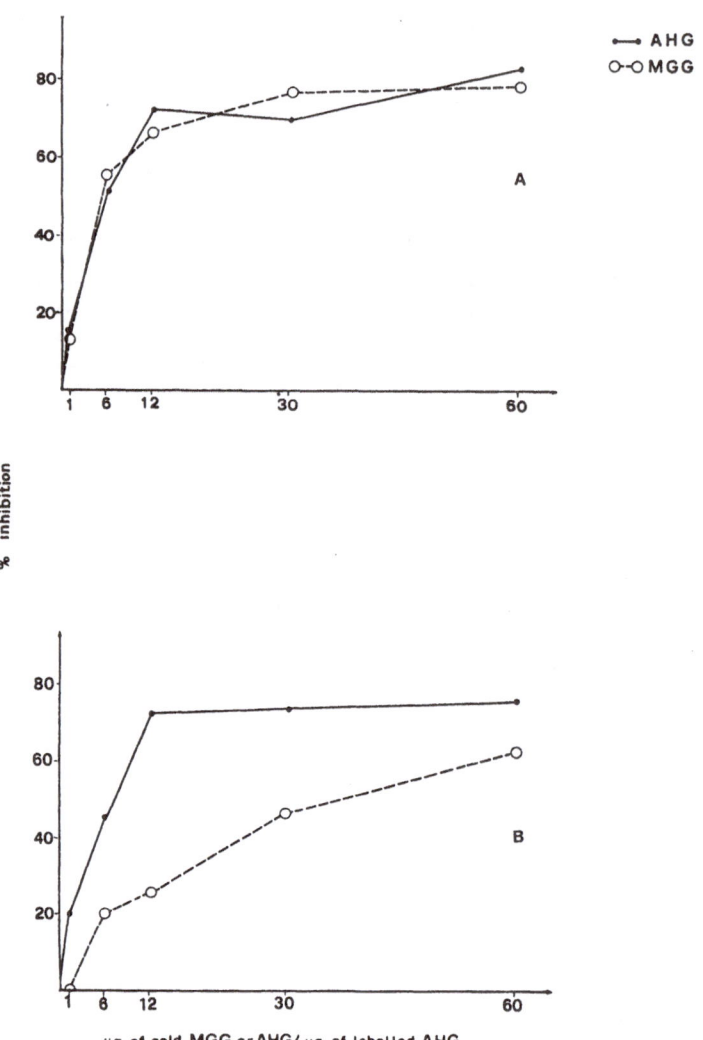

Fig 6.   Inhibition by cold MGG (dashed line) or by cold AHG (full
line) of binding to labelled AHG of the products of clones
PB 3.2 and PB 4.1.

fact, the direct binding assay seemed to indicate a considerable
increase of affinity upon aggregation, which failed however to be
confirmed by the inhibition assay. Conversely, it should be mentioned that the sensitivity of the latter is severely affected by
the concentration of monoclonal antibodies in the test: as shown
in Figure 7, the difference, as far as competition for clone product
4.1 is concerned, between MGG and AHG vanished when the antibody was
diluted.

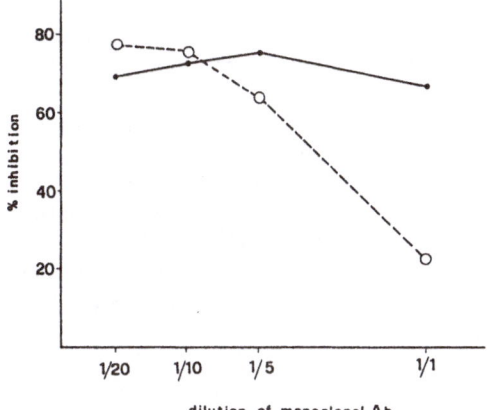

dilution of monoclonal Ab

Fig. 7. Variations, upon dilution of the reaction mixture, of the
inhibitory effect exerted, upon the binding of clone
product PB 4.1 to labelled AHG, by a 12-fold excess of
cold MGG (dashed line) or cold AHG (full line). The undiluted sample was the same in Figure 6b.

As a conclusion, some monoclonal antibodies appear to be able
to detect the formation, in aggregated antigen, of higher affinity
determinants.

Even in the most favourable cases, such increases in affinity
are however of limited extent, no really "new" determinant having
been found within the very limited number of anti-AHG monoclonal
antibodies investigated.

REFERENCES

1. H. Metzger, Effect of antigen binding on the properties of
    antibody, Adv.Immunol. 18:169-207 (1974).
2. J. Schlessinger, I.Z. Steinberg, D. Givol, J. Hichman, and I.
    Pecht, Antigen-induced conformational changes in antibodies
    and their Fab fragments studied by circular polarization of
    fluorescence, Proc.Natl.Acad.Sci. U.S.A. 72:2775-2779 (1975).
3. D. Lancet and I. Pecht, Kinetic evidence for hapten-induced
    conformational transition in immunoglobulin MOPC 460, Proc.
    Natl.Acad.Sci. U.S.A. 73:3549-3553 (1976).
4. V.P. Zav'yalov, V.M. Abramov, V.G. Skortzov, and G.V. Troitsky,
    Conformational transitions of antibodies induced by hapten
    binding. Biochim.Biophys.Acta. 493:359-366 (1977).
5. V.P. Timofeev, I.V. Dudich, Yu.K. Sykulev, R.S. Nezlin and F.
    Franek, Slow conformational change in anti-dansyl antibody as
    a consequence of hapten binding, FEBS Letters 102:103-106 (1979).

6. R. Zidovetzki, Y. Blatt, C.P.J. Glandemans, B.N. Manjula, and
    I. Pecht, A common mechanism of hapten binding to immunoglobulins
    and their heterologous chain recombinants, Biochemistry 19:
    2790-2795 (1980).

7. H. Metzger, The effect of antigen on antibodies: recent studies,
    Contemp.Tp.Mol.Immunol. 7:119-152 (1978).

8. J.K. Wright, J. Tschopp, J.C. Jatan, and J. Engel, Dimeric,
    trimeric and tetrameric complexes of immunoglobulin G fix
    complement, Biochem.J. 187:775-780 (1980).

9. K. Jahnke and W.S. Scholtan, in:"Die blut eiweisskörperin der
    ultracentrifuge" p.47 Geory Thieme Verlag, Stuttgart, (West
    Germany) (1960).

10. J.K. Wright, J. Tschopp, and J.C. Jaton, Preparation and
    characterization of chemically defined oligomers of rabbit
    immunoglobulin G molecules for the complement binding studies.
    Biochem.J. 187:767-774 (1980).

11. G. Koehler and C. Milstein, Continuous cultures of fused cells
    secreting antibody of predefined specificity, Nature 256:495
    (1975).

12. S.J. Singer, Structure and function of antigen and antibody
    proteins, in:"The Proteins", vol.3 H. Neurath, ed., p.269-357
    (1965).

13. A. Gandolfi and R. Strom, Analysis of binding curves in multi-
    valent antigen-heterogeneous antibody systems, J.Theor.Biol.
    92:57-84 (1981).

II: STRUCTURE/FUNCTION INTERRELATIONSHIPS IN IgM AND IgG

INTRODUCTION

For IgM a strictly associative model cannot be correct since it appears that a single molecule of IgM acquires the ability to bind and activate Cl when the IgM becomes attached to a cell surface. Indeed, IgM has the advantage of being a "built-in" aggregate, so that the degree of intermolecular crosslinking is not a variable altered by attachment. Since IgM is already an aggregated structure, this implies that sites on the Fc responsible for the activation of complement cannot be exposed or preformed to a significant extent prior to the interaction of the combining sites on the Fab with antigen; thus a conformational change in the structure of the Fc region seems essential. However, there does appear to be some interaction between unliganded IgM in solution and Clq. If we accept the concept that the structure of a protein in solution is not static but rather a dynamic equilibrium between all of its kinetically accessible forms, then the sites which interact with complement may flicker into and out of existence prior to the interaction of IgM with antigen. In this case, there would be a small background rate of complement binding and, perhaps, activation in solutions of un-aggregated and unliganded IgM. Then, when IgM binds to a cell surface, this dynamic equilibrium would be much shifted in favor of the forms possessing the sites which interact with complement, and in consequence the rate would be greatly enhanced. This shift in the dynamic equilibrium of protein structures would then represent the conformational change.

What is the nature of the conformational change which occurs when IgM binds to antigens? This is the question explored in the first two contributions to Topic II. Arnold Feinstein and his colleagues develop the concept, originally proposed by them, that this conformational change is induced by the multivalent binding of IgM to a polyvalent antigen causing a distortion from a planar structure into a "staple" form. They now present evidence that a simple steric model in which movement of the $(Fab')_2$ arms exposes sites on the unaltered $(Fc)_5$ is <u>not</u> correct, but that a rearrangement of the Fc disk must also take place. And they discuss the possibility that the $C1r_2C1s_2$ tetramer need bind to an additional site on IgM for activation to occur.

In contrast to the distortive mechanism just discussed, Crossland and Koshland summarize the case for an allosteric mechanism for IgM activation, referring to previously published studies indicating that while activation of the Fc function was proportional to the <u>number</u> of combining sites occupied, it was not proportional to the <u>degree</u> of <u>aggregation</u> of the antigen. In an attempt to resolve the differences between their results and the results of others indicating a requirement for polyvalent antigen, Crossland and Koshland report new and interesting experiments using an anti-phenylarsenate system, which apparently bind ligands through a strong charge, charge interaction without properly filling the combining site. Even multivalent binding to a polyvalent antigen does not result in significant complement fixation in this system, and they conclude that "a major determining factor in the expression of IgM effector function appears to be the ability of the antigen to fit approximately within the IgM active site and induce the propagation of conformational changes."

In contrast to the undeveloped sites on IgM, the C1q binding sites on IgG appear to be fully exposed to the solution in the intact antibody. It seems quite possible that a model for the structure of the C1q binding site is to be found on the surface of the models for the Fc region developed from X-ray studies of IgG fragments and from sequence conservation analyses. Dennis Burton and his colleagues describe in detail the evidence which has led them to propose a particular charged region on the surface of IgG, which may be involved in C1q binding. In addition, they have extended their analysis of solvent accessibility and sequence conservation to the other constant domains on IgG (that is, $C\gamma3$, $C\gamma1$ and C_L). Hydrophobic patches and exposed surface residues showing a high or specific pattern of conservation are described in detail, for these are likely to be sites of protein-protein interactions involved in other aspects of the immune response.

In the last paper of Topic II by R. Bragado and colleagues, the location is reported of the acidic residues responsible for the loss of complement activation capacity upon chemical modification of IgG. Their results identify a few glutamic acid residues located in or near the X face of the $C\gamma2$ domain which appear to be essential, and they discuss the role of charge, charge interactions for IgG-C1q binding.

IMMUNOGLOBULIN M CONFORMATIONAL CHANGE

IS A SIGNAL FOR COMPLEMENT ACTIVATION

A. Feinstein, N.E. Richardson,
B.D. Gorick and N.C. Hughes-Jones

Immunology Department
ARC Institute of Animal Physiology
Babraham, Cambridge CB2 4AT, England

In line with the theme of this meeting, this contribution will confine itself to a consideration of how the IgM molecule could alter its conformation on binding to antigens, so as to provide a signal for the binding and subsequent activation of C1, the first component of the classical complement system. A model for the IgM molecule (based on electron microscopy and other evidence) which we have found useful is shown in Figure 1.[1]. In each of the five subunits a pair of $C\mu2$ domains is present where a hinge region would be found in the IgG molecule.

In the case of IgG, Metzger[2] has reviewed the evidence that there need be no requirements for a conformational change to occur when IgG interacts with antigens to form lattices, since the aggregation alone of IgG molecules allows polyvalent binding of C1 to the Fc regions of IgG molecules. However, in the case of IgM, some type of conformational change must be involved when IgM is bound to particulate antigens, since a single molecule of IgM can activate a C1 molecule[3,4,5].

We proposed[6,7,8] that the IgM molecule was activated by multivalent binding to antigens. Comparing electronmicrographs of IgM, free and bound to particulate antigens, it appeared that the free IgM molecule exists in a planar form, with its $F(ab')_2$ regions in the same plane as the central $(Fc)_5$ disc. Binding to antigens caused dislocation of the $F(ab')_2$ units out of the plane, resulting in a distortion which allowed C1 binding to the IgM, leading to C1 activation. The new stabilized conformation of IgM was described as a 'staple' form, and our recent use of freeze-etching[9] confirmed the nature of the attached IgM conformation seen earlier in negatively stained preparations. There appeared to be relatively little flexibility within $F(ab')_2$ units.

DIAGRAM OF THE ARRANGEMENT OF THE DOMAINS
IN THE IgM MOLECULE.

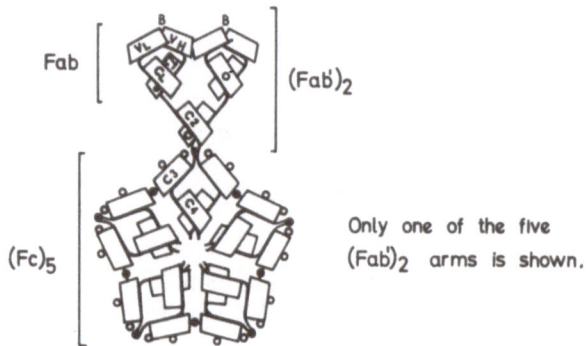

Only one of the five
(Fab')$_2$ arms is shown.

- Interchain disulphide bridge
- Carbohydrate
- Antigen-binding site
- Regions of polypeptide chains folded into domains.

Fig. 1.

Interesting results supporting this model were obtained[10] using a bivalent ligand of appropriate length to cross-link F(ab')$_2$-regions intramolecularly. The model is in general supported by the demonstration by many groups that optimal IgM/antigen ratios exist for fixing and activating complement. Complexes made in excess antigens are not active, demonstrating that the mechanism is not allosteric. An apparent exception to this was the finding that the binding of monovalent ligands by IgM antibody, (where IgM distortion by cross-linking would appear to be impossible) could activate the antibody to fix complement. However, Metzger[2] calculated that only 0.02% of the antibody sites would have been occupied in these experiments, and the matter remains unresolved.

Further support for the distortional model was obtained in experiments showing that as the hapten density on liposomes was increased, complement fixation to bound IgM was enhanced, presumably due to an increase in the extent of multiple binding of IgM molecules[13].

Recent studies[14] have also indicated that varying hapten density on red cells affects the binding of C1 to attached human IgM antibody molecules.

In our own complement studies we have investigated several human and mouse IgM preparations for their ability to bind C1q and to bind and activate C1. The studies on C1 activation have been carried out both by C1 fixation[15]; Reid, Richardson and Feinstein, in preparation, and by the use of ^{125}I-C1 where the rate of cleavage of C1r-C1s has been determined[16]. In the interpretation of the data, two separate phenomena have to be investigated. First, the nature of the alteration in the IgM molecule which follows binding to antigen and which results in the subsequent binding and activation of C1. The two possibilities to be considered are the allosteric and distortive models[2]. In the allosteric model changes in antigen binding sites result in the appearance of a distal C1 binding site; in the distortive model, movement of the F(ab')$_2$ arms in relation to the Fc piece results in exposure of a binding site. Limited removal of Fab arms from pig IgM antibody produces molecules which will agglutinate antigen but not fix complement[17].

Secondly, there is the mode of activation of the C1r-C1s tetramer. Two possibilities may be considered. 1) That one C1 molecule bound to a single IgM molecule is sufficient to bring about activation. The mechanism involved may be either a) through binding of C1q alone which results in a conformational change lead to C1r-C1s activation, or b) that the C1r-C1s tetramer must also bind the IgM before activation can be initiated. 2) That two C1 molecules bound to a single IgM molecule may be necessary for activation. This model envisages an interaction between C1 molecules, such as the necessity for C1r-C1s to span between two C1q molecules. The requirement for the IgM molecules to be close neighbours for C1 activation (the associative model[2] for IgG) is unlikely since the evidence that IgM molecules activate independently of each other is well-established[3,4,5].

Human myeloma IgM (Qu). Two species of (Fc)$_5$ fragments were prepared. Treatment of IgM with trypsin at 60°C results in a fragment consisting mainly of the Cµ3 and Cµ4 domains but carrying also a portion of the Cµ2 domain with an oligosaccharide attached. Subsequent treatment of this tryptic fragment with either soluble or insoluble preactivated papain resulted in removal of the Cµ2 fragment. The first preparations of these (Fc)$_5$ species were tested for C1 fixation and it was found that the tryptic (Fc)$_5$ had no activity whereas the papain (Fc)$_5$ had several-fold higher C1-fixing activity than the original non-complexed IgM (Table 1). The papain (Fc)$_5$ was prepared using insolubilized papain and it was subsequently shown that this resulted in the presence of a significant quantity of oligomeric (Fc)$_5$ molecules. Preparations enriched with oligomeric species were found to have a greater fixing ability than preparations depleted of oligomers[15]; Reid, Richardson and Feinstein in preparation. Later preparations were made using soluble papain and contained a much lower proportion of oligomers. These preparations did not bind C1q to any greater extent than the original IgM preparation (Table 2), all functional affinity constants (K) falling in the range 3-6 x 10^5 M^{-1}. Moreover, using ^{125}I-C1, neither trypsin nor papain (Fc)$_5$

Table 1. Human Cl fixation to human IgM and fragments
          (non-complexed)

Material tested	Moles $\times 10^{11}$ giving 50% Cl fixation	Functional affinity constant* K (M^{-1})
Human IgM (Qu)	45	0.03×10^8
Tryptic (Fc)$_5$	Nil fixing at 50	0.03×10^8
Papain (Fc)$_5$	1	1.5×10^8
IgG heat aggregated (control)	3	0.5×10^8

*These values were calculated assuming that every molecule
 was capable of binding Cl.

activated Cl any faster than did the original IgM preparation
(Figure 2). The interpretation of the binding constants must be
made in the light of the results obtained with IgG antibody. The
balance of evidence is in favour of IgG molecules having a preformed
Clq-binding site on the Fc piece and that this site binds with a
value of K of the order of $5 \times 10^5 M^{-1}$; there is about a 1000-fold
increase in the value of K when Clq binds by two of its heads to
two adjacent IgG molecules[18]. The value of K obtained for monomer
IgM in solution is about 10-fold higher than that obtained for monomer
IgG. This could indicate that there is one binding site on an IgM
molecule for Clq and that this bond is firmer than that with IgG.
Following binding to antigen the value of K rises 100-fold (see later
section) indicating that other Clq-binding sites have become available
on the molecule. If the suggestion that there is only one site available for binding when the pentameric IgM or (Fc)$_5$ molecules is in
solution is correct, then the presence of only one site on a pentameric molecule would indicate that the molecule is asymmetrical.
Our finding that oligomeric (Fc)$_5$ will fix Cl efficiently may be explained in that the aggregation has brought about either association
of the weaker binding sites on the (Fc)$_5$, or has caused distortion.

Table 2. Functional affinity constants for the binding
          of ^{125}I Clq with human IgM and fragments
          (non-complexed).

Material tested	Amount used	K (M^{-1})
Human IgM (Qu)	2.5mg	6.0×10^5
	2.5mg	5.0×10^5
Tryptic (Fc)$_5$	1.0mg	3.0×10^5
	0.5mg	3.0×10^5
Papain (Fc)$_5$	1.0mg	3.5×10^5
	0.5mg	3.0×10^5

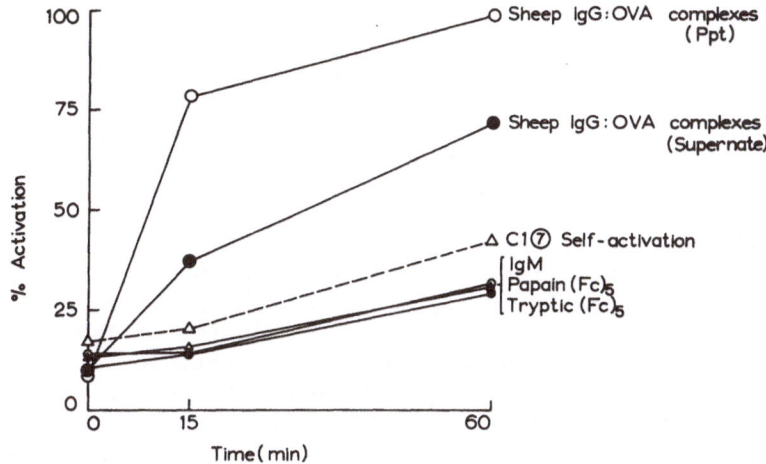

Fig. 2. Activation of ^{125}I-C1 by human IgM (Qu) and (Fc)$_5$ fragments (non-complexed). The spontaneous rate of ^{125}I-C1 self-activation is shown and was not increased by the presence of intact IgM or papain and tryptic (Fc)$_5$ fragments. The upper two curves are positive controls using sheep IgG - ovalbumin immune complexes. The extent of activation of ^{125}I-C1 was determined by measurement of the rate of cleavage of the C1r-C1s tetramer.

Human IgM anti-I. We have also examined the C1-fixing and lytic activities of a cold agglutinin human IgM(Pa) and its F(ab')$_2$ and IgMs derivatives[15]; Reid, Richardson and Feinstein, in preparation. The IgMs (7S) subunits bound to rabbit erythrocytes were as effective as the original pentameric IgM in C1-fixation but were much less effective than the pentamers in lytic activity. An F(ab')$_2$ preparation had no fixing activity. These results indicated that the C1q fixation site is present on the Cμ3-Cμ4 region of IgM and that the site is available on IgMs subunits when these are bound to antigen. Studies on the binding of C1q to the IgMs anti-I subunits were not carried out but the IgMs subunits from the human myeloma IgM when free in solution were found to bind only weakly to C1q with the same order of binding constant as the free intact IgM (K = 5-7 x 10^5M^{-1}). The failure of IgMs to bring about the lysis of cells despite the binding of C1 could be explained on the basis that a site is also required on the IgM for C1r-C1s binding and that these are not available on IgMs.

Mouse MOPC 104E IgM. We have examined the behaviour of the mouse myeloma antidextran MOPC 104E in relation to C1 binding and activation. We found that this IgM can be bound to dextran in saturating amounts to give complexes containing IgM:dextran in a ratio of 4:1 by weight[19] and that these complexes will efficiently

bind C1q, there being one C1q molecule bound for every molecule of
IgM (Table 3) and that such complexes will bind and activate C1
(Figure 3). On the other hand, IgM combined to dextran in the ratio
0.05:1 binds C1q much less efficiently (1 C1q molecule per 10 complexed IgM molecules) but with a binding constant of $3 \times 10^7 M^{-1}$, a
value similar to that found for 4:1 ratio complexes. In addition,
the 0.05:1 ratio complexes did not activate C1 (Figure 3). When we
examined these IgM dextran complexes by electron microscopy, we
found that those containing a high ratio of IgM to dextran appeared
to contain almost exclusively IgM molecules bound to single particles
of dextran in the 'staple' form, whereas complexes formed in low
weight ratios of IgM:dextran contained a majority of IgM molecules
involved in cross-linking between different particles of dextran.
These latter findings appear to be analogous to our previous findings
using IgM from various species of IgM antibody to Salmonella flagella
where again the ratio conditions gave either IgM molecules attached
in the 'staple' form to a single flagellum or cross-linking different
flagella. The value of the functional affinity constant for the
binding between C1q and free IgM in solution was found to be $5 \times
10^5 M^{-1}$. In contrast binding of C1q to IgM complexed to dextran was
increased 30- to 60-fold (Table 3). These results are in contrast
to the finding of no difference in binding C1q between free and complexed equine IgM[20].

In the experiments with the 4:1 ratio complexes where activation
took place, the IgM was in about 100-fold molar excess over the C1
and hence the chance of two C1 molecules being present on a single
IgM molecule must have been very low. It is thus more likely that
a single C1 molecule on one IgM is sufficient for activation. The
failure of the 0.05:1 complexes to activate C1 following binding
could be consistent with the possibility that the C1r-C1s tetramer
requires an additional binding site on Ig and that none is available
in the cross-linked configuration of IgM.

Table 3. Human ^{125}I-C1q fixation to Mouse IgM/Dextran
          complexes.

Material tested	Functional affinity constant $K(M^{-1})$	*Sites/IgM bound
Mouse IgM (non-complexed)	0.05×10^7	–
4 IgM/1 Dextran	1.6×10^7	1:1
1 IgM/1 Dextran	1.25×10^7	0.5:1
0.05 IgM/1 Dextran	3.0×10^7	0.1:1

*This value represents binding sites on IgM for C1q
 determined by Scatchard plot analysis.

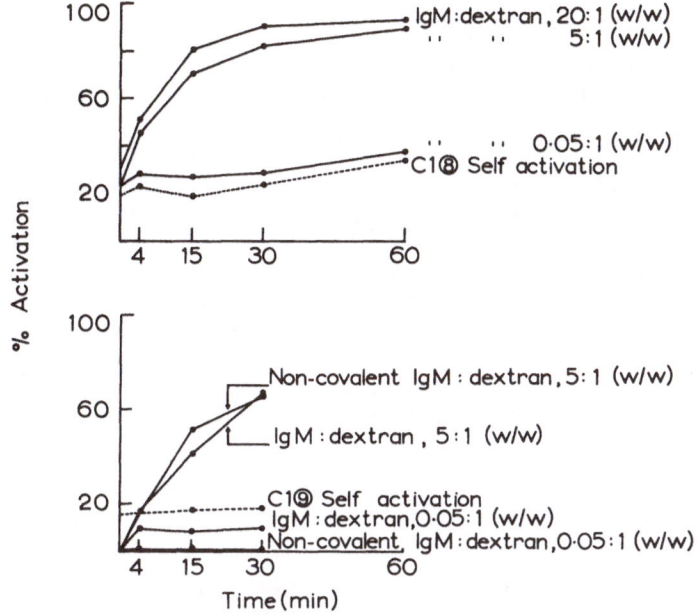

Fig. 3. Activation of ^{125}I-Cl by mouse IgM:dextran complexes. Top: rate of activation of ^{125}I-Cl by complexes made at various ratios of IgM:dextran; rate of spontaneous activation of ^{125}I-Cl is also shown. Bottom: rate of activation of ^{125}I-Cl by either native or reduced non-covalent pentameric IgM complexed at various ratios with dextran.

We have also carried out experiments using mouse IgMs subunits and F(ab')$_2$ fragments complexed with dextran. IgMs subunits in solution do not bind Clq to any greater extent than intact IgM but our preliminary results show that when complexed with dextran at the same weight ratio as with pentamer IgM (2:1) they will fix Clq, with similar functional affinity constants, but with only 10% of the number of Clq binding sites. In parallel experiments with ^{125}I-Cl, the Cl bound to the IgMs-dextran complexes but was not activated. These results are compatible with those obtained with the human IgMs anti-I, which fixed Cl well, but gave very little complement mediated hemolysis, which again suggests the requirement for a Clr-Cls binding site on intact IgM molecules.

Interestingly, both the IgMs subunits and F(ab')$_2$ preparations bound to dextran with association constants of 3×10^6 and $0.7 \times 10^6 M^{-1}$ respectively showing that their binding was bivalent, (the association constant of IgM binding monovalently to the hapten (nigerosyl-α (1→3)-nigerose) is $3.6 \times 10^4 M^{-1}$)[21].

Reduced mouse and human pentamer IgM

We have discovered conditions (unpublished) where we can reduce the inter-Fab' disulphide bridge in mouse or human IgM resulting in non-covalently pentameric IgM molecules which will dissociate into covalent subunits in SDS. The mouse non-covalent pentamers precipitate with dextran to form complexes similar to those obtained with native mouse IgM and we have found that their behaviour as regards C1q binding (Figure 4) and C1 binding and activation (Figure 3) is unaltered relative to native IgM. We obtained essentially the same results as regards C1 fixation with non-covalent pentamers produced from the cold agglutinin IgM in that when these molecules were bound to rabbit erythrocytes they were almost as effective as the native IgM in fixing 50% of the offered C1. This is in contrast with the loss of activity of rabbit and human IgG on reduction of the inter-heavy chain bridges.

Conclusions

1. The available evidence indicates that the C1q-binding site is present on the Cμ3-Cμ4 domains of IgM, although there is no definite evidence against the suggestion that there may be a site present on the Cμ2 domain.

2. The evidence obtained from binding constant values indicates that there maybe one binding site for C1q on the IgM molecule and that further sites become available following binding to antigen. No evidence was obtained that removal of the $F(ab')_2$ arms revealed any additional binding sites. This indicates that a simple steric model in which movement of $F(ab')_2$ arms exposes sites on the unaltered $(Fc)_5$ is not correct. The $(Fc)_5$ must undergo further conformational change. It is proposed then that the binding of more than one IgMs subunit is required, to bring about a distortional change corresponding to 'staple' appearances seen in the electron microscope[6] which we have previously proposed were the active forms. However, the active distortional form appears to involve a rearrangement of the $(Fc)_5$ disc. This mechanism has the advantage that a variety of attached distorted forms could be active.

   It would be difficult experimentally to rule out a proposal that although allosteric effects involving the binding antigen sites are not sufficient, this may make a necessary contribution to the activation of the complexed IgM molecule. However, it is hardly conceivable that such a mechanism could transmit a signal across the various forms of the dislocated attached IgM molecules.

3. IgMs subunits bound to antigens have binding sites for C1q but have a considerably lower binding capacity than intact IgM when

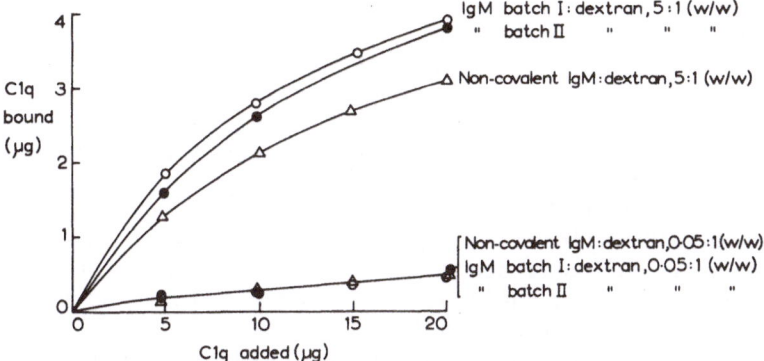

Fig. 4. Binding of human ^{125}I-Clq by either native or reduced
non-covalent pentameric mouse IgM complexed at various
ratios with dextran.

compared on a molar basis.

4. Non-covalent IgM in which the intrasubunit disulphide bridge
is reduced is as effective as native IgM in both binding Clq
and Cl activation.

5. Our observation that the conformation of IgM present in complexes
formed very low IgM/dextran ratios (ie. mainly cross-linking
IgM) binds Cl without activating it can be explained in either
of two ways:

(a) The spatial arrangement of sites for binding the globular
heads of Clq is inappropriate for the activation of
Clr-Cls.

(b) The Clr-Cls tetramer needs to bind to an additional site
on IgM for activation to take place, and this site is not
available under these conditions.

Acknowledgements

We are grateful to Dr. D. Beale for helpful discussion,
Dr. M.E. Slodki for α-1→3 dextran and Miss A. Froggatt, Mr. N.R.
Brown and Mr. N. Buttress for their technical assistance.

REFERENCES

1. A. Feinstein, Conclusions. An IgM model, in:"Prog.Immunol.(II)"
L. Brent and J. Holborrow, eds., North Holland, Amsterdam 115
(1974).

2. H. Metzger, The effect of antigen on antibodies: Recent studies
    in:Cont.Topics Molecular Imm. 7. R.A. Reisfeld and F.P. Inman,
    eds., Plenum Press, New York. 119. (1978).
3. T. Borsos, and H.J. Rapp, Hemolysin titration based on fixation
    of the activated first component of complement. Evidence that
    one molecule of hemolysin suffices to sensitize an erythrocyte.
    J.Immun. 95:559 (1965).
4. T. Borsos and H.J. Rapp, Complement fixation on cell surfaces
    by 19S and 7S antibodies, Science N.Y. 150:505 (1963).
5. T. Ishizaka, T. Tada, and K. Ishizaka, Fixation of C' and C'la
    by rabbit γG and γM antibodies with particulate and soluble
    antigens, J.Immun. 100:1145 (1968).
6. A. Feinstein, E.A. Munn, and N.E. Richardson, The three-dimen-
    sional conformation of γM and γA globulin molecules, Ann.N.Y.
    Acad.Sci. 190:104 (1971).
7. D. Beale and A. Feinstein, Structure and function of the constant
    regions of immunoglobulins, Q.Rev.Biophys. 9:135 (1976).
8. A. Feinstein and D. Beale, Models of immunoglobulins and antigen-
    antibody complexes, in:"Immunochemistry 'An advanced textbook',"
    L.E. Glynn and M.W. Steward, eds., Wiley, Chichester. 263 (1977).
9. E.A. Munn, L. Bachmann, and A. Feinstein, Structure of hydrated
    immunoglobulins and antigen-antibody complexes. Electron mi-
    croscopy of spray-freeze-etched specimens, Biochim.Biophys.Acta.
    625:1 (1980).
10. F. Karush, M-M Chua, and J.D. Rodwell, Interaction of a bivalent
    ligand with IgM anti-lactose antibody, Biochemistry 18:2226
    (1979).
11. J.C. Brown and M.E. Koshland, Activation of antibody Fc function
    by antigen-induced conformational changes, Proc.Natl.Acad.Sci.
    U.S.A. 72:511 (1975).
12. H-C. Chiang and M.E. Koshland, Antigen induced conformational
    changes in IgM antibody. I. The role of the antigenic deter-
    minant, J.Biol.Chem. 254:2736 (1979).
13. G.M.K. Humphries and H.M. McConnell, Membrane-controlled de-
    pletion of complement activity by spin label-specific IgM,
    Proc.Natl.Acad.Sci.U.S.A. 74:3537 (1977).
14. T. Borsos, R.M. Chapuis and J.J. Langone, Distinction between
    fixation of C1 and the activation of complement by natural IgM
    antihapten antibody: effect of cell surface hapten density,
    Mol.Immunol. 18:863 (1981).
15. A. Feinstein and N.E. Richardson, Structure and Activity of
    IgM, in:"Immunology 1978," J. Gergely, G.A. Medgyesi, and S.R.
    Hollan, eds., Akademiai Kiado, Budapest. 257 (1978).
16. E.J. Folkerd, B. Gardner, and N.C. Hughes-Jones, The relation-
    ship between the binding ability and the rate of activation
    of the complement component C1, Immunology 41:179 (1980).
17. D. Beale and J.K. Fazakerley, The action of pepsin on porcine
    immunoglobulin M and its effect on biological activity.
    Biochem.J. 191:183 (1980).

18. N.C. Hughes-Jones, Functional affinity constants of the reaction between ^{125}I-labelled C1q and C1q binders and their use in the measurement of plasma C1q concentrations, Immunology 32:191 (1977).
19. A. Feinstein and N.E. Richardson, Tertiary structure of the constant regions of immunoglobulins in relation to their function, Monogr.Allergy. 17:28 (1981).
20. R.C. Siegel and R.E. Cathou, Conformation of immunoglobulin M. III. Structural requirements of antigen for complement fixation by equine IgM, J.Immunol. 125:1910 (1980).
21. N.M. Young, I.B. Jocius, and M.A. Leon, Binding properties of a mouse immunoglobulin M myeloma protein with carbohydrate specificity, Biochemistry 10:3457 (1971).

EXPRESSION OF Fc EFFECTOR FUNCTION IN HOMOGENEOUS MURINE ANTI-ARS IgM

Kathryn D. Crossland and Marian E. Koshland

Department of Microbiology and Immunology
University of California
Berkley, California 94720

INTRODUCTION

The essential role of protein conformational changes in translating ligand binding into an effective signal or response has been established in many enzyme-substrate[1,2], receptor-ligand[3,4], and other protein-ligand systems[5]. Yet its role in carrying out antibody effector functions remains controversial. The possible roles of allosteric, distortive, and associative mechanisms in antibody function have been reviewed elsewhere[6,7], but it is probable that the relative role of each of these mechanisms may vary significantly with antibody class, and between soluble and membrane-bound antibody molecules.

IgM exists as both a monomeric membrane receptor on the B cell surface, and as a soluble pentameric antibody. As a membrane receptor it is involved in transmission of a signal for the proliferation and differentiation of the immuno-competent B cell; and as the first secreted antibody product of that cell, pentameric IgM provides significant protection by its efficiency in activating the complement sequence. Despite the importance of these functions, the mechanisms

This investigation was supported by Grant AI 07079 from the National Institute of Allergy and Infectious Diseases. K. D. C. was a trainee on Grant CA-09179 to the Department of Microbiology and Immunology.

by which antigen binding induces their expression remains to be
elucidated.

Studies to date have focussed on the secreted pentamer IgM,
since it is both easier to obtain and has the advantage of being a
"built-in" aggregate so that the variable of intermolecular cross-
linking is not a factor. Ligand-induced changes have been detected
in both the Fab and Fc regions of IgM by spectral[8] and depolarization
of fluorescence studies[9,10]. Perhaps the strongest evidence for an
allosteric mechanism of signal transmission in IgM has come from
studies with rabbit anti-phenyl-β-D-lactoside (anti-Lac) IgM[11-14].
Binding of monovalent as well as multivalent antigen was found to
increase the exposure of J chain determinants in the $(Fc)_5$ portion
of pentameric IgM[12]. In addition, by using defined monovalent Lac
antigens, it was shown that monovalent antigen was almost as effective
as multivalent antigen in induction of complement fixation by the
rabbit IgM. The possibility of cross-linking or aggregation by a
small proportion of multi-substituted antigen was eliminated by
coupling of the haptenic determinant to the single active site cysteine of papain[13]. These results indicated that activation of Fc
function was proportional to the number of sites occupied and not
the degree of aggregation or ability of the antigen to crosslink.
Moreover, from measurements of the affinity of the anti-Lac IgM for
the mono-Lac-papain (K_a of 5 x 10^5), it was calculated that maximal
activation was achieved by the binding of 2 moles antigen/mole IgM,
a value in good agreement with that obtained for the requirement for
IgM-activated lysis of haptenated sheep red cells[15].

Studies in other hapten-IgM systems, on the other hand, have
given different results. Siegel and Cathou have found that equine
anti-dansyl IgM antibody activates complement efficiently only in
combination with a large, highly polymerized antigen, such as dansyl-
Ficoll, while other smaller dansyl derivatives are ineffective even
though antigen-antibody complexes are formed[16]. Other studies have
demonstrated a similar requirement for multivalent interaction in
activation of IgM Fc function[17,18].

In an attempt to resolve these differences, we have undertaken
studies of IgM antibodies with a very different specificity from the
neutral Lac determinant. An anti-phenylarsonate system was chosen
because (a) the arsonate hapten, like the dansyl group, is negatively
charged and (b) anti-Ars IgM antibodies have been found both in our
and other laboratories to be relatively poor at complement fixation.
Homogenous IgM antibodies were prepared by somatic cell hybridization
techniques in order to rule out effects due to heterogeneity in
binding affinity. In this paper the complement-fixing properties
of these hybridoma IgM antibodies are analyzed with respect to binding
affinity, specificity of binding, and antigen crosslinking.

MATERIALS AND METHODS

Antibody Preparation

Five hybridomas producing pentamer anti-Ars IgM (2-2, 2-4, 2-5, 2-12, and 2-15) were the generous gift of Maxwell Slomich. These hybridomas were produced by fusion of spleen cells from an Ars-KLH (keyhole limpet hemocyanin) immunized Balb/c mouse with NS-1 myeloma cells, according to standard procedures[19,20]. Hybrids were screened for production of anti-Ars antibody by their ability to lyse Ars-modified sheep red blood cells in the presence of guinea pig complement. Positive hybrids were cloned and maintained in continuous cell culture in RPMI 1640 containing 20% fetal calf serum, in an atmosphere of 10% CO_2-in-air. The anti-Ars hybridoma 36-54.3 was the generous gift of Dr. Vernon Oi and Dr. L. Herzenberg. Anti-Ars IgM was purified from cell culture supernatants by either affinity purification with hapten elution[21], or ammonium sulfate precipitation and gel filtration[20]. Antibody class and concentration were confirmed by a liquid-phase double antibody radioimmunoassay specific for murine IgM[20]. Sodium dodecyl sulfate polyacrylamide gel electrophoresis (SDS-PAGE) was used to evaluate size and purity[20].

The MOPC 104E myeloma, producing an IgM with anti-α (1→3) dextran activity was carried as either an ascites or subcutaneous tumor in Balb/c mice. IgM myeloma protein was purified from serum or ascites by ammonium sulfate precipitation and gel filtration on Sepharose 6B, or by affinity purification on a dextran B1355-poly-acrylamide column[22] with elution by 0.2M glycine-HCl, pH 2.3 at 4°C.

Antigen Preparation

A variety of multi-substituted antigens, including multi-Ars-BSA, multi-Ars-KLH, and multi-Ars-succinylated-papain, were prepared according to standard diazotization procedures[20]. An estimation of the Ars groups/mole was made spectrophotometrically[23,24] and protein concentration determined by the Bradford (BioRad) protein-dye binding assay[25].

Coupling of Ars to Ficoll was accomplished by an adaptation of a procedure used for preparation of dinitrophenol-Ficoll[26]. In a representative preparation, 1.0g Ficoll (T-400, Pharmacia) was dissolved in 4.0ml carbonate buffer, pH 9.7. To the Ficoll solution was added 0.6g cyanuric chloride dissolved in 2.0ml dimethylformamide, and the mixture stirred two minutes at room temperature. Two grams of arsanilic acid, dissolved in 30ml water and adjusted to pH 9.7, were added rapidly, and the mixture reacted for 16 hours with stirring at room temperature. The coupled product was dialyzed extensively to remove excess hapten.

Nigeran oligosaccharides were prepared by mycodextranase diges-
tion of nigeran, a linear glucan of alternating $\alpha(1\to4)$ and $\alpha(1\to3)$
linkages[27]. Multivalent antigens were synthesized by cyanoborohydride
coupling of the isolated oligosaccharides to BSA[28]. Dextran fraction
S from Leuconostoc mesenteroides NRRL B-1355 was a generous gift of
Dr. Allene Jeanes of the USDA Agricultural Research Service.

Association Constant Determination

Radioiodinated p-iodoazobenzenearsonate hapten was synthesized
from its corresponding diazonium compound following a procedure developed by Dr. Leon Wofsy (unpublished), adapted from Lucas and Kennedy[29].
Briefly, a diazonium solution was prepared by dissolving 15mg
arsanilic acid in 0.6ml ice cold 0.2 N HCl. To the ice cold solution
was added 0.3ml of 8.5mg/ml $NaNO_2$. After reaction for thirty minutes
on ice, 2mg of urea was added to quench excess nitrous acid. An
iodide solution containing 5.2mg KI and $1.0mC_i$ of $Na^{125}I$ in 0.1ml
water was prepared and flushed with N_2. The diazonium solution was
added to the iodide solution, flushed with N_2, stoppered, and reacted
for one hour at $80°C$. After cooling, the reaction mixture was
chromatographed on a 1.5 x 35cm Bio-Gel P-2 column to separate the
^{125}I-hapten from free iodine and other reaction products. Concentration of the ^{125}I-hapten was calculated on the basis of cpm/ml,
based on the ratio of cold/hot iodide in the original solution.

The Beckman Airfuge Ultracentrifuge was utilized to determine
the association constants of the anti-Ars hybridoma IgM proteins by
the method of Schachman[30], as adapted by Brown and Koshland[11].

Antibody Characterization

The purified hybridoma proteins were run on two-dimensional gels
essentially as described in Mishell and Shiigi[20], modified from the
procedure of O'Farrell[31]. Binding specificity was assessed in a one-
step solid-phase radioimmunoassay[20], and by quantitative precipitin
assays using ^{125}I-labelled IgM. Complement fixing ability of the
hybridoma proteins was measured by the ability of antigen-antibody
complexes to inhibit complement lysis of sensitized sheep red cells
in a quantitative microcomplement fixation assay system[32], with the
modifications noted in Brown and Koshland[11].

RESULTS

Complement-fixing Activity of Murine Hybridoma Anti-Ars IgM

When five hybridoma anti-Ars IgM antibodies were examined for
their capacity to activate complement in solution, they were all

Table 1.

IgM	Complement Fixation			Association Constant	
	Antigen	Amount IgM at Maximum Fixation	Estimated determinant/ IgM ratio at Maximum Fixation	Antigen	K_a (M^{-1})
2-4 Anti-Ars hybridoma	Ars-BSA	7000ng	200	iodophenyl-arsonate	3.5×10^5
	Ars Ficoll	5000ng	100		
MOPC 104 Myeloma	dextran B1355	75ng	20[a]	nigeran tetra-saccharide	3.6×10^4
	nigeran oligo-saccharide-BSA	300ng	1000		
Normal rabbit anti-Lac	Lac-BGG	90ng	20	p-amino-phenyl-β-lactoside	1.0×10^5

[a]Young, N. M., I. B. Jocius, and M. A. Leon, 1971. Biochem. 10:3457.
[b]Chiang, H.-C., and M. E. Koshland (unpublished).

found to be very inefficient. The two most active preparations required the addition of at least 200Ars determinant mol/mol of IgM (Table 1). The three other IgM hybridomas did not even produce 50% fixation over the entire range of antigen and antibody concentrations tested. These results were dramatically different from the complement-fixing activities evidenced by the murine anti-dextran myeloma IgM, MOPC 104E, and by the rabbit anti-phenyl-β-lactoside IgM. With each of these antibodies, maximum fixation was achieved at significantly lower determinant/IgM ratios, 20 to 50mol/mol, and the 50% endpoints at saturation were 75 to 90ng, almost 100-fold less than that found for the best of the anti-Ars IgM preparations (Table 1).

The inefficiency of the anti-Ars IgM antibodies could not be attributed to structural aberrations. They appeared to be

pentameric molecules by standard biochemical and immunological criteria including SDS polyacrylamide gel electrophoresis, radioimmunoassay, and J chain content. Two-dimensional gel electrophoresis
showed that each of the preparations was homogeneous with respect
to the component μ and light chains and there was no evidence for
heterogeneous molecules that had incorporated the light chain synthesized by the fusion partner. Moreover, it was unlikely that the
hybridoma antibodies represented a non-complement-fixing subclass
of IgM because analyses at the genomic level have established the
presence of a single μ gene per haploid genome[33]. On the basis of
these findings, the poor complement-fixing capacity of the anti-Ars
hybridomas appeared to be related to the interaction of antigen at
the active site, that is, either to the extent of antigen cross-
linking or to the affinity and specificity of antigen binding.

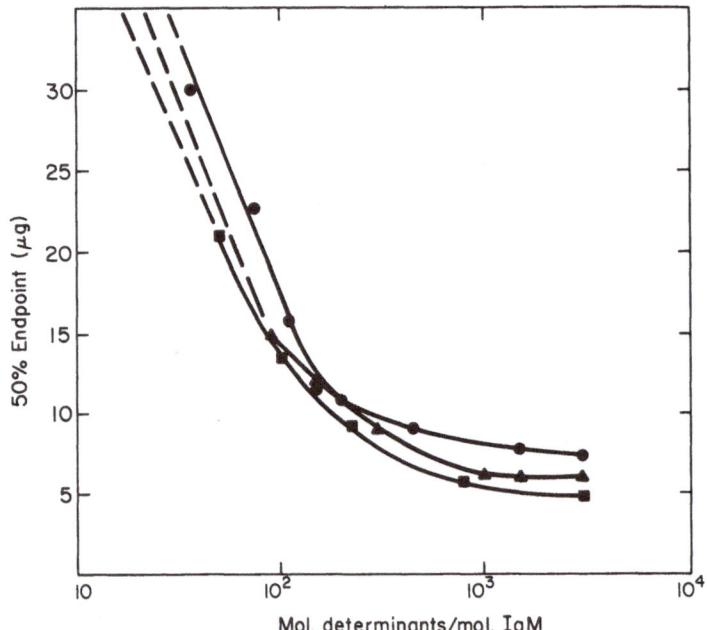

Fig. 1.   Complement fixation of 2-4 anti-Ars IgM with multivalent
          Ars antigens. A microcomplement fixation assay using
          2.0µg of IgM and a range of Ars antigen concentrations
          was performed, and the 50% endpoints were plotted for
          each determinant/IgM molar ratio. Antigens: Ars-BSA (▲);
          Ars-KLH (●); Ars-Ficoll (■). BSA was substituted with 20
          groups/mol; KLH with 40 groups/10^5 daltons; and Ficoll
          with 50 groups/10^5 daltons.

Relationship Between Antigen Crosslinking and the Complement-Fixing of Hybridoma Anti-Ars IgM

The possibility that inadequate crosslinking by antigen might be responsible for the behavior of the anti-Ars hybridomas was investigated by synthesizing a series of multivalent antigens and testing their effectiveness in the micro-complement fixation assay. The results obtained are shown in Figure 1. It is evident that the extent of complement fixation was independent of the size and structure of the carrier used and the degree of hapten substituion. Neither the slope of the complement fixation curve nor the 50% endpoint at saturation was significantly affected whether the Ars hapten was coupled to BSA at a ratio of 20-25 groups/mol, to KLH with an average of 40 groups per 10^5 dalton subunit, or to Ficoll in amounts ranging from 10-15 groups per 10^5 daltons. All these Ars antigens were shown to be functionally multivalent; in quantitative precipitin assays with radioiodinated hybridoma IgM, they gave maximum precipitation in the expected range of 10-200 determinant mol/mol IgM.

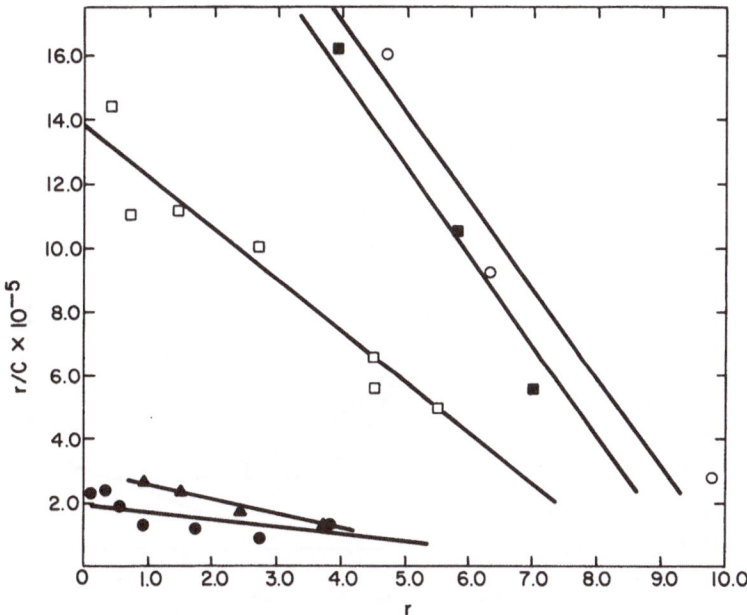

Fig. 2. Association constants of hybridoma anti-Ars IgM antibodies. Measurements of the binding of anti-Ars IgM to ^{125}I-labelled Ars hapten were made in a Beckman airfuge and results expressed on Scatchard plots. Hybridoma anti-Ars IgM: 2-2 (●); 2-4 (○); 2-5 (□); 2-12 (▲); 2-15 (■).

Relationship Between the Binding Affinities of Anti-Ars Hybridomas and Complement-Fixing Activities

The association constant of each of the hybridoma proteins was determined from ultracentrifugation measurements with the ^{125}I-labeled hapten, p-iodophenylarsonate. As expected for homogeneous IgM antibodies, the binding plots obtained (Figure 2) were linear and extrapolated to an IgM valence of 9–10. The anti-Ars IgM from the 2–4 and 2–15 hybridomas was found to have the highest affinity with a K_a of 3.5 x 10^5 1/mol. The IgM from the 2.5 hybridoma had an intermediate K_a of 1.4 x 10^5 1/mol and the IgM from the 2–2 and 2–12 hybridomas gave the lowest K_a of approximately 2.5 x 10^4 1/mol. These binding affinities correlated roughly with the capacity of the anti-Ars antibodies to fix complement. Thus, the 2–4 and 2–15 IgM preparations which had the highest affinity gave a measurable response, whereas the lower affinity 2–5, 2–2, and 2–12 preparations gave a barely detectable response. However, when the association constants of the anti-Ars hybridomas were compared with those of MOPC 104E or anti-phenyl-β-lactoside IgM, it was clear that the

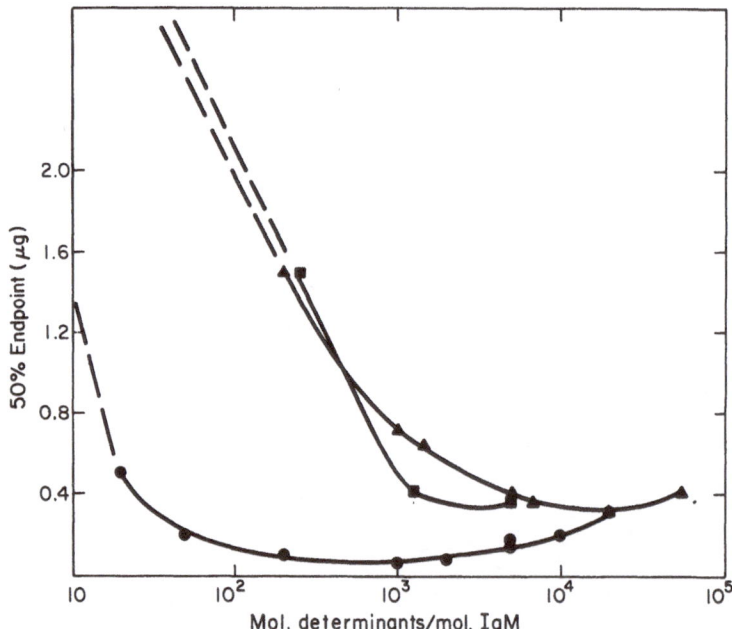

Fig. 3.   Complement fixation of MOPC 104E IgM and nigeral oligo-saccharide-BSA. A standard microcomplement fixation assay was performed using 2.0μg MOPC 104 IgM and varying amounts of antigen. Antigens giving significant complement fix-ation: octasaccharide-BSA (▲); tetrasaccharide-BSA (■); and dextran B1355 (●).

binding strengths of the anti-Ars IgM antibodies could not account for the inefficiency of their complement response. As shown in Table 1, the 2-4 hybridoma anti-Ars IgM bound the Ars hapten with a K_a tenfold higher than that of MOPC 104E IgM for the nigeran hapten[34] and threefold higher than that of rabbit anti-phenyl-β-lactoside IgM for its homologous hapten[11]. Yet the 2-4 preparation was 10-100 times less effective than the anti-CHO IgM antibodies at fixing complement. The difference is illustrated by the complement fixation response of the MOPC 104E IgM shown in Figure 3. This anti-dextran myeloma protein binds the heterologous $\alpha(1\rightarrow3)$ nigeran hapten relatively weakly with a K_a of 3.6×10^4 sites with a nigeran-BSA antigen approached the maximum level of fixation obtained with dextran.

Relationship Between the Specificity of Anti-Ars IgM Hybridomas Complement-Fixing Activity

When the specificity of the anti-Ars IgM hybridomas was examined in a solid phase binding assay, the preparations were found to react strongly with antigens containing negatively charged groups such as succinate and citrate. Competition experiments (Table 2) showed that succinate and citrate haptens were equivalent to the homologous Ars hapten in inhibiting the binding of multivalent Ars antigen to the 2-4 hybridoma. In contrast, an uncharged hapten control, lactose, was unable to inhibit the interactions. This unusual binding site specificity was not confined to the five hybridomas selected from a single fusion experiment. The anti-Ars IgM from an independently derived hybrid cell line, 36-54.3, demonstrated a similar pattern of succinate and citrate cross-reactivity (Table 2).

Table 2.

Hapten[a]	%Inhibition of Precipitation[b] of ^{125}I-Anti-Ars IgM Hybridomas	
	2-4.4	36-54.3
Ars	57.9	65.4
Succinate	57.6	20.6
Citrate	54.6	32.3
Lactose	2.8	0

[a]Hapten added to give 5×10^5 mol hapten/mol Igm.
[b]Ars-BSA added at a ratio of 100 mol Ars determinants/mol IgM.

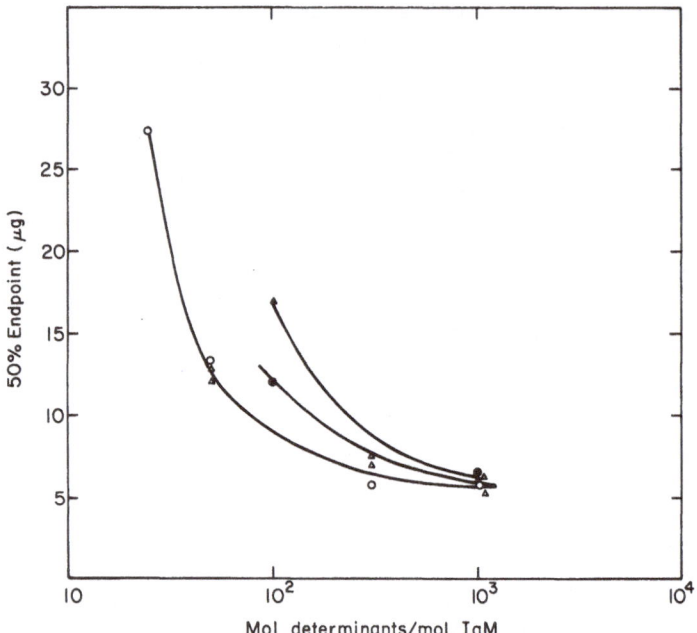

Fig. 4. Complement fixation of hybridoma anti-Ars IgM with Ars-BSA
         (20 groups/mol) succinylated BSA (55 groups/mol). 2-4 with
         Ars-BSA (O); 2-4 with succ-BSA (●); 36-54.3 with Ars-BSA
         (△); 36-54.3 with succ-BSA (▲).

     Moreover, succinylated antigens were almost equivalent to Ars
antigens in the extent to which they induced complement fixation
by the anti-Ars hybridoma antibodies (Figure 4). The shape of the
fixation curves and the saturation endpoints were relatively similar
with both the Ars-substituted and the succinylated antigens despite
the structural dissimilarity of their determinant groups. These
findings suggested that binding in the anti-Ars hybridoma antibodies
involves primarily a charge interaction which is not sufficient to
induce significant expression of the Fc complement reaction. Thus,
the unusual binding site specificity of the anti-Ars IgM hybridomas
appeared to be the major factor responsible for their poor complement
fixing ability.

DISCUSSION

     Previous studies have shown that both the hapten and carrier
portions of a ligand are necessary for antigen triggering of IgM
effector function. Hapten structure plays an important role in
developing the binding affinity which determines site occupancy and
thus the extent of Fc reaction[13], whereas the carrier serves a

separate, non-specific amplifying function[14]. The present studies define a third major factor in the antigen triggering process, the ability of the hapten determinant to fit appropriately within the IgM active site and induce the propagation of effective conformational changes. The evidence was obtained by analyzing the complement response of a series of homogeneous murine IgM antibodies with specificity for the anti-phenylarsonate group. Multivalent Ars antigens were found to elicit only a minimal response from all the antibodies examined. These results could not be attributed to an inadequate binding affinity or an inadequate carrier structure. However, the inefficiency of the triggering could be related to the lack of complementarity between the hapten determinant and the IgM active sites. The finding that such structurally different determinants as arsonate and succinate were bound to the IgM antibodies with almost equivalent affinity indicated that the interaction was based primarily on charge and did not involve any extensive fit between the ligand and antibody groups. Thus, in addition to developing binding affinity, hapten structure also appears to be critical for generating active site changes that contribute to the initiation of the Fc signal.

The existence of homogeneous IgG immunoglobulins that bind structurally diverse antigens has been reported previously (see 35 for review). In general such multiple binding has involved hydrophobic ligands and only rarely have IgG molecules been found to bind dissimilar haptens, each with a reasonable interaction energy, ie. greater than 1×10^4/mol. In contrast, the anti-Ars IgM antibodies described here represent one of the few examples of multiple binding in which high affinity for a charged group is the common feature. Moreover, the pattern of charged group crossreactivity appeared at a high frequency among the anti-Ars IgM hybridomas examined. These characteristics suggest that the IgM response to Ars antigen selects for the expression of germ-line V region genes[36] with broad specificity and high affinity for negatively charged determinants.

The minimal response of the anti-Ars IgM antibodies to antigen triggering indicates that, like many other protein-ligand systems, there can be non-productive as well as productive binding at the active site. If these results with pentamer IgM can be extrapolated to the monomeric IgM membrane receptor, then the question arises as to whether productive and non-productive interactions serve a regulatory function in B cell activation. It is clear that such a regulatory mechanism does not operate in the early stages of B cell differentiation to pentamer IgM secretion; otherwise IgM antibodies defective in their complement-fixing capacity would not be produced. Moreover, there is evidence that the crosslinking of membrane IgM receptors by anti-Ig antibody, in combination with T cell factors, is sufficient to induce pentamer IgM secretion[37], indicating that ligand interaction at the active site is not critical to the initial differentiation signal. However, it may be that later stages of the B cell response, such as the class switch and affinity maturation, do depend upon a productive ligand interaction with the membrane Ig receptor that relays an internal signal via Fc conformational changes. In this way, the stimulus to class switch would result in the selection of antibodies best able to interact with antigen in a productive fashion. This hypothesis is supported by the finding that the normal anti-Ars IgG population lacks succinyl crossreactivity and displays a normal complement fixation response (manuscript in preparation).

ABSTRACT

The activation of IgM Fc function was investigated by measuring the capacity of homogeneous anti-phenylarsonate (anti-Ars) antibodies to fix complement in the presence of multivalent Ars antigen. Only a minimal response was elicited from all the anti-Ars IgM hybridomas examined. The inefficiency of the triggering could not be explained by a low binding affinity for the antigen or by inadequate cross-linking by antigen. The inefficiency could be explained, however, by the unusual binding site specificity of the anti-Ars antibodies. The preparations were found to crossreact strongly with antigens containing unrelated negatively charged groups, such as succinate and citrate. Such haptens not only effectively inhibited the binding of homologous Ars antigens, but succinylated antigens also induced complement fixation to essentially the same extent. The lack of structural similarity among these ligands indicated that their binding to anti-Ars IgM reflects a charge interaction that does not result in significant complement fixation. Thus, a major determining factor in the expression of IgM effector function appears to be the ability of the antigen to fit appropriately within the IgM active site and induce the propagation of conformational changes.

REFERENCES

1. C. Y. Hu, G. J. Howlett and H. K. Schachman, Spectral alterations associated with the ligand-promoted gross conformational change in aspartate transcarbamylase, J. Biol. Chem. 256:4998 (1981).
2. B. Birdsall, A. S. Burgen, D. E. Rodrigues, J. Miranda and G. C. Roberts, Cooperativity in ligand binding to dihydrofolate reductase, Biochem. 17:2102 (1978).
3. A. S. V. Burgen, Conformational changes and drug action, Fed. Proc. 40:2723 (1981).
4. P. A. Insel, Membrane-active hormones: Receptors and receptor regulation, in:"Biochemistry and Mode of Action of Hormones II, Vol. 20, H. V. Richenberg, ed., University Park Press, Baltimore, USA (1978).
5. M. F. Perutz, Regulation of oxygen affinity of hemoglobin, Ann. Rev. Biochem. 48:327 (1979).

6. H. Metzger, Effect of antigen binding on the properties of anti-
    body, Adv. Immunol. 18:169 (1974).
7. H. Metzger, The effect of antigen on antibodies: recent studies,
    in:"Comtemporary Topics in Molecular Immunology," Vol. 7,
    pp. 119, F. P. Inman and R. A. Reisfeld, eds., Plenum,
    New York (1978).
8. H. K. Forster and M. Sela, Conformational changes induced in
    antipoly (L-prolyl) antibodies by oligoproline haptens of
    different sizes, Immunology 16:651 (1979).
9. D. A. Holowka and R. E. Cathou, Conformation of immunoglobulin
    M. I. Characterization of anti-epsilon-1-dimethylamino-5-
    naphthalenesulfonyl-L-lysine immunoglobulin M antibodies from
    horse, pig, and shark, Biochem. 15:3373 (1976).
10. D. A. Holowka and R. E. Cathou, Conformation of immunoglobulin M.
    II. Nanosecond fluorescence polarization analysis of segmen-
    tal flexibility in anti-ε-1-dimethylamino-5-naphthalenesul-
    fonyl-L-lysine IgM-immunoglobulin from horse, pig, and shark,
    Biochem. 15:3379 (1976).
11. J. C. Brown and M. E. Koshland, Activation of antibody Fc func-
    tion by antigen-induced conformational changes, Proc. Natl.
    Acad. Sci. USA 72:5111 (1975).
12. J. C. Brown and M. E. Koshland, Evidence for a long-range confor-
    mational change induced by antigen binding to IgM antibody,
    Proc. Natl. Acad. Sci. USA 74:5682 (1977).
13. H. C. Chiang and M. E. Koshland, Antigen-induced conformational
    changes in IgM antibody. I. The role of the antigenic deter-
    minant, J. Biol. Chem. 254:2736 (1979).
14. H. C. Chiang and M. E. Koshland, Antigen-induced conformational
    change in IgM antibody. II. The role of the carrier, J. Biol.
    Chem. 254:2742 (1979).
15. T. Borsos, R. M. Chapuis and J. J. Langone, Distinction between
    fixation of C1 and the activation of complement by IgM anti-
    hapten antibody: effect of cell surface hapten natural den-
    sity, Mol. Immun. 18:863 (1981).
16. R. C. Siegel and R. E. Cathou, Conformation of immunoglobulin M.
    III. Structural requirements of antigen for complement fix-
    ation by equine IgM, J. Immunol. 125:1910 (1980).
17. C. Fewtrell, M. Geier, A. Goetze, D. Holowka, D.E. Isenman,
    J. F. Jones, H. Metzger, M. Navig, D. Sieckmann, E. Silverton
    and K. Stein, Mediation of effector functions by antibodies:
    Report of a workshop, Mol. Immunology 16:741 (1979).
18. F. Karush, M.-M. Chua and J. D. Rodwell, Interaction of a bi-
    valent ligand with IgM anti-lactose antibody, Biochem. 18:2226
    (1979).
19. G. Kohler and C. Milstein, Continuous cultures of fused cells
    secreting antibody of predefined specificity, Nature 256:495
    (1975).
20. B. B. Mishell and S. S. Shiigi, eds., "Selected Methods in
    Cellular Immunology," W. H. Freeman, San Francisco (1980).
21. L. Wofsy and B. Burr, The use of affinity chromatography for
    the specific purification of antibodies and antigens, J.
    Immunol. 103:380 (1969).

22. S. Carrel and S. Barandun, Protein-containing polyacrylamide
    gels: Their use as immunoabsorbents of high capacity,
    Immunochem. 8:39 (1971).
23. M. Tabachnick and H. Sobotka, Azoproteins. I. Spectrophotometric
    studies of amino acid azo derivatives, J. Biol. Chem. 234:1726
    (1959).
24. M. Tabachnick and H. Sobotka, Azoproteins. II. A spectrophoto-
    metric study of the coupling of diazotized arsanilic acid
    with proteins, J. Biol. Chem. 235:1051 (1960).
25. M. Bradford, A rapid and sensitive method for the quantitation
    of microgram quantities of protein utilizing the principle
    of protein-dye binding, Anal. Biochem. 72:248 (1976).
26. P. R. B. McMaster, R. J. Bachvaroff and F. T. Rapaport, Mechan-
    isms of antibody formation. II. Use of DNP-Ficoll in studies
    of hapten-specific B-cell responses, Cell. Immunol. 16:949
    (1975).
27. K. K. Tung and J. H. Nordin, An enzymic method for the large-
    scale preparation of nigerose and nigeran oligosaccharides,
    Anal. Biochem. 38:164 (1970).
28. G. R. Gray, Antibodies to carbohydrates: Preparation of antigens
    by coupling carbohydrates to proteins by reductive amination
    with cyanoborohydride, Meth. Enz. Vol. L, p. 155 (1979).
29. H. J. Lucas and E. R. Kennedy, Iodobenzene, Org. Syntheses Coll.
    Vol. II, p. 351 (1942).
30. M. A. Bothwell, G. J. Howlett and H. K. Schachman, A sedimen-
    tation equilibrium method for determining molecular weights
    of proteins with a tabletop high speed air turbine centrifuge,
    J. Biol. Chem. 253:2073 (1978).
31. P. H. O'Farrell, High resolution two-dimensional chromatography
    of proteins, J. Biol. Chem. 250:4007 (1975).
32. E. Wasserman and L. Levine, Quantitative micro-complement fix-
    ation and its use in the study of antigenic structure by
    specific antigen-antibody inhibition, J. Immunol. 87:290
    (1960).
33. J. Rogers, P. Early, C. Carter, K. Calame, M. Bond, L. Hood and
    R. Wall, Two mRNAs with different 3' ends encode membrane-
    bound and secreted forms of immunoglobulin μ chain, Cell 20:
    303 (1980).
34. N. M. Young, I. B. Jocius and M. A. Leon, Binding properties of
    a mouse immunoglobulin myeloma protein with carbohydrate
    specificity, Biochem. 10:3457 (1971).
35. F. F. Richards, W. H. Konigsberg, R. W. Rosenstein and
    J. M. Varga, On the specificity of antibodies, Science 187:130
    (1975).
36. J. D. Rodwell and F. Karush, Restriction in IgM expression. I.
    The V_H regions of equine anti-lactose antibodies, Mol. Immunol.
    17:1553 (1980).
37. D. C. Parker, D. C. Wadsworth and G. B. Schneider, Activation
    of murine B lymphocytes by anti-immunoglobulin is an induc-
    tive signal leading to immunoglobulin secretion, J. Exp. Med.
    152:138 (1980).

THE LOCALIZATION OF EFFECTOR SITES ON IMMUNOGLOBULIN G

Dennis R. Burton, Raymond A. Dwek, Jiří Novotný

Department of Biochemistry
University of Sheffield, Sheffield, UK
Department of Biochemistry
University of Oxford, Oxford, UK
Molecular and Cellular Research Laboratories
Massachusetts General Hospital, Boston, USA

The triggering event for a number of immunological effector functions is the recognition of a site on the Fc part of the IgG molecule by a molecule forming a component of the effector function. Whereas high resolution crystallographic structures are now available for pooled human Fc[1] and rabbit Fc[2], very little information is available on the interacting effector function molecules. These are often cell surface molecules who have not yet been isolated in pure form. The best characterised of these molecules is C1q which is a subcomponent of the first component of the complement cascade.

Given our detailed knowledge of the molecular structure of Fc and information on the interaction involved, a number of conclusions about the localization of the effector function binding site on Fc can generally be drawn. In the absence of definitive crystallographic data on complexes, experimental data will hopefully accumulate to enable as precise a localization as possible. In this article, we shall briefly consider how this localization process is proceeding for the C1q and cell receptor binding sites on IgG.

THE C1q BINDING SITE

A number of suggestions have been made for the site on Fc of interaction with C1q. There is general agreement that this site is to be found in the $C\gamma 2$ domain as the Facb fragment[3] and isolated $C\gamma 2$ domains[4] both bind C1q with an affinity similar to that for Fc. Much interest has centered on Trp 277 not least because a peptide corresponding to the sequence flanking this tryptophan (Lys 274 to Gly 281) is reported to inhibit C1 binding to IgG[5]. However both Allan and Isliker[6] and ourselves[7] find that tryptophan modification of Fc has little effect on C1q binding. Furthermore the crystal structure of Fc indicates that Trp 277 is essentially a buried residue with very little solvent accessibility. Boackle et al.[5] have raised the possibility of a "cryptic" C1q site and in a similar vein it has been suggested that a tryptophan residue normally "sequestered" from the surface of the IgG molecule may become exposed on antibody interaction with antigen[8]. If this tryptophan were Trp 277 this would seem to require a fairly substantial conformational change in the Fc region on antigen binding. Available experimental evidence in our view argues against the occurrence of such a change[9].

Work in a number of laboratories indicates that the interaction of C1q and IgG can be inhibited by a variety of molecules[4,7,10-12]. This has been attributed[7] to the presence in these molecules of charged groups since charge is thought to be an important component in the IgG-C1q interaction (see below). As for example Dnp-ethyl phosphate inhibits the interaction at 10^{-3}M, only one order of magnitude less effectively than monomeric IgG ($K_{50} \sim 10^{-4}$M) with some polyions being more effective eg. polylysine ($K_{50} < 10^{-4}$M), the use of peptide inhibitors to mimic an IgG binding sequence is seen to be limited. The peptide inhibitor no.98 described by Boackle[5] is extraordinarily effective as an inhibitor of C1 consumption ($K_{50} \sim 10^{-7}$M compared to 10^{-4}M for monomeric IgG) which brings into question its mode of inhibitory action.

Two C1q binding sites have recently been proposed on the Cγ2 domain. The first by Brunhouse and Cebra[13] involves residues in an extended chain region Lys 290 to Glu 295 in human IgG1 and the second by Burton et al[7] involves residues on the last two antiparallel β-strands of the Cγ2 domain. Both sites contain a number of charged residues and evidence has accumulated in the literature to indicate the importance of charged groups in the IgG-C1q interaction. Our own work[7] has to date probably most thoroughly investigated this phenomenon when we concluded that of the order of 12 ions were released into solution when 1 molecule of C1q bound to an immune aggregate. Although there is a natural tendency, because of the involvement of charge, to describe the C1q-IgG interaction as "ionic", this statement should be treated with caution. Calculations based on our data indicate that roughly 50% of the interaction free energy is electrostatic in origin and 50% non-electrostatic. These figures are subject to very large errors as they involve a substantial extrapolation from data at low to higher salt concentration but do indicate that the interaction between C1q and IgG probably involves both ionic and other, eg. hydrophobic, contributions.

The two most recently proposed hypotheses for the C1q binding sites on Fc draw greatly on comparison of Cγ2 sequences. Brunhouse

and Cebra[13] note that no residue or set of adjacent residues can be
found which absolutely correlate with the ability to bind complement
of the component IgG isotopes. However they find a region of extended chain which is not invariant but shows more conservative
substitutions than might be expected for exposed residues. This
region is from Lys 290 to Glu 295 (human γ1) and a comparison of
available γ sequences in this region reveals the following:

	290					295
human γ1	lys	pro	arg	glu	gln	gln
human γ2	lys	pro	arg	glu	glu	gln
human γ3	lys	pro	arg	glx	glx	glx
human γ4	lys	pro	arg	glu	glu	gln
mouse γ1	gln	pro	arg	glu	glu	gln
mouse γ2a	thr	his	thr	arg	gln	asn
mouse γ2b	gln	thr	his	arg	glu	asp
guinea pig γ1	asx	val	leu	glx	glx	glx
guinea pig γ2	lys	pro	arg	val	glu	gln
rabbit	pro	leu	arg	glu	gln	gln

The generally accepted non-complement fixing molecules are human
IgG4, mouse IgG1 and guinea pig IgG1. Since Fc4 binds C1q although
IgG4 does not[14], it seems that in this case loss of C1q binding
probably reflects Fab arm obstruction of a site rather than disappearance of this site from the IgG4 sequence. For mouse IgG1 and
guinea pig IgG1 this mechanism may also operate although loss of the
non-functional C1q binding site may have then followed in evolution,
precluding the unambiguous use of these sequences in any comparison.
Concentrating on those IgGs known to bind C1q then this proposed site
requires either species specificity in C1q or the ability of C1q to
recognise a variety of structures (compare human vs. mouse γ2a, γ2b
vs. rabbit sequences). Some support for this site has recently come
from chemical modification studies[15].

Burton[7] proposed their site on the basis of a combination of
residue accessibility and sequence comparison analyses, inhibitor
studies and chemical modification studies. They looked for accessible highly conserved residues pointing to the considerable cross-
species reactivity of C1q and IgG and located the following:

	318	320	322	331	333	335	337
human γ1	glu	lys	lys	pro	glu	thr	ser
human γ2	glu	lys	lys	pro	glu	thr	ser
human γ3	glu	lys	lys	pro	glu	thr	ser
human γ4	glu	lys	lys	ser	glu	thr	ser
mouse γ1	glu	lys	arg	pro	glu	thr	ser
mouse γ2a	glu	lys	lys	pro	glu	thr	ser
mouse γ2b	glu	lys	lys	pro	glu	thr	ser
guinea pig γ1	-	-	-	thr/ala	thr	thr	ser
guinea pig γ2	glu	lys	lys	pro	glu	thr	ser
rabbit	glu	lys	lys	pro	glu	thr	ser

- = unsequenced

These residues form a contiguous roughly planar area on the
Y-face of the Cγ2 domain. Inhibitor studies support the importance
of charged groups and reported chemical modification studies lend
support to this site.

While on the subject of charged groups, it is interesting to
note the conservation of two carboxylate amino acid residues at positions 269 and 270 at the N-terminal extremities of the Cγ2 domain.
These two residues are within roughly 20Å of their counterparts on
the other Cγ2 domain. Their involvement in Clq binding we have considered unlikely at least in part because for antibody recognition
of cell surface determinants, Clq binding to this region would be
under extreme, if not prohibitive steric restrictions from the cell.

Another suggested possibility as the Clq binding region is the
Cγ2 carbohydrate. Winkelhake[16] has enzymatically removed the carbohydrate fro IgG and found decreased Clq binding affinity. In
another approach to this problem Leatherbarrow and Dwek in Oxford
are producing aglycosylated monoclonal antibodies and assaying their
functional activity. A major problem is of course to show that
aglycosylation does not radically alter IgG conformation - preliminary results indicate that it does not[17]. Burton et al.[7] find that
a variety of monosaccharides do not inhibit the IgG-Clq interaction.
It seems unlikely that the carbohydrate alone forms the Clq binding
site on IgG, although some involvement cannot be ruled out at this
stage.

In conclusion, interest in possible Clq binding sites on the
Cγ2 domain in IgG seems to have moved somewhat from the Trp 277
region to two regions of relatively high density of charged residues.
One of these regions is highly conserved, the other not. Unequivocal
identification of the site will most likely require either crystallographic studies on a Clq head·Fc complex or a series of chemical
modifications combined with peptide mapping studies.

THE CELL Fc RECEPTOR SITE(S)

There are a number of different cell types possessing surface
Fc receptors capable of binding IgG including monocytes/macrophages,
B- and T-lymphocytes, neutrophils, trophoblasts and natural killer
cells. It is possible that all of these cell types possess receptors
binding to different sites on Fc. This seems intuitively unlikely
and in the human system is argued against by similarities in binding
constant and subclass specificity patterns for human IgG interacting
with Fc receptors on the various cell types (see below).

However in contrast to Clq and IgG, there would seem to be
species specificity in the interaction of cell Fc receptors and IgG.
For instance Leslie[18] has shown that rabbit IgG but not human IgG1

and IgG3 compete with guinea pig IgG2 for the receptor on guinea pig peritoneal macrophages. Also a number of workers[19] have shown the probable existence of different receptors for different mouse IgG subclasses. In contrast human IgG subclasses would appear to compete for the same receptor on human monocytes[20].

There is considerable controversy as to whether Fc receptor binding is to a site on the Cγ2 or Cγ3 domain. This may of course vary between species and IgG subclass. Perhaps the least equivocal report is that of Diamond et al.[21] who find that a variant of mouse IgG2b lacking the Cγ3 domain competes as effectively as IgG2b for the mouse macrophage Fc receptor in a rosetting assay. This clearly implicates Cγ2 as the binding domain in this instance.

A detailed solvent accessibility and sequence conservation analysis of the Cγ2 domain has been presented[7] as described earlier. We have carried out a similar analysis of the Cγ3 domain of human Fc using the coordinates of Huber et al.[1], details of which will appear elsewhere but a few points of interest will be taken up here. The Cγ3 domain of human Fc has two large exposed hydrophobic patches. The first ($\sim$450Å^2) is on the inside of the domains, taking in both Cγ3 domains, and between the two Cγ2 domains. The second patch ($\sim$300Å^2) is on the outside of the Cγ3 domains and again takes in both domains. Both patches are present in all human subclasses and to varying degrees in other IgGs. Only 10 Cγ3 residues are solvent accessible and identically conserved in all available Cγ3 sequences compared to 22 in the Cγ2 domain. The conservation of most of these residues can be understood in terms of Cγ3/Cγ2 contact, general domain architecture or salt bridge formation. There is no region of neighbouring solvent accessible highly conserved residues in Cγ3 as found for Cγ2 and proposed as a potential Clq binding site.

It is interesting to concentrate on the binding of human IgG to human cells when considering possible receptor binding sites on Fc. The principal features of human IgG binding to human monocytes emerging from a number of laboratories[22-28] can be summarised as follows:

1. $K_{ass} \sim 10^6 - 10^8 M^{-1}$ for monomeric IgG1, IgG3 and pooled IgG.

2. IgG1 $\sim$ IgG3 > IgG4 > IgG2 in K_{ass}. Same receptor.

3. pFc$^\prime$: $K_{ass} \sim 2.10^4 M^{-1}$ (pFc$^\prime$1, pFc$^\prime$3 > pFc$^\prime$2, pFc$^\prime$4).

4. Cγ2 : $K_{ass} < 2.10^4 M^{-1}$

5. Hinge sensitivity : K_{ass} decreased by factor $\sim$10 by interheavy S-S reduction and in hinge-deleted Dob.

6. Ciccimarra et al peptide residue nos.(407-416, Cγ3) inhibits IgG binding at $\sim$1μM.

For human IgG binding to human placental (trophoblastic) Fc receptors the following features emerge[29-32]:

1. $K_{ass} \sim 10^6\text{-}10^7$ M^{-1} for monomeric IgG1, IgG3 and pooled IgG.

2. IgG1 $\sim$ IgG3 > IgG4 > IgG2 in K_{ass}.

3. Fc (pool) $\sim$ Fc2 $\gg$ IgG2 (29) Fc1 $\gg$ Fc2 < IgG2 (30) in K_{ass}.

4. pFc' : $K_{ass} \ll 10^5$.

5. Cγ2 : $K_{ass} \ll 10^5$.

6. Hinge sensitivity: K_{ass} decreased by a factor $\sim$ 20 by interheavy S-S reduction and by more in hinge-deleted Dob.

7. Binding sensitive to ionic strength and pH.

Broadly similar features have been observed for the binding of human IgG to human neutrophils and lymphocytes although some differences are found.

Concentrating on subclass specificity, we can find no convincing sequence differences to explain the observed specificity. The solvent accessible part of the peptide in[6] residues 413-416 is identical in the four human subclasses. If this sequence were involved in receptor binding then subclass specificity would require either interference of the Fab arms with this distant region (which at least at first consideration seems unlikely) or a rather complex mechanism involving a nearby loop (residues 384-390) on which there are some minor subclass differences. The major difference between the subclasses is however in the hinge and we would like to speculate that the pattern of subclass specificity observed in cell receptor binding is related to involvement of the hinge directly or indirectly by eg. interfering with critical residues on Cγ2.

pFc'((Cγ3)$_2$) has been reported to inhibit human IgG binding to human monocytes with a subclass specificity the same as that for native IgG[25]. This has been taken as evidence for the importance of the Cγ3 domain in monocyte receptor binding. However it should be noted that pFc' is only effective as an inhibitor at orders of magnitude higher concentrations than IgG (or Fc). Further pFc' is not observed to show any similar inhibitory ability with respect to Fc receptors on human trophoblasts or neutrophils.

In conclusion, identification of cell receptor sites on IgG would seem to present a more complex problem than that of Clq with the likelihood of variations between species and subclasses. Clearly a valuable approach is to isolate a pure receptor molecule and undoubtedly progress in the murine system will be greatly advanced by the work of Unkeless[19] in isolating monoclonal antibody against the IgG2b Fc receptor on mouse macrophages which should permit isolation and molecular characterisation of this receptor. Presumably this approach in the human system will follow and combined with more detailed knowledge on the interaction of human IgG subclasses and the various receptor-bearing cell types allow localisation of the receptor site(s).

REFERENCES

1. R. Huber, J. Deisenhofer, P. M. Colman, M. Matsushima and W. Palm, Crystallographic studies of an IgG molecule and an Fc fragment, Nature 264:415 (1976).
2. B. Sutton and D. C. Phillips, Private communication.
3. M. Colomb and R. R. Porter, Characterisation of a plasmin-digest fragment of rabbit immunoglobulin gamma that binds antigen and complement, Biochem. J. 145:177 (1975).
4. D. Yasmeen, J. R. Ellerson, K. J. Dorrington and R. H. Painter, Structure and function of immunoglobulin domains. IV. The distribution of some effector functions among the Cγ2 and Cγ3 homology regions of human immunoglobulin G, J. Immun. 110: 1706 (1976).
5. R. J. Boackle, B. J. Johnson and G. B. Caughman, An IgG primary sequence exposure theory for complement activation using synthetic peptides, Nature 282:742 (1979).
6. R. Allan and H. Isliker, Studies on the complement-binding site of rabbit IgG-II. The reaction of rabbit IgG and its fragments with Clq, Immunochem. 11:243 (1974).
7. D. R. Burton J. Boyd, A. D. Brampton, S. E. Easterbrook-Smith, E. J. Emanuel, J. Novotný, T. W. Rademacher, M. R. van Schravendijk, M. J. E. Sternberg and R. A. Dwek, The Clq receptor site on immunoglobulin G, Nature 288:338 (1980).
8. R. H. Painter, D. B. Foster, B. Gardner and N. C. Hughes-Jones, Functional affinity constants of subfragments of IgG for Clq, Hoppe-Seylers Z, Physiol. Chem. 362:23 and references therein (1981).
9. C. Fewtrell, M. Geier, A. Goetze, D. Holowka, D. E. Isenman, J. F. Jones, H. Metzger, M. Navia, D. Sieckmann, E. Silverton and K. Stein, Mediation of effector functions by antibodies. Report of a workshop, Molec. Immunol. 16:741 (1979).
10. G. R. Sledge and D. H. Bing, Binding properties of the human complement protein Clq, J. Biol. Chem. 248:2818 (1973).
11. T. Y. Lin and D. S. Fletcher, Interaction of human Clq with insoluble immunoglobulin aggregates, Immunochemistry 15:107 (1978).
12. N. C. Hughes-Jones and B. Gardner, The reaction between the complement subcomponent Clq, IgG complexes and polyionic molecules, Immunology 34:459 (1978).
13. R. Brunhouse and J. C. Cebra, Isotypes of IgG: Comparison of three pairs of isotypes which differ in their ability to activate complement, Mol. Immunol. 16:907 (1979).

14. D. E. Isenman, K. J. Dorrington and R. H. Painter, The structure and function of immunoglobulin domains. II. The importance of interchain disulphide bonds and the possible role of molecular flexibility in the interaction between immunoglobulin G and complement, J. Immunol. 114:1726 (1975).

15. R. Bragado, J. A. Lopez de Castro, C. Juarez, J. P. Albar, A. Garcia-Pardo, F. Ortiz and F. Vivanco, Chemical modification of carboxyl groups in human Fc fragment: II. Location of acidic residues involved in complement activation, Mol. Immunol. in press.

16. J. L. Winkelhake, T. J. Kunicki, B. M. Elcombe and R. H. Aster, Effects of pH treatments and deglycosylation of rabbit immunoglobulin G on the binding of C1q, J. Biol. Chem. 255:2822 (1980).

17. R. J. Leatherbarrow, Private communication.

18. R. G. Q. Leslie and A. H. Niemetz, Species specificity in the binding of IgG to macrophages, Immunology 37:835 (1979).

19. J. C. Unkeless, I. S. Mellman, M. McGettigan and H. Plutner, in:"Purification and characterisation of a mouse macrophage Fc receptor in Heterogeneity of Mononuclear Phagocytes," Forster and Landy, eds., Academic Press, New York, p.92 and references therein (1981).

20. M. D. Alexander, J. A. Andrews, R. G. Q. Leslie and N. J. Wood, The binding of human and guinea pig IgG subclasses to homologous macrophage and moncyte Fc receptors, Immunology 35:115 (1978).

21. B. Diamond, B. K. Birshstein and M. D. Schaff, Site of binding of mouse IgG2b to the Fc receptor on mouse macrophages, J. Exp. Med. 150:721 (1979).

22. H. Huber and H. H. Fudenberg, Receptor sites of human monocytes for IgG, Int. Arch. Allergy 34:18 (1968).

23. N. Abramson, E. W. Gelfand, J. H. Jandl and F. S. Rosen, The interaction between human monocytes and red cells. Specificity for IgG subclasses and IgG fragments, J. Exp. Med. 132:1207 (1970).

24. G. Holm, E. Engwal, S. Hammarström and J. B. Natvig, Antibody-induced hemolytic activity of human blood monocytes. The role of antibody class and subclass, Scand. J. Immunol. 3:173 (1974).

25. G. O. Okafor, M. W. Turner and F. C. Hay, Localisation of monocyte binding site of human immunoglobulin G, Nature 248:228 (1974).

26. F. Ciccimarra, F. S. Rosen and E. Merler, Localisation of the IgG effector site for monocyte receptors, Proc. Nat. Acad. Sci. USA 72:2081 (1975).

27. D. E. Barnett-Foster, K. J. Dorrington and R. H. Painter, Structure and function of immunoglobulin domains. VIII. An analysis of the structural requirements in human IgG1 for binding to the Fc receptor of human monocytes (1980).

28. M. Klein, N. Haeffner-Cavaillon, D. E. Isenman, C. Rivat,
    M. A. Navia, D. R. Davies and K. J. Dorrington, Expression of
    biological effector functions by immunoglobulin G molecules
    lacking the hinge region, Proc. Natl. Acad. Sci. USA 78:524
    (1981).
29. R. Matre, Similarities of Fc receptors on trophoblasts and
    placental endothelial cells, Scand. J. Immunol. 6:953 (1977).
30. J. A. van der Muelen, T. C. McNabb, N. Haeffner-Cavaillon,
    M. Klein and K. J. Dorrington, The Fcγ receptor on human
    placental plasma membrane I. Studies on the binding of homolo-
    gous and heterologous immunoglobulin G, J. Immunol. 124:500
    (1980).
31. P. J. Brown and P. M. Johnson, Fcγ-receptor activity of isolated
    human placental syncytiotrophoblast plasma membrane,
    Immunology 42:313 (1981).
32. M. Niezgódka, J. Mikulska, M. Ugorski, J. Boratynski and
    J. Lisowki, Human placental membrane receptor for IgG. I.
    Studies on properties and solubilisation of the receptor, Mol.
    Immunol. 18:163 (1981).

HUMAN Fcγ FRAGMENT: LOCATION OF ACIDIC RESIDUES INVOLVED IN

COMPLEMENT ACTIVATION*

R. Bragado, J. A. López de Castro
F. Ortiz, F. Vivanco

Department of Immunology
Fundación Jiménez Díaz, Madrid 3

INTRODUCTION

Classical activation of the complement system is initiated by
the binding of Cl to immune complexes or aggregated IgG[1]. It is
generally assumed that the binding site for the complement subcomponent Clq resides on the Cγ2 domain. However, a convincing picture of the precise topography and chemical nature of the complement
binding site is still lacking.

The participation of ionic side chains in the IgG-Clq interaction is suggested by the capacity of a wide variety of ionic substances either to bind to Clq or to inhibit IgG-Clq binding[2-4].

Chemical modification studies also support the idea that ionic
side chains may be important for IgG-Clq interaction. Preservation
of positive charge in the IgG molecule has been shown to be essential
for complement activation[5,6] although it appears not to be so for
Clq binding[7]. The location of the groups involved has not been established.

The importance of negatively charged groups in the Fc portion
of IgG for complement activation has been documented in a previous
report[8]. The potential of this approach for the topographical description of the Clq effector site is based on that: (a) modification
of a few groups results in a complete loss of complement activating
capacity; (b) this appears to be primarily due to chemical modifi-

*The data presented in this communication have been originally
published in Molecular Immunology[12].

cation itself and not to secondary conformational changes; (c) it
is possible to define precise reaction conditions to achieve selective loss of anticomplementary activity. This is attained at 5
minutes reaction; at 1 minute, significant modification occurs which
is hardly, if at all, accompanied by a decrease in complement activation ability.

In this communication, the exact location of the modified acidic
goups in the Cγ2 is reported, as well as an estimate of the extent
to which individual groups get modified at the point of functional
inactivation. The results aim at providing a quantitative insight
into our understanding of the nature of immunoglobulin–Clq interaction.

MATERIALS AND METHODS

Radioactive Chemical Modification

It was performed at 1 and 5 minutes, as previously described[8].
Briefly, 5mg of non-aggregated Fcγl were dissolved in a radioactive
solution of 1M glycine-ethyl-ester (1mCi of 44.7mCi/mmole, NEN) to
give a final specific activity of 200μCi/mmole. Once the pH was
adjusted to 4.75, solid EDC was added to a final concentration of
0.1M. During the time of reaction the pH was maintained at 4.75 by
means of the addition of 0.5M HCl; 2M sodium acetate pH 4.75 was
added to stop the reaction. Unincorporated radioactivity was removed
by exhaustive dialysis against 0.01M NH_4HCO_3. Samples were lyophilized until their use.

Reduction and S-Carboxymethylation

Each of the radioactively modified Fcγl samples (about 10mg/ml)
was reduced under N_2, with 0.01M dithiothreitol in 6M guanidine
hydrochloride, 0.5 Tris/HCl and 0.002M EDTA buffer, pH 8.0, at 37°C.
After 2hrs, iodoacetamide was added to a final concentration of 25mM;
the samples were left 30 minutes in the dark and then immediately
applied onto a column (1.6 x 100cm) of Sephadex G-75 equilibrated
with 1M acetic acid. Fractions containing the reduced and alkylated
Fcγl fragment (MW 25000) were pooled and lyophilized.

Tryptic Cleavage

Protein samples dissolved (5-10mg/ml) in NH_4HCO_3 pH 8.2, were
digested with TPCK-treated trypsin (Merck) at a ratio 1:1000 for
6hr at 37°C. The reaction was stopped by freeze-drying the solution.

Purification of Tryptic Peptides

Microbore ion-exchange chromatography and HPLC were used as
the main procedure for peptide purification in preparative amounts.
Chromatographic conditions were as described elsewhere[9,10,12].

Peptide detection was performed by a fluorescamine assay as
described by Nakai[11]. Radioactive peptides were detected by liquid
scintillation counting of 50-100μl aliquots and by TLC in butanol
acetic acid:water:pyridine, 15:3:12:10, (20 x 20 cellulose plates,
Merck) followed by autoradiography (X-Omat S fims, Kodak).

Assignment of Radioactive Acidic Residues

1. Automated sequence analysis. Automated Edman degradations
were performed in a Beckman 890C sequencer using polybrene[13] and a
0.1M Quadrol program. A blank cycle was run before the addition of
peptide sample.

2. Pth characterization and identification of radioactive peptides. Routinely, 20% of each Pth-amino acid, was employed for
identification and quantitation by HPLC[12]. The remaining portion
(80%) was desiccated and its radioactivity was determined by β-
scintillation counting after dissolving in 2ml of Aquasol.

RESULTS

Purification of Peptides

Figure 1a shows the chromatographic profile of a tryptic digest
of chemically modified Fcγ1 fragment. The peptide maps were identical after 1 or 5 minutes reaction.

The radioactive peptide profile (Figure 1b) suggests that the
same peptides get labelled at 1 or 5 minutes. The amount of label
incorporated in most peptides increased with longer times but at
least T6 and T7 showed no increase in radioactive uptake beyond 1
minute of modification.

The flowthrough peak (T1) of the ion-exchange column contained
a complex mixture of peptides that were resolved by HPLC (Figure 2).
Seven radioactive peptides were obtained. T1-1 and T1-2 as well as
T1-4 and T1-7 in Figure 2 showed to be identical by amino acid compo-
sition and sequence analysis. In addition, T2 (Figure 1) and T1-3
(Figure 2) were identical as judged by the same criteria.

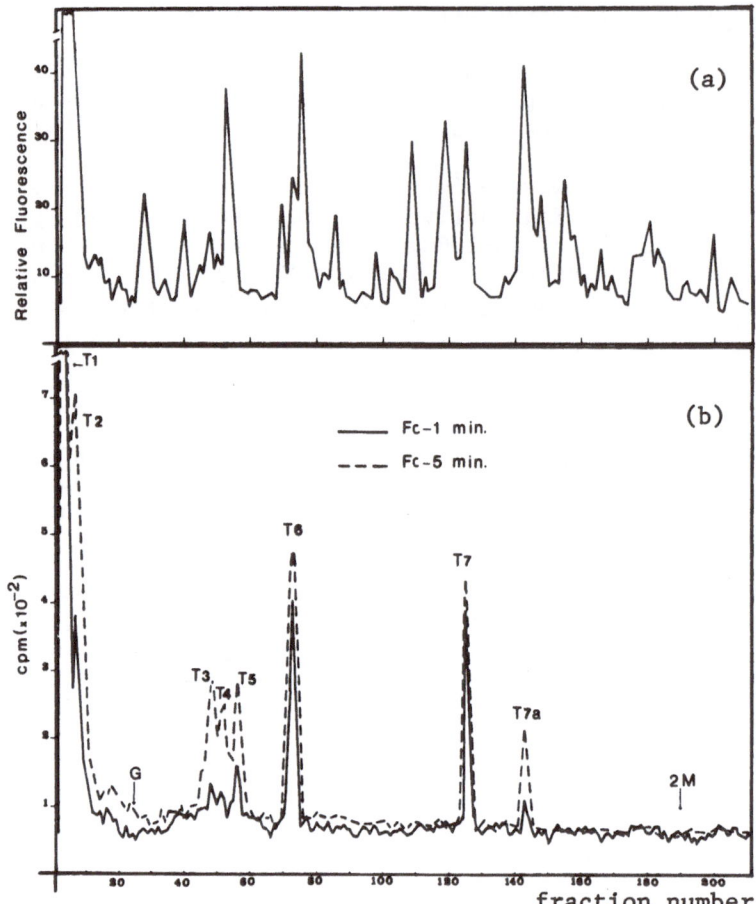

Fig. 1. Purification of tryptic peptides. Ion-exchange chromatography of tryptic peptides from modified Fcγl fragment at
1 and 5 minutes. (a) Fractions were monitored by fluorescamine reaction. The chromatographic profile was identical for both Fc-1' and Fc-5'. (b) Peptide map of radiolabelled tryptic peptides. Peptides identified in each
peak are stated. Peptide T7a was shown to be identical to
T7.

        Amino acid sequence of all radioactive peptides from both Fc-1'
and Fc-5' was determined in order to determine their position in
the Fc molecule and to establish the nature of the modified residues
as well as the extent of their modification. The results are shown
in Table 1. Five peptides belonged to the C 2 domain. A total of
13 residues, of which only six were located in the C 2 region, were
modified at both one and five minutes reaction. The possible
labelling of sialic acid in the CHO moiety was not determined.

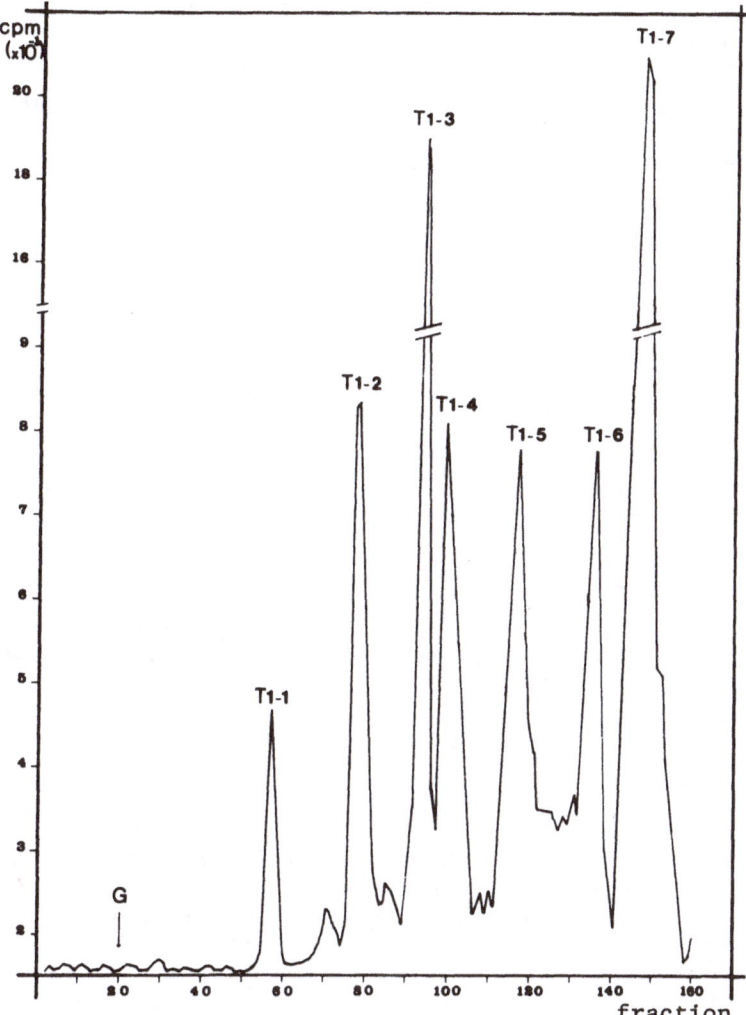

Fig. 2. HPLC fractionation of radiolabelled peptides contained in
peak T1 (Figure 1b).

Quantitation of Labelling

     As suggested by the chromatographic profile (Figure 1b), amino
acid sequence analysis of the radioactive peptides obtained after
1 or 5 minutes reaction showed that the same residues were modified
at either time, differences between both sets being only quantitative
as measured by the specific activity of labelled residues. Thus,
an estimate of the relative extent of modification of each reaction
time was judged to be essential, considering that the modification
that occurs between 1 and 5 minutes has drastic effects on the

Table 1.   Amino Acid Sequence Analysis of Radiolabelled Peptides
           from Modiefied Fcγ1 Fragment.

PEPTIDE	SEQUENCE	MODIFIED RESIDUES
T4	249 ＊ 255 Asp-Thr-Leu-Met-Ile-Ser-Arg	Asp(249)
T1-5	256 ＊ 274 Thr-Pro-Glu-Val-Thr-Cys-Val-Val-Val-Asp-Val-Ser-His-Glu-Asp-Pro-Gln-Val-Lys	Glu(258) & Glu(269)
T1-1 & T1-2	293 ＊ 301 Glu-Gln-Gln-Tyr-Asn-Ser-Thr-Tyr-Arg	Glu(293)
T7 & T7a	318 ＊ 320 Glu-Tyr-Lys	Glu(318)
T3	327 ＊ 334 Ala-Leu-Pro-Ala-Pro-Ile-Glu-Lys	Glu(333)
T1-3 & T2	345 ＊ 349 Glu-Pro-Gln-Val-Tyr	Glu(345)
T5	356 ＊ 360 Glu-Glu-Met-Thr-Lys	Glu(356)
T1-6	371 ＊ ＊ 392 Gly-Phe-Tyr-Pro-Ser-Asp-Ile-Ala-Val-Glu-Trp-Glu-Ser-Asn-Gly-Gln-Pro-Glu-Asn-Asn-Tyr-Lys	Asp(376), Glu(380) & Glu(382)
T6	410 ＊ 414 Leu-Thr-Val-Asp-Lys	Asp(413)
T1-4 & T1-7	440 ＊ Ser-Leu-Ser-Leu-Ser-Pro-Gly	Gly(446)

Arrows underline residues which were directly identified by HPLC.
Asterisks denote radiolabelled residues. (a) The proposed sequence
of peptide T1-6 is in agreement with that reported previously by
Hofman and Parr[21].

capacity of Fcγ to activate complement (see Introduction).   The
specific activity of labelled residues after both 1 and 5 minutes
is shown in Table 2.

        Two types of modified residue may be distinguished:

(a) Those in which the extent of labelling does not increase beyond
    1 minute. They are: Glu (269) and Glu (318) in Cγ2 and Asp
    (413) in Cγ3.
(b) Those showing a significant increase in specific activity between
    1 and 5 minutes reaction. They include: Asp (269), Glu (258)
    Glu (293) and Glu (333) in Cγ2; Glu (345), Asp (376), Glu (380),
    Glu (382) and the C-terminal residue Gly (446) in Cγ3.

Table 2.   Specific Activity of Radiolabelled
Residues from Fc-1' and Fc-5'.

Modified residues	S (μCi/mmole)		Relative increase S_5/S_1
	1 minute	5 minutes	
Asp (249)	11	32	2.9
Glu (253)	17	34	2.0
Glu (269)	52	56	1.1
Glu (293)	40	92	2.3
Glu (318)	53	66	1.2
Glu (333)	3	10	3.3
Glu (345)	13	65	5
Glu (356)	16	36	2.3
Asp (376)	14	43	3.1
Glu (380)	20	45	2.3
Glu (382)	14	27	1.9
Asp (413)	47	51	1.1
Gly (446)	38	120	3.2
Total	338	712	2.1

(a) Absolute values of specific activity of individual residues significantly underestimate the extent of modification as measured by the specific activity of the undigested protein. Relative increase of labelling is not affected[12].

DISCUSSION

The coupling of nucleophiles with water-soluble-carbodiimides is the mildest method for modifying acidic residues in proteins. Although several side reactions may take place during the modification of carboxyl groups, this reaction appears to be very specific in the Fcγ fragment[8].

The chemical modification of acidic side chains, under precise reaction conditions, abrogates its capacity to activate complement and this inactivation occurs in the absence of significant conformational changes, as judged by CD spectroscopy and other criteria[8]. Thus, the location of residues modified under these conditions might give direct information as to the topography of the effector site for activation.

In evaluating the contribution of individual modified groups to the loss of anticomplementary capacity it is important to take

into account the extent of modification of each group at the time
when no functional alteration occurs (1 minute) and at the time when
loss of complement activating capacity takes place (5 minutes).
This is especially so since the number of modified residues at both
reaction times is the same.

Out of thirteen modified residues, six are located in the Cγ2
domain. This region of the molecule is believed to be the one to
which Clq binds[14],[15]. Likely, therefore, it is the modification of
acidic residues on Cγ2 which is directly responsible for loss of
anticomplementary capacity.

Glu (269) and Glu (318) do not increase their extent of modification after 1 minute of reaction (Table 2). This suggests that
both groups react essentially to completion in 1 minute. Since at
this time of modification no significant functional alteration is
observed, it must be concluded that either the acidic side chains
of these residues are not involved in the Clq effector site or that
their modification is not sufficient to abrogate complement activating capacity. The possibility that they may participate in a
more extensive network of atomic interactions involving larger areas
of both IgG and Clq can not be discarded.

Glu (333) is not extensively labelled even after 5 minutes.
Solvent accessibility analysis[12], based on the atomic coordinates
of Fc 1, showed that although the side chain of this residue is
exposed to solvent, it is likely involved in an ionic pair with Lys
(320).

The side chain of Asp (249) is relatively unexposed to solvent
as judged from accessibility analysis. It is in close proximity to
to the guanidinium group of Arg (255) with which it possibly forms
an ionic pair[16]. Thus, although label uptake by this residue significantly increases between 1 and 5 minutes, its participation in
complement activation seems unlikely.

Both Glu (258) and Glu (293) significantly increase their extent
of modification with time of reaction up to 5 minutes. Their side
chains are exposed to solvent. Glu (258) is in close proximity to
the carbohydrate moiety and it is possibly forming a hydrogen bond
with a galactose residue[16]. The side chain of Glu (293) is pointing
outwards from the surface of the molecule. It seems likely that
modification of one or both of these two residues may be responsible
for functional inactivation of Cγ2 upon modification of acidic side
chains. However, it must be considered that this modification adds
to the previous one affecting Glu (269) and Glu (318). Thus, it is
possible that functional inactivation may result from a cumulative
cancellation of negative charges rather than from modification of
a single group.

Glu (258), Glu (269) and Glu (293) are on or close to the X face of the Cγ2 domain[16,17]. Thus, it is likely that some area of this face may be primarily involved in activating C1 through contacts which may involve the establishment of salt bridges with basic residues in C1q. IgG-C1q interaction probably involves a large number of atomic contacts. In addition to charge interactions, participation of hydrophobic bonds has been suggested by a number of reports[18,19]. It is conceivable that charge "long range" interactions may play an important role in a first step favouring a precise orientation of the IgG-C1q complex. The stabilization of this complex would be accomplished through the establishment of non polar "short range" interactions and the increased stabilization of salt bridges after hiding from solvent exposure. Presumably, this second step would allow the activation of the C1 component.

The precise area on Cγ2 to which C1q binds, remains controversial. Burton et al.[7] proposed the β-bend spanning residues Gly (316) to Lys (340). This area includes both Glu (318) and Glu (333) which, as discussed above, appear not to play a significant functional role in our experimental system. A short segment including residues Lys (290)-Gln (295) has been proposed on the basis of comparative analysis of amino acid sequences as involved in C1q binding[20]. Interestingly, this segment includes Glu (293), which as mentioned above, may be involved in complement activation. However, the search for a restricted segment as the putative site may be an idle question. It is conceivable that IgG-C1q interaction may involve relatively large areas of the Cγ2 domain so that abrogation of complement activation may be accomplished by altering areas relatively distant from each other. One such area could be the one surrounding Glu (293).

The binding of 2,5-diaminotoluene, an inhibitor of C1q binding, to the stretch of Cγ2 including residues Ser (324)-Leu (326), (Sutton, personal communication), suggest that this area, corresponding to the turning point of the β-bend suggested by Burton et al.[7], may also be important. Glu (269) is in close proximity to this area. Although, as discussed above, modification of this residue does not induce significant functional inactivation, it may still be involved in C1q binding.

In summary, the data presented in this communication suggest that a few side chain carboxyl groups which are located in or near the X face of the Cγ2 domain are essential for complement activation by human IgG. The data are compatible with the involvement of relatively distant areas of the domain surface which may include the area around Glu (293) and perhaps the one around Glu (258). They do not exclude the participation of an area close to Glu (269) but this residue does not constitute by itself an essential contact. A working hypothesis may be proposed for IgG-C1q binding as being primarily driven by charge interactions. The interactions would be

further stabilized by the establishment of additional nonpolar bonds
and by solvent exclusion in the contact areas.

SUMMARY

    Acidic residues of human Fcγ seem to play a role in complement
activation by the classical pathway[8]. The location of the acidic
residues which are responsible for the abrogation of complement
activating capacity upon chemical modification, has been now established. Out of thirteen residues modified at the point of functional
inactivation only six are located on the Cγ2 region.

    Glu (333) and Asp (249) appear to be only partially labelled.
They are possibly involved in ionic pairs[16], so that their contribution to complement activation is dubious. However, an extensive
modification occurs in Glu (258) and in Glu (293), which adds to
the more rapid one of Glu (318) and Glu (269), resulting in loss of
anticomplementary capacity.

    From the data it can be hypothesized that formation of inter-
molecular salt bridges between acidic residues of Fc and basic residues of C1q may contribute as an essential part of the mechanism of
complement activation. Moreover, they also help to determine the
topography of the C1q effector site.

REFERENCES

1. R. R. Porter and K. B. M. Reid, Activation of the complement
        system by antibody-antigen complexes. The classical pathway,
        Advances in Protein Chemistry 33:1-71 (1979).
2. C. Hughes-Jones and B. Gardner, The reaction between the comp-
        lement C1q, IgG complexes and polyionic molecules, Immunology
        34:459-463 (1978).
3. T. Lin and D. S. Fletcher, Interaction of human C1q with in-
        soluble immunoglobulin aggregates, Immunochemistry 15:107-117
        (1978).
4. R. Allan, M. Rodrick, H. R. Knobel and H. Isliker, Inhibition
        of the interaction between the complement component C1q and
        immune complexes, Int. Archs. Allergy Appl. Immunol. 58:140-
        148 (1979).
5. S. Cohen and E. L. Becker, The effect of carbamylation and
        amidination of rabbit IgG-antibody on its ability to fix
        complement, J. Immunol. 100:395-402 (1968).
6. S. Cohen and E. L. Becker, The effect of benzylation or sequen-
        tial amidination and benzylation on the ability of rabbit
        IgG-antibody to fix complement, J. Immunol. 100:403-406
        (1968).

7. D. R. Burton, J. Boyd, A. D. Brampton, S. B. Easterbrook-Smith,
    E. J. Emanuel, J. Novotny, W. T. Rademacher,
    M. R. van Schravendijk, M. J. E. Sternberg and R. S. Dwek,
    The C1q receptor site on immunoglobulin G, Nature 288:338-344
    (1980).
8. F. Vivanco, R. Bragado, H. P. Albar, C. Juárez and F. Ortiz,
    Chemical modification of carboxyl groups in human Fc fragment:
    Structural role and effect on the complement fixation, Mol.
    Immunol. 17:327-336 (1980).
9. W. Macheidt, J. Otto and E. Wachter, Chromatography on microbore
    columns, Methods in Enzymology 47:210-220 (1977).
10. J. A. López de Castro, H. T. Orr, R. J. Robb, T. G. Kostyk,
    D. L. Mann and J. L. Strominger, Complete amino acid sequence
    of a papain solubilized human histocompatibility antigen HLA
    B-7. Isolation and amino acid composition of fragments and
    of tryptic and chymotryptic peptides, Biochemistry 18:5704-
    5711 (1979).
11. N. Nakai, C. Y. Lai and L. Horecker, Use of fluorescamine in
    the chromatographic analysis of peptides from proteins, Anal.
    Biochem. 58:563-570 (1974).
12. R. Bragado, J. A. López de Castro, C. Juarez, J. P. Albar,
    A. García Pardo, F. Ortíz and F. Vivanco, Chemical modifi-
    cation of carboxyl groups in human Fc fragment: II. Location
    of acidic residues involved in complement activation, Molec.
    Immunol. (in press).
13. D. G. Klapper, C. E. Wilde and J. D. Capra, Automated amino acid
    sequence of small peptides utilizing polybrene, Anal. Biochem.
    85:126-131 (1978).
14. M. Colomb and R. R. Porter, Characterization of a plasmin digest
    fragment of rabbit immunoglobulin gamma that binds antigen
    and complement. Biochem. J. 145:177-183 (1975).
15. D. Yasmeen, J. R. Ellerson, K. J. Dorrington and R. H. Painter,
    The structure and function of immunoglobulin domains. IV.
    The distribution of some effector functions among the Cγ2 and
    Cγ3 homology regions of human immunoglobulin G. J. Immunol.
    116:518-526 (1976).
16. J. Deisenhofer, Crystallographic refinement and atomic models
    of a human Fc-fragment, and its complex with fragment B of
    protein A from Staphylococcus Aureus at 2.9Å and 2.8Å resol-
    ution, Biochemistry 20:2361-2370 (1981).
17. D. Beale and A. Feinstein, Structure and function of the constant
    regions of immunoglobulins, Quart. Rev. Biophys. 9:135-180
    (1976).
18. R. Allan and H. Isliker, Studies on the complement binding site
    of rabbit immunoglobulin G. II. The reaction of rabbit IgG
    and its fragments with C1q, Immunochemistry 11:243-248 (1974).
19. D. E. Isenman, J. R. Ellerson, R. H. Painter and K. J. Dorrington,
    Correlation between the exposure of aromatic chromophores at
    the surface of the Fc domains of immunoglobulin G and their
    ability to bind complement, Biochemistry 16:233-240 (1977).

20. R. Brunhouse and J. J. Cebra, Isotypes of IgG: Comparison of
    the primary structures of three pairs of isotypes which differ
    in their ability to activate complement, Mol. Immunol. 16:907-
    917 (1979).
21. T. Hofman and D. M. Parr, A note on the amino acid sequence of
    residues 381-391 of human immunoglobulin gamma chains, Mol.
    Immunol. 16:923-925 (1979).

III: STRUCTURE OF C1, INTERACTIONS BETWEEN ITS SUBUNITS AND

WITH OTHER MACROMOLECULES

INTRODUCTION

In order to understand at the molecular level how the immunological signal, generated upon interaction of antibody with
antigen, can result in the cleavage of a peptide chain in C1r and,
thereby, initiate the proteolytic cascade among the early complement
components of the classical pathway, it is essential to know how
C1 is assembled from its subunits, that is, how the quaternary
structure of the complex is formed, and what is the value of the
association constant for the interaction between C1q and $C1r_2C1s_2$.
Moreover, as part of understanding how different immune complexes
activate the classical pathway, it is important to know how tightly
the binding sites located on the C1q heads interact with immunoglobulins of different classes and subclasses, and with immunoglobulin aggregates of different sizes. Finally, it appears that a
number of fibrous macromolecules, notably fibronectin and heparin,
can compete for the site on C1q to which the $C1r_2C1s_2$ tetramer binds,
and it seems probable that similar interactions between C1q and
fibrous macromolecules play a role in the immune defense.

In the first paper of Topic III, Schumaker and Poon describe
the ultrastructure of the first component of complement, as seen
with the electron microscope, in which the flexible $C1r_2C1s_2$ tetramer
apparently binds among the C1q arms, separated from the C1q heads
but beneath the central bundle. They also report the equilibrium
constant for the reversible binding which occurs when the C1q subunit and the $C1r_2C1s_2$ tetramer are mixed together in solution to
form C1. The binding constant between the unactivated tetramer and
C1q is reported by them to be about 14nM, and becomes an order of
magnitude weaker upon activation.

In the second paper, Karoly Cseh and colleagues present data indicating that fibronectin binds to IgG3 and IgG1, as well as confirming that fibronectin binds to Clq. They suggest this binding could play a role in the fixation of Clq to cryoglobulins in the course of activation of the classical complement system.

In the third paper, David Bing extends the observation that mucopolysaccharides, and in particular heparin, bind to Clq tightly, and that the binding is pH and temperature independent but ionic strength dependent. Two molecules of heparin bind to each Clq. The binding appears to be to the collagenous portion, and does not occur with Cl. Thus, the binding of heparin to Clq appears similar in many respects to the binding of fibronectin. The binding between Clq and a variety of immunoglobulins are summarized as well in this paper, including binding to IgM and IgG, different IgG subclasses and crosslinked and aggregated IgG.

In the final paper, Arlaud and colleagues report sequence studies on the b chain of Clr, including the relative positioning of the 8 cyanogen bromide peptides and identification and placing in sequence of about two-thirds of the residues. The charge relay system common to the "trypsin-like" enzymes is confirmed, but two invariant cysteine residues, found in other mammalian serum proteases, except Cls, are lacking. The authors speculate on what this structural abnormality may imply.

ULTRASTRUCTURE OF THE FIRST COMPONENT OF HUMAN COMPLEMENT

Verne N. Schumaker, Pak H. Poon

Department of Chemistry and Biochemistry
and the Molecular Biology Institute
University of California
Los Angeles, CA 90024

Although a substantial amount is known about the structures
and functions of some of the components of the first component of
complement, very little is known about how these are assembled
or how they interact to express the Cl function. What is known includes, first, the stoichiometry, which finally appears to have been
clarified[1-3] as shown in Table 1. Two molecules of Clr and two molecules of Cls spontaneously reassemble at neutral pH to give the
Clr_2Cls_2 tetrama[4], and then one tetramer and a single Clq come
together to form the 16S Cl. Moreover, Ziccardi and Cooper[5,6] have
provided evidence that it is this 16S component which also exists in
sera, although a loose association between it and some other serum
protein which would not survive sucrose gradient centrifugation
cannot be ruled out.

Second, the activation sequence seems clear, that is, the first
step evidently involves the multivalent binding between the Clq heads
and a cluster of certain IgG's or IgM bound as a soluble immune complex or on a cell surface[7,8]. However, we do not know how this
binding of the Clq heads to the clustered Fc regions causes the activation signal to be sent to the Clr_2Cls_2 tetramer. Clq does not
appear to possess enzymatic activity; thus, the immunological signal
probably involves a protein conformational change. The nature of
this conformational change is, perhaps, the central outstanding
question yet to be answered concerning the mechanism of activation
of the classical complement pathway. When the signal is received

This work was supported by National Institutes of Health Grant GM
13914 and by National Science Foundation Grant PCM 80-21368.

Table 1. Stoichiometry, Molecular Weights, and
          Sedimentation Properties of C1 and its
          Subunits

Subunits	$s_{20,w}$	Number	Mol.Wt.
$C1r_2$	6.7	2	170,000
C1s	4.3	1	85,000
$C1r_2C1s_2$	8.5	4	340,000
C1q	10.2	1	410,000
C1	16	5	750,000

by the tetramer, it initiates a sequence which includes the autolytic
activation of C1r by peptide chain cleavage, and then the activated
C1r cleaves the polypeptide chain of the proenzyme C1s, which, in
turn, becomes a protease possessing the specific enzymatic function
of C1, which is to cleave C4 and C2, and, thus, initiate the classical complement pathway.

Third, a substantial amount is known concerning the detailed
structural features of C1q following the elegant work of Reid and
Porter who sequenced the collagenous portion of C1q and proposed the
model shown in Figure 1 which corresponds to the ultrastructure of
C1q as seen with the electron microscope[9], and which resembles "a
bunch of tulips". Various portions of the C1 molecule have been
provided with appropriate names as well. Thus the tulips are called
the C1q heads, and the 15A fibers leading to the heads are often
called the collagenous arms. That rod-like portion of the molecule
composed of the six collagen helicies after they are joined together,
has not yet received a commonly accepted name. We propose the term
"central bundle" for this rod-like portion. Then the term "C1q
stalk" is reserved for that part of the molecule remaining after the
C1q heads have been digested away with pepsin, and includes the central bundle together with the collagenous arms. At the point where
the collagenous arms join to the central bundle there occurs a break
in the typical collagen-type amino acid sequence, which has been proposed to correspond to the sharp bend clearly visualized to occur at
this point[9]. We propose that the location of this bend is also a
semi-flexible joint[10], as will be discussed next.

In Figure 2 is shown an electron micrograph of C1q molecules including a few of the bunch-of-tulip or profile views, as indicated by
some of the arrows. Most of the molecules have landed on the grid in
a different manner, however, presenting a different aspect which we
call a "top view". In these top views the six C1q heads are spread

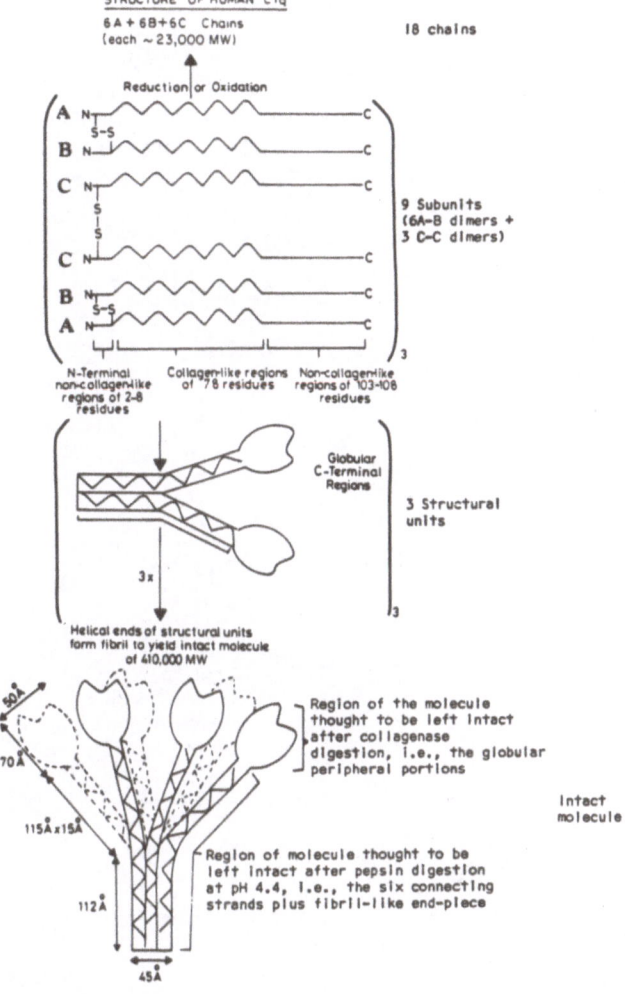

Fig. 1. Primary, secondary and quaternary structure of Clq as suggested from the work of Reid and Porter is shown in this drawing. The three different types of polypeptide chains, A, B, and C, have been sequenced, and the collagenous portions are shown as wavy lines denoting the regions suggested to be in the typical collagenous triple helix; the non-collagenous portions form the Clq heads. A short stretch of residues at the N-terminal end also has a non-collagen sequence, and in this region the disulfide bonds are found joining the A and B chains and also adjacent C chains. Thus, pairs of Clq arms are joined together by a a disulfide bond. Three such pairs then assemble without further covalent linkage to form the six armed Clq structure which resembles a bunch of tulips in profile. (Taken from Porter and Reid, 1979, with permission of the authors).

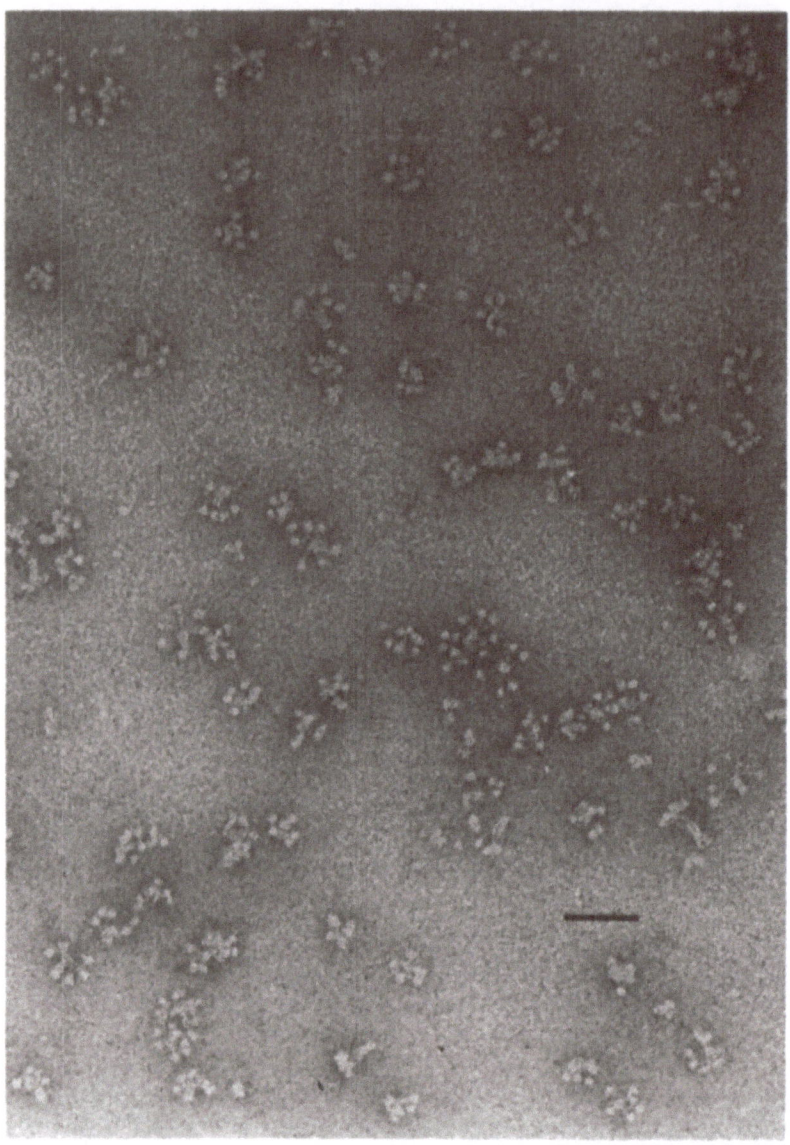

Fig. 2. Electron microscopy of the Clq subunit of the first
         component of complement. Both profile views, with the
         characteristic "bunch of tulips" appearance, and top views
         are seen in this photograph. Arrows point to molecules
         clearly representative of both classes, and careful inspec-
         tion shows that most of the molecules on this grid can be
         identified as belonging to one or the other configuration.
         In this negatively stained preparation, the molecules them-
         selves are not seen but rather the three-dimensional cast
         of the molecule embedded in a thin layer of stain. The
         length of the horizontal bar is 500Å.

out to form a circle with the central bundle forming a seventh dot
in the center of the circle; we believe the central bundle is pointing
straight up from the grid and, thus, is seen in cross section as a
central dot rather than as a rod. One particularly striking top view
is identified by one of the arrows of Figure 2 for which the arms,
only 15A in diameter, may be clearly visualized radiating out from
the central dot to the Clq heads located on the peripheral circumfer-
ence. For that particular top view, the angle made by the arms to
the central bundle must be nearly 90°; for the remaining top views
the angles are less, and often the heads and central bundle are found
to be clustered closely together. By fitting a circle through the
heads it is possible to estimate the angle made by the collagenous
arms as they leave the central bundle as shown in Figure 3. An additional assumption may be inferred from inspection of this figure:
For the molecules forming these top view images, the collagenous arms
are sufficiently strong to withstand the forces of drying and protrude
upwards lifting both themselves and the central bundle off the grid
toward the observer; a distribution of sizes of these top view images
is seen, corresponding to the angular distribution of the supporting
arms. Two such samplings of this angular distribution made by the
Clq arms with respect to an axis through the central bundle are shown
in Figure 4, according to whether rigid or loose criteria are used
to select the top view images, as described in reference[10]. All
angles from 20°, where the clustered heads would touch one another,
to 90°, where the molecule is flattened on the grid, seem to be possible, however, the distribution peaks at about 55°, leading to the
concept of a favored angle, and thus, the joint is not freely flexible. We have used the term "semi-flexible" joint to describe this
limited pattern of flexibility; evidently the restoring forces become
progressively greater as the angle of distortion is increased from
a favored angle of about 55° toward either extreme position. Thus,
Clq may be said to possess segmental flexibility, at least as seen
on the electron microscope grid.

Segmental flexibility seems to be a characteristic of the early
components of the immune system, including at least most IgG and
IgM[11,12] and Clq. For the immunoglobulins, the reason for this flexibility seems clear, it is required if these molecules are to recognise
and bind to a cluster of antigenic sites arranged in an irregular
pattern. And for Clq segmental flexibility is needed if the heads
are to bind to the corresponding irregular cluster of Fc regions.

In Figure 5 is shown an electron micrograph of Clr_2Cls_2 in the
activated form. Similar structures are seen for the unactivated
tetramer. This tetrameric molecule is a long rod-like structure with
a poorly defined length, ranging between 55 to 65nm, and domain
structure seems to be present. Moreover, the molecules appear to
be flexible, for a variety of configurations are apparent upon inspection of the various images present on the grid. This immediately
raises an interesting question: whereas it is apparent why antibody

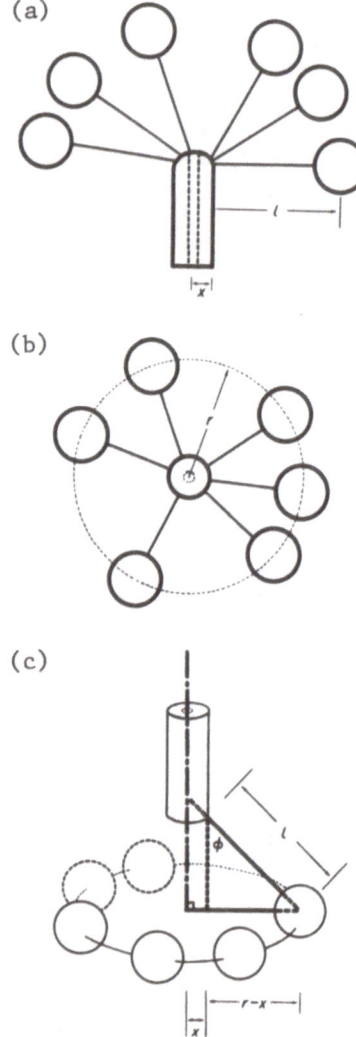

Fig. 3. In this cartoon of the top and profile views of Clq is indicated the two measurements which we take on the images seen
in the previous photograph of Figure 2, and the trigono-
metric calculation used to determine the cone angle made
by the Clq arms with an axis through the central bundle.
The distribution of these angles from 20 to 90° is shown
in Figure 4. (Taken from reference[10]).

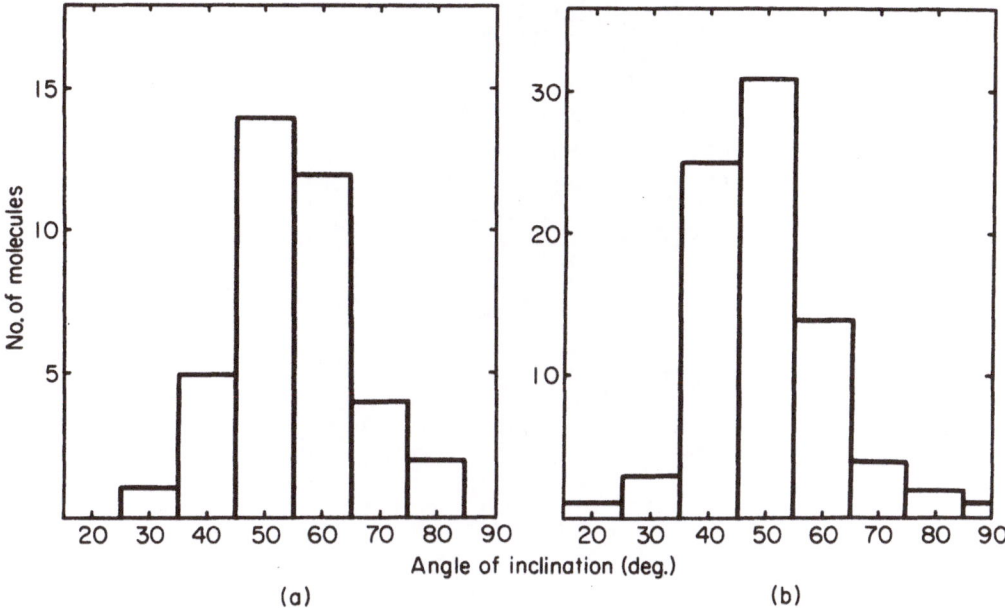

Fig. 4. Distribution of angles calculated from the top views of Clq. These are the angles formed between the arms and an axis through the central bundle of Clq, as shown in Figure 3. The molecule seems quite flexible with all angles between 20 and 90° represented, although the distribution peaks at a most probable angel of 55°. The existence of this most probable angle has led us to postulate a semi-flexible joint or hinge at the location where the arms form the central bundle. (Taken from reference[10]).

must be flexible to permit binding to the antibody, it is not immediately apparent what purpose is served by the flexibility of the $C1r_2C1s_2$ tetramer. This is a point to which we shall return. Moreover, we do not yet know the sequence of the Cls and the Clr subunits along the length of the rod-like $C1r_2C1s_2$ tetramer, although it has been plausibly suggested, on the basis of what is known about the stability of pairs of subunits, that the order is ClsClrClrCls[4].

When the $C1r_2C1s_2$ tetramer and Clq are mixed in solution, they spontaneously reassociate to form the 16S Cl. The equilibrium constant for this reassociation has a value readily measured by physical techniques, and the hydrodynamic method which we have employed is described in Figure 6. This is a new adaptation of sucrose gradient ultracentrifugation in which the slower sedimenting component, in this case the 8.55 $C1r_2C1s_2$ tetramer, is initially distributed uniformly throughout the gradient, while a very dilute solution of faster component, the 12S ^{125}I-Clq, is layered in a zone on top of

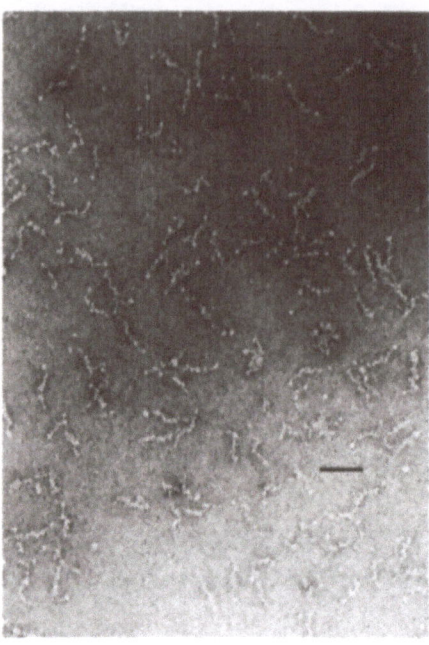

Fig. 5. Electron micrograph of the activated tetrameric subunit,
         $C1r_2C1s_2$. Similar structures are seen for the unactivated
         tetramer. These long, rod-like molecules exhibit a limited
         degree of flexibility and are often seen in a reverse "S"-
         shaped configuration. Domain structure seems to be present,
         although it does not appear consistent from molecule to
         molecule. Length variability also seems excessive, as
         though these tetrameric complexes exhibit extensive local
         golding and unfolding along the length of the flexible rod.
         The length of the horizontal bar is 500Å.

the gradient. The zone of C1q also contains $C1r_2C1s_2$ at the same
concentration as found below in the gradient. To understand how
the dissociation constant is measured, first consider the case where
the concentration of the $C1r_2C1s_2$ tetramer is much lower than the
numerical value of the dissociation constant. In this case upon
centrifugation the zone of C1q moves down the tube with the migration
rate of unassociated C1q, that is, 10.2S. In contrast, if the concentration of $C1r_2C1s_2$ is much higher than the numerical value of the
dissociation constant, then the zone sediments at 16S, characteristic of the completely reassociated C1. At intermediate concentrations of $C1r_2C1s_2$, the zone will move with a sedimentation rate
characteristic of intermediate degrees of association, as shown in
Figure 7. Thus, each point in Figure 7 represents a separate experiment, and a smooth "S-shaped" curve is produced when sedimentation
rate is plotted as a function of the logarithm of the $C1r_2C1s_2$ concentrations. The midpoint of this curve, that is, the concentration

ULTRACENTRIFUGATION OF A ZONE OF Cl THROUGH A 5–20% SUCROSE GRADIENT CONTAINING Clr_2Cls_2

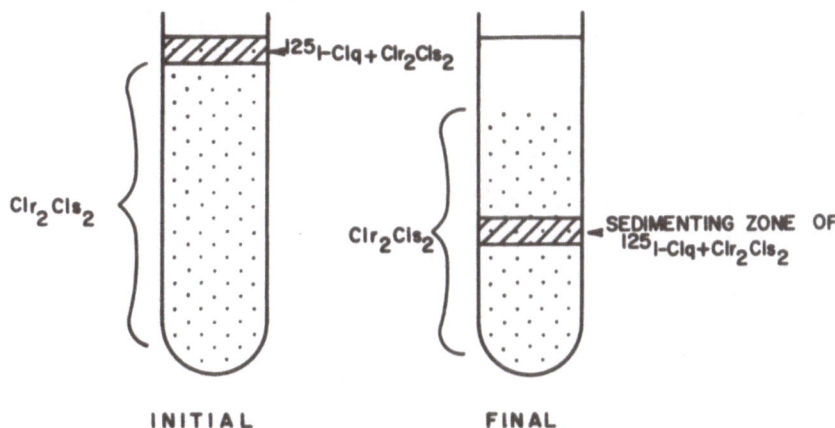

Fig. 6.   In this cartoon is illustrated our new technique of zone centrifugation through a uniform concentration of a slower macromolecular species, which we have employed to measure the binding constants between different macromolecules. Thus, a zone of radio-labelled Clq is placed on top of a sucrose gradient containing a uniform concentration of Clr_2Cls_2. The sedimentation coefficient of the zone will then reflect the fraction of time each molecule of Clq spends in the bound and in the unbound states with the Clr_2Cls_2 tetramer. Since the zone remains intact, this result also indicates that the dynamic equilibrium is occurring at a rate much faster than the tendency of the zone to split into two zones representing the two types of species that it contains.

of Clr_2Cls_2 at which the zone moves at 13.1S, is numerically equal to the thermodynamic dissociation constant. This is found to be 14 nM for Clr_2Cls_2 and Clq in equilibrium with unactivated Cl, and 140nM for Clr_2Cls_2 and Clq in equilibrium with activated Cl[13]. The dissociation constant will be, in general, a function of temperature, salt concentration and type of salt; these experiments were performed at 5°C in Tris-buffered saline with an ionic strength of 0.165M, and 5mM CaCl . In addition to providing a measure of the association constant, these experiments show that the equilibrium must be dynamic, otherwise the sedimenting zone would become two zones, one composed of the 10.2S unassociated Clq and the other of the 16S Cl.

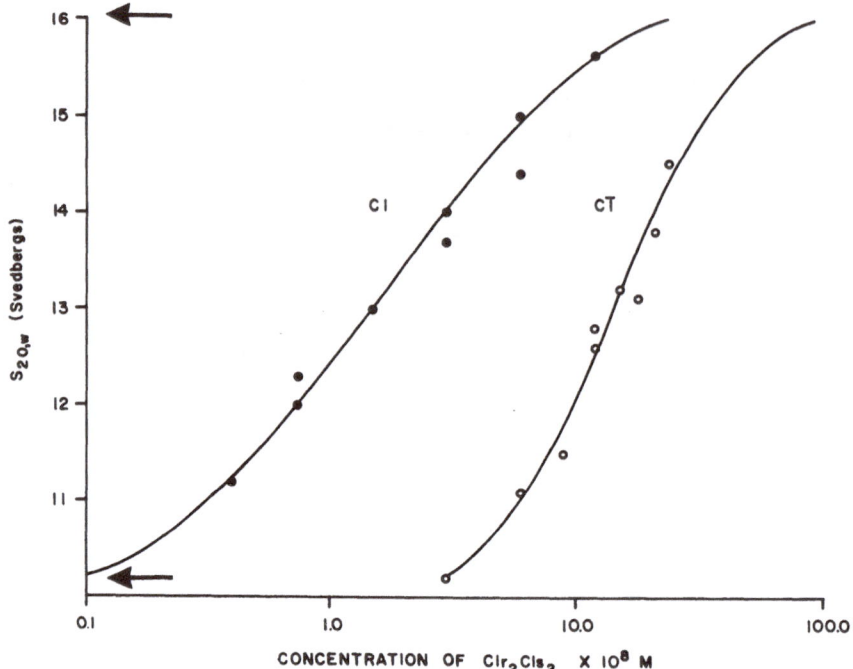

Fig. 7. In this graph are shown the data obtained from the technique
of zone centrifugation througy a uniform concentration of
a macromolecular ligand depicted in the previous illustration of Figure 6. The two curves represent the binding
of Clr_2Cls_2 and of activated Clr_2Cls_2 to Clq. The sedimentation coefficient of the zone is plotted on the vertical
axis, and the logarithm of the concentration is plotted on
the horizontal axis. The abscissal value of the midpoint
of each curve is equal to the dissociation constant of the
complex and can be seen to be 14nm for Clr_2Cls_2 and 140nm
for Clr_2Cls_2.

The initial attempts to visualize the reassociated Cl with the
electron microscope failed; evidently the molecule came apart when
it touched the carbon grid, for only the unassociated Clq and the
rod-like Clr_2Cls_2 tetramer could be found. Therefore, it was necessary to crosslink the complex, and this was readily accomplished using
a water-soluble carbodiimide which forms covalent peptide bonds between adjacent amino and carboxyl groups[14]. Since electron micro-
scopists tend to choose among the available images to present evidence for a favored model, we shown in Figure 8 two fields of images
showing the many aspects of Cl from which have selected 91 molecules
as evidence for the model of crosslinked Cl presented below. About
10% of the molecules are usable, and as can be seen from this figure,
the initial impression is a jumble of confused forms with amorphous,

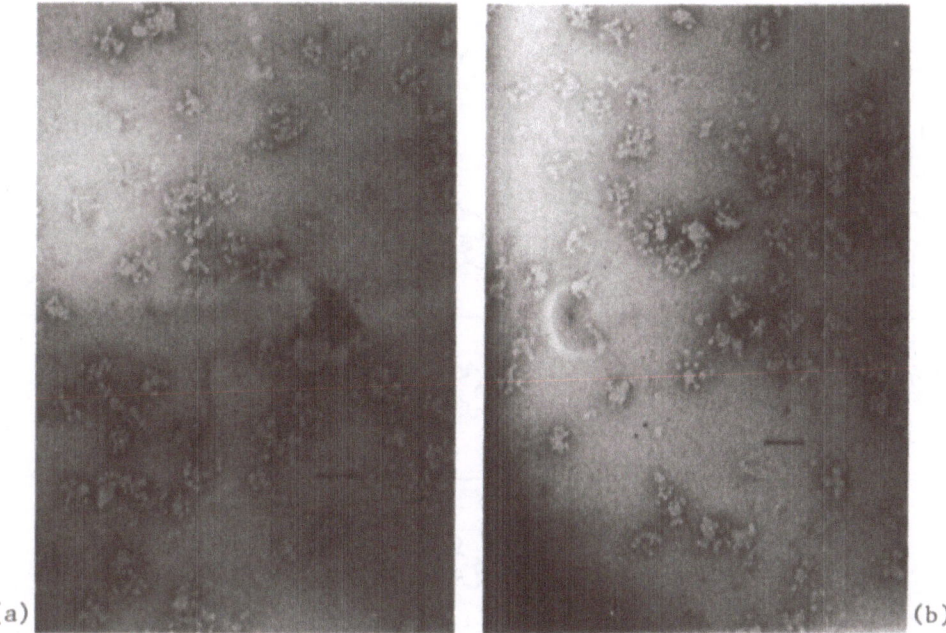

(a)   (b)

Fig. 8.   a and b. Electron microscopy of chemically cross-linked
          Cl shows both top views and profile views, as indicated
          by the arrows. Clq heads are clearly visible as well as
          a short portion of the collagenous arms, indicating that
          the $C1r_2C1s_2$ must be centrally located. The rod-like form
          of $C1r_2C1s_2$ as shown in the previous electron micrograph
          of Figure 6, cannot be found in this illustration, instead
          the tetramer must be folded into a compact form. Some of
          these can be seen rather clearly as indicated by the arrows.
          From the top the $C1r_2C1s_2$ can be seen to be a compact, disk-
          like mass centered about the central bundle and obscuring
          about one-half of the lengths of the collagenous arms.   In
          profile, this mass can be seen to be located among the arms
          just beneath the position where the central bundle flares
          out to form the collagenous arms. The lengths of the hori-
          zontal bars are 500Å.

overlapping shapes. A clearly defined model hardly springs forth.
But on closer inspection, two things become apparent. First, the
Clq heads are plentiful throughout the grid, and second, there are
no images of the rod-like $C1r_2C1s_2$ tetramers which had been clearly
visualized in the uncrosslinked preparations. Evidently the $C1r_2$-
$C1s_2$ tetramers are folded in some compact form associated with the
Clq subunits. Next, as indicated by arrows, a few distinct top
views and profile views can be resolved. In one clearly resolved
top view in the upper field shown in Figure 8, the Clq heads and

Fig. 9.   Twenty four tracings of profile views of Cl are shown in
           this illustration. A comparison of the length and width
           of the central bundle of these images with the length and
           width of the central bundle for Clq shows these to be the
           same, at least along 90% of the length, suggesting that
           the Clr_2Cls_2 tetramer is entirely folded among the arms
           and does not overlap the central bundle to any significant
           extent.

even a portion of the collagenous arms can be seen circumscribing a
dense central region which must represent the folded Clr_2Cls_2 tetramer
obscuring the central bundle. And in two profile views shown in the
lower portion of Figure 8, both the Clq heads as well as the central
bundle may be visualized while the folded Clr_2Cls_2 tetramer may be
located among the collagenous arms as a poorly resolved mass separated
both from the Clq heads and from most, if not all, of the central
bundle as well. These initial impressions are reinforced by the
twenty-four tracings of profile views shown in Figure 9. The entire
Clq heads and roughly one-half of the collagenous arms appear to be
entirely free of the folded tetramer. Moreover, comparison of the
dimensions of the central bundle of Cl with corresponding measurements
of the dimensions of Clq shows that they are the same, at least along
90% of the length and, perhaps, along the entire length of the central
bundle. This observation came as a surprise, for it has often been
assumed, at least in published cartoons[15,16], that the Clr and Cls
subunits were attached to the central bundle, thus concealing that

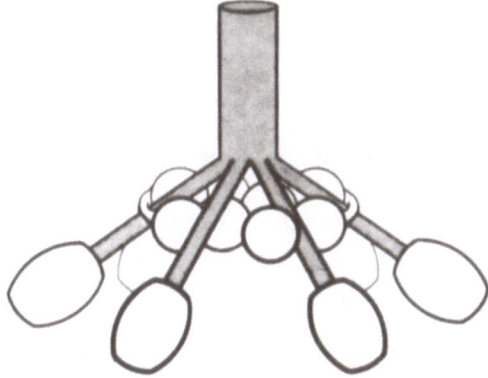

Fig. 10. A plausible model consistent with the electron microscopic
          and ultracentrifugal evidence is shown in this drawing.
          The Clr_2Cls_2 is folded among and around the arms of Clq
          forming an interlaced structure. In this illustration the
          Clq heads seem oversized, but conform to the dimensions
          suggested by Porter and Reid who took them from electron
          micrograph images. We have modeled the Clr_2Cls_2 as a
          string of 8 equally sized beads connected by short peptide
          strands, with a total volume consistent with an anhydrous
          molecular weight of 340,000 daltons. (Taken from refer-
          ence 14.)

portion of the molecule from putative Clq receptors on macrophages[17,18] or other cells. It has not been possible for us to tell
from these electron micrographs whether the Clr_2Cls_2 molecule is
entirely contained within the cage formed by the Clq arms or whether
it is wrapped around the outside of the arms like a chain of popcorn
decorating a holiday tree. When models are constructed, however,
it becomes apparent that there is insufficient space inside the cage
for the entire, folded tetramer if 50% of the collagenous arms are
to remain visible. We rather prefer the model in which the tetramer
is folded, intertwined among the arms with some of the mass protruding to the outside, as shown in Figure 10.

     A word of caution is required at this point. The crosslinked
Cl does not appear to be hemolytically active. This has been tested
by adding Cl or crosslinked Cl back to Cl-depleted serum and then
using various dilutions for the lysis of antibody-coated sheep erythrocytes, as shown in Figure 11. Reassociated Cl shows good activity
in this test, but after crosslinking for 1 hour at 37° with 0.1M
water-soluble carbodiimide, hemolytic activity appears to be completely lost. Not shown are other experiments which demonstrate
that Clq which has been treated with the water-soluble carbodiimide
still reforms a 16S Cl-like molecule upon the addition of the

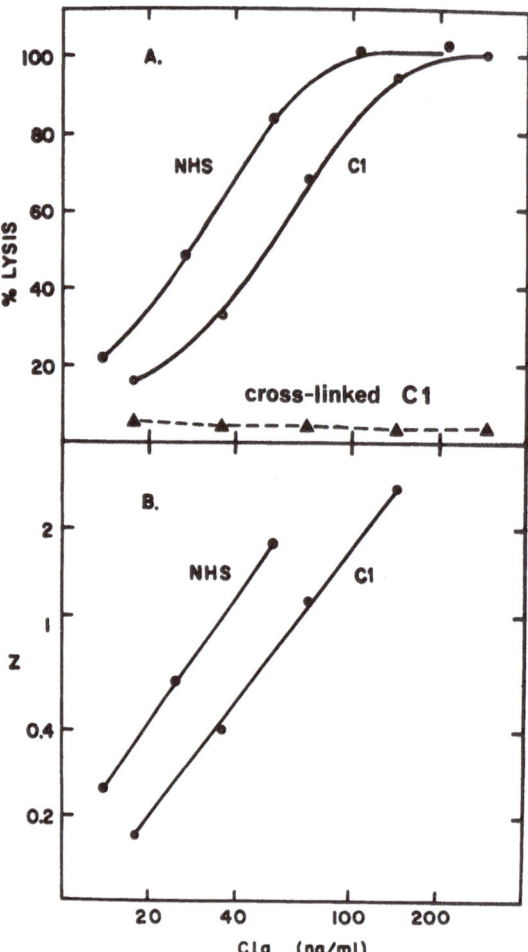

Fig. 11. The chemically crosslinked Cl is not hemolytically active,
as shown in this figure in which serum Cl and reconstituted
Cl are compared with Cl which has been crosslinked. For
this assay the reconstituted and the crosslinked reconstituted Cl were added back to Cl depleated serum, and then
used in a standard assay to cause lysis of sheep red blood
cells coated with antigen and antibodies.

Clr_2Cls_2 tetramer. But if the Clr_2Cls_2 is treated first with the
reagent and then added to untreated Clq, there is no reformation of
a 16S species.

We have now examined 65 top views of the chemically crosslinked
Cl, and some of these are shown in Figure 12. From these top views,
it is possible to circumscribe the heads to determine the cone angle
as was previously described for Clq. The distribution of such angles

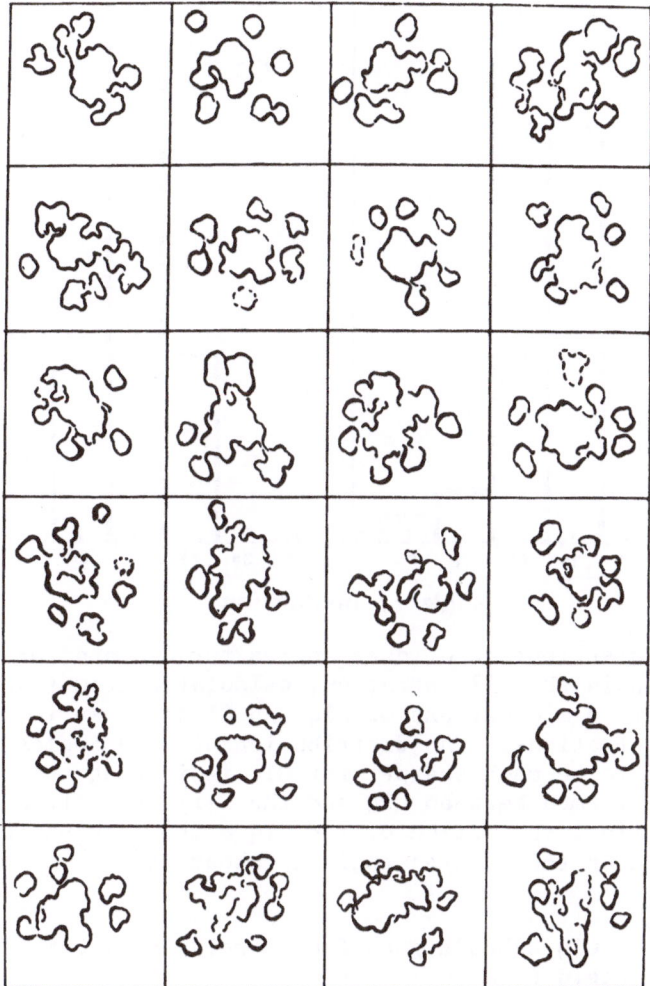

Fig. 12.   Some of the 65 top views of C1 which we have now examined
           are shown in this illustration. The $C1r_2C1s_2$ can be seen
           to obscure the central bundle and a portion of the col-
           lagenous arms. It also appears to lock the arms in place.
           Thus, by placing a circle through the arms we can determine
           by trigonometry the cone angle, as was previously done for
           C1q as illustrated in Figure 3. The distribution of these
           cone angles is shown in the next figure.

is shown in Figure 13, where it is compared with a corresponding
distribution of C1q. The distribution is much sharper, indicating
that when the folded tetramer forms among the C1q arms they become
locked in place, and the molecule loses much of its segmental flexibility. In the vocabulary of protein conformational change, the

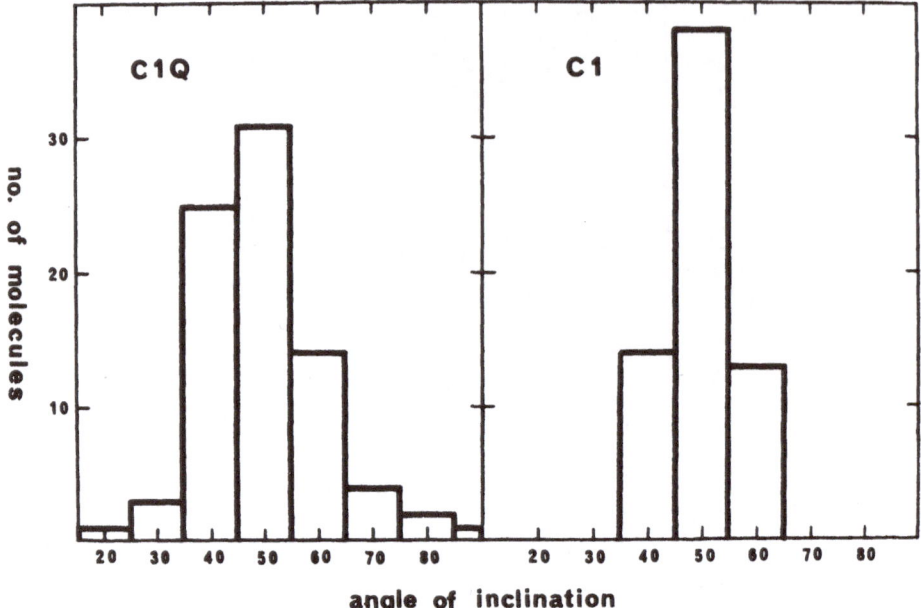

Fig. 13.   The distribution of cone angles for Clq and for Cl is
           shown in this illustration, calculated from tracings of
           electron micrographs of Clq and Cl as shown in previous
           illustrations. The distribution of cone angles can be
           seen to be much sharper for Cl, indicating that the com-
           plex formed between Clq and the Clr_2Cls_2 tetramer greatly
           restricts the motion of the Clq arms, locking them into
           place at the favored angle of about 55°.

structure of the Clq subunit goes from a relaxed, flexible form to
a taut, almost rigid form.

     At this point it is tempting to sepeculate, at least to a lim-
ited extent, concerning a model for activation based upon a protein
conformational change as being the immunological signal causing the
initiation of the activation sequence. We will propose that the Cl
molecule exists in continuous dynamic equilibrium between a taut
relatively rigid configuration and a relaxed, more flexible one.
It may be imagined that some of the looser interactions between
portions of the folded Clr_2Cls_2 and the Clq subunits are intermittently breaking and reforming under the impulsive forces of random
thermal bombardment by energetic solvent molecules. Perhaps the
molecule would spend only one-percent or less of its time in this
relaxed configuration. Yet while it is in this relaxed configuration, we suggest that two things may occur:

(1) the activation sequence may initiate, that is, partially bound
    to Clq, the relaxed tetramer is in a spontaneously activatable
    state, and

(2) the Clq heads may bind to an irregular cluster of antibody mol-
    ecules located on a soluble or cell-bound immune complex.

But if the relaxed molecule does bind to such an irregular cluster,
then it is held open in this relaxed configuration, and, thus, the
lifetime of the relaxed state and in consequence, the rate of activation, would be increased one-hundred to one-thousand fold.

    This hypothesis may have the virtue of being susceptible to
experimental test. Thus, a freely mobile cluster of antibody molecules would not be expected to cause activation, for there would
exist no forces generated by the binding of relaxed Cl to such a
cluster to prevent spontaneously return to the taut configuration.
Indeed, this seems to be the case for concentrated solutions of un-
aggregated immunoglobulins which bind readily to Clq, but which do
not activate Cl, at least not in plasma where free immunoglobulin
is found in enormous abundance. It is not clear that the same can
be said for completely fluid surface distributions of antibody molecules. The viscosity of such surface layers is a thousand times
higher than the viscosity in aqueous solution and this enhanced viscosity would exert a viscous drag slowing greatly the return to the
taut configuration. This theory would predict that certain symmetrical, regular clusters of Fc regions would be arranged properly to
bind Cl tightly without exerting any stressful forces on the molecule
to distort it from the taut configuration. Indeed, Loos, Rapp and
Borsos[19] present good evidence for a non-self activating population
of Cl bound to EAC4 cells, a population which is susceptible to activation by a gentle treatment with trypsin. They present another
explanation for the existence of this population involving the dissociation of a portion of Clq. We would like to suggest alternatively that these non-self-activatable Cl molecules may be those
which provide no distortion of the Cl molecule and thus capture it
but do not initiate the classical complement sequence.

REFERENCES

1. I. Gigli, R. R. Porter and R. B.Sim, The unactivated form of
    the first component of human complement, Cl, Biochem. J. 157:
    541 (1976).
2. R. J. Ziccardi and N. R. Cooper, The subunit composition and
    sedimentation properties of human Cl, J. Immunol.118:2047
    (1977).
3. R. C. Siegel, V. N. Schumaker and P. H. Poon, Stoichiometry and
    sedimentation properties of the complex formed between the
    Clq and Clr_2Cls_2 subcomponents of the first component of
    complement, J. Immunol. 127:2447 (1981).
4. J. Tschopp, W. Villiger, H. Fuchs, E. Kilchherr and J. Engel,
    Assembly of subcomponents Clr and Cls of first component of
    complement: Electron microscope and ultracentrigugal studies,
    Proc. Natl. Acad. Sci. USA 77:7014 (1980).

5. R. J. Ziccardi and N. R. Cooper, Direct demonstration and quanti-
    tation of the first complement component in human serum,
    Science 199:1080 (1978).
6. R. J. Ziccardi and N. R. Cooper, Demonstration and quantitation
    of the first component of complement in human serum, J. Exp.
    Med. 147:385 (1978).
7. H. Metzger, The effect of antigen on antibodies: Recent studies,
    Contemp. Topics Molec. Immunol. 7:119 (1978).
8. S. K. Dower and D. M. Segal, Clq binding to antibody-coated
    cells: Predictions from a simple multivalent binding model,
    Molec. Immunol. 18:823 (1981).
9. R. R. Porter and K. B. M. Reid, Activation of the complement
    system by antibody-antigen complexes: The classical pathway,
    Adv. Prot. Chem. 33:1 (1979).
10. V. N. Schumaker, P. H. Poon, G. W. Seegan and C. A. Smith, Semi-
    flexible joint in the Clq subunit of the first component of
    human complement, J. Mol. Biol. 148:191 (1981).
11. J. Yguerabide, H. F. Epstein and L. Stryer, Segmental flexibility
    in an antibody molecule, J. Mol. Biol. 51:573 (1970).
12. R. E. Cathou, Solution conformation and segmental flexibility
    of immunoglobulins, Comprehensive Immunol. 5:37 (1978).
13. R. C. Siegel, C. J. Strang, M. L. Phillips, P. H. Poon and
    V. N. Schumaker, Hydrodynamics and electron microscopy of Cl
    and Cl, Fed. Proc. Am. Soc. Exp. Biol. 40:4152a (1981).
14. C. J. Strang, R. C. Siegel, M. L. Phillips, P. H. Poon and
    V. N. Schumaker, Ultrastructure of the first component of
    human complement: Electron microscopy of the crosslinked
    complex, Proc. Natl. Acad. Sci. USA (in press).
15. M. Mayer, The complement system, Scientific American 229:5:54
    (1973).
16. R. R. Porter, The biochemistry of complement, Biochem. Soc.
    Trans. 5:1659 (1977).
17. A. J. Tenner and N. R. Cooper, Identification of types of cells
    in human peripheral blood that bind Clq, J. Immunol. 126:1174
    (1981).
18. J.-L. Wautier, H. Souchon, K. B. M. Reid, A. P. Peltier and
    J. P. Caen, Studies on the mode of reaction of the first
    component of complement with platelets: Interaction between
    the collagen-like portion of Clq and platelets, Immunochem.
    14:763 (1977).
19. M. Loos, T. Borsos and H. J. Rapp, The first component of comp-
    lement in serum: Evidence for a hitherto unrecognized factor
    in Cl necessary for internal activation, J. Immunol. 110:205
    (1973).

BINDING OF HUMAN I[125] FIBRONECTIN TO C1q AND TO HUMAN MONOCLONAL MYELOMA IgG SUBCLASSES

J. Cseh, L. Jakab
T. Pozsonyi and J. Fehér

III. Dept. Med.
Semmelqwis Univ. Med. School
Budapest, Hungary

INTRODUCTION

Plasma fibronectin seems to be a component of mixed polyclonal cryoglobulins[1,6]. According to our previous studies, mixed cryoglobulins, consisting of monoclonal and polyclonal immunoglobulins contain fibronectin (unpublished results). In vivo activation of the complement system was demonstrated in all of our patients with essential mixed cryoglobulinaemia. Moreover, in vitro studies showed that isolated cryoglobulins activated the complement system via the C1q. The purpose of this study was to analyze the binding of purified human fibronectin to potential components of the cryoglobulins, to monoclonal immunoglobulins and to C1q, in vitro. This was performed by binding of I[125] labelled fibronectin to human C1q, to human C1 complex, to human C1q bound to immune complexes and to human myeloma IgG subclasse.

MATERIALS AND METHODS

Human plasma fibronectin was isolated from fresh normal human plasma by affinity chromatography according to the method of Vuento and Vaheri[7] on gelatin-Sepharose 4B (gelatin Sigma type I, Sepharose 4B Pharmacia Fine Chemicals) and on L arginine-Sepharose 4B columns (L arginine Reanal, Budapest). Human and rabbit C1q was prepared according to the method of Liberty[2]. Human C1 complex and monoclonal human myeloma proteins, IgG1, IgG2, IgG3 and IgG4 subclasses were kindly supplied by Gy. Füst and Gy. Medgyesi (National Institute of Hematology and Blood Transfusion, Budapest). Plasma fibronectin, 1mg/ml was labelled with I[125] (100 uCi) by Cloramine T (25ug), incu-

115

bated for 2 minutes. The reaction was stopped by sodium metabi-
sulfite (25ug). Unbound I^{125} was removed by dialysis. Solid phase
radioimmunoassay was performed on plastic micro RIA plates (Falcon,
5052, 12 x 75mm). Polystyrene tubes were coated with human and
rabbit Clq, ovalbumin-rabbit antiovalbumin immunoglobulin immune
complexes (OV/aOV) and myeloma proteins, each 10ug, by incubation
at 37°C for 2 hours and at 4°C overnight. Tubes were washed 3 times
with PBS+BSA. The remaining isotope activity was counted in a well
type scintillation spectrometer (Gamma 300, Beckman Instruments).

Antiovalbumin immunoglobulin was produced by immunising rabbits
with ovalbumin (Sigma) and immunoglobulins were purified on oval-
bumin-Sepharose 4B immunosorbent[3]. Immune complexes were made by incubation of ovalbumin with antiovalbumin immunoglobulin at 37°C for
2 hours and at 4°C overnight, in equivalence and at ovalbumin-antiovalbumin immunoglobulin ratios of 1:2 and 1:4. Clq was bound to
the complexes by incubation at 37°C for 2 hours and at 4°C overnight,
at molar ratios of 1:1, 1:2 and 1:4. Complexes were separated by
centrifugation and washed 3 times with PBS.

RESULTS

Human and rabbit Clq, human C1 complex and Clq complexed with
ovalbumin-antiovalbumin immunoglobulin immune complexes all bound purified human I^{125} fibronectin. Human Clq bound more I^{125} fibronectin than human C1 complex or the rabbit Clq (Figure 1). I^{125}

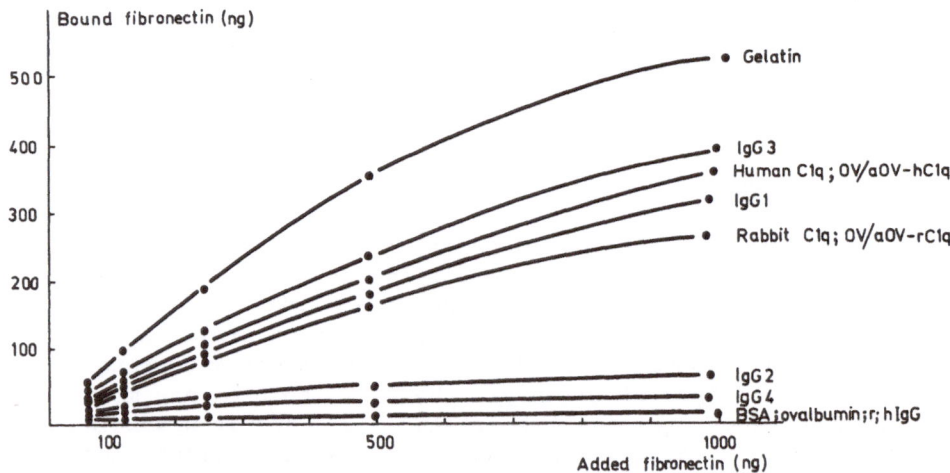

Fig. 1.   Binding of I^{125} fibronectin to different proteins
          (protein, each 10ug).

fibronectin was bound to human monoclonal myeloma proteins of IgG1 and IgG3 subclasses. However, monoclonal myeloma proteins of IgG2 and IgG4 subclasses did not bind substantial amounts of I^{125} fibronectin. I^{125} fibronectin was not bound to human or rabbit polyclonal IgG and to ovalbumin-antiovalbumin immunoglobulin immune complexes not containing C1q.

DISCUSSION

It has been established that plasma fibronectin can bind to the collagen-like part of human C1q[4,5]. In our assay, human C1 complex and C1q complexed with ovalbumin-antiovalbumin immunoglobulin immune complexes also bound I^{125} fibronectin. According to these results it seems possible that plasma fibronectin can also bind in vivo to the immune complexes or cryoglobulins containing C1q. Since human monoclonal myeloma IgG3 and IgG1 also bound I^{125} fibronectin, we suppose that plasma fibronectin can bind in vivo to certain monoclonal immunoglobulin components of the cryoglobulins. C1q was not demonstrated by double immunodiffusion in the mixed monoclonal-polyclonal cryoglobulins. Therefore we suggest that fibronectin binds to one of the immunoglobulin components of the cryoglobulins. Binding of plasma fibronectin to C1q may play a role in fixation of the collagen-like part of C1q to the cryoglobulins in the course of activation of the complement system via the C1q.

REFERENCES

1. B. Anderson, M. Rucker, R. Entwistle, F. R. Schmid and G. N. Wood, Plasma fibronectin is a component of cryoglobulins from patients with connective tissue and other diseases, Ann. Rheum. Dis. 40:50 (1981).
2. A. Liberty, D. M. Bausch, R. D. Baillie, The incorporation of high pH euglobulin precipitation in the isolation of C1q, J. Immunol. Methods 40:243 (1981).
3. S. C. March, I. Parikh and P. Cuatrecasa, A simplified method for cyanogen bromide activation of agarose for affinity chromatography, Anal. Biochem. 60:149 (1974).
4. E. J. Menzel, J. S. Smolen, L. Liotta and K. B. M. Reid, Interaction of fibronectin with C1q and its collagen like fragment, FEBS Letters 129:1:188 (1981).
5. E. Ruoshlaihi, E. Engvall and E. G. Hayman, Fibronectin: Current concepts of its structure and functions, Coll. Res. Vol.1:95 (1981).
6. G. Wood, M. Rucker, J. W. Davis, R. Entwistle and B. Anderson, Interaction of plasma fibronectin with selected cryoglobulins, Clin. Exp. Immunol. 40:358 (1980).
7. M. Vuento and A. Vaheri, Purification of fibronectin from human plasma by affinity chromatography under non-denaturing conditions, Biochem. J. 183:331 (1979).

BIOCHEMICAL SPECIFICITY OF HUMAN COMPLEMENT COMPONENT C1q*

David H. Bing**

Center for Blood Research
Boston, Massachusetts 02115

The serum complement system has an important role in host defense and haemostasis, and inflammatory processes[1]. It consists of 9 components, 2 factors and 9 regulatory processes[2,3]. With the exception of factor D, the complement proteases circulate <u>in vivo</u> as zymogens which are activated to a system of interacting transiently active proteases by antigen-antibody complexes and/or by both cationic and anionic biopolymers. The latter include DNA, acidic proteoglycans, acidic polysaccharides and polymers of basic amino acids. There are at least three inactivator proteins which exert internal control on the complement system by interacting with the activated forms of various components[2,3].

That complement activity can be regulated at the level of the first component of complement, C1, was recognized with the discovery of an inhibitor of C1, C1 INH[4] and the inherited condition, hereditary angioneuroticedema in which C1 INH is lowered or present in non-functional form[5]. Activation of C1 to proteolytically active complex with limited substrate specificity occurs following binding via the C1 subcomponent C1q to polymerized Ig (eg. antigen-antibody aggregates, dimers, trimers and tetramers of IgG or heat-aggregated IgG). The activated C$\bar{1}$ is an unstable complex from which the same protease subcomponents of C$\bar{1}$, C1$\bar{r}$ and C1$\bar{s}$, readily dissociate to react with C1 INH[6-9]; C1$\bar{r}$-C1$\bar{s}$-C1 INH complexes can in fact be detected in plasma[10]. The observed neutralization of C1$\bar{r}$ and C1$\bar{s}$ by C1 INH which only occurs upon dissociation of C1 explains the in-

---

*This work was supported in part by NIH grants AM 17351 and HL 24856.
**Established Investigator of the American Heart Association.

ability of Cl INH to inactivate the bound Cl complex[11]. The fate of
the third Cl subcomponent, Clq, once the Cl$\bar{r}$ and Cl$\bar{s}$ subunits dissociate is not known. It has been suggested that upon binding Ig Clq
itself dissociates into dimers, tetramers, etc. leading to a non-
functional form[12], but the work of Liberti et al.[13] in which it was
shown that active Clq can reversibly associate with EA without losing
activity argues against this hypothesis. Studies on human disease
states demonstrate that Cl activity is self limiting[14]. Thus, following dissociation and neutralization of Cl$\bar{r}$ and Cl$\bar{s}$ _in vivo_, the
newly exposed Clq is prevented from initiating a new round of complement activation by _either_ binding additional immunoglobulin leading
to formation of a complement activating locus _or_ binding and activating additional Clr and Cls via exchange with circulating Cl. As
Clq has no proteolytic activity _per se_, a possible mechanism by which
ing species which can be in the circulation or on cell surfaces.
Clq inhibitors[15-17] and cellular receptors for Clq have been described[18-23] but the role of these in complement activation and regulation has yet to be determined.

The structure of human Clq has been studied in detail in
physical-chemical methodology and electron microscopy. The proposed
model for Clq based on such studies is summarized as follows[24]:
Clq is a 410,000M_r protein containing 6 disulfide linked subunits
which are comprised of 3 polypeptide chains A, B and C of M_r =
24,000, 23,000 and 22,000, respectively[25]. Examination of the entire
molecule reveals a unique structure of six globular units which are
arranged around a fibrous bundle which consists of amino acid sequences similar to that found in collagen[26]. Clq binds to the C_H2
domain of immunoglobulin (Ig) via its globular units and the Clq
subcomponents Clr and Cls bind in Ca^{+2}-dependent complex to the
collagenous portion of Clq[27-30]. Binding of Cl via Clq to immunoglobulin initiates classical pathway complement activation via a
conformationally induced conversion of Clr to an active serine protease[31]. The Clr-Cls binding site on Clq is in the collagenous
region and electron microscopic studies suggest that in Cl the Clr
and Cls are interlaced in the extended portion connecting the globular units to the central collagenous bundle[32]. The importance of
the collagenous portion of Clq is further emphasized by the work of
Schultz and coworkers who have found in ant venom a low molecular
weight polysacharide which can activate the Cl complex via interaction with this portion of Clq[33-36] and studies which have shown
that the collagenous portion of Clq interacts with platelets[37-39].

Other than antigens, the ability to bind specifically to immunoglobulin is shared by only a few kinds of molecules. Clq stands
out as being a normal circulating plasma protein which will bind to
a specific site in the Fc region of immunoglobulin. Quantitative
measurements of the binding affinity of Clq for IgG and IgM have
been made and the K_d range from 235μM for pooled monomeric IgG[40] to
a K_d=0.3nM for tetramers of IgG[41] and from K_d=4.3μM for (Fc) 5μ to 1nM

for a Sepharose bound Waldenstrom macroglobulin[42] (Table 1). The increase in binding affinity of C1q for aggregated immunoglobulin is consistent with the concept that binding and activation of C1 occurs optimally in the presence of antigen-antibody complexes[43,44]. The binding of C1 or C1q to immunoglobulin can be reversed by alkyl or aryl diamines as well as other polyionic molecules[42,45-47]. As high ionic buffers (0.5 to 1M NaCl) will also elute C1q and C1 from Sepharose bound IgG, collectively these data suggest that binding of C1q to immunoglobulins is predominantly ionic in nature[48,49]. A specific binding site in the C_2H domain has been proposed[50] and a peptide has been isolated which fulfills the criteria for being part of this site[51]. The exact nature of the binding sites on C1q and IgG have yet to be fully described, but it is known that the intrinsic binding affinity of C1q for immunoglobulin can be enhanced as a result of the multivalent nature of each of the two molecules for each other and that the functional K_d can approximate the physiological concentrations of each binding species[52].

Our laboratory has obtained quantitative data on the binding properties of C1q with molecules other than immunoglobulin. Earlier studies[42,53,54] demonstrated that alkyl and aryl diamines would bind to C1q and inhibit binding to immunoglobulin. More recently, we have measured the binding constants of plasma and cellular fibronectin and heparin and heparin-like mucopolysaccharides for purified C1q. The results of these studies are summarized in Table 1.

Our studies on fibronectin have used human plasma fibronectin (cold insoluble globulin) and cellular fibronectin derived from a hamster fibroblast cell line. The experiments with plasma fibronectin have employed a solid phase fibronectin (fibronectin adsorbed to plastic tubes). Both ELISA[55,56] and radiobinding assays with ^{125}I-C1q[57] have been used to investigate the binding of C1q to fibronectin. These studies have shown that C1q binds to solid phase fibronectin with a K_d of 60 to 80nM. Studies on the mode of binding have revealed that the collagenous portion of C1q will competitively inhibit this interaction (K_i=59nM) but the globular regions will not[57]. This is further confirmed by experiments in which reconstitution of C1 by addition of C1r + C1s inhibited binding of C1q and heating C1q at 52°, which uncoils the triple coiled collagen structure[58], increases the fibronectin binding activity. Examination of the fibronectin binding activity of the isolated A, B and C polypeptide chains have shown that the isolated A chain had as much binding activity as equimolar mixtures of the A, B and C chains[56]. This is reminiscent of the reactivity of C1q with platelets where the A chain also appeared to play a prominent role[38].

The interaction of heparin or heparin-like material (HLM) with complement was noted 50 years ago[59]. Studies from 1940 to the present have indicated that heparin reversibly inhibits complement activity in a time dependent fashion[60] and that one site of action of

Table 1. Summary of Binding Properties of Human C1q

Species	Binding Constant	Method	Parameters	Ref.
Monomeric Ig				
Pooled Ig	$K_i = 110\mu M$	Inhibition of Binding to IgM Sepharose	Alkyl and Aryl diamines inhibit, binds to globular portion of C1q.	42
Pooled IgG	$K_d = 20\mu M$	Inhibition of Binding of ^{125}I-C1q to Glutaraldehyde RBC		67
Subclasses IgG				
IgG$_1$	$K_d = 85\mu M$			
IgG$_2$	$K_d = 85\mu M$	Ultracentrifugation		40
IgG$_3$	$K_d = 34\mu M$			
IgG$_4$	$K_d = 229\mu M$			
Monoclonal IgM	$K_i = 6.4\mu M$	Inhibition of Binding to IgM Sepharose		42
Aggregated Ig				
Sepharose Bound IgM	$K_d = 704 nM$	Binding of ^{125}I-C1q to Sepharose-IgM		42
Crosslinked IgG	$K_d = 10 nM$	Binding of ^{125}I-C1q to IgG Aggregates		68
IgG Dimer	$K_d = 1\mu M$			

Ligand	K_d	Method	Description	Ref.
Trimer	K_d=100nM	Complement Fixation		41,69
Tetramer	K_d=.33nM			
Heat-aggregated IgG	K_d= 20 to .4nM	Inhibition Binding ^{125}I-Clq to Glutaraldehyde-fixed RBC		67
Heparin	K_d=76.6nM	Radioassay with ^{125}I-heparin	Binds to Clq Collagenous Region. pH and Temperature Independent, Ionic Strength Dependent. 2 Moles bound/mole Clq.	70,71
Fibronectin	K_d=60 to 80nM	Binding ^{125}I-Clq to Solid Phase Fibronectin	Binds to Clq Collagenous Region. pH Independent, Ionic Strength Dependent	55-57

HLM is the heat labile fraction (eg. C1) of complement[61]. Addition
of polyanions in serum lead to activation of C1 and it was suggested
this occurred as a result of formation of complexes with serum proteins such as CRP and possibly others[62-64]. Lipopolysaccharides and
polyanionic polymers inhibited binding of activated C1 via the C1q
subcomponent to EA or EAC4[65,66]. All of these studies were based
on inhibition of the C1 dependent lysis of antibody sensitized sheep
red blood cells and used 10 to 1000 molar excess of polyionic species
over C1/C1q. Previous results could thus reflect displacement of
bound complement component by HLM, HLM inhibition of complement via
interaction with complement regulatory proteins, or impurities in
the complement reagents. Previously reported studies on the interaction of C1q with heparin and synthetic polymers indicated that
there were a limited number of binding sites on C1q (2 to 6), the
inhibition was competitive, and the K_d was less than μM[45,68,69].

Our laboratory has made quantitative measurements of the interaction of human complement subcomponent C1q with heparin and a variety
of mucopolysaccharides. These studies have been described[70,71] and
are summarized as follows (see Table 1). It was shown that 300nM
heparin preferentially inhibited the ability of 1.4nM C1$\bar{r}$ and 1.6nM
C1$\bar{s}$ to recombine with 0.7nM C1q to form a hemolytically active C$\bar{1}$
complex. The binding of C1 to heparin was quantitatively determined
in an assay which used an ^{125}I-labeled low molecular weight heparin
fraction (M_r=8500). Two classes of binding sites were detected.
In the first, 2.02 moles of heparin were bound per mole to C1q with
a K_d of 76.6nM. The second class bound 12 moles of heparin with a
K_d of 1.01μM. That the binding site for the heparin was the collagenous portion of C1q was demonstrated in separate binding experiments which used the isolated collagenous position of C1q with approximately 2.2 moles of heparin bound/mole C1q collagenous region
and a K_d=381nM. In contrast the isolated C1q globular region did
not bind to ^{125}I-heparin. The binding of heparin to the collagenous
region of C1q was further demonstrated by the failure of ^{125}I-labeled
heparin to bind the first component of complement reconstituted from
purified C1q, C1$\bar{r}$ and C1$\bar{s}$. A variety of mucopolysaccharides were
able to inhibit interaction of C1q with ^{125}I-labeled heparin, the
most effective being of chondroitin-6-sulfate and a porcine heparin
which was affinity fractionated to remove the fraction with a high
affinity for antithrombin III.

Fibronectin and heparin-like materials are widely distributed
on the surface of cells and possibly in the circulation. Based on
the magnitude of binding constants which have been measured, it can
be hypothesized that these molecules or ones similar in structure
might be able to exert control on the complement system via interaction with C1/C1q. From the work of others as well as our laboratory, there appears to be at least four possible mechanisms by which
this regulation might occur. These are: (i) the ability to inhibit
directly the interaction of C1q with a receptor which could be on
an immunoglobulin molecule or another complement component; (ii) the
enhancement of the rate and extent of inhibition of activated comp-

lement C1$\bar{\text{r}}$ and C1$\bar{\text{s}}$ components by C1 INH[72]; (iii) modulation of the accessibility of an activated complement component to complement regulatory proteins[74,75]; and (iv) the activation of the complement system directly[62-66]. Whether or not any or all of these mechanisms are operative under physiological conditions will require experimentation to demonstrate that the reactions readily demonstrated with purified proteins have counterparts which occur at the cellular level.

ACKNOWLEDGEMENTS

The author acknowledges the advice and helpful discussions of Ms. Sheri Almeda, Mr. Richard Laura, Dr. Robert D. Rosenberg and Dr. Henri Isliker, and the help of Ms. Rachelle Rosenbaum in preparing this manuscript.

REFERENCES

1. H. J. Müller-Eberhard, Complement, <u>Ann. Rev. Biochem.</u> 44:697-724 (1975).

2. R. D. Schreiber and H. J. Müller-Eberhard, Complement and renal disease, <u>in</u>:"Contempory Issues in Nephrology," C. B. Wilson, B. M. Brenner and J. H. Stein, eds., 3:67-105 (1979).

3. D. T. Fearon and K. F. Austen, Activation mechanisms of the alternative complement pathway, <u>in</u>:"Chemistry and Physiology of the Human Plasma Proteins," D. H. Bing, ed., Pergamon Press, N.Y. pp. 229-254 (1979).

4. O. Ratnoff and I. H. Lepow, Some properties of an esterase derived from preparations of the first component of complement, <u>J. Exp. Med.</u> 106:327-342 (1957).

5. V. H. Donaldson, Serum C1 inhibitor (C$\bar{\text{I}}$ INH), <u>in</u>:"Chemistry and Physiology of the Human Plasma Proteins," D. H. Bing, ed., Pergamon Press, N.Y. pp. 369-383 (1979).

6. R. J. Ziccardi and N. R. Cooper, Demonstration and quantitation of activation of the first component of complement in human serum, <u>J. Exp. Med.</u> 147:385-364 (1978).

7. R. J. Ziccardi and N. R. Cooper, Modulation of the antigenicity of C1$\bar{\text{r}}$ and C1$\bar{\text{s}}$ by C1 inactivator, <u>J. Immunol.</u> 121:2148-2153 (1978).

8. R. J. Ziccardi and N. R. Cooper, Active disassembly of the first component of complement component C$\bar{\text{I}}$ by C$\bar{\text{I}}$ inactivator, <u>J. Immunol.</u> 123:788-792 (1978).

9. R. B. Sim, G. J. Arlaud and M. Columb, C$\bar{\text{I}}$ inhibitor-dependent dissociation of human complement component C$\bar{\text{I}}$ bound to immune complexes, <u>Biochem. J.</u> 179:449-457 (1979).

10. A-B. Laurell, U. Johnson, U. Martensson and A. G. Sjoholm, Formation of complexes composed of C1r, C1s and C1 inactivator in human serum on activation of C1, <u>Acta Pathol. Microbiol. Scand. (C)</u> 86:299-304 (1978).

11. I. Gigli, S. Ruddy and K. F. Austen, The stoichiometric measure-
    ment of the serum inhibitor of the first component of comp-
    lement by inhibition of immune hemolysis, J. Immunol. 100:
    1154-1164 (1968).

12. T. Borsos, M. Loos, R. M. Chapuis, R. Medicus and H. Isliker,
    A novel way of relating the structure of C1q to the hemolytic
    activity of the first component of complement, Mol. Immunol.
    17:1415-1424 (1980).

13. P. A. Liberti, R. D. Baillie, K. Milligan and D. Bausch, On the
    interaction of rabbit C1q with sheep and rabbit immune com-
    plexes, J. Immunol. 123:2212-2219 (1979).

14. S. Ruddy, I. Gigli and K. F. Austen, The complement system of
    man, N. Engl. J. Med. 287:489-495 (1972).

15. J. D. Conradie, J. E. Volanakis and R. M. Stroud, Evidence for
    a serum inhibitor of C1q, Immunochemistry 12:967-971 (1969).

16. B. Ghebrehiewet and H. J. Müller-Eberhard, Lysis of C1q coated
    chicken erythrocytes by human lymphoblastoid cell lines, J.
    Immunol. 120:27-32 (1978).

17. L. Silvestri J. R. Baker, L. Roden and R. M. Stroud, The C1q
    inhibitor in serum is a chondroitin-4-sulfate proteoglycan,
    J. Biol. Chem. 256:7383-7387 (1981).

18. H. B. Dickler and H. J. Kunkel, Interaction of aggregate γ
    globulin with B-lymphocytes, J. Exp. Med. 136:191-196 (1972).

19. A. T. Sobel and U. A. Bokish, Receptors for C4b and C1q on human
    peripheral lymphocytes and lymphoblastoid cells, in:"Membrane
    Receptors of Lymphocytes," M. Seligman, J. L. Preud-Homme and
    K. M. Knovilsk, eds., North Holland, Amsterdam p. 151 (1975).

20. A. J. Tenner and N. R. Cooper, Analysis of receptor-mediated C1q
    binding to human peripheral blood mononuclear cells, J.
    Immunol. 125:1658-1664 (1980).

21. B. Ghebrehiwet, C1q inhibitor (C1q INH): Functional properties
    and possible relationship to a lymphocyte membrane associated
    C1q precipitin, J. Immunol. 126:1837-1842 (1981).

22. E. Linder, Binding of C1q and complement activation by vascular
    endothelium, J. Immunol. 126:648-657 (1981).

23. B. S. Andrews, M. Shadforth, P. Cunningham and J. S. Davis IV,
    Demonstration of a C1q receptor on the surface of human
    endothelial cells, J. Immunol. 127:1075-1080 (1981).

24. R. R. Porter and K. B. M. Reid, Activation of the complement
    system by antibody-antigen complexes: The classical pathway,
    Adv. Protein Chem. 33:1-71 (1979).

25. K. B. M. Reid, Isolation, by partial pepsin digestion, of the
    three collagen-like regions present in subcomponent C1q of
    the first component of human complement, Biochem. J. 155:5-17
    (1976).

26. K. B. M. Reid, A collagen-like amino acid sequence in a polypep-
    tide chain of human C1q (a subcomponent of the first com-
    ponent of complement), Biochem. J. 141:189-203 (1979).

27. H. R. Knobel, C. Heusser, M. L. Rodrick and H. Isliker, Enzymatic
    digestion of the first component of human complement (C1q),
    J. Immunol. 112:2094-2101 (1974).

28. N. C. Hughes-Jones and B. Gardner, Reaction between isolated globular subunits of the complement component C1q and IgG complexes, Mol. Immunol. 16:697-701 (1979).

29. K. B. M. Reid, R. B. Sim and A. P. Faiers, Inhibition of the reconstitution of the haemolytic activity of the first component of human complement by a pepsin derived fragment of subcomponent C1q, Biochem. J. 161:239-245 (1977).

30. R. R. Porter, Activation mechanism of the classical pathway of complement, in:"Chemistry and Physiology of the Human Plasma Proteins," D. H. Bing, ed., Pergamon Press, N.Y. pp. 281-288 (1979).

31. N. R. Cooper, Activation of the complement system, Contemp. Top. Immunol. 3:155-183 (1973).

32. V. N. Schumaker and Pak H. Poon, Ultrastructure of the first component of complement, This symposium, 99-116 (1981).

33. D. R. Schultz and P. I. Arnold, Venom of the ant Pseudonyrmex sp.: further charaterization of two factors that affect human complement proteins, J. Immunol. 119:1690-1699 (1977).

34. D. R. Schultz, P. I. Arnold, M.-C. Wu, T. M. Lo, J. E. Volanakis and M. Loos, Isolation and partial characterization of a polysaccharide in ant venom (Pseudomyrmex sp.) that activates the classical complement pathway, Mol. Immunol. 16:253-264 (1979).

35. D. R. Schultz, M. Loos, F. Bub and P. I. Arnold, Differentiation of hemolytically active fluid-phase and cell-bound human C1q by an ant venom-derived polysaccharide, J. Immunol. 124:1251-1257 (1980).

36. D. R. Schultz and P. I. Arnold, The first component of human complement: on the mechanism of activation by some carbohydrates, J. Immunol. 126:1994-1998 (1981).

37. J. P. Cazenave, S. N. Assimeh, R. N. Painter, M. A. Packham and J. F. Mustard, C1q inhibition of the interaction of collagen with human platelets, J. Immunol. 116:162-163 (1976).

38. E. A. Suba and G. Csako, C1q (C1) receptor on human platelets: Inhibition of collagen-induced platelet aggregation by C1q (C1) molecules, J. Immunol. 117:304-309 (1976).

39. J.-L. Wautier, H. Souchon, K. B. M. Reid, A. P. Peltier and J. P. Caen, Studies on the mode of reaction of the first component of complement with platelets: Interaction between the collagen-like portion of C1q and platelets, Immunochemistry 14:763-766 (1977).

40. V. N. Schumaker, M. A. Calcott, H. L. Spiegelberg and H. J. Müller-Eberhard, Ultracentrifuge studies on the binding of IgG of different subclasses to the C1q subunit of the first component of complement, Biochemistry 15:5175-5181 (1976).

41. J. K. Wright, J. Tschopp, J. C. Jaton and J. Engle, Dimeric, trimeric and tetrameric complexes of immunoglobulin G fix complement, Biochem. J. 187:775-780 (1980).

42. C. R. Sledge and D. H. Bing, Binding properties of human complement protein C1q, J. Biol. Chem. 248:2818-2823 (1973).

43. A. W. Dodds, R. R. Sim, R. R. Porter and M. A. Kerr, Activation of the first component of human complement (C1) by antibody-antigen aggregates, Biochem. J. 175:388-390 (1978).

44. J. Tschopp, T. Schulthess, J. Engel and J. T. Jaton, Antigen-independent activation of the first component of complement C1 by chemically crosslinked rabbit IgG oligmers, FEBS Letters 112:152-154 (1980).

45. N. C. Hughes-Jones and B. Gardner, The reaction between complement subcomponent C1q, IgG and polyionic molecules, Immunology 34:459-463 (1978).

46. E. Raepple, H.U. Hill and M. Loos, Mode of interaction of different polyions with the first (C1, CĪ), the second (C2) and fourth (C4) component of complement. I. Effect on fluid phase CĪ and CĪ bound to EA and EAC4, Immunochemistry 13:251-253 (1976).

47. D. S. Fletcher and T.-Y. Lin, Inhibition of immune complex-mediated activation of complement, Inflammation 4:113-123 (1980).

48. W. P. Kolb, L. M. Kolb and E. R. Podack, C1q: Isolation from human serum in high yield by affinity chromatography and development of a highly sensitive hemolytic assay, J. Immunol. 122:2103-2111 (1979).

49. R. Medicus and R. M. Chapuis, The first component of complement. I. Purification and properties of native C1, J. Immunol. 125:390-395 (1980).

50. D. R. Burton, J. Boyd, A. D. Brampton, S. B. Easterbrook-Smith, E. J. Emanuel, J. Novotny, M. R. Rademacher, M. J. E. van Schravendyk-Sternberg and R. A. Dwek, The C1q receptor site on immunoglobulin G, Nature 288:338-344 (1980).

51. J. P. Lee and R. H. Painter, Complement binding properties of two peptides from the $C\gamma 2$ region of IgG_1, Mol. Immunol. 17:1155-1162 (1980).

52. S. K. Dower and D. M. Segal, C1q binding to antibody coated cells: predictions from a simple mutivalent model, Mol. Immunol. 18:823 (1981).

53. G. Wirtz, Influence of aliphatic amines on the interaction of human C'1 with EAC'4, Immunochemistry 2:95-162 (1965).

54. J. R. Sledge and D. H. Bing, Purification of human complement component C1q by affinity chromatography, J. Immunol. 111:661-666 (1973).

55. H. Isliker, D. H. Bing and R. O. Hynes, Interactions of fibronectin with C1q, a subcomponent of the first component of complement, in:"The Immune System," Vol. 2, C. M. Steinberg and I. Lefkowitz, eds., S. Karger, Basel pp.231-238 (1981).

56. H. Isliker, D. H. Bing and R. O. Hynes, Fibronectin interacts with C1q, a subcomponent of the first component of complement, Immunological Comm. in press (1981).

57. D. H. Bing, S. Almeda, H. Isliker, J. Lahav and R. O. Hynes, Fibronectin binds to the C1q component of complement, Proc. Natl. Acad. Sci. USA, submitted (1981).

58. B. Brodsky-Doyle, K. R. Leonard and K. B. M. Reid, Circular dichroism and electron microscopy studies on human C1q before and after limited proteolysis by pepsin, Biochem. J. 159:279-286 (1976).

59. E. E. Ecker and P. Gross, Anticomplementary power of heparin, J. Infect. Dis. 44:250-253 (1929).

60. P. J. Wising, The complement-fixing properties of heparin salts, Acta Med. Scand. 91:550-554 (1937).

61. E. E. Ecker and L. Pillimer, Anticoagulants and complementary activity, an experimental study, J. Immunol. 40:73-80 (1941).

62. R. Rent, N. Entel, R. Eisenstein and H. Gewurz, Complement activation by interaction of polyanions and polycations. I. Heparin-protamine induced consumption of complement, J. Immunol. 114:120-124 (1975).

63. B. A. Fiedel, R. Rent, R. Myhrman and H. Gewurz, Complement activation by interaction of polyanions and polycations. II. Precipitation and role of IgG, C1q and C1 INH during heparin-protamine induced consumption of complement, Immunology 30: 161-169 (1976).

64. D. R. Claus, J. Siegel, K. Petras, D. Skor, A. P. Osmand and H. Gewurz, Complement activation by interaction of polyanions and polycations. III. Complement activation by interaction of multiple polyanions and polycations in the presence of C-reactive protein, J. Immunol. 118:83-87 (1977).

65. M. Loos, D. Bitter-Suermann and M. Dierich, Interaction of the first (C1), the second (C2) and the fourth (C4) component of complement with different preparations of bacterial lipo-polysaccharides and with lipid A, J. Immunol. 112:935-940 (1974).

66. M. Loos and D. Bitter-Suermann, Mode of interaction of different polyanions with the first (C1, C1), the second (C2) and the fourth component of complement. IV. Activation of C1 in serum by polyanions, Immunology 31:931-934 (1976).

67. N. C. Hughes-Jones, Function affinity constants of the reaction between ^{125}I-labeled C1q and C1q binders and their use in the measurement of plasma C1q concentrations, Immunology 32: 191-198 (1977).

68. T-Y. Lin and D. S. Fletcher, Interaction of human C1q with in-soluble immunoglobulin aggregates, Immunochemistry 15:107-117 (1978).

69. J. Tschopp, W. Villiger, A. Lustig, J. C. Jaton and J. Engel, Antigen independent binding of IgG dimers to C1q as studied by sedimentation equilibrium, complement fixation and electron microscopy, Eur. J. Immunol. 10:529-535 (1980).

70. D. H. Bing, S. Almeda and R. D. Rosenberg, The binding properties of human C1q; Interaction with mucopolysaccharides. Complement Workshop, J. Immunol., in press (1982) (Abs).

71. S. Almeda, R. D. Rosenberg and D. H. Bing, The binding properties of human complement C1q; Interaction with mucopolysaccharides, J. Biol. Chem., submitted (1981).

72. R. Rent, B. Myhrman, A. Fiedel and H. Gewurz, Potentiation of C1-esterase inhibitor activity by heparin, Clin. Exp. Immunol. 23:264-271 (1971).

73. K. Nagaki and S. Inai, Inactivator of the first component of complement (CĪ INA). Enhancement of CĪ INH activity against CĪs by acidic mucopolysaccharides, Int. Arch. All. Appl. Immunol. 50:172-180 (1976).

74. J. M. Weiler, R. W. Yurt, D. T. Fearon and K. F. Austen, Modulation of the formation of the amplification convertase of complement C3b, Bb by native and commercial heparin, J. Exp. Med. 147:409-421 (1978).

75. M. D. Katzatchkine, D. T. Fearon, J. E. Silbert and K. F. Austen, Surface-associated heparin inhibits zymosan-induced activation of the human alternative complement pathway by augmenting the regulatory action of the control proteins on particle bound C3b, J. Exp. Med. 150:1202-1215 (1979).

SEQUENCE STUDIES ON HUMAN COMPLEMENT SUBCOMPONENT C$\overline{1}$r

Gérard J. Arlaud[1], Jean Gagnon and Rodney R. Porter

DRF/BMC, C.E.N.-G., 85 X, 38C41 Grenoble, France[1]
MRC Immunochemistry Unit, Department of Biochemistry
Oxford University, South Parks Road
Oxford OX1 3QU UK

C1, the first component of the classical complement pathway, is a calcium-dependent complex consisting of three distinct glycoproteins, subcomponents C1q, C1r and C1s. Binding of C1 to activators is mediated mainly by C1q and induces activation of proenzyme C1r to its proteinase form C$\overline{1}$r, which in turn cleaves and activates proenzyme C1s. Subcomponent C$\overline{1}$s is the proteinase responsible for specific cleavage by C$\overline{1}$ of complement components C2 and C4[1]. Activation of C1 is triggered by interaction with the antibody moiety of antibody-antigen complexes, but can also be induced by antigen-independent, direct C1 binding to activators such as bacteria[2,3] and viruses[4]. Thus, the C1 complex is not only the activation unit of the classical pathway of complement, but can also fulfil a direct recognition function.

The first proteolytic event of C1 activation is currently attributed to C1r itself. This is supported by the experiments of Dodds[5], who proposed that C1r is autocatalytically activated when incorporated within the C1 complex bound to antibody-antigen aggregates. An intramolecular autocatalytic mechanism has also been proposed to account for activation of isolated C1r[6]. However, there are conflicting reports that isolated C1r by itself has[6,7,8] or has not[5,9] the capacity to autoactivate. The main question is therefore to know whether the signal which is transmitted to the C1 complex upon binding to activators induces "de novo" the appearance of the C1r autoactivating capacity, or only modulates a property inherent in the C1r molecule, but normally repressed inside the C1 complex in vivo. Activated C$\overline{1}$r is a highly specific serine proteinase, as its biological function in C$\overline{1}$ is thought to be restricted to a single cleavage in proenzyme C1s. This double feature of C1r: autoacti-

vating property of the proenzyme, and very restricted specificity
of the active form, prompted us to investigate the primary sequence
of this C1 subcomponent, in search of a structural basis of its
unique properties.

C̄1̄r was purified from human serum[10] and, after reduction and
alkylation, the a- and b-chains (apparent molecular weights 58000
and 35000, respectively) were separated by high-pressure gel-
permeation chromatography. CNBr cleavage was performed on the catalytic b chain and yielded eight major peptides (CB Ia, CB Ib, CB II,
CB III, CB IV, CB Va, CB Vb and CB Vc), as expected from the methionine content of the chain (7 residues/mole). The CNBr peptides were
purified by gel filtration on Sephadex G-50, then by high-pressure
reversed-phase chromatography on a μ Bondapak C 18 column[11]. As
determined from the sum of their amino acid compositions, these peptides accounted for a minimum molecular weight of 28000, close to
the value 29100 calculated from the whole b chain.

N-terminal sequence determinations of C̄1̄r b chain and its CNBr-
cleavage peptides allowed the identification of about two-thirds of
the amino-acid residues of the chain[11]. N-terminal sequence of residues 1-20 of the whole b chain was found in agreement with previous
results[12] and allowed the alignment of peptides CB IV and CB Ib
(Figure 1). The alignment of peptides CB Vb and CB II was provided
by the sequence of the overlapping peptide, whereas peptide CB Va,
the only one lacking homoserine, was identified as the C-terminal
peptide of C̄1̄r b chain. These findings allowed the relative alignment of five peptides in the order:

(CB IV) - (CB Ib) / (CB Vb) - (CB II) / (CB Va)

From homology with other serine proteinases[13], peptides CB Ia,
CB Vc and CB III were located in this order between peptides CB Ib
and CB Vb (Figure 1). Thus, the complete alignment proposed for
the eight major CNBr-cleavage peptides of C̄1̄r b chain is the following:

(CB IV) - (CB Ib) - (CB Ia) - (CB Vc) - (CB III) - (CB Vb) -
(CB II) - (CB Va)

Positioning of the eight peptides in this order shows extensive
homology with the conserved residues of the catalytic chain of serine
proteinases (Figure 1). The residues forming the "charge-relay"
system of the active site of serine proteinases (His-57, Asp-102 and
Ser-195 in the chymotrypsinogen numbering) are found at position 39
of C̄1̄r b chain, position 27 of peptide CB Ia, and position 17 of
peptide CB II, in a linear distribution that is in accordance with
that found in other serine proteinases[13]. The presence of an as-
partic acid residue at position 11 of peptide CB II indicates that
C̄1̄r belongs to the family of "trypsin-like" enzymes, which is con-

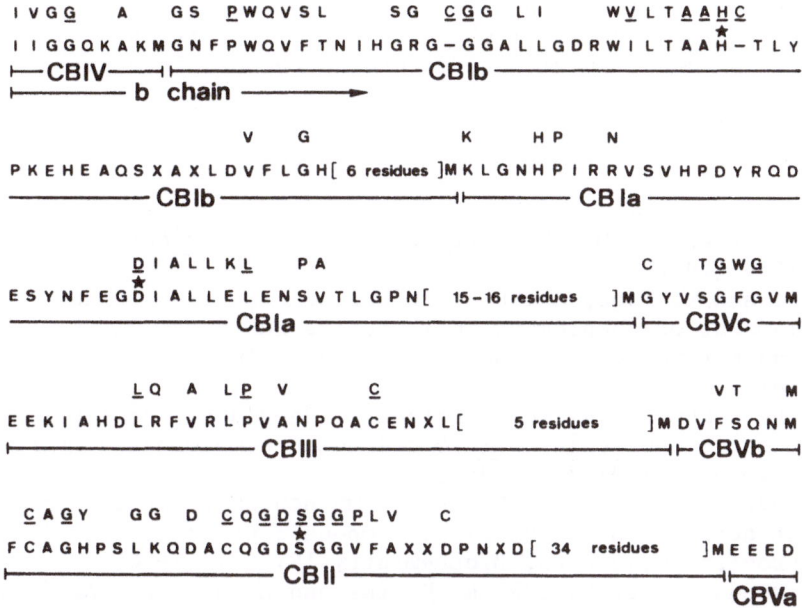

Fig. 1. Alignment and N-terminal amino acid sequences of the major
         CNBr-cleavage peptides from human C1̄r b chain. The histi-
         dine, aspartic acid and serine residues involved in the
         charge-relay system are indicated by a star (★). Two gaps
         (-) were left at positions were are normally located the
         two half-cystine residues forming the "histidine-loop"
         disulphide bridge in other mammalian serine proteinases.
         The conserved residues indicated above the C1̄r sequence
         are from Young[13], those underlined being invariant in the
         known serine proteinases sequences[13]. The single-letter
         code of amino acid residues has been used.

sistent with the proteolytic and esterolytic specificities of C1̄r[14].
The sequence found at positions 15-20 of peptide CB II (Gly-Asp-Ser-
Gly-Gly-Val) is homologous with the primary binding site of serine
proteinases[13], with the exception of residue at position 20, which
is invariably a proline in other known serine proteinases (Figure 1).

     The N-terminal part of C1̄r b chain structure lacks two half-
cystine residues invariant in other mammalian serine proteinases,
in which they are linked to form a disulphide bond called the
"histidine-loop"[13]. These two residues, usually located 15 positions
before the histidine residue of the charge-relay system, and immediately after it (Figure 1), have also been found missing in the corresponding part of C1̄s b chain structure[15]. Given that proenzyme C1s,
in contrast to proenzyme C1r, is not self-activatable, and that
activated C1̄s has a wider esterolytic specificity than C1̄r, the

absence of the "histidine-loop" disulphide bridge cannot be directly
related to precise functional characteristics. Further studies are
currently being done, in view to determine the complete structure
of C$\overline{1}$r b chain.

REFERENCES

1. R. R. Porter, The biochemistry of complement, Biochem. Soc. Trans.
   5:1659 (1977).
2. F. Clas and M. Loos, Antibody-independent binding of the first
   component of complement (C1) and its subcomponent C1q to the
   S and R forms of Salmonella minnesota, Infect. Immun. 31:1138
   (1981).
3. S. J. Betz and H. Isliker, Antibody-independent interactions
   between Escherichia coli J5 and human complement components,
   J. Immunol. 127:1748 (1981).
4. R. M. Bartholomew and A. F. Esser, Mechanism of antibody-independent activation of the first component of complement (C1) on
   retrovirus membranes, Biochemistry 19:2847 (1980).
5. A. W. Dodds, R. B. Sim, R. R. Porter and M. A. Kerr, Activation
   of the first component of human complement (C1) by antibody-
   antigen aggregates, Biochem. J. 175:383 (1978).
6. G. J. Arlaud, C. L. Villiers, S. Chesne and M. G. Colomb,
   Purified proenzyme C1r. Some characteristics of its activation and subsequent proteolytic cleavage, Biochim. Biophys.
   Acta. 616:116 (1980).
7. R. J. Ziccardi and N. R. Cooper, Activation of C1r by proteolytic
   cleavage, J. Immunol. 116:504 (1976).
8. S. N. Assimeh, R. M. Chapuis and H. Isliker, Studies on the precursor form of the first component of complement - II.
   Proteolytic fragmentation of C1r, Immunochemistry 15:13
   (1978).
9. J. Bauer, E. Weiner and G. Valet, Mechanism of C1r activation,
   J. Immunol. 124:1514 (Abstr.) (1980).
10. G. J. Arlaud, R. B. Sim, A.-M. Duplaa and M. G. Colomb, Differen-
    tial elution of C1q, C$\overline{1}$r and C$\overline{1}$s from human C1 bound to immune
    aggregates. Use in the rapid purification of C$\overline{1}$ subcompon-
    ents, Mol. Immunol. 16:445 (1979).
11. G. J. Arlaud, J. Gagnon and R. R. Porter, The catalytic chain
    of human complement subcomponent C$\overline{1}$r. Purification and N-
    terminal amino acid sequences of the major cyanogen bromide-
    cleavage fragments, Biochem. J. 201:49 (1982).
12. R. B. Sim, R. R. Porter, K. B. M. Reid and I. Gigli, The struc-
    ture and enzymatic activities of the C1r and C1s subcomponents
    of C1, the first component of human complement, Biochem. J.
    163:219 (1977).
13. C. L. Young, W. C. Barker, C. M. Tomaselli and M. O. Dayhoff,
    Serine proteases, in:"Atlas of Protein Sequence and Structure"
    M. O. Dayhoff, ed., Vol. 5, suppl. 3:73 (1978).

14. R. B. Sim, The human complement system serine proteases C$\overline{1}$r and
    C$\overline{1}$s and their proenzymes, <u>Methods Enzymol</u>. 80, in press (1981).
15. G. J. Arlaud and J. Gagnon, C$\overline{1}$r and C$\overline{1}$s subcomponents of human
    complement : two serine proteinases lacking the "histidine-
    loop" disulphide bridge, <u>Bioscience Reports</u> 1:779 (1981).

IV: THE ACTIVATION OF THE FIRST COMPONENT OF COMPLEMENT

INTRODUCTION

Can a similar conceptual framework, including the allosteric, distortive, and associative models, which have proven so useful in efforts to understand the generation of the immunological signal during antibody antigen interaction, also be employed as an aid toward the design of experiments to explore the molecular mechanism by which the immunological signal is transmitted from the complex formed between the Fc domains and the Clq heads to the Clr_2Cls_2 tetramer to trigger its activation? The problem is similar in some respects, for example, it involves signal transmission over a long distance. Since Clq is flexible, a distortive model seems possible; thus, a distortive model could involve a relative motion between the Clq arms, mechanically transmitting the activating signal over a distance of 50 to 100A to the Clr, thought to be located near the hinge of the central bundle as previously described in this volume by Schumaker and Poon. Alternatively, an allosteric model might involve signal transmission through allosteric change in the dynamic structure of the collagenous arms of Clq, resulting in the propagation of activating signals which ripple along the length of collagenous helix from the Clq heads to the central bundle. Finally, two different associative models could be developed involving either (a) interaction between antibodies and two or more molecules of Cl, or (b) physical contact between the Clr subunits and the complex formed between the Clq heads and the Fc portion of the immunoglobulins, or some other portion of the immunoglobulins.

Whatever the mechanism, it must be compatible with the kinetics observed for the activation of Cl. In the first paper of this Topic by Kilchherr and coworkers, it is shown that the kinetics are sig-

137

moidal, which the authors interpret in terms of several slow steps and the build-up in concentration of intermediate species. Moreover, they show that the soluble activating system employed in these studies is sensitive to concentration, and that this is a consequence of the dissociation of the C1 complex. From their kinetic studies, they report a dissociation constant of 50nM, which is intermediate between the values found for the unactivated and activated C1 by Schumaker and Poon.

A study of the fluid phase activation of proenzyme C1r is reported in the second paper by Colomb and his colleagues. In this presentation it is shown that spontaneous activation of C1r occurs in the presence of DFP or C1 inhibitor at 37° for 90 minutes, and activation is abolished at pH 5 and strongly inhibited in the presence of Ca^{++}. These workers also show that the pattern of iodination is markedly changed upon activation of C1r, suggesting that a dramatic conformational change takes place when the polypeptide is cleaved. The implications of these results toward understanding the activation of C1 is discussed by the authors, and a distortive model for the activation mechanism is proposed by them.

The final paper in this Topic concerns the activation of mammalian complement by chicken C1q. Else Marie Nicolaisen reports that human $C1r_2C1s_2$ can be activated by chicken C1q bound to chicken antibodies. It is remarkable that the binding site of C1q for $C1r_2C1s_2$ has been conserved in such evolutionarily distant and diverse species of higher animals as chickens, frogs and mammals. The mixed complement reaction is inhibited by a normally occurring chicken protein, HEF, which may be homologous with fibronectin or human C1q inhibitor.

DISSOCIATION OF C1 AND CONCENTRATION DEPENDENCE OF ITS ACTIVATION

KINETICS

Erich Kilchherr, Harald Fuchs
Jurg Tschopp, and Jurgen Engel

Department of Biophysical Chemistry
Biozentrum der Universitat Basel
Basel, Switzerland
Department of Molecular Immunology
Scripps Clinic and Research Foundation
La Jolla, CA 92037, USA

The activation of the zymogen C1s to the enzyme C1s in human C1 complex $(C1q(C1rC1s)_2)$ was studied as a function of the concentrations of $(C1rC1s)_2$ and of C1q which was saturated with oligomers of rabbit IgG (chemically crosslinked, defined size). A large concentration dependence of the sigmoidal kinetics was observed in the 2 to 180nM concentration range (Figure 1). This was explained by association-dissociation equilibria between the antibody-saturated C1q and various forms of the $(C1rC1s)_2$ complex (unactivated to activated).

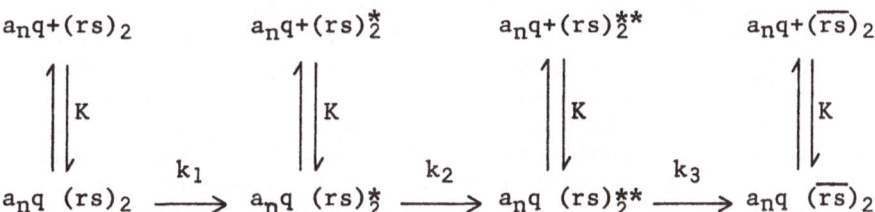

The establishment of these equilibria (binding constants $K=2 \times 10^7 M^{-1}$) was assumed to be fast as compared to the rates of the activation steps (rate constants $k_1=10^{-3}s^{-1}$ and $k_2=k_3=10^{-2}s^{-1}$ at 30°C). The fast re-equilibration of the C1 complex explains the finding that small amounts of antibody saturated C1q catalyzed the activation of large amounts of C1s. The interpretation of the kinetic results was supported by a direct demonstration of the dissociation properties of reconstitued C1 and C1 isolated from serum.

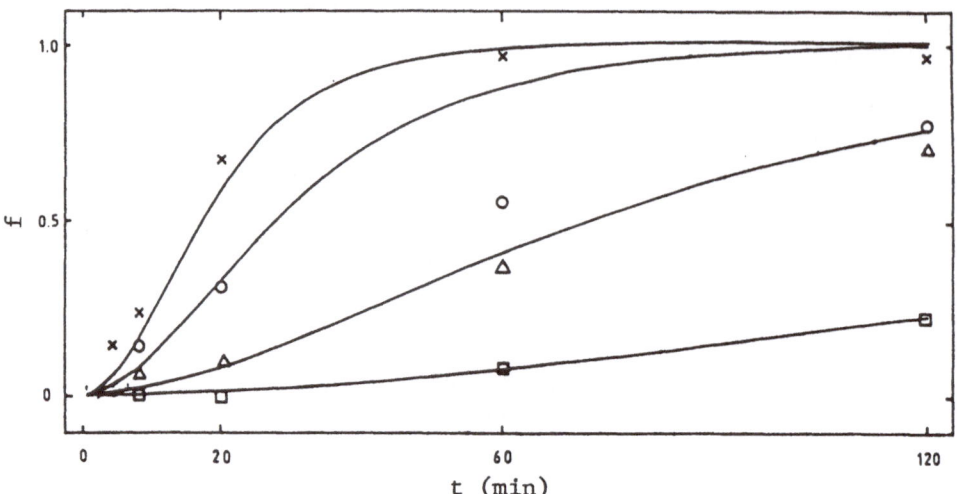

Fig. 1. Time courses of C1s activation at various concentrations
         of C1q and (C1rC1s)$_2$. The concentration of C1q was
         [C1q] =180nM (X), 40nm (O), 8nm (Δ) and 2nM and the molar
         ratio of C1q to (C1rC1s)$_2$ was 2 in all experiments. The
         activation was initiated by addition of IgG-dimers to a
         final concentration of 1.2μM. The degree of activation f
         of C1s at 30°C in 10nM Tris HCl, pH 7.4, 150nM NaCl con-
         taining 5mM Ca^{++} was followed by the Cooper and Ziccardi
         test. The curves were calculated from the mechanism
         described in the text.

REFERENCES

1. J. K. Wright, J. Tschopp, J.-C. Jaton and J. Engel, Dimeric,
    trimeric and tetrameric complexes of immunoglobulin G fix
    complement, Biochemical J. 187:775 (1980).
2. J. Tschopp, W. Villiger, A. Lustig, J.-C. Jaton and J. Engel,
    Antigen-independent binding of immunoglobulin G-dimers to
    C1q as studied by sedimentation equilibrium, complement fix-
    ation and electron microscopy, Eur. J. Immunol. 10:529 (1980).
3. J. Tschopp, T. Schulthess, J. Engel and J.-C. Jaton, Antigen-
    independent activation of the first component of complement
    C1 by chemically cross-linked rabbit IgG-oligomers, FEBS Lett.
    112:152 (1980).
4. J. Tschopp, W. Villiger, H. Fuchs, E. Kilchherr and J. Engel,
    Electron microscopic and ultracentrifugal studies on the
    assemble of the subcomponents C1r and C1s of the first com-
    ponent (C1), Proc. Natl. Acad. Sci. USA 77:7014 (1980).

FLUID PHASE ACTIVATION OF PROENZYMIC C1r

Maurice G. Colomb, Cristian L. Villers
Gérard J. Arlaud

Equipe de Recherche "Immunochimie -
Système Complémentaire" du DRF-G et de l'USM-G
associée au CNRS (ERA No 695) et a l'INSERM (U. No 238)
Laboratoire de Biologie Moléculaire et Cellulaire
Centre d'Etudes Nucléaires de Grenoble, 85 X
38041 Grenoble, Cedex, France

C1r is the first protein in the complement sequence to activate through proteolysis[1]: hydrolysis of a single peptide bond in monochain proenzymic C1r leads to an activated molecule consisting of two chains, a and b, the latter bearing the active site responsible for subsequent proteolysis of C1s inside C1. Activation of C1r in C1 results from various interactions expressed either at the level of subcomponent C1q (immune complexes with IgG or IgM, aggregated immunoglobulins)[1], or at the level of both subcomponents C1q and C1s (viral membranes)[2], or even directly at the level of C1r (plasmin, Hageman factor fragment)[3,4]. In order to evaluate the relevance of these different interactions to the same mechanism of activation it appears necessary to study first the basal activation of isolated C1r.

The activation of C1r has been a controversial problem, with results against[5,6] or in favour[7-10] of an autocatalytic process. As most discrepancies may be due to traces of contaminant proteases

---

Abbreviations:
DFP, diisopropyl phosphorofluoridate.
SDS, sodium dodecyl sulphate.
PAGE, polyacrylamide gel electrophoresis.
The nomenclature of the components of complement is that recommended by World Health Organization (1968). Activation of a component is indicated by a bar.

in C1r, a new protocol for C1r purification was established, based
on a one-step affinity chromatography, from human citrated plasma,
without any exogenous inhibitor[11]. This paper reports the main
features of C1r activation. Current views concerning the location
of C1r in C1 are discussed.

The purity of C1r was assessed, on SDS-PAGE, from protein staining or after labelling of proenzymic C1r with tritiated DFP: after
incubation for 60 minutes at 37°C with 1mM [^{3}H]-DFP, C1r was analysed
by SDS-PAGE in non reducing and reducing conditions. Radioactivity
was detected only at the level of 91000 daltons (C$\overline{1}$r) or 35000
daltons (b chain of C$\overline{1}$r) respectively (Figure 1), reflecting the
only labelling of C$\overline{1}$r present at time zero of the incubation or C1r
activated during the 60 minute incubation at 37°C.

Total activation of isolated C1r was achieved within 30 to 60
minutes at 37°C. C1q, C1s and C$\overline{1}$s were individually without effect
on this activation; in contrast, the rate of activation was increased
in the presence of activated C$\overline{1}$r (Figure 2a): total activation,
achieved in 60 minutes in the absence of activated C$\overline{1}$r, was completed in 15 minutes in the presence of 66% (w/w) C$\overline{1}$r. This acti-

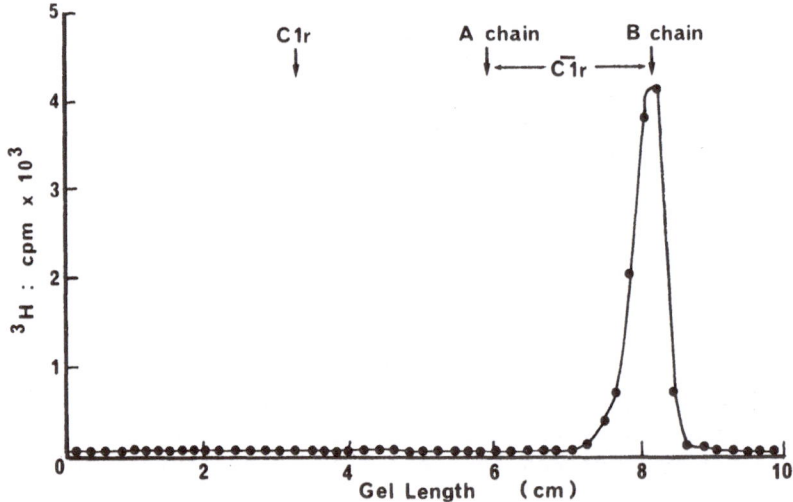

Fig. 1.   Labelling of proenzymic C1r by [1,3-^{3}H] DFP and subsequent
          activation. Proenzymic C1r (0.4mg/l) was incubated in 1mM
          [1,3-^{3}H] DFP/145mM NaCl/5mM triethanolamine-HCl (pH 7.4)
          for 60 minutes at 37°C. After exhaustive dialysis against
          the above medium, but without DFP, 12μg protein were re-
          duced, alkylated and submitted to SDS-PAGE. Radioactivity
          was estimated on gel slices by scintillation counting.

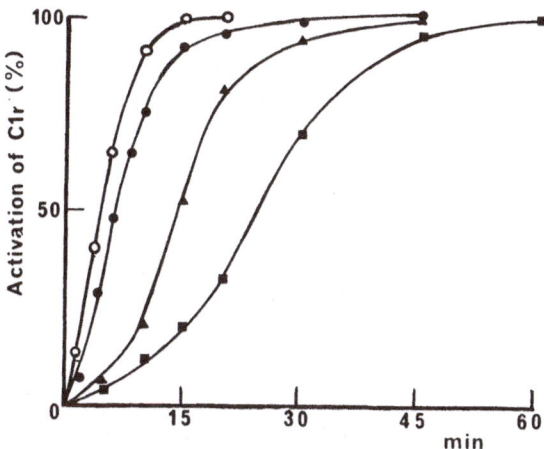

Fig. 2a.   Effect of activated C1̄r on C1r activation. Samples of
C1r (12μg) were incubated with different amounts of C1̄r
in 50μl of 145mM NaCl/5mM triethanolamine-HCl (pH 7.4),
for different periods, at 37°C. After reduction and alkylation, samples were analysed by SDS-PAGE. The amount
of activated C1̄r was obtained by scanning the gels after
Coomassie Blue staining. C1̄r: C1r molar ratios = 0
(—■—), 0.04 (—▲—), 0.40 (—●—), 0.66 (—○—).

vating effect of C1r was abolished when activated C1̄r was blocked
by C1̄ Inh or DFP and, in this last case, submitted to exhaustive
dialysis, before incubation with proenzymic C1r (Figure 2b), indicating thus a catalytic effect of activated C1̄r on C1r activation.

The direct effect of DFP and C1̄ Inh on the activation of C1r
was also checked: the rate of C1r activation was decreased by 30%
and 40% after incubation for 90 minutes at 37°C in 5mM and 10mM DFP
respectively (Figure 3). As the only protein able to bind DFP was
activated C1̄r (Figure 1), the decrease in the rate of activation
was due to the inhibition of C1̄r formed during the process of activation. A similar inhibition was found for C1̄ Inh: C1̄ Inh added to
C1r in a 1:1 molar ration inhibited C1r activation by 48%, upon incubation for 45 minutes at 37°C. In this case also the inhibition was
due to a blocking of nascent activated C1̄r as separate experiments
showed that there was no interaction between C1̄ Inh and proenzymic
C1r.

Activation of C1r was not affected by an increase in ionic
strength up to 0.5M, but progressively inhibited at higher values.
In contrast, calcium was found to inhibit C1r activation very efficiently, probably through a promotion of C1r aggregation. The
influence of pH on C1r activation was checked at pH 5.0, 7.4 and

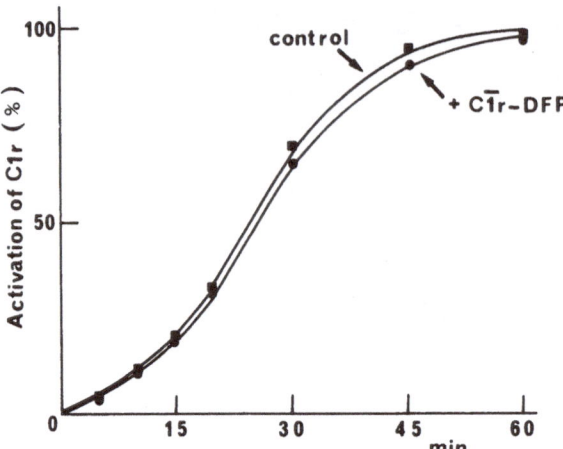

Fig. 2b. Effect of DFP-blocked C̄1r on C1r activation. C̄1r
          (436μg/ml) was incubated in 5mM DFP/145mM Nacl/50mM tri-
          ethanolamine-HCl (pH 7.4) for 30 minutes at 37°C; 5mM DFP
          was then added before a second incubation for 30 minutes
          at 37°C; 5mM DFP was then added before a second incubation
          for 30 minutes at 37°C. After two extensive dialyses
          during 24 hours against 2 liters of 145mM NaCl/5mM tri-
          ethanolamine-HCl (pH 7.4), C̄1r-DFP was incubated with pro-
          enzymic C1r as described in Figure 2a (DFP-C̄1r: C1r
          molar ratio of 0.40).

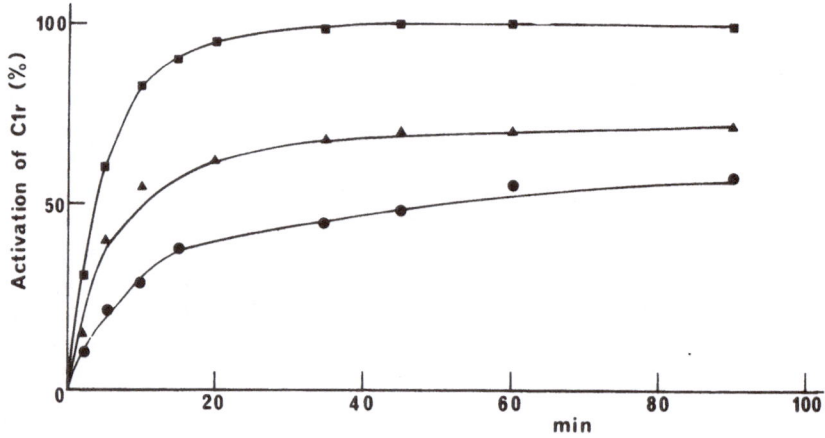

Fig. 3. Activation of proenzymic C1r in the presence of DFP. In-
         cubation was carried out as described in Figure 2a, in the
         absence of added C̄1r, in the presence of 5mM (——▲——) or
         10mM (——●——) DFP; control without DFP (——■——).

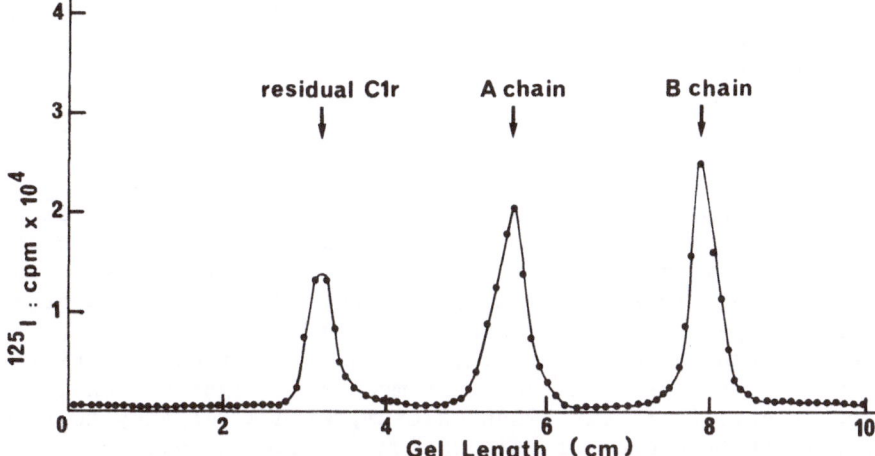

Fig. 4. Scan of purified C1r iodinated in the presence of lacto-
         peroxidase. Lactoperoxidase-catalysed iodination of C1r
         (0.4mg/ml) was at 0°C in 145mM NaCl/5mM triethanolamine-
         HCl (pH 7.4). After extensive dialysis, 15μg protein in
         50μl were activated by incubation for 30 minutes at 37°C,
         reduced, alkylated and analysed by SDS-PAGE. ^{125}I was
         counted directly on gel slices.

10.5: at pH 5.0, activation was abolished, due either to the mon-
merization of C1r or to a direct pH effect on the pro-site of the
zymogen[10]; activation increased slightly from pH 7.4 to pH 10.5.
A similar increase in the rate of activation by exogenous activated
$\overline{\text{C1r}}$ was also observed at this last pH, a finding in favour of the
coexistence of the two activation mechanisms at the different pH.

     Gross structural differences between proenzymic and activated
C1r were investigated by surface iodination of tyrosines in the presence of lactoperoxidase. Iodination of proenzymic C1r at 0°C,
followed by activation of C1r during 45 minutes at 37°C, led to a
distribution of 48% of the label in the a chain a 52% in the b chain
(Figure 4). Parallel labelling of activated $\overline{\text{C1r}}$ led to fixation
of 20% and 80% of ^{125}I in the a chain and the b chain respectively.
These findings illustrate profound alterations of C1r conformation
upon activation of the molecule, at contrast with parallel experiments on C1s failing to detect net alterations in C1s conformation
upon activation.

DISCUSSION

     Spontaneous activation of purified C1r in the fluid phase was
described previously[7-10]. This activation was rapid (full activation

in 10-15 minutes at 37°C[7,9] and was inhibited by calcium. According to Ziccardi and Cooper[7] this inhibition was totally prevented by 10mM DFP; the results presented here show that DFP was able to decrease the rate of activation but not to block it even at 10mM. Similarly C$\overline{1}$ Inh was also able to partially inhibit C1r activation, which confirmed an auto-catalytic activation mediated by the pro-site of C1r[10]. Addition of activated C$\overline{1}$r to proenzyme C1r clearly increased its rate of activation; this observation adds to the above data on DFP and C$\overline{1}$ Inh inhibition, indicating that a second mechanism of activation of C1r was triggered by nascent activated C$\overline{1}$r formed in the first autocatalytic reaction. This second mechanism appeared as an autocatalytic interdimer proteolysis; it was not found by some authors[7], while others[5,6] tend to attribute the initial activating potential to contaminant proteases. In fact this mechanism might be confined to purified soluble C1r: C1r binds calcium, with a tendency to form aggregates; in the presence of this cation its autoactivation was drastically reduced at 37°C; furthermore, intermolecular contacts between C1r inserted in different C1 complexes are very unlikely. This interdimer proteolysis confirms the large exposure of C1r to extrinsic activating proteases such as plasmin[3] or a fragment of Hageman factor[4].

Surface iodination of proenzymic and activated C1r indicated a net redistribution of the peripheral tyrosines upon activation[11,12]; these findings seem to reflect profound structural alterations upon C1r activation, in contrast to C1s which appears less affected by activation[10]. These structural changes in C1r illustrate the key role of C1r inside C1, converting a structural signal into a proteolytic activity. Activation of C1r seems to be under control from its intimate neighbour C1s as well as from the more loosely associated C1q subunits. The ribbon-like structure of C$\overline{1}$r$_2$C$\overline{1}$s$_2$ observed in electron microscopy and ultracentrifugation[13,14] appears compatible with the structure proposed recently for C1[15], in which C1r$_2$-C1s$_2$ might be intricated within the collagen-like arms of C1q, not far from the globular heads. From iodination data on C1r, either in C1r$_2$-C1s$_2$ or in C1[11], from the easy dissociation of C1 by calcium-chelating agents and also from recent data on the affinity between immunoglobulins aggregates-C1q and C$\overline{1}$r$_2$-C$\overline{1}$s$_2$[16] we propose that C1r$_2$-C1s$_2$ is bound loosely at the outer surface of C1q collagenous arms. Thus a small change in the relative position of the globular heads of C1q, upon binding to an activator, is likely to trigger C1r activation, with a slight opening of the ring-like structure assumed by C1r$_2$-C1s$_2$ around C1q[15]. This opening might concern the two distal C1s in the C1s-C1r-C1r-C1s complex, as it was found in separate experiments that tetrameric C1r$_2$-C1s$_2$ could be formed from dimeric C1r and monomeric C1s. Monomerization of C1r observed upon binding of C$\overline{1}$ Inh to the active site of C$\overline{1}$r, is responsible for a disruption of C1 structure, resulting in the unveiling of the collagenous stem of C1q, available then either for further C1r$_2$-C1s$_2$ binding or interaction with C1q receptors and/or C1q inhibitor.

REFERENCES

1. K. B. M. Reid and R. R. Porter, The proteolytic activation systems of complement, Ann. Rev. Biochem. 50:433 (1981).
2. R. M. Bartholomew and A. F. Esser, Mechanism of antibody-independent activation of the first component of complement (C1) on retrovirus membranes, Biochemistry 19:2847 (1980).
3. N. R. Cooper, L. A. Miles and J. H. Griffin, Effects of plasma kallikrein and plasmin on the first complement component, J. Immunol. 124:1517 (Abstr) (1980).
4. B. Ghebrehiwet, M. Silverberg and A. P. Kaplan, Activation of the classical pathway of complement by Hageman factor fragment, J. Exp. Med. 153:665 (1981).
5. A. W. Dodds, R. B. Sim, R. R. Porter and M. A. Kerr, Activation of the first component of human complement (C1) by antibody-antigen aggregates, Biochem. J. 175:383 (1978).
6. J. Bauer and G. Valet, Mechanism of C1r activation, J. Immunol. 124:1514 (Abstr) (1980).
7. R. J. Ziccardi and N. R. Cooper, Activation of C1r by proteolytic cleavage, J. Immunol. 116:504 (1976).
8. K. Takahashi, S. Nagasawa and J. Koyama, Partial purification and autocatalytic activation of the subunit of the first component of human complement, C1r, FEBS Lett. 65:20 (1976).
9. S. N. Assimeh, R. M. Chapuis and H. Isliker, Studies on the precursor form of the first component of complement. II Proteolytic fragmentation of C1r, Immunochemistry 15:13 (1978).
10. G. J. Arlaud, C. L. Villiers, S. Chesne and M. G. Colomb, Purified proenzyme C1r. Some characteristics of its activation and subsequent proteolytic cleavage, Biochim. Biophys. Acta 616:116 (1980).
11. C. L. Villiers, A. M. Duplaa, G. J. Arlaud and M. G. Colomb, Fluid phase activation of proenzymic C1r purified by affinity chromatography, Biochim. Biophys. Acta (in press) (1981).
12. J. Bauer and G. Valet, Conformational changes of the subunits C1q, C1r and C1s of human complement component C1 demonstrated by ^{125}I labeling, Biochim. Biophys. Acta 670:129 (1981).
13. J. Tschopp, W. Villiger, H. Fuchs, E. Kilchherr and J. Engel, Assembly of subcomponents C1r and C1s of first component of complement : electron microscopic and ultracentrifugal studies, Proc. Natl. Acad. Sci. USA 77:7014 (1980).
14. V. N. Schumaker, P. H. Poon, G. W. Seegan and C. A. Smith, Semiflexible joint in the C1q subunit of the first component of human complement, J. Mol. Biol. 148:191 (1981).
15. V. N. Schumaker, C1 : Its structure and mechanism of activation, This volume, p99 (1982).
16. E. Kilchherr, H. Fuchs, J. Tschopp and J. Engel, Dissociation of C1 and concentration dependence of its activation kinetics, This volume, p141 (1982).

ACTIVATION OF MAMMALIAN COMPLEMENT BY CHICKEN C1q

Else Marie Nicolaisen

Institute for Experimental Immunology
University of Copenhagen
Norre Alle 71, Copenhagen Ø, DK 2100 Denmark

INTRODUCTION

Activation of chicken complement by the classical pathway is still debatable. Both positive and negative evidence of antibody-dependent lysis has been reported. The components from C3 to C9 function since activation via the alternative pathway will lyse mammalian erythrocytes in the absence of Ig. Actually the discrepancy in the literature concerning classical activation might be explained by the fact that heterologous erythrocytes were used in most studies. On the other hand, absence of lysis of chicken erythrocytes coated with specific chicken antibodies might be due to the strong anti-complementary effect ascribed to chicken serum.

Mammalian complement is not activated by chicken antibodies, due to incompatibility between the chicken antibody and mammalian C1. It is however possible to obtain lysis of chicken erythrocytes, sensitized with alloantibodies, with a mixture of normal chicken serum and guinea pig serum, and it has been suggested, that chicken C1 activates guinea pig C4 and C2[1]. This mixed complement (MC) reaction was used for studies of the initial step of classical activation in chicken. In the experiments reported here human serum was used instead of guinea pig serum.

One purpose for the investigation was to test whether the Fc-binding chicken serum protein, HEF, (Haemagglutination Enhancing Factor), demonstrated earlier[2] is at all involved in the first step of classical activation in chicken. In vitro the HEF protein precipitates in a quantitative manner with chicken IgG in immune complexes, and it enhances the titre of haemagglutinating allo-antibodies[3]. HEF is neither an immunoglobulin or a lipoprotein, it is

149

a carbohydrate containing betaglobulin with a high molecular weight
(to be published). HEF is probably the same protein as described
by Ivanyi & Tempelis[4]. This chicken serum protein is a euglobulin,
which binds to immune complexes, migrates as a betaglobulin and has
a sedimentation coefficient of 15.6S.

RESULTS AND DISCUSSION

Normal chicken serum was fractionated by standard protein separation methods, such as preceipitation, gel filtration and ion-
exchange, and fractions were tested in parallel for ability to enhance haemagglutinating titre, HEF-activity, and ability to activate
human complement, MC-activity. The functional assays are described
in detail elsewhere[5].

The initial experiments indicated that two different molecules
are responsible for these two effects, and that the MC-active molecule has properties in common with mammalian Clq, but it turned
out to be very difficult reproducibly to obtain one effect without
contamination with the protein causing the other. Also, since both
assays for detection are based on Fc-binding, competition for Fc
might occur. Consequently, competition for Fc sites might eventually
determine whether lysis or agglutination is observed.

Eventually, the best separation was obtained by isotachophor-
esis, ITP. The ITP was performed in a flat-bed with granulated gel
as supporting material, and the obtained fractions were analysed
for activity. Methodological details will be published elsewhere.
ITP applied to whole chicken plasma shows that HEF migrates with
the main part of proteins between the leading ion and the terminating
ion, while the MC-active component is found in the terminating zone.
This result shows that HEF is at least not the molecule that activates human complement, but does it interfere with the activation?
This point will be discussed below.

Meanwhile, for comparison human plasma was separated by ITP.
It was found with monospecific antisera, that human Cls migrates
between the leading and the terminating ions, while human Clq is
found in the terminating zone, in agreement with the conclusion
above, that the active chicken protein in the MC reaction is the
chicken Clq.

The ITP fractions from the terminating zone were concentrated
and analysed on SDS-PAGE. The human Clq preparation shows bands of
IgG and albumin contamination, and for the Clq A-B and C-C chains
are found at 59,000 and 52,000 MW positions, The corresponding
chicken MC-active preparation shows also contamination with IgG and
albumin and two bands at 58,000 and 53,000.

After reduction and alkylation, human C1q shows A, B and C chains at 32,000. 28,500 and 25,000. There may however be a third band also from chicken C1q. When more concentrated MC-active preparations, obtained by other methods than ITP, are analysed, a third band at 32,000 is seen. Unfortunately these preparations contain impurities, and the SDS-PAGE has still to be re-run with a more concentrated ITP fraction. The fact that the 32,000 band is not seen in the first run, could be a result of weak staining, due to high carbohydrate content. It is also possible that two chains migrate to the same position under the conditions used here.

No bands of any C1r of C1s like molecules have ever been observed in the MC-active preparations. Both from this negative finding and from the resemblence with human C1q, it is concluded that the only chicken complement protein needed to activate human complement is the chicken C1q. This means that the part of the C1q molecule, where $C1r_2, C1s_2$ are bound, has been more strongly conserved during evolution than the part responsible for Fc-binding. This is in accordance with results presented on frog complement by Alexander and Steiner[6]. They found that frog C1q could activate human C1r, C1s and furthermore that the whole of frog C1 activates human C4 and C2. It is possible that also the chicken C1 complex can activate human C4 and C2; this possibility is not excluded from the results presented here.

It is very puzzling that the chicken serum contains a C1q molecule that can be activated by binding to chicken antibodies in the sense that it is then able to activate human C1r, C1s, and still chicken serum alone will not result in lysis of sensitized chicken erythrocytes. One possibility would be that the chicken C1q is in some way inhibited. An obvious candidate for such inhibitory effect is the HEF molecule. When increasing amounts of HEF is added to a fixed concentration of an MC-active fraction lysis will be reduced, and at a certain point only agglutination is obtained.

The mechanism of this in vitro inhibition could be that HEF competes with C1q for binding-sites on Fc because both molecules bind to the same or neighboring sites. But it is tempting to speculate, since it is so difficult to separate the two molecules, that HEF may compete with $C1r_2, C1s_2$ for binding to C1q, and if that is the case HEF exerts its agglutination enhancing effect via bound C1q. If this is correct neither HEF or the C1q alone should be able to enhance agglutination. In human serum a C1q inhibitor has been demonstrated[7], and recent evidence suggests that this molecule might be the same as the C1q receptor found on human peripheral blood leucocytes, and that C1q is bound through the collagen-like region[8].

In work on this C1q inhibitor no agglutination enhancing effect has been described. However, a widely used method for C1q detection is agglutination of Ig-coated latex particles[9]. When this method

was first published the authors discussed the possible existence of
a Clq inhibitor influencing the reactivity of Clq. Later, in the
paper demonstrating a Clq inhibitor[7], it was shown that the Clq inhibitor is found together with Clq in euglobulin precipitates, a
method usually employed for separation of Clq.

So it can not be excluded that the agglutination seen with human
Clq is actually an effect which depends on the simultaneous presence
of the Clq inhibitor.

When the available physico-chemical data on the human Clq inhibitor and chicken HEF are compared, there are several similarities;
they are both euglobulins which are heat stable and have high molecular weight. The only difference so far known is the mobility,
where HEF migrates in the beta-region while the human Clq inhibitor
is found to be extremely anionic. In functional terms the human
Clq inhibitor only inhibits free Clq, and not Clq bound to Clr_2,
Cls_2.

Recently it has been reported that human serum fibronectin has
the same property as human Clq inhibitor, ie. it binds to the collagen-like region of free Clq and thereby impairs the activity of
free Clq, while the native Cl is unaffected[10].

The function of soluble fibronectin in plasma is not clarified.
It is supposed to act as an opsonizing protein, which facilitates
phagocytosis of bacteria, cellular debris, coagulation products,
microaggregates and immune complexes. Fibronectin (or "cold-
insoluble globulin") has a molecular weight of 440,000 and contains
two similar covalently linked subunits with carbohydrate attached.

Purified plasma fibronectin has a tendency to form filamentous
protein polymers, especially when stored at 4°C. The polymerization
seems to be through a self-assemble process, it is non-covalent and
can be reversed. This was shown by gel filtration on Sepharose 4B,
the K_{av} was found to be 0.47[11].

HEF, too, forms reversible aggregates and has a K_{av} of 0.47 on
Sepharose 4B; it is not dissociated in 8M urea. Analysis on SDS-PAGE
of highly purified preparations of HEF shows for the unreduced molecule five bands between 220,000 and 75,000 and one band at 20,500.
After reduction and alkylation five bands from 200,000 to 51,000 and
one band at 19,500 are found. HEF was purified from serum and the
SDS-PAGE pattern shows similarities with human fibronectin after
plasmic digest[12]. It has to be tested, whether HEF purified from
plasma will give the bands reported for chicken fibronectin: one
band between 250,000 and 200,000 for the unreduced molecule, and
two bands near 220,000 for the reduced and alkylated molecule[13].
So far it is still possible that HEF is chicken fibronectin, if
that is the case, it will be interesting to see whether HEF and Clq
form complexes in vivo or the enhancing effect seen in vitro is an
artefact.

Conclusion

Chicken C1q is found to be the only chicken serum protein needed to activate human C1r,C1s. The chicken C1q freshly isolated by ITP is very similar to the mammalian homologue in basic subunit structure. It was not tested whether whole chicken C1 activates human C4 and C2. The mixed complement reaction is inhibited by a normally occurring chicken serum protein, HEF. The possible homology of this protein with human serum C1q inhibitor and human serum fibronectin is discussed. Whether HEF impairs the activation of chicken C1q and thereby inhibits classical activation of chicken complement is in need of further investigation.

REFERENCES

1. H. N. Benson, H. P. Brumfield and B. S. Pomeroy, Requirement of avian C'1 for fixation of guinea pig complement by avian antibody-antigen complexes, J. Immunol. 87:2 (1961).
2. E. M. Nicolaisen, C. Koch, K. Hála and M. Simonsen, Antigen-antibody complex binding serum protein in the chicken, Foila Biol. (Praha) 23:426 (1977).
3. E. M. Nicolaisen and C. Koch, Haemagglutination-enhancing factor protein in chicken serum, Scand. J. Immunol. 13:11 (1981).
4. J. Ivanyi and C. Tempelis, Genetic polymorphism of a serum euglobulin of chickens which binds to antigen-antibody complexes, Immunogenetics 10:83 (1980).
5. E. M. Nicolaisen and M. Simonsen, Elution of chicken Ig from fixed target cells, J. Immunol. Meth. 29:139 (1979).
6. R. J. Alexander and L. A. Steiner, The first component of complement from the bullfrog Rana catesbiana: functional properties of C1 and isolation of subcomponent C1q, J. Immunol. 124:1418 (1980).
7. J. D. Conradie, J. E. Volanakis and R. M. Stroud, Evidence for a serum inhibitor of C1q, Immunochemistry 12:967 (1975).
8. B. Ghebrehiwet, C1q inhibitor (C1qINH): functional properties and possible relationship to a lymphocyte membrane-associated C1q precipitein, J. Immunol. 126:1837 (1981).
9. R. W. Ewald and A. F. Schubart, Agglutinating activity of the complement component C1q in the FII latex fixation test, J. Immunol. 97:100 (1966).
10. H. Isliker, D. H. Bing and R. O. Hynes, Interactions of fibronectin with C1q, a subcomponent of the first component of complement, in:"The Immune System," Karger, Basel, 2:231 (1981).
11. M. Vuento, T. Vartio, M. Saraste, C. von Bonsdorf and A. Vaheri, Spontaneous and polyamine-induced formation of filamentous polymers from soluble fibronectin, Eur. J. Biochem. 105:33 (1980).

12. F. Jilek and H. Hörmann, Cold-insoluble globulin, II. Plas-
    minolysis of cold-insoluble globulin, Hoppe-Seyler's Z.
    Physiol. Chem. 358:133 (1977).
13. K. M. Yamada and D. W. Kennedy, Fibroblast cellular and plasma
    fibronectins are similar but not identical, J. Cell. Biol.
    80:492 (1979).

V: THE IMPORTANCE OF ANTIGEN CONFORMATION FOR ANTIGENICITY AND IMMUNOGENICITY

INTRODUCTION

The existence of two major classes of antigenic determinants, sequential and conformation-dependent, has been established in the classic paper by Sela and collaborators some 22 years after Karl Landsteiner had described the poor crossreactivity of native and denatured proteins. More recently, further distinction was made by Atassi , who observed that in certain cases sequential determinants ("continuous") would also be dependent on the conformation of the protein, while the discontinuous ones composed by amino acids belonging to non-contiguous sections of a polypeptide chain or to different chains in oligomers are always conformation-dependent.

Obviously the target of these determinants is the antibody receptor of the B cell whose specificity can be indirectly studied by looking at the antibody paratope. This structure can accommodate a cluster of 3-5 amino acids, and it does not see whether they are or are not in sequence. An elegant demonstration of this blindness was reported recently by Atassi who synthesized a continuous peptide mimicking a discontinuous determinant of lysozyme and found that it displayed crossreactive antigenicity.

Conformation-dependent determinants constitute the basis for immunologic recognition of native molecules, for the probing of conformation by antibodies and for the antibody-mediated stabilization of activation of defective molecules.

In this topic V, Ruth Arnon (paper 15) discusses the identification of conformational determinants on a number of antigens and

155

reports on interesting and successful attempts to reproduce syntheti-
cally, key protein determinants both as antigens and as immunogens.
Berzofsky et al. (paper 16), determines the specificity of 6 monoclonal antibodies raised against spermwhale myoglobin (some recognizing conformational sites) and finds that a well established immunogenic dominance of certain zones of the molecule does not apply to
hybridomas.

Van Regenmortel, working with 2130-mer tobacco mosaic virus
capsid and with monoclonal antibodies, distinguishes (a) cryptotypes
(determinants facing inward in the assembled capsid) (b) exposed
determinants on the surface recognized both on the virus and its
subunits, and (c) neotopes (determinants dependent upon the quaternary structure of the polymer). Ivanyi describes the topography of
antigenic determinants of human growth hormone and of chorionic
somatomammotropin on the basis of crossinhibition reactions using a
series of hybridomas. Zabin et al. describe the use of antibodies to
probe for the three-dimensional structure of a protein where crystallographic analysis is still wanting. By the use of anti-peptide
(anti-sequential), anti-wild type tetramer and anti-dimer antibodies,
it has been established that some peptides contain cryptotopes and
that a number of dominant determinants of the wild type enzyme are
dependent on the tetrameric form. Goldberg also uses antibodies
directed against the native antigen (tryptophan-synthetase) and
against its fragments to understand the changes of conformation that
the B subunits undergo when they assemble to form the complete
protein. It is predicted that a further study of this model will
reveal antibody-induced changes of conformation, which provides a
connection with the next topic.

REFERENCES

1. M. Sela, B. Schechter, I. Schechter, and F. Borek. Cold Spring
        Harbor Symposia on Quantitative Biology, 32, p.537. (1967).
2. K. Landsteiner, The Specificity of Serological reactions,
        p.42. Howard Univ. Press, Cambridge, Mass. (1945).
3. M.Z. Atassi, The complete antigenic structure of myoglobin:
        approaches and conclusions for antigenic structures of
        proteins, In: "Immunochemistry of Proteins II" (ed. M.Z.
        Atassi), Plenum Press, New York, p.77 (1977).
4. M.Z. Atassi, Precise determination of protein antigenic
        structures has unravelled the molecular immune recognition of
        proteins and provided a prototype for synthetic mimicking of
        other protein binding sites, Mol.Cell.Biochem. 32:21 (1980).

CONFORMATIONAL ANTIGENIC DETERMINANTS IN PROTEINS

Ruth Arnon

Department of Chemical Immunology
The Weizmann Institute of Science
Rehovot, Israel

INTRODUCTION

The decisive role of conformation in determining the antigenic specificity of protein antigens is widely recognized. A large body of experimental evidence indicates that antigenic properties change drastically upon denaturation of native proteins (by heat or chemical modification) or upon unfolding of their polypeptide chains[1,2]. The denatured or unfolded proteins are usually still immunogenic, but their antigenic specificity is totally different from that of the corresponding native proteins. In many instances, more subtle conformational alterations in a protein are also accompanied by a change in antigenic reactivity. One example of this phenomenon is the change in the antigenicity associated with the removal of the heme group from sperm-whale myoglobin: In the conversion of metmyoglobin to apomyoglobin the loss of heme is associated with only a small structural change and the antigenic properties of the two molecular species are not much different. The precipitate formed between metmyoglobin and antiapomyoglobin, however, is colorless and does not contain the ferriheme group[3]; thus the antibodies specific to the heme-free molecule must have induced a conformational change in the metmyoglobin and released the heme group from it during the antigenantibody interaction.

Not all conformational alterations have a measurable effect upon antigenicity and vice versa. There are, for example, several aberrant hemoglobins for which X-ray crystallography shows distortion in tertiary structure, and yet they cannot be distinguished antigenically from hemoglobin A[4]. On the other hand, α and β chains of human hemoglobin do not show antigenic resemblance in spite of the remarkable similarity in their conformations[5]. These are probably exceptions,

157

however, to the general rule that there is an intimate relationship
between conformation and antigenicity.

For a better assessment of the role of conformation in determining the antigenic properties of a protein molecule, the structural
features that affect antigenic specificity should be analyzed.
Antibodies elicited in response to immunization with protein antigens
are reactive with various antigenic determinants and may be directed
against one or more of the structural aspects of the protein, including the primary, secondary and tertiary structures, as well as the
quaternary structure which results from specific association of several polypeptide chains to form a multi-subunit protein. The antigenic determinants, therefore, can be divided theoretically into two
broad categories according to whether their specificity is due only
to stretches of amino acid sequences in the protein (sequential) or
to other structural features (conformational). Conformational determinants may contain amino acid residues that, although remote in the
unfolded polypeptide chain, are probably juxtaposed in the native
structure. There are a few cases in which sequential determinants
have been clearly demonstrated, such as the terminal segments of
collagen [6] or silk fibroin[8]. In most cases, however, antibodies to
protein are directed mainly towards conformation-dependent determinants[8].

DEFINED ANTIGENIC DETERMINANTS IN PROTEINS

Attempt to identify antigenic determinants of proteins involve
either the splitting of the molecule by limited proteolysis and/or
chemical cleavage and screening of the resultant fragments for antigenic activity, or alternatively, the synthesis of peptides analogous
to defined regions in the molecule. In both cases the fragments
delineate the antigenic determinants of the protein in terms of their
localization and conformation in the native structure. A few
examples:

1) In tobacco mosaic virus protein (TMVP), a protein of 158 amino
acid residues, an icosapeptide comprising of residues 93-112 of the
protein has been demonstrated to possess the entire antigenic
reactivity of a peptide digest[9]. Moreover, most of this activity was
shown to reside in a tripeptide Ala-Thr-Arg[10], pinpointing to a very
small and defined antigenic determinant.

2) In staphylococcal nuclease three determinants have been located in
regions of the molecule that occupy the corner of the folded peptide
chain[11] and consequently possess conformational specificity.
3) Sperm whale myoglobin is a protein for which a full antigenic
mapping has been accomplished[12]. The five antigenic determinants
which have been identified are localized at the corners of the molecule that separate the helical regions, and although composed of

short sequences (5-7 residues each), they are all exposed on the surface of the molecule.

4) In hen egg white lysozyme two independent antigenic regions have been identified, both are conformation-dependent. One consists of the two terminal segments (residues 1-27 and 122-129) linked by a disulfide bond[13] and the second, comprising residues 60-88, contain an intrachain disulfide bond and is denoted "loop"[14]. The latter one has been investigated in detail in our laboratory and was clearly demonstrated to be a conformational determinant.

SYNTHETIC ANTIGENS WITH PROTEIN SPECIFICITY

When the chemical structure of a protein antigenic determinant is known, it is possible to synthesize it and if necessary to attach it to a carrier (synthetic or protein) for the preparation of appropriate immunogenic conjugates. This approach has been employed in the case of several protein antigens and has led to the production of synthetic antigens capable of eliciting antiprotein specific immune response. A few examples illustrate this point:

The first example is vertebrate collagen, a molecule composed of three polypeptide chains which are supercoiled in the form of a characteristic triple helix[15]. Collagen as such is a weak immunogen, leading mainly to antibodies directed against the N- and C-terminal nonhelical regions, which show considerable inter-species differences[6]. The helical region has been conserved during evolution to a greater extent than other parts, yet some interspecies differences have been observed in this region as well, and they are expressed in some conformation-dependent antigenic determinants. This assumption was confirmed by the synthetic approach: a synthetic periodic polypeptide (ProGlyPro), which was shown to have a collagen-like triple-helical structure[16], was found to be immunogenic in guinea pigs[17]. Immunization with this co-polymer elicited antibodies that, in addition to their reaction with the homologous polymer, reacted also with native collagens of several species[18]. This cross-reaction is by virtue of the triple-helical conformation, which is common to the the synthetic material and the various collagens.

The second example is hen egg white lysozyme. As mentioned, one of its defined antigenic determinants is the "loop" (residues 60-83), which when attached to a macromolecular carrier elicits antibodies with restricted heterogeneity that react with native lysozyme and show conformation-dependent specificity[14]. In later studies it has been reported[19] that the "loop" peptide can be synthesized chemically. Attachment of the resultant peptide to a synthetic carrier yielded a completely synthetic antigen which elicited antibodies that were reactive with native lysozyme and could still recognize the conformational determinant of the intact molecule. By using various

synthetic analogs of the loop peptide, in which one amino acid at a
time was replaced by alanine, it was possible to determine the contribution of each of these residues to its antigenic properties[20].

The last example is a protein of great interest, namely the
carcinoembryonic antigen (CEA) of the colon. A glycoprotein of about
200,000 mol wt, it is characteristic of many types of cancer tissues[21]. Antibodies against it are used in a radioimmunoassay for
detection of cancer. The synthetic approach has been employed in
this case as well: A peptide corresponding to the 11 amino acid
residues of the NH -terminal portion of the sequence of CEA has been
synthesized and chemically attached to a synthetic carrier and to
bovine serum albumin. Both macromolecular conjugates provoked anti-
peptide antibodies in rabbits[22]. Investigation of the specificity of
this immunological system by two different techniques has revealed
that the peptide reacted not only with its homologous antibodies but
also with antiCEA sera. Likewise, the antiserum against the peptide
reacted with intact CEA.

ANTIVIRAL RESPONSE INDUCED BY SYNTHETIC ANTIGENS

As seen from the above mentioned examples, synthetic antigens
that contain immunoreactive region(s) of protein molecules can give
rise to an immune response towards the intact protein. There is no
reason why this should be limited to purified proteins or enzymes.
It should be feasible to use a similar approach for components of
viruses and bacteria, with the purpose of inducing antibodies reactive with the intact agent and hopefully leading to its neutralization.

As a model system for testing this possibility, we chose the
coliphage MS-2, a RNA-containing virus with icosahedral symmetry.
The viral capsid consists of 180 identical coat protein subunits with
a molecular weight of 13,700 daltons each, and a single molecule of
another protein, denoted protein A. The coat protein consists of a
single polypeptide chain with 129 amino acid residues, which have
been sequenced[23]. Antisera against it were found to be as efficient
as anti-phage sera in neutralizing the MS-2 phage. To locate the
region(s) that participate in the neutralization process, we cleaved
the coat protein with cyanogen bromide, to yield three fragments,
possessing the sequences 1–88 (P_1), 89–108 (P_2), and 109–129 (P_3),
respectively[24]. The peptide P_1 was separated from the mixture of P_2
and P_3 by gel filtration on Sephadex G-25.

The mixture of peptides P_2 and P_3 was found capable of inhibiting the neutralization of the phage by antiserum to the whole
MS-2. Since the mixture of these two peptides could not be separated
because of their similarity in size and electrical charge, the peptides corresponding to P_2 and P_3 were synthesized. The synthetic P_3,

corresponding to the carboxyterminal 21 amino acid residues in the sequence of the coat protein, had no activity related to the neutralization of MS-2. On the other hand, the synthetic P2 peptide was very efficient in inhibiting the inactivation of the phage by the antiserum against the phage. Furthermore, a synthetic antigen prepared by attachment of P2 covalently to a synthetic carrier induced antiserum in rabbits, that was capable of neutralizing MS-2 activity almost as efficiently as the antiserum prepared in rabbits against the intact coat protein[24].

This study provided the first reported evidence that a synthetic peptide corresponding to a region which is involved in viral neutralization can be utilized for eliciting anti-viral activity. It is logical to assume that the P2 peptide assumes in P2-A--L a steric conformation close to the one which this peptide region has in the intact coat protein, and that this conformation is conserved also in the intact phage.

More recently we have been investigating this chemical approach in a system of an animal virus, and chose the influenza virus as a model[25]. We have synthesized a peptide that corresponds to a region in the influenza hemagglutinin which, according to the predicted structure, should have comprised a β-bend folded region. This peptide consisted of the residues 91-108 of the hemagglutinin subunit HA1, which is a segment common in its sequence to at least nine strains of influenza subtype (A). The peptide was attached to purified tetanus toxoid and the conjugate served for the immunization of rabbits and mice. The elicited antibodies reacted in the radioimmunoassay not only with the peptide or its conjugates, but also with intact influenza virus subtype A, either denatured or live. Moreover, the antibodies from both rabbits and several mice strains were also capable of inhibiting the capacity of the hemagglutinin of the relevant influenza strains to agglutinate chicken red blood cells.

A more interesting finding was the capacity of the antiserum to neutralize the viability of the virus, as measured by the inhibition of the virus plaque formation in tissue culture in vitro. Under the experimental condition used, the extent of inhibition of the virus was up to 65%, with effectivity of the antiserum up to a dilution of 1:32.

The mouse strain C3H/DiSn was selected for checking the potential of the synthetic conjugate to provide protection against viral challenge. The preliminary results of such experiments, indicated that immunization of the mice with the peptide conjugate, resulted in their effective protection against infection with the A/Texas mouse-adapted influenza virus. The protective effect was demonstrated by the difference in the incidence of the infection in the mice, as well as by the lower titer of virus particles in the lungs of the immunized mice as compared to control groups. In view of the fact that the synthetic peptide used is of a sequence common to several influenza strains, the conjugate should in principle elicit antibodies equally effective with the different strains, and possibly leading to cross-strain protection.

The intention of such studies is not to prepare the most effective commercial vaccine against a particular virus such as influenza, but rather to prove the feasibility of the approach. The results presented here demonstrated that it is indeed feasible.

CONCLUDING REMARKS

In this manuscript evidence has been presented for the possibility of identifying and localizing antigenic determinants of proteins that demonstrate conformation-dependent specificity. When the structure of such determinants is established, it is possible to synthesize them and prepare synthetic conjugates which elicit immune response towards the intact native proteins. Moreover, if the protein is a component of a virus the immune response may lead to antiviral activity. This opens the road for the future development of synthetic vaccines.

REFERENCES

1. R. Arnon, Confirmation-dependent antigenic determinants in proteins and synthetic polypeptides. In: Peptides Polypetpides and Proteins. N. Lotan, M. Goodman and E. R. Blout eds., John Wiley and Sons, p.538 (1974).
2. R. Arnon and B. Geiger, Molecular basis of immunogenicity and antigenicity. In: Immunochemistry, Glynn ed., John Wiley and Sons, p.307 (1978).
3. M. J. Crumpton and J. M. Wilkinson, Conformational changes in sperm-whale myoglobin due to combination with antibodies to apomyoglobin. Biochem.J. 100,223 (1966).
4. A. Reichlin, Localizing antigenic determinants in human hemoglobin with mutants: Molecular correlation of immunological tolerance. J.Mol.Biol. 64,485 (1972).
5. M. Reichlin, E. Bucci, C. Fronticelli, J. Wyman, E. Antonini, C. Luppolo and A. Rossi-Fanelli, The properties and interactions of the isolated - and -chains of human hemoglobin. IV. Immunological studies involving antibodies against the isolated chains. J.Mol.Biol. 17, 18 (1966).
6. U. Becker, R. Timpl and K. Kuhn, Carboxyterminal antigenic determinants of collagen from calf skins. Localization within discrete regions of the nonhelical sequence. Eur.J.Biochem. 28,221 (1972).
7. J. J. Cebra, Studies on the combining sites of the protein antigen silk fibroin. I. Characterization of the fibroin-rabbit antifibroin system. J.Immunol. 86,190 (1961).

8. M. Sela, Antigen design and immune response. In: The Harvey Lectures, series 67, Academic Press, N.Y. p. 214 (1973).
9. E. Benjamini, J. D. Young, W. T. Petersen, C. Y. Leung, and M. Shimizu, Immunochemical studies on the tobacco mosaic virus protein, II. The specific binding of a tryptic peptide of the protein with antibodies to the whole protein. Biochemistry 4, 2081 (1969).
10. E. Benjamini, M. Shimizu, J. D. Young, and C. Y. Leung, Immunochemical studies on tobacco mosaic virus protein. IX. Investigations of binding and antigenic specificity of antibodies to an antigenic area of tobacco mosaic virus protein. Biochemistry 8,2242 (1969).
11. D. Sachs, A. N. Schechter, A. Eastlake, and C. B. Anfinsen, Antibodies to a distinct antigenic determinant of staphylococcal nuclease. J.Immunol. 109,1300 (1972).
12. M. Z. Atassi, Antigenic structure of myoglobin: The complete immunological anatomy of a protein and conclusions relating to antigenic structures of proteins. Immunochemistry, 12,423 (1975).
13. H. Fujio, M. Imanishi, K. Nishioka, and T. Amano, Antigenic structure of hen egg white lysozyme. II. Significance of the N- and C-terminal region as an antigenic site. Biken J. 11,207 (1968).
14. R. Arnon, and M. Sela, Antibodies to a unique region in lysozyme provoked by a synthetic antigen conjugate. Proc.Nat.Acad.Sci. 62,163 (1969).
15. K. A. Piez, H. A. Bladen, J. M. Land, E. J. Miller, P. Bornstein, W. T. Buther, and A. H. Kang, Comparative studies on the chemistry of collagen utilizing cyanogen bromide cleavage. Brookhaven Symp.Biol. Vol. V, No. 21,345 (1968).
16. J. Engel, J. Kurtz, E. Katchalski, and A. Berger, Polymers of tripeptides as collagen models. II. Conformational changes of poly(L-prolyl-glycyl-L-prolyl) in solution. J.Mol.Biol. 17,255 (1966).
17. F. Borek, J. Kurtz, and M. Sela, Immunological properties of a collagen-like synthetic polypeptide. Biochim.Biophys.Acta, 188,314 (1969).
18. A. Moaz, S. Fuchs, and M. Sela, On immunological crossreactions between the synthetic ordered polypeptide (L-Pro-Gly-L-Pro) and several collagens. Biochemistry, 12,4246 (1973).
19. R. Arnon, E. Maron, M. Sela and C. B. Anfinsen, Antibodies reactive with a native lysozyme elicited by a completely synthetic antigen. Proc.Nat.Acad.Sci., 68,1450 (1971).
20. E. Teicher, E. Maron and R. Arnon, The role of specific amino acid residues in the antigenic reactivity of the loop peptide of lysozyme. Immunochemistry, 10,265.
21. J. Krupey, P. Gold and S. O. Freedman, Physicochemical studies of the carinoembryonic antigens of the human digestive system. J.Exp.Med. 128,387 (1968).

22. R. Arnon, M. Bustin, E. Calef, S. Chaitchik, J. Haimovich,
    N. Novik, and M. Sela, Immunological cross-reactivity of
    antibodies to a synthetic undecapeptide of carcinoembryonic
    antigen with the intact protein and human sera. Proc.Nat.
    Acad.Sci. 73,2123 (1976).
23. U. Y. Lin, Ch. M. Tsung, and H. Fraenkel-Conrat, The coat protein
    of the RNA Bacteriophage MS-2. J.Mol.Biol. 24,1 (1967).
24. H. Langbeheim, R. Arnon and M. Sela, Antiviral effect on MS-2
    coliphage obtained with a synthetic antigen. Proc.Natl.Acad.
    Sci. 73, 4636, (1976).
25. G. Müller, M. Shapira, and R. Arnon, Anti-influenza response
    achieved by immunization with a synthetic conjugate. Proc.
    Nat.Acad.Sci. in press (1981).

TOPOGRAPHIC ANTIGENIC DETERMINANTS DETECTED BY

MONOCLONAL ANTIBODIES TO MYOGLOBIN

Jay A. Berzofsky, Gail K. Buckenmeyer, Gloria Hicks,
Douglas J. Killion, Ira Berkower, Yoichi Kohno,
Margaret A. Flanagan, Mark R. Busch, Richard J. Feldmann,
John Minna and Frank R. N. Gurd

NCI and DCRT, NIH, Bethesda
Maryland and Indiana University
Bloomington, Indiana

We have been interested in studying the structure of immunological receptors on T and B lymphocytes for natural protein antigens, and the nature of the antigenic determinants recognized by
these. As a model protein antigens, we have used mammalian myoglobins, as we have also been studying the H-2-linked Ir gene
control of the T and B lymphocyte responses to these antigens in
mice (Berzofsky, 1978a, b; Berzofsky et al., 1979; Richman et al.,
1980; Berzofsky and Richman, 1981; Kohno and Berzofsky, 1982;
Berzofsky, submitted for publication). As a model of B lymphocyte
receptors for sperm whale myoglobin, we prepared monoclonal hybridoma
antibodies in A.SW mice, which have an antigen-specific H-2linked Ir
gene for high responsiveness to sperm whale myoglobin, (Berzofsky,
1978), as well an a non-H-2linked gene which results in higher serum
antibody responses to many antigens (Dorf et al., 1974; Berzofsky et
al., 1977; Pisetsky et al.,1978. We hoped that these antibodies
would allow us to extend the knowledge of antigenic determinants on
globular proteins which had been obtained with heterogeneous immune
serum antibodies (Crumpton, 1974; Reichlin, 1975; Atassi, 1975;
Hurrell et al., 1977a; East et al., 1980), and also the knowledge of
antibody combining sites, which, because of the requirement for
homogeneous antibody preparations, was limited to myeloma proteins
which bind small molecules (haptens) and carbohydrates (Potter, 1977;
Kabat, 1978).

The monoclonal antibodies were prepared by fusing spleen cells
from hyperimmune A.SW mice with NS.1, a non-secreting, hypoxanthine-
guanosine phosphoribosyl transferase deficient variant of MOPC 21

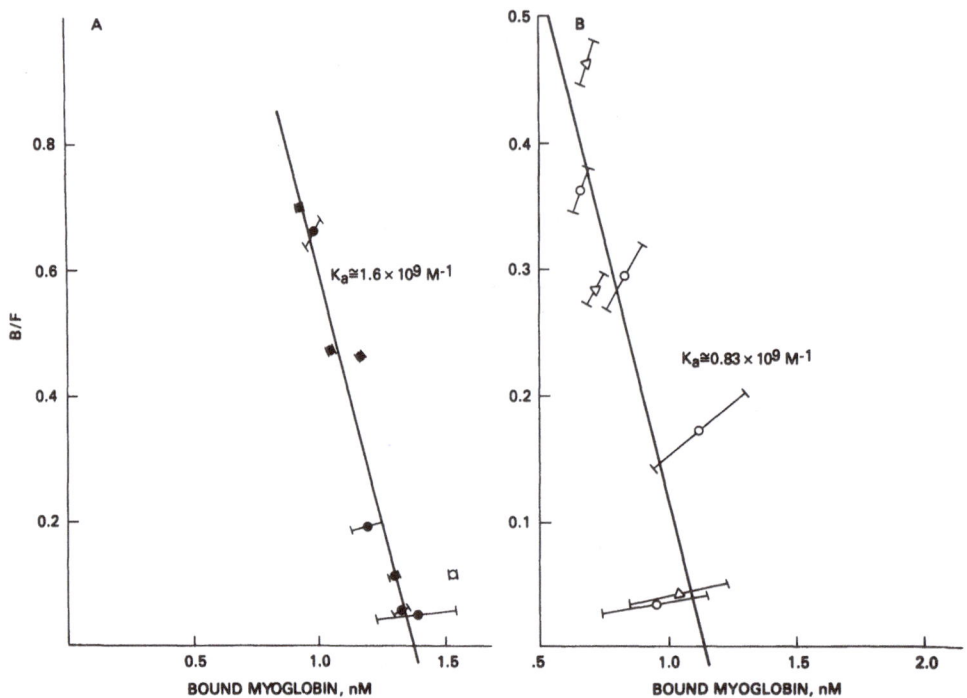

Fig. 1.   Scatchard analysis of hybridoma monoclonal antibodies
             specific for sperm whale myoglobin. A, clone 5, 1:50 di-
             lution of culture supernatant; B, clone 6, 1:50 dilution
             of culture supernatant. Increasing concentrations of
             [3H]sperm whale myoglobin were added to a constant con-
             centration of monoclonal antibody. After equilibrium was
             achieved, at pH 7.6 and 4°C, polyethylene glycol (MW
             6000, final concentration 10% W:W) was added to precipi-
             tate all the immunoglobulin and bound myoglobin, leaving
             free myoglobin in the supernatant. The bound/free
             [3H]myoglobin ratio is plotted vs the concentration of
             bound myoglobin. Modified from Berzofsky et al., 1980,
             with permission of the publisher.

plasmacytoma cell line (Berzofsky et al., 1980). The fusion technique
was essentially that of Köhler and Milstein (1975) except that polyethylene glycol was used instead of Sendai to induce fusion. Hybrid
colonies were screened both by a solid phase radioimmunoassay and by
a solution phase radiobinding assay modified from that of Desbuquois
and Aurbach (1971). Colonies secreting antimyoglobin antibodies were
cloned by limiting dilution and soft agar methods.

     The binding of these antibodies to myoglobin was studied in
aqueous solution in equilibrium at pH 7.6-8.0 at 4°C, using [3H]sperm
whale myoglobin, labeled preferentially at the N-terminal alpha amino
group, and using polyethylene glycol (MW 6000, 10% W:W final concen-

tration) (Desbuquois and Aurbach, 1971) to precipitate all of the immunogloubulin together with any bound myoglobin, leaving free myoglobin in the supernatant solution (Berzofsky et al., 1980). All of the monoclonal antibodies studied manifested linear Scatchard plots (Figure 1) indicative of a single affinity as expected for a homogeneous antibody. Association constants determined from the Scatchard analyses were all in the range 0.2 to 2.0 x $10^9 \underline{M}^{-1}$. These uniformly high affinities may have resulted from the immunization and the selection procedure in screening for myoglobin-specific hybrid clones, in contrast to myeloma proteins, for which no selection took place. The latter have been found to have affinities rarely above $10^6 M^{-1}$. (Potter, 1977).

To determine the fine specificity of these monoclonal antibodies we studied six of them (all IgG_1, Kappa except for clone 2, which was IgG_{2a} Kappa) for binding to cyanogen bromide cleavage fragments of myoglobin. However, none of these six (as well as six additional antimyoglobin monoclonals) bound radiolabeled fragments (1-55) or (132-153). Moreover, unlabeled fragments (1-55) and (56-131) failed to compete for binding to labeled native myoglobin, even out to over a thousand-fold excess (8000 nM) (Figure 2). These data suggested that the affinities of these antibodies for fragments (if any) must be less than $10^5 M^{-1}$. This result could not be attributed solely to the conformational equilibria of the peptides, since Hurrell et al. (1977b) had determined the K_{conf} (Sachs et al., 1972) value for fragment (132-153) to be about 100, i.e., about 1% was in a native-like conformation at any time. Therefore, since these three fragments span the whole linear sequences of myoglobin, we postulated that the antigenic determinants recognized by some of these antibodies might not be represented in any contiguous stretch of primary sequence, but rather might consist of residues close together on the surface of the native molecule, but far from each other in the sequence and possibly on different CNBr cleavage fragments.

We did not use smaller peptides to identify the antigenic determinants because they would be less likely to react than the larger CNBr cleavage fragments, and because the two synthetic peptides we made failed to react with even conventional immune antimyoglobin sera raised in goats, a sheep, or high responder mice. We used the rapid modification (Corley et al., 1972) of the Merrifield (1965) method to synthesize peptides corresponding to residues 56-63 and 93-102 of sperm whale myoglobin. These sequences had previously been reported to be two of the five sites which were said to account for all of the antigenicity of myoglobin (Atassi, 1975). The peptides were purified by gel filtration and ion exchange chromatography, demonstrated to be essentially homogeneous by two dimensional chromatography/electrophoresis, and had the predicted amino acid analysis. The peptides were labeled with $K^{14}CNO$ under conditions which predominantly label the N-terminal alpha amino group, not lysines, and in the case of peptide (93-102), the labeling was accomplished before the lysine residues were deblocked after synthesis, to insure that only the N-terminal

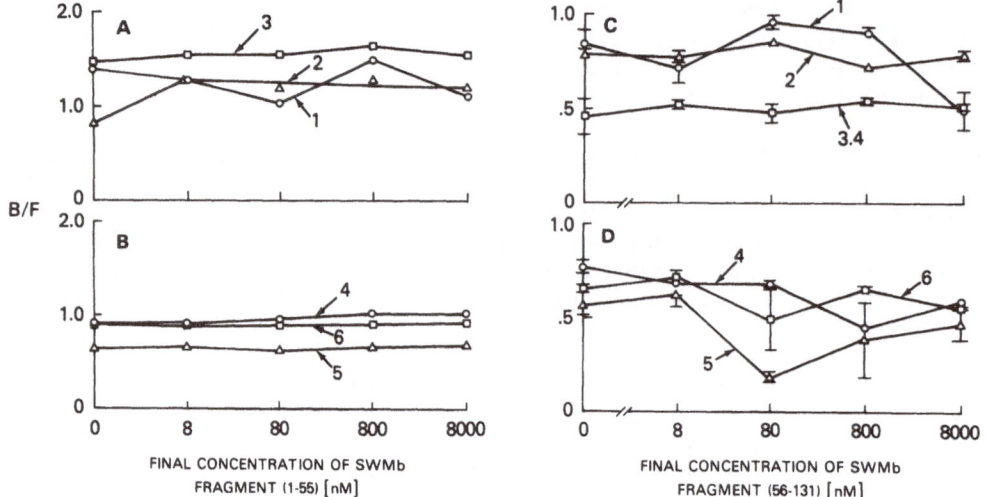

Fig. 2.   Failure of CNBr cleavage fragments of myoglobin to
            inhibit interaction of [³H]myoglobin with monoclonal
            antimyoglobin antibodies. Curves for antibodies of
            clones 1 through 6 are numbered accordingly. Labeled
            sperm whale myoglobin was held at a constant 7nM, and
            each monoclonal antibody concentration was held constant
            at 4-6nM, except clone 3.4 which was tested at 12nM.
            Equilibrium conditions and assay were as in Figure 1.
            Reproduced from Bersofsky et al., (1982).

alpha amino group was modified. Nevertheless, no binding of radioactivity from either peptide could be demonstrated to either early or
hyperimmune antimyoglobin antisera from two goats, a sheep, and
several high responder strains of mice (Table 1) even out to large
antibody excess. All of these sera bound both labeled native myoglobin and labeled CNBr cleavage fragments of myoglobin. Since the
activity originally reported for these peptides had been based on
inhibition of quantitative precipitin reactions of goat and rabbit
antisera (Atassi, 1975), we tested the unlabeled synthetic peptides
for any activity using this assay. Even out to 1000-fold molar
excess, we could not detect inhibition of goat or sheep antisera
(Figure 3), although the peptides were previously reported to achieve
maximal inhibition at a 150-200-fold molar excess (Koketsu and
Atassi, 1974; Pai and Atassi, 1975). We concluded that these two
sites on myoglobin, reported to be antigenic for antisera raised in
a few goats and rabbits, may not be antigenic for antisera raised in
other goats, much less sheep and mice.

    Therefore, we decided to analyze the antigenic determinants
recognized by these monoclonal antibodies de novo, without making any
assumptions about previously reported sites. Furthermore, to avoid
the conformational problems of cleavage fragments and synthetic

Table 1. Failure of Labeled Synthetic Peptides to Bind to Antimyoglobin Antisera

Serum[a]		$[^{14}C]$Myoglobin[d]		$[^{14}C]$56-63[e]		$[^{14}C]$93-102[f] Prep 1		$[^{14}C]$93-102[g] Prep 2	
Species	Bleed[b]	Cpm[c] Free	Cpm Bound	Cpm Free	Cpm Bound	Cpm Free	Cpm Bound	Cpm Free	Cpm Bound
B10.D2 Mouse	Pre	7155	613	1663	49	3056	57	5965	69
	3°	215	7912	1601	61	3133	57	6280	89
A.SW Mouse	Pre	8178	427	1724	54				
	2°	138	8381	1739	55				
Goat 1023	Pre	4230	231	2128	63	2757	73	940	57
	3°	1004	4005	2007	74	2756	86	900	57
	6°	821	3917	2086	64				
	7°	805	3917			2769	84		
Sheep 1364	Pre	4084	212	2115	61	2790	77		
	3°	260	4333	2117	64	2797	102		
	6°	197	4693	2099	69				
	7°	189	4488			2784	76		

[a]10µl of neat serum was used in a final reaction volume of 50µl. Doubling the serum to 20µl or halving to 5µl gave similar results.
[b]Number of immunizations before serum obtained.
[c]Cpm are uncorrected means of radioactivity in supernatant and precipitate from duplicate assays. Machine background was 40 cpm.
[d]Final concentration of $[^{14}C]$sperm whale myoglobin, 0.87µM.
[e]Final concentration of $[^{14}C]$peptide (56-63) was 0.33µM.
[f]Peptide (93-102) labeled preferentially at N-terminal amino group after deblocking and purification. Final concentration, 1.1µM.
[g]Peptide (93-102) labeled exclusively at N-terminal amino group with $K^{14}CNO$ before lysines were deblocked, then purified after deblocking. Final concentration 1µM in the case of B10.D2 antibodies and 0.14µM in the case of goat antibodies.

polypeptides and to allow for the possibility that some determinants might be present only on the surface of the intact molecule folded in its native conformation, we carried out the subsequent studies using only native myoglobins.

A series of native mammalian myoglobins with known amino acid substitutions relative to sperm whale myoglobin were used as competi-

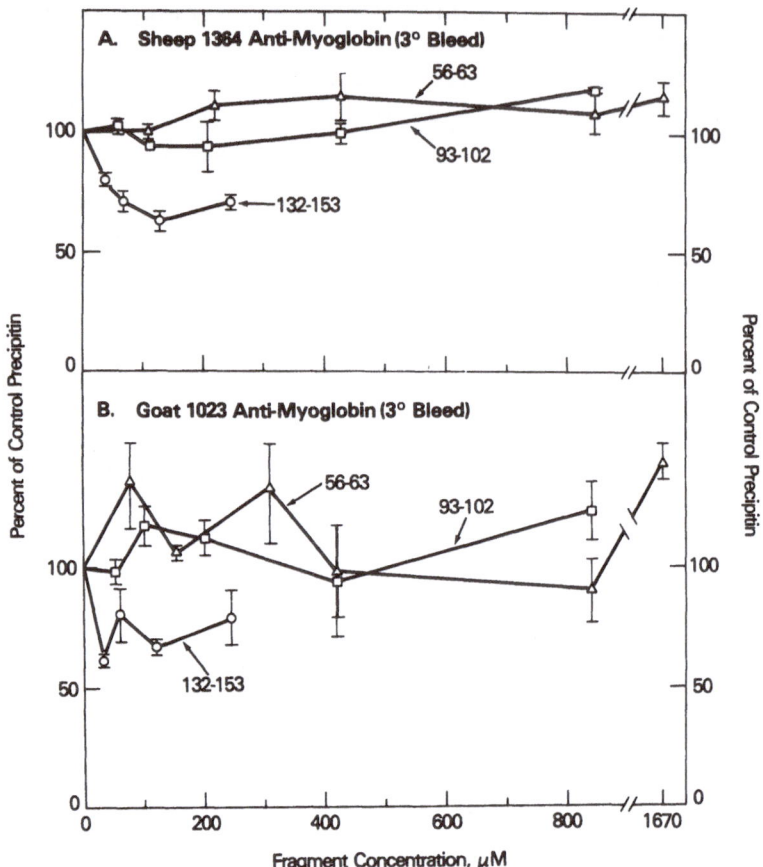

Fig. 3. Failure of synthetic peptides corresponding to putative
         antigenic sites to inhibit sheep or goat antimyoglobin
         quantitative precipitin. Final dilution of serum, 1:2;
         final concentration of sperm whale myoglobin 3.3μM for the
         sheep antimyoglobin curves and 1.6μM for the goat antimyo-
         globin curves. Peptides corresponding to sequences 56–63
         and 93–102 were purified solid–phase synthetic peptides as
         described in the text, corresponding to sites 2 and 3 of
         Atassi (1975). Similar synthetic peptides were reported to
         inhibit quantitative precipitation by Koketsu and Atassi
         (1974) and Pai and Atassi (1975). The significant in-
         hibition by CNBr cleavage fragment (132–153) is shown as a
         positive control. Similar results were obtained with serum
         from later bleeds and serum from another goat.

tive inhibitors of the binding of [³H]sperm whale myoglobins to
monoclonal antibody in a solution phase equilibrium radioimmunoassay.
The relative affinities of the various myoglobins were then compared

by comparing the concentrations required to produce 50% inhibition.
We reasoned that since all of the molecules of monoclonal antibody
could bind to only a single site on myoglobin, an amino acid substitution within this site should perturb the affinity. Subsitutions
very near the site may or may not affect the affinity, but substitutions far from the site should not alter the affinity unless they
significantly perturbed the tertiary structure of the molecule.
Thus, the only assumption necessary for this analysis was that the
three dimensional conformations of all the mammalian myoglobins
tested were nearly the same. This assumption appeared justified since
the three dimensional structures of the three myoglobins studied by
x-ray crystallography were virtually identical (Takano, 1977;
Bradshaw et al., 1969; Scouloudi and Baker, 1978; Scouloudi, 1978).
Correlation of the sequence differences with relative affinities for
14 myoglobins has allowed us, thus far, to identify critical residues
recognized by three of the six antibodies studied. Additional myoglobins will be required to resolve ambiguities in the assignments of
the other antibodies.

Two of the antibodies were found to recognize determinants
consisting of residues which are far apart in the primary sequence
but close together on the surface of the molecule in its native
three-dimensional conformation, i.e., topographic determinants. All
of the myoglobins with high or low affinity for clone 3.4 antibody
could be accounted for by one and only one substitution, that of Asp
for Glu at position 4 (Table 2). Why should Glu and Asp, which have
the same carboxylic acid functional group and differ by only one
methylene group in the length of the side chain, produce such a
marked difference in affinity? The answer became apparent when we
examined the three-dimensional structures (Figure 4). Glu 4 forms a
very tight ionic bond or salt bridge with Lys 79, in the E-F interhelical bend. By computer analysis, the closest approach possible
between the carboxylic oxygen and amino nitrogen is 1.84 Å (Although,
admittedly, they cannot really get quite so close) for Glu 4-Lys 79,
but is 3.2 Å for Asp 4-Lys 79. Substitution of Asp for Glu at position 4 results in a weaker bond between the shorter Asp and Lys 79
than could be formed by the longer Glu, and produces a large decrease
in affinity for the antibody. Computer analysis of electrostatic
potential maps demonstrates that Asp 4 and Lys 79 neutralize each
other's charge significantly less (at the pH 8 of the experiments)
than do Glu 4 and Lys 79. Presumably, this difference leads to
unfavorable contacts with the complementary site on the antibody.
Substitution of Asn for His at position 12 also lowers the affinity.
Since Glu 4 and Lys 79 are on different CNBr cleavage fragments, the
determinant is not represented intact on any fragment, accounting for
the failure of the antibody to bind fragments.

A second topographic determinant is recognized by antibody from
clone 1. In this case, replacement of Glu 83 by Asp without any
other changes lowers the affinity. (Compare killer whale and pilot

Table 2.   Inhibition of Monoclonal Antibody 3.4 Binding by Myoglobin
           Variants

Inhibitor Myoglobin	Residue Number		Concentration for 50% Inhibition nM
	4	12	
Sperm Whale	Glu	His	8
Dwarf Sperm Whale	Glu	His	46
Dall Porpoise	Glu	Asn	333
Goosebeaked Whale	Glu	His	10
Bottlenosed Dolphin	Asp	Asn	8,000
Minke Whale	Asp	Asn	>40,000
Killer Whale	Asp	Asn	>40,000
California Sea Lion	Asp	Asn	3,200
Human	Asp	Asn	>>800
Dog	Asp	Asn	>40,000
Horse	Asp	Asn	>40,000
Beef	Asp	Asn	>40,000

Modified from Berzofsky et al., (1982) with permission

whale myoglobins in Table 3). However, in contrast to clone 3.4,
this substitution does not account for all of the low affinity myo-

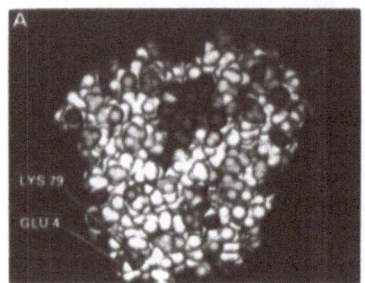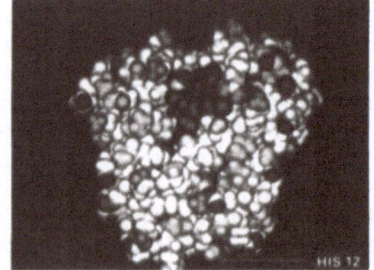

Fig. 4.   Stereoscopic views of a computer-generated space-filling
          molecular model of sperm whale myoglobin, based on the
          Takano (1977) x-ray diffraction coordinats. This orien-
          tation is arbitrarily designated the "front view." The
          computer method was described by Feldmann et al., (1978).
          The heme and aromatic carbons are shaded darkest, followed
          by carboxyl oxygens, then other oxygens, then primary amino
          groups, then other nitrogens, and finally side chains of
          aliphatic residues. The backbone and the side chains of
          nonaliphatic residues, except for the functional groups, are
          shown in white.

Table 3.   Inhibition of Monoclonal Antibody 1 Binding by Myoglobin
           Variants

| | Residue Number | | | | Concentration for 50% Inhibition |
| | 83 | 140 | 144 | 145 | nM |
Inhibitor Myoglobin					
Sperm Whale	Glu	Lys	Ala	Lys	10
Dwarf Sperm Whale	Glu	Lys	Ala	Lys	13
Goosebeaked Whale	Glu	Lys	Ala	Lys	3
Dog	Glu	Asn	Ala	Lys	5
Horse	Glu	Asn	Ala	Lys	3
Pilot Whale	Glu	Lys	Ala	Lys	24
Killer Whale	Asp	Lys	Ala	Lys	240
California Sea Lion	Asp	Asn	Ala	Lys	160
Human	Glu	Lys	Ser	Asn	470
Dall Porpoise	Asp	Lys	Thr	Lys	500
Beef	Glu	Asn	Glu	Lys	>40,000
Sheep	Glu	Asn	Ala	Gln	>40,000

Modified from Berzofsky et al., (1982) with permission.

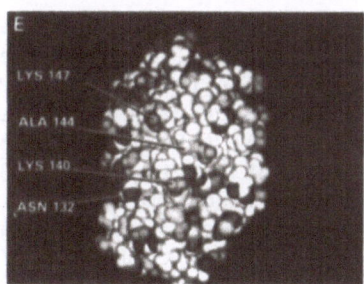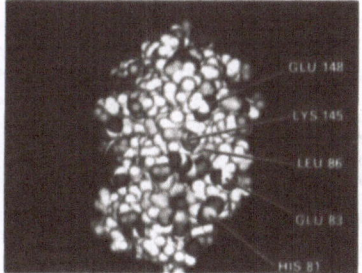

Fig. 5.   Stereoscopic view of computer-generated space-filling model
          of the left view of sperm whale myoglobin. Methods and
          shading are as indicated in Figure 4.

globins. To account for human, beef, and sheep myoglobins we must
find another substitution or group of substitutions nearby on the
protein surface of these myoglobins which are not present in any of
the high affinity myoglobins. The only substitutions that meet these
criteria are in positions 144 (Ala) and 145 (Lys) (Table 3). In
fact, nonconservative substitutions of Glu for Ala at position 144 in
beef or of Gln for Lys at position 145 in sheep myoglobin result in
the lowest affinities of all, despite the fact that they are coupled
with no change at position 83. Thus, the cluster of residues 83,
144, and 145 was the only group of adjacent surface residues which

Table 4.   Inhibition of Monoclonal Antibody 5 Binding by Myoglobin
           Variants

Inhibitor Myoglobin	Residue Number			Concentration for 50% Inhibition nM
	140	144	145	
Sperm Whale	Lys	Ala	Lys	17
Dwarf Sperm Whale	Lys	Ala	Lys	26
Dall Propoise	Lys	Thr	Lys	10
Goosebeaked Whale	Lys	Ala	Lys	32
Human	Lys	Ser	Asn	30
Minke Whale	Lys	Ala	Lys	21
Killer Whale	Lys	Ala	Lys	55
California Sea Lion	Asn	Ala	Lys	>20,000
Dog	Asn	Ala	Lys	>40,000
Horse	Asn	Ala	Lys	6,400
Beef	Asn	Glu	Lys	>40,000
Sheep	Asn	Ala	Gln	>40,000

Modified from Berzofsky et al., (1982) with permission.

could account for the whole pattern of relative affinities (Table 2
and Figure 5). These residues are also far apart in the primary
sequence and not present together on any of the three CNBr cleavage
fragments, but are close together on the surface of the native three-
dimensional conformation of the molecule (11.7 Å between Glu 83
carboxyl oxygen and Ala 144 β carbon, and 7.1 Å between Glu 83 oxygen
and Lys 145 ε-amino nitrogen). Thus, these also represent a topographic antigenic determinant. Between residues 83 and 144/145 is a
cavity into which the convex part of the antibody combining site
might fit. Although antibody combining sites studied thus far,
primarily those of myeloma proteins which bind haptens or carbohydrates, have concavities which engulf the antigen, one might expect
that the reverse could sometimes be true for globular protein antigens, i.e., that the site on the antigen would be concave and that on
the antibody convex. It is interesting to speculate that this may be
the first example of such a reverse topological relationship between
the complementary sites on antigen and antibody.

     The third monoclonal antibody, that of clone 5, distinguishes
Lys 140 from Asn 140 (Table 4). Although no other residues could be
assigned in this analysis, presumably this antibody also recognizes
other residues nearby on the surface of myoglobin, since it also does
not bind the CNBr cleavage fragments. In this case, residues 144 and
145 can be excluded from the site. For instance, Dall porpoise myoglobin has a substitution at position 144 and human myoglobin has
substitutions at both positions 144 and 145, and yet both myoglobins,

which have a Lys —→Asn substitution at position 140, resulting in
low affinity for clone 5 antibody, have a high affinity for clone 1
antibody (Tables 3 and 4). Thus, the determinant recognized by clone
5 antibody does not appear to overlap with that recognized by clone 1
antibody.

The prediction that these two determinants, although close to
gether on what we have called the left view of the myoglobin surface
(Figure 5), do not overlap, was tested by attempting to bind both
antibodies to the same myoglobin molecule simultaneously. This
approach was feasible since myoglobin is a monomer, without any
repeating sequences. Plastic microtiter wells were coated with clone
5 antibody. Monomeric native sperm whale myoglobin was allowed to
bind to the attached clone 5 myoglobin. The wells were then incubated
with alkaline-phosphatase conjugated monoclonal antibody of clone 1
or clone 5, and the amount bound, after extensive washing, was determined by addition of substrate (p-nitrophenyl phosphate) and monitoring of the absorbance change at 405mm with time (Figure 6). As a
control, neither antibody bound directly to the clone 5 antibody-
coated plates if myoglobin was left out and both antibodies bound
well to plates coated directly with myoglobin in random orientation
(data not shown). As expected, clone 5 antibody could not bind to a
myoglobin molecule already bound to another molecule of the same
antibody. In contrast, clone 1 antibody bound very well to myoglobin
molecules already bound by clone 5 antibody. Thus, the two antibodies
can bind simultaneously to the same molecule of myoglobin-proof that
the determinants recognized by these two monoclonal antibodies do not
overlap.

In summary, the determinants recognized by these three
monoclonal antibodies differ from previously reported antigenic sites
on sperm whale myoglobin both in that they are topographic, rather
than sequential, and in that all but one of the seven residues
identified for the three determinants are outside the previously
reported sites. Similar findings have recently been obtained for two
monoclonal antibodies to human myoglobin (East et al., 1982) and for
a monoclonal antibody to lysozyme (Smith-Gill et al., 1982). In both
cases, the monoclonal antibodies bound topographic determinants
involving residues outside the antigenic sites previously identified
for myoglobin (Atassi, 1975) and lysozyme (Atassi, 1978) by use of
conventional antisera and small synthetic peptides. Taken together,
these other studies of monoclonal antibodies and our own (reported
here, and Berzofsky et al., 1982), plus the studies of Hurrell et
al., (1977a) and East et al., (1980) using conventional antisera,
lead us to conclude that there are far more antigenic determinants on
these globular protein antigens than previously suggested. Probably
most of the surface of the molecule can be immunogenic depending on
the species immunized (P.E.E. Todd and S.J. Leach, personal
communication), although some regions may tend to be more immunogenic
than others. These immunogenic regions, or domains (Hurrell et al.,

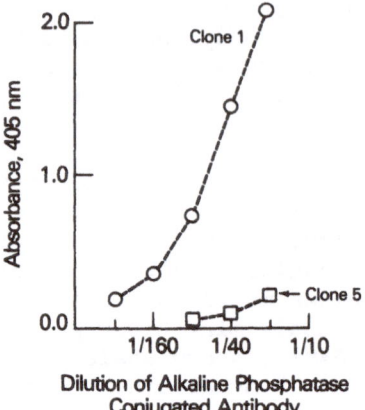

Fig. 6.   Monoclonal antibodies of clone 1 and clone 5 can bind
          simultaneously to the same molecule of sperm whale myo-
          globin.   Polystyrene plates were coated with purified clone
          5 antibody overnight, washed, and reacted with 50nM myo-
          globin at 20°C for 1hr, and washed again.   Then alkaline-
          phosphatase-conjugated antibody of clone 1 or clone 5 was
          reacted for another hour at 20°C.   After extensive washing,
          the amount of enzymelabeled antibody bound was determined by
          cleavage of substrate, monitored at 405nM.

1977a), probably consist of many overlapping determinants, so that no
single antibody will recognize a whole domain. Moreover, these
antigenic domains, as well as individual determinants, must be viewed
as topographic entities on the protein surface. There is no a priori
reason that we can see why residues connected together by a short
stretch of peptide backbone should be any more likely to be part of
the same determinant than any other residues which are equally close
on the surface topography, even if they are at opposite ends of the
primary sequence. This fundamental principle should apply as long as
the antibodies in question, whether conventional or monoclonal, were
raised against the pure native protein. If the immunogen contains
denatured or fragmented material, then, of course, residues
contiguous on the backbone will have more chance of being seen

together than residues which are neighbors only in the native conformation.

This study also demonstrates the exquisite sensitivity of monoclonal antibodies to such subtle change as Glu vs. Asp in an antigenic site. Such sensitivity would likely be lost due to heterogeneity in a conventional antiserum.

Finally, we would like to study the primary and three-dimensional structure of these monoclonal antibody combining sites to understand the complementary interaction between two globular proteins, antigen and antibody, which leads to such high affinities. One example of what we might find was the suggestion above that the combining site of clone 1 antibody may contain a convex surface complementary to the concavity on the antigen surface. This would be the opposite of what one usually envisions from studies with small haptens which fit into a cavity on the antibody. We are also studying the idiotypes of these monoclonal antibodies, and have found that some which recognize different antigenic determinants may nevertheless share idiotopes, some even in the combining site (Kohno, Berkower, Minna, and Berzofsky, 1982).

SUMMARY

As models of B cell receptors for a defined globular protein antigen under H-2-linked Ir gene control, we have studied several high affinity ($\sim 10^9 M^{-1}$) IgG$_1$, K murine monoclonal antibodies specific for sperm whale myoglobin. None of 6 tested reacted with any of the three CNBr cleavage fragments which span the whole sequence of myoglobin. Also, two synthetic peptides corresponding to sequences previously reported to be antigenic sites for some goat and rabbit antisera were found not to react with goat, sheep, or mouse antimyoglobin. Therefore, since we could not determine the fine specificity with fragments and peptides, we used a series of native myoglobins with very similar three dimensional structure and known amino acid substitutions relative to sperm whale myoglobin. By correlating relative affinities with known sequence differences, we identified residues involved in antigenic determinants recognized by 3 of the monoclonal antibodies. Clone 3.4 antibody recognized Glu 4 on the A helix and Lys 79 on the E-F interhelical segment. Clone 1 antibody recognized Glu 83 on the E-F interhelical segment and Ala 144 and Lys 145 on the H helix. Clone 5 antibodies recognized Lys 140 on the H helix, but were not affected by substitutions at 144 and 145, while clone 1 antibodies were not affected by substitutions at position 140. Therefore we predicted that the determinants recognized by these two antibodies, although both involving the H helix, did not overlap. This prediction was confirmed by the demonstration that both clone 1 and clone 5 antibodies could bind simultaneously to the same molecule of myoglobin. These determinants are of a new type for the case of myoglobin in that 1) for clones 3.4 and 1 they are topographic, that is, they involve residues which are far apart in the primary sequence but are brought together on the surface by the folding of the molecule in its native conformation; and 2) 6 of the 7 residues are outside of previously reported determinants. We have also found that monoclonal antibodies can distinguish such subtle changes as Asp for Glu in a complex protein determinant. The structures of the antibody combining sites are now under study.

REFERENCES

Atassi, M. Z., 1975, Antigenic structure of myoglobin: The complete immunochemical anatomy of a protein and conclusions relating to antigenic structures of proteins. Immunochemistry 12,423-438.

Atassi, M. Z., 1978, The precise and entire antigenic structure of lysozyme: Implications of surface-simulation synthesis and the molecular features of protein antigenic sites. Adv.Exp.Med.Biol. 98, 41-99.

Berzofsky, J. A. 1978a, Genetic control of the immune response to myoglobins in mice. I. More than I-region gene in H-2 controls the antibody response. J. Immunol. 120,360.

Berzofsky, J. A., 1978b, Genetic control of the antibody response to sperm whale myoglobin in mice. Adv.Exp.Med.Biol. 98,225.

Berzofsky, J. A., Richman, L. K., 1981, Genetic control of the immune response to myoglobins. IV. Inhibition of determinant-specific Ir gene-controlled antigen presentation and induction of suppression by pretreatment of presenting cells with anti-Ia antibodies. J.Immunol. 126, 1898.

Berzofsky, J. A., Schechter, A. N., Shearer, G. M., and Sachs, D. H., 1977, Genetic control of the immune response to staphylococcal nuclease. III. Time-course and correlation between the response to native nuclease and the response to its polypeptide fragments. J.Exp.Med. 145, 111.

Berzofsky, J. A., Richman, L. K., and Killion, D. J., 1979, Distinct H-2-linked Ir genes control both antibody and T cell responses to different determinants on the same antigen, myoglobin. Proc.Nat.Acad.Sci.U.S.A. 76,4046.

Berzofsky, J. A., Hicks, G., Fedorko, J., and Minna, J., 1980, Properties of monoclonal antibodies specific for determinants of a protein antigen, myoglobin. J.Biol.Chem. 255, 11188-11191.

Berzofsky, J. A., Buckenmeyer, G. K., Hicks, G., Gurd, F. R. N., Feldmann, R. J., and Minna, J., 1982, Topographic antigenic determinants recognized by monoclonal antibodies to sperm whale myoglobin, J.Biol.Chem. 257, 3189-3198.

Bradshaw, R. A., Kretsinger, R. H., and Gurd, F. R. N., 1969, Comparison of myoglobins from harbor seal, porpoise, and sperm whale. J.Biol.Chem. 244, 2159-2166.

Corley, L., Sachs, D. H., and Anfinsen, C. B., 1972, Rapid solid-phase synthesis of bradykinin. Biochem.Biophys.Res.Comm. 47, 1353.

Crumpton, M. J., 1974, Protein antigens: the molecular bases of
    antigenicity and immunogenicity. In The Antigens, (edited by
    Sela, M.), Vol. 2, pp. 1-79. Academic Press, New York.

Desbuquois, B., and Aurbach, G. D., 1971, Use of polyethlene glycol
    to separate free and antibody-bound peptide hormones in radio-
    immunoassays. J.Clin.Endocrinol.Metab. 33, 732.

Dorf, M. E., Dunham, E. K., Johnson, J. P., and Benacerraf, B., 1974,
    Genetic control of the immune response: The effects of non-H-2-
    linked genes on antibody production. J.Immunol. 112,1329.

East, I. J., Todd, P. E., and Leach, S. J., 1980, On topographic
    antigenic determinants in myoglobins. Molecular Immunol. 17,
    519-525.

East, I. J., Hurrell, J. G. R., Todd, P. E. E., and Leach, S. J.,
    1982, Antigenic specificity of monoclonal antibodies to human
    myoglobin. J.Biol.Chem. 257, 3199-3202.

Feldmann, R. J., Bing, D. H., Furie, B. C., and Furie, B., 1978,
    Interactive computer surface graphics approach to study of the
    active site of bovine trypsin. Proc.Nat.Acad.Sci.U.S.A. 75,
    5409-5412.

Hurrell, J. G. R., Smith, J. A., Todd, P. E., and Leach, S. J.,
    1977a, Crossreactivity between mammalian myoglobins: Linear vs.
    spatial antigenic determinants. Immunochemistry 14, 283-288.

Hurrell, J. G. R., Smith, J. A., and Leach, S. J., 1977b,
    Immunological measurements of conformational motility in regions
    of the myoglobin molecule. Biochemistry 16, 175-185.

Kabat, E. A., 1978, The structural basis of antibody complementarity.
    Adv.Protein Chem. 32, 1-76.

Kohler, G., and Milstein, C., 1975, Continuous cultures of fused
    cells secreting antibody of predefined specificity. Nature 256,
    495-497.

Kohno, Y., and Berzofsky, J. A., 1982, Genetic control of immune
    response to myoglobin: Ir gene function in genetic restriction
    between T and B lymphocyte. J.Exp.Med. 156, 1486-1501.

Kohno, Y., Berkower, I., Minna, J., and Berzofsky, J. A., 1982,
    Idiotypes of anti-myoglobin antibodies: Shared idiotypes
    among monoclonal antibodies to distinct determinants of sperm
    whale myoglobin. J.Immunol. 128, 1742-1748.

Koketsu, J., and Atassi, M. Z., 1974, Immunochemistry of sperm-whale
    myoglobin. XIX. Accurate delineation of the single reactive
    region in sequence 54-85 by immunochemical study of synthetic
    peptides. Biochim.Biophys.Acta 324,21.

Merrifield, R. B., 1965, Automated synthesis of peptides. Science
    150, 178.

Pai, R. C., and Atassi, M. Z., 1975, Immunochemistry of sperm-whale
    myoglobin. XX: Accurate delineation of the single reactive
    region in sequence 80-103 by immunochemical studies of synthetic
    peptides. Immunochemistry 12, 285.

Pisetsky, D. S., Berzofsky, J. A., and Sachs, D. H., 1978, Genetic
    control of the immune response to staphylococcal nuclease. VII.
    Role of non-H-2-linked genes in the control of the anti-nuclease
    antibody response. J.Exp.Med. 147, 396.

Potter, M., 1977, Antigen-binding myeloma proteins of mice. Adv.Immunol. 25, 141-211.

Reichlin, M., 1975, Amino acid substitution and the antigenicity of globular proteins. Adv.Immun. 20,71-123.

Richman, L. K., Strober, W., and Berzofsky, J. A., 1980, Genetic control of the immune response to myoglobin. III. Determinant-specific, two Ir gene phenotype is regulated by the genotype of reconstituting Kupffer cells. J.Immunol. 124, 619.

Sachs, D. H., Schecter, A. N., Eastlake, A., and Anfinsen, C. B., 1972, An immunological approach to the conformational equilibria of polypeptides. Proc.Natl.Acad.Sci.U.S.A. 69, 3790-3794.

Scouloudi, H., 1978, A preliminary comparison of metmyoglobin molecules from seal and sperm whale. J.Mol.Biol. 126, 661-671.

Scouloudi, H., and Baker, E. H., 1978, X-ray crystallographic studies of seal myoglobin: The molecule at 2.4 Å resolution. J.Mol.Biol 126, 637-660.

Smith-Gill, S. J., Wilson, A. C., Potter, M., Prager, E. M., Feldmann, R. J., and Mainhart, C. R., 1982, Mapping the antigenic epitope for a monoclonal antibody against lysozyme. J.Immunol. 128, 314-322.

Takano, T., 1977, Structure of myoglobin refined at 2.0 Å resolution. I. Crystallographic refinement of metmyoglobin from sperm whale. J.Mol.Biol. 110, 537-568.

ROLE OF CONFORMATION ON THE ANTIGENIC DETERMINANTS

OF TOBACCO MOSAIC VIRUS PROTEIN

Marc H.V. Van Regenmortel, Danièle Altschuh
and Jean-Paul Briand

Institut de Biologie Moléculaire et Cellulaire
15 rue Descartes
Strasbourg, France

The majority of viruses have a fairly simple architecture con-sisting of a number of identical protein subunits assembled into a capsid with icosahedral or helical symmetry. The polymerization of subunits leads to the effective disappearance of certain surfaces of the protein that are turned inward, a feature common to all macro-molecular structures built up from monomers. Epitopes situated on these hidden surfaces have been called cryptotopes[1]; they become antigenically expressed only when the capsids have been disassembled.

Antigenic studies performed with all major groups of viruses have shown that viral capsids also possess specific epitopes that are not present in the constituent protein subunits[2-6]. Such epi-topes, which have been called neotopes[2] arise either by conformational changes induced by the intersubunit bonds, or by the creation of a new antigenic structure through the juxtaposition of amino acid resi-dues from neighboring subunits. In the latter case, the neotope would correspond to a type of "discontinuous" determinant[7] or "topographical" determinant[8] in which the juxtaposed residues orig-inate from separate polypeptide chains. In some cases a total lack of antigenic cross-reactivity between protein monomer and capsid is observed, while in others a limited cross-reactivity is found to exist. In the case of certain animal viruses, antibodies directed to neotopes appear to play an important role in virus neutralization. This means that antibodies obtained as a result of immunization with monomeric subunits might fail to neutralize the virus efficiently. This is of considerable importance in view of current attempts to produce subunits or synthetic viral vaccines[9-12]. It has been shown that immunization with the product of a limited reassociation of subunits can increase the neutralizing activity of the vaccine com-pared with the poor efficacy of isolated monomers[13]. Since relatively

181

small aggregates may already give rise to a neotope-like antigenic
configuration, it is clear that a knowledge of the factors that control neotope formation is of considerable importance for ensuring
the success of the new types of viral vaccines. Tobacco mosaic virus
(TMV) is a particularly good model for studying the role played by
tertiary and quaternary protein structure in viral antigenicity. The
antigenic properties of TMV have been studied more extensively than
those of any other virus[14-15] and since the three-dimensional structure of its protein subunits has been established by X-ray crystallography[16] it is possible to interpret the immunochemical data in
terms of the location of each residue.

ANTIGENIC STRUCTURE OF TMV PROTEIN

     TMV is a rod-shaped particle, 300nm long, which consists of 2130
identical protein subunits arranged as a helix around an RNA molecule
of molecular weight 2×10^6. The peptide chain contains 158 amino
acid residues (Table 1) and its folding was determined from electron
density maps obtained at a resolution of 2.8 Å. The central part of
the subunit consists of two pairs of helices which comprise about 60
residues.

     Although the individual epitopes that are present at the surface
of the virus are expressed 2130 times in each particle, only 780
antibody molecules are able to bind simultaneously to the surface of
one virion. This figure represents the antigenic valence of the
virus and applies to IgG antibody molecules that bind univalently to
the particle surface. When both combining sites of the antibody
molecules are bound to neighboring subunits on the same virion, a
maxiumum of 400 IgG molecules can attach to the virus surface[17].
The latter type of binding, known as monogamous bivalent binding, is
characterized by approximately a twentyfold higher avidity than that
found when the same antibody molecules bind monovalently. The flexibility of the two Fab arms must be considerable to allow the two
paratopes of an IgG molecule to bind to neighboring epitopes on the
virus surface. Although there is a large thermodynamic advantage in
bivalent binding[18,19], there must also be considerable constraint
when the two Fab arms that are presented in opposite orientation
bind to identical epitopes that all face the same way on the virus
surface. In the case of cellular antigens lying in various orien-
tations in the membrane, the constraints of monogamous bivalent binding of IgG are likely to be less, which agrees with the much larger
differential in avidity observed between monovalent and bivalent
binding on cellular surfaces[20].

     Although it has been suggested by some workers that the TMV sub-
unit possesses only one or two epitopes (see ref.[3]) subsequent work
has demonstrated the presence of at least eight antigenic regions in
the protein monomer. The following experimental approaches have been
used to study the epitopes of TMV protein:

1)   Inhibition experiments with peptides obtained by degradation

Peptides obtained by exzymatic degradation of the subunit were tested for their ability to inhibit the precipitation, complement fixation and ELISA (enzyme-linked immunosorbent assay) reactions between virus or viral protein and antibody[21-24]. These studies demonstrated antigenic activity in three short regions of the molecule corresponding to residues 47 - 61, 62 - 68 and 153 - 156 (Figure 1). In addition, the results pointed to the presence of two other epitopes in tryptic peptide 1 (residues 1 - 41) and of one epitope in tryptic peptide 8 (residues 93 - 112).

Table 1.   Amino sequence of tobacco mosaic virus protein

```
SYSITTPSQFVFLSSAWADPIELINLCTNALGNQFQTQQA
      10            20            30            40

RTVVQRQFSEVWKPSPQVTVRFPDSDFKVYRYNAVLDPLV
      50            60            70            80

TALLGAFDTRNRIIEVENQANPTTAETLDATRRVDDATVA
      90           100           110           120

IRSAINNLIVELIRGTGSYNRSSFESSSGLVWTSGPAT
     130           140           150          153
```

---

2)   Inhibition experiments with synthetic peptides

In a series of classical experiments with synthetic peptides corresponding to various parts of tryptic peptide 8, Benjamini and coworkers showed that the shortest peptide that possessed demonstrable antigenic activity was the pentapeptide 108 - 112[25]. As shown by the location of this region in the central hole of the virus rod (Figure 1), residues 108 - 112 correspond to a cryptotope of TMV. Recent unpublished experiments from this laboratory showed that synthetic peptides corresponding to residues 34 - 39, 72 - 77 and 142 - 147 also possessed inhibitory activity in ELISA experiments. Since it has been shown that tryptic peptide 1 probably contains two epitopes[23], it seems likely that in addition to the activity found in residues[34-39], there is another active region in the N-terminal half of this peptide, presumably in the vicinity of residues 1 - 10.

3)   Study of the role of single amino acid exchanges on antigenic reactivity

The antigenic properties of a large number of TMV mutants with a limited number of amino acid exchanges have been compared with those of the wild strain. Mutations that altered residues 65, 66, 107, 136, 138, 140, 148 and 156 were found to change the antigenic properties of the virus, whereas exchanges at position 20, 21, 25, 33, 46, 59, 63, 81, 97, 99, 126, 129 did not[26,27]. These experiments

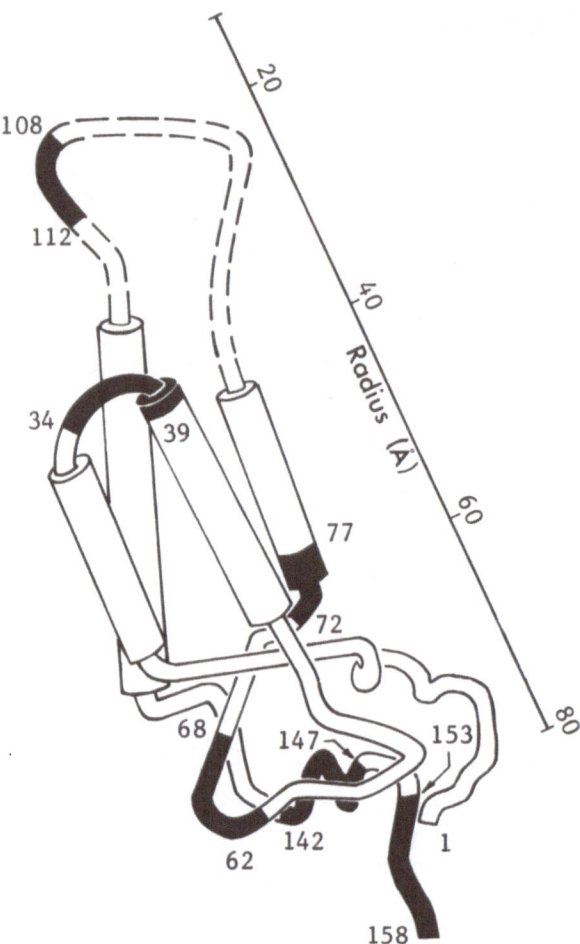

Fig. 1.   Folding of the TMV protein molecule, based on the X-ray
          crystallographic data of Bloomer et al.[16]. The helices
          are represented as cylinders. The radius scale starts
          at the center of the hole in the assembled capsid.
          Blackened portions correspond to sequences with demon-
          strable antigenic activity. Two additional antigenic
          regions are located in the vicinity of residues 1 - 20
          and 47 - 61.

performed fifteen years ago made use of classical antisera and it is
likely that subtle changes in the epitopes induced by certain mutations may have been overlooked. In several cases it was indeed
found that antisera directed to the wild type virus did not dis-
tinquish between the immunogen and the mutant and that the antigenic
difference between them only became apparent when an antiserum
prepared against an heterologous strain was used for the comparison.

This question is at present being reinvestigated in our laboratory by means of monoclonal antibodies. Preliminary results indicate that single exchanges are able to completely abolish the ability of the virus to react in ELISA with a particular monoclonal antibody. It is a well-recognized advantage of monoclonal antibodies that they usually accentuate any difference in cross reactivity between two antigens. In a classical antiserum that contains an heterogenous mixture of antibodies recognizing various regions of the same antigen, any difference in reactivity pertaining to one epitope is obviously diluted out by the remaining cross-reactivities due to the unchanged epitopes. This phenomenon is illustrated in Table 2 for the mutant with an exchange of position 65. When the wild type and mutated capsids were compared, or when the inhibitory activities of the wild type and mutated peptides 62 - 68 were compared, an antiserum clearly picked up the antigenic difference. However, when the wild type and mutated monomeric subunits were compared, the change in cross-reactivity was masked by the reaction of the remaining seven unaltered epitopes present on the subunit[28].

When an amino acid substitution in a mutant protein is found to affect the antigenic properties, the simplest inference is that the particular residue contributes to the structure of an epitope. However, without additional information one cannot be sure that the mutated residue is located within the boundaries of an epitope, since it is also possible that the substitution influences antigenic reactivity by altering the conformation of a distal antigenic region of the molecule. For instance in the case of the TMV mutant that presented an exchange at position 107, this substitution situated far away from the virus surface (see Figure 1) nevertheless altered the antigenic properties of the capsid[26].

DETECTION OF HETEROSPECIFIC ANTIBODIES

Heterospecific (or heteroclitic) antibodies are antibodies that are better recognized by another antigen than the one used for immunization. The two antigens are structurally related and cross-react with the antibody, the only oddity being that the reaction with the heterologous antigen is stronger than with the immunogen. Depending on the operational constraints of the serological test used it may happen that the weak reactivity of heterospecific antibodies with the immunogen is below detection level. In such a case it would seem as if the antibody had been elicited by an antigen with which it is unable to react. Such a situation has been found with several mutants and strains of TMV[27,29-31]. When TMV antisera were fully absorbed by intragel absorption with the homologous TMV, they were invariably found to contain heterospecific antibodies capable of reacting (in ELISA and precipitin tests) with mutated antigens where a PRO → LEU exchange had occurred at position 156 of the peptide chain. This exchange leads to a conformational change that

Table 2. Influence of amino acid exchanges in the TMV protein
          subunit on the antigenic properties of capsids, monomeric
          subunits and tryptic peptides.[a]

Mutant	Location of exchange	Location of tryptic peptide	Antigenic difference revealed in		
			capsid [b]	Subunit [c]	Tryptic [d] peptide
414	65	62 – 68	+	–	+
Ni 118	20	1 – 41	–	+	–
Ni1927	156	142 – 158	+	+	+

a) From Milton et al.[28]
b) The presence (+) or absence(-) of serological differences between
    wild type and mutant capsids was determined by precipitin tests.
c) Serological differences were determined in inhibition assays
    with radiolabeled TMV protein using unlabeled mutant proteins.
d) Wild type and mutant peptides were compared in inhibition assays
    with labeled TMV protein.

alters a neotope specificity of the virus; this is borne out by the
fact that the heterospecific antibodies do not react in ELISA and
precipitin tests with the monomeric subunit protein derived from
the mutated virus.

      It seems likely that heterospecific antibodies are more common
than is usually thought. There is no reason to suppose that the
clonal selection that triggers antibody production by a specific B
cell clone corresponds necessarily to a situation where the best
possible fit between epitope and paratope occurs with the immunogen.
When an antigen with a better fit is discovered by chance (usually
a close structural relative of the immunogen) the reactivity of the
antibody with the immunogen may then subjectively appear to be a
weak "cross-reactivity".

      The usual antibody heterogeneity of antisera is likely to mask
the presence of any heterospecific antibodies; however, in the case
of monoclonal antibodies the superior reactivity with an antigen
other than the immunogen may be much more clear-cut. Unpublished
experiments from this laboratory have shown that monoclonal antibodies with heterospecific activity are readily detected when hybridoma screening tests for antibody activity are done with TMV
mutants instead of with the wild type immunogen. Depending on the
sensitivity of the ELISA procedure used for screening, such antibodies may appear totally unreactive with TMV.

      Since monoclonal antibodies are being increasingly used for
the analysis of antigenic structures, it is clear that heterospecific

antibodies will be encountered more frequently in future. Whenever monoclonal antibodies are used as tools for probing changes in protein conformation, it is important to consider the possibility that altered antigenic configurations may sometimes react better with the antibody than the original one present in the immunogen.

USE OF ANTIBODIES FOR PROBING CONFORMATIONAL CHANGES

Immunochemical comparisons between the members of several groups of related proteins have demonstrated that there is a strong correlation between degree of sequence difference and degree of antigenic difference. From the extent of correlation it has been inferred that about 80% of the amino acid substitutions that have accumulated during the evolution of monomeric globular proteins are antigenically detectable[32]. Such a high incidence of antigenic changes induced by mutation is in agreement with the view that substitutions outside the surface epitopes are able, by some kind of allosteric mechanism, to produce conformational changes that alter the antigenicity. This means that the antigenic reactivity of globular proteins is modulated by long-range interactions at the level of secondary and tertiary structure, and that it is impossible to describe fully the structure of an epitope in terms of a short peptide sequence. The fact that short peptides are able to inhibit the reaction between antibodies and the complete antigen is due to the high molar excesses of peptide usued as well as to the existence of a variety of random conformations in these short fragments. It is generally assumed that conformations that approximate the native one can be generated by spontaneous and reversible folding of the fragments. Results obtained in the TMV system indicate that even conformations specific for the protein quaternary structure may be induced in certain peptides. This was shown by the ability of the C-terminal tryptic peptide (residues 142 - 158) to inhibit the reaction between TMV and neotope-specific antibodies[23]. Such neotope-specific antibodies recognize the quaternary structure of the capsid and are obtained by removing from a TMV antiserum all the antibodies capable of reacting with monomeric protein subunits[31].

Another useful approach for unraveling the role of conformation in antigenic reactivity consists in combining the study of mutants presenting single substitutions with that of the inhibitory activity of short fragments of the antigen[28]. It was found for instance that the protein subunit of a mutant with an exchange at position 20 was distinguishable from TMV protein, although the wild type and mutant tryptic peptides that contained the substitution did not differ in their inhibitory capacities (Table 2). It seems that residue 20 is not itself part of an epitope of TMV protein but that the exchange PRO → LEU at this position affects the conformation of an epitope elsewhere on the subunit surface. This substitution was found not to affect the antigenic reactivity of the viral capsid, indicating

that it is a cryptotope of the protein that has been affected by
the mutation.

In conclusion, the data obtained in the TMV system clearly
illustrate the value of combining various approaches for elucidating
the antigenic structure of a polymerized protein. Several epitopes
appear to be unique to the protein monomer (eg. those in the vicinity
of residues 34 - 39, 72 - 77 and 108 - 112), others are expressed
in both capsid and depolymerized subunit whereas some exist only by
virtue of the conformational constraints of the quaternary structure
of the protein.

SUMMARY

Several immunochemical techniques have been applied to the
elucidation of the antigenic structure of TMV protein. Inhibition
experiments with tryptic as well as synthetic peptides have identified
eight antigenic regions in the molecule. Some of them are situated
on surfaces of the subunit that are turned inward in the capsid, and
they correspond therefore to cryptotopes of the virus. Other antigenic regions are present on the surfaces of the subunits that are
exposed at the periphery of the virion and those induce antibodies
that recognize both the virus and the depolymerized subunits. In
addition, antisera prepared against virions also contain antibodies
that cannot be removed by absorption with depolymerized subunits.
Such antibodies, called neotopes, recognize a unique conformation
found only in the quaternary structure of the viral protein.

Mutants of the virus presenting a single amino acid exchange
in the protein have also been studied with respect to their antigenic properties. It was found that certain exchanges did not affect
in the same way the antigenicity of the corresponding tryptic peptides,
of the subunit and of the virion. Several monoclonal antibodies
prepared against TMV failed to react with certain mutants. On the
other hand, some of the hybridomas produced antibodies that failed
to recognize TMV (the immunogen) but reacted with certain mutants.
These monoclonal antibodies are similar to the previously described
heterospecific antibodies found in TMV antisera.

REFERENCES

1. N.K. Jerne, Immunological speculations, Annu.Rev.Microbiol.
   14:341 (1960).
2. M.H.V. Van Regenmortel, Plant virus serology, Adv.Virus Res.
   12:207 (1966).
3. M.H.V. Van Regenmortel, Serology and Immunochemistry of Plant
   Viruses, 302 pp. Academic Press, New York. (1982).

4.   A.R. Neurath and B.A. Rubin, Viral Structural Components As
     Immunogens of Prophylactic Value, 87 pp. S. Karger Publ.,
     Basel. (1971).

5.   E. Norrby and G. Wadell, The relationship between the soluble
     antigens and the virion of adenovirus type 3. VI. Further
     characterization of antigenic sites available at the surface
     of virions. Virology 48:757 (1972).

6.   D.A. Eppstein and J.A. Thoma, Characterization and serology of
     the matrix protein from a nuclear-polyhedrosis virus of tricho-
     plusiani before and after degradation by an endogenous protein-
     ase, Biochem.J. 167:321 (1977).

7.   M.Z. Atassi and J.A. Smith, A proposal for the nomenclature
     of antigenic sites in peptides and proteins, Immunochemistry
     15:609 (1978).

8.   I.J. East, P.E. Todd, and S.J. Leach, On topographic antigenic
     determinents in myoglobin, Molec.Immunol. 17:519 (1980).

9.   T.J. Wiktor, E. György, H.D. Schlumberger, F. Sokol, and H.
     Koprowski, Antigenic properties of rabies virus components,
     J.Immunol. 110:269 (1973).

10. W.P. Parks and F. Rapp, Prospects for herpesvirus vaccination-
     safety and efficacy considerations, Prog.Med.Virol. 21:188
     (1975).

11. H.L. Bachrach, D.M. Moore, P.D. McKercher, and J. Polatnick,
     An experimental protein vaccine of foot-and-mouth disease,
     in:"Perspectives in Virology" (M. Pollard, ed.), Vol.10, pp.
     147-159. Raven, New York. (1978).

12. R. Arnon, Chemically defined antiviral vaccined, Annu.Rev.
     Microbiol. 34:593 (1980).

13. B. Morein, A. Helenius, K. Simons, R. Pettersson, L. Kääriäinen,
     and V. Schirrmacher, Effective subunit vaccines against an
     enveloped animal virus, Nature (Lond.). 276:715 (1978).

14. I. Rappaport, The antigenic structure of tobacco mosaic virus,
     Adv.Virus Res. 11:223 (1965).

15. M.H.V. Van Regenmortel, Tobamoviruses, in:"Handbook of Plant
     Virus Infection" (E. Kurstak, ed.), pp. 541-564. North Holland
     Publ., Amsterdam. (1981).

16. A.C. Bloomer, J.N. Champness, G. Bricogne, R. Staden, and A.
     Klug, Protein disk of tabacco mosaic virus at 2.8 Å resolution
     showing the interactions within and between subunits, Nature
     (Lond.). 276:362 (1978).

17. M.H.V. Van Regenmortel, and G. Hardie, Immunochemical studies
     of tobacco mosaic virus. II. Univalent and monogamous bivalent
     binding of IgG antibody, Immunochemistry 13:503 (1976).

18. E.D. Day, Advanced Immunochemistry, Williams and Wilkins,
     Baltimore, Maryland. (1972).

19. F. Karush, The affinity of antibody: Range, variability, and
     the role of multivalence, in:"Immunoglobulins" (G.W. Litman
     and R.A. Good, eds.), pp. 85-116. Plenum, New York. (1978).

20. C.L. Greenburry, D.H. Moore, and L.A.C. Nunn, The reaction with
     red cells of 7 S rabbit antibody: Its subunits and their re-
     combinations, Immunology 8:420 (1965).

21.   F.A. Anderer, Versuche zur Bestimmung der serologisch deter-
      minanten Gruppen des Tabakmosaikvirus, Z.Naturforsch. 18b:1010
      (1963).

22.   E. Benjamini, J.D. Young, M. Shimizu and C.Y. Leung, Immuno-
      chemical studies on the tobacco mosaic virus protein. I. The
      immunological relationship of the tryptic peptides of tobacco
      mosaic virus protein to the whole protein, Biochemistry 3:1115
      (1964).

23.   R.C. de L. Milton and M.H.V. Van Regenmortel, Immunochemical
      studies of tobacco mosaic virus. III. Demonstration of five
      antigenic regions in the protein subunit, Mol.Immunol. 16:179
      (1979).

24.   D. Altschuh and M.H.V. Van Regenmortel, Localization of anti-
      genic determinants of a viral protein by inhibition of enzyme-
      linked immunosorbent assay (ELISA) with tryptic peptides,
      J.Immunol.Methods, in press. (1982).

25.   E. Benjamini, Immunochemistry of the tobacco mosaic virus
      protein, in:"Immunochemistry of Proteins" (M.Z. Atassi, ed.),
      Vol.2, pp. 265-310. Plenum, New York. (1977).

26.   P. Sengbusch, Aminosäureaustausche und Tertiarstruktur eines
      Proteins. Vergleich von Mutanten des Tabakmosaikvirus mit
      serologischen und physikochemischen Methoden, Z.Vererbungsl.
      96:364 (1965).

27.   M.H.V. Van Regenmortel, Serological studies on naturally
      occurring strains and chemically induced mutants of tobacco
      mosaic virus, Virology 31:467 (1967).

28.   R.C. de L. Milton, S.C.F. Milton, M.V. Von Wechmar, and M.H.V.
      Van Regenmortel, Immunochemical studies of tobacco mosaic
      virus. IV. Influence of single amino acid exchanges on the
      antigenic activity of mutant coat proteins and peptides, Mol.
      Immunol. 17:1205 (1980).

29.   P. Sengbusch and H.G. Wittmann, Serological and physicochemical
      properties of the wild strain and two mutants of tobacco mosaic
      virus with the same amino acid exchange in different positions
      of the protein chain, Biochem.Biophys.Res.Commun. 18:780 (1965).

30.   F. Loor, On the existence of heterospecific antibodies in sera
      from rabbits immunized against tobacco mosaic virus determinants,
      Immunology 21:557 (1971).

31.   M.H.V. Van Regenmortel and N. Lelarge, The antigenic specificity
      of different states of aggregation of tobacco mosaic virus
      protein, Virology 52:89 (1973).

32.   T.J. White, I.M. Ibrahimi, and A.C. Wilson, Evolutionary sub-
      stitutions and the antigenic structure of globular proteins,
      Nature (Lond.). 274:92 (1978).

MONOCLONAL ANTIBODIES MAY BLOCK STERICALLY OR CONFORMATIONALLY THE ANTIGENIC DETERMINANTS OF HUMAN GROWTH HORMONE

J. Ivanyi

Department of Experimental Immunobiology
The Wellcome Research Laboratories
Beckenham, Kent, United Kingdom

Elucidation of the structural specificity of antigenic determinants of proteins has been a subject of extensive studies for various reasons. Although the principle of B-cell-T-cell co-operation in response to hapten- and carrier-like determinants has been firmly established[1], the functional allocation of the respective determinants within natural proteins was attempted only in few instances[2-5]. The specificity of the B-cell response alone is also a subject for dispute in terms of antigenic structures which are immunogenic for B-cells and those which bind to secreted antibodies[6]. The production of heteroclitic antibodies binding to related rather than homologous antigen[7] and the original antigenic sin[8,9] are other examples indicating the difficulties in defining the structural basis of specificity and immunogenicity. It is not clear, whether antigenicity is determined primarily by those structures which differ in sequence from self molecules[10-13] or whether antigenicity is influenced mainly by topographical localization of determinants in the molecule, irrespective of taxonomic distance from the immunized host[14,15]. Recently it was proposed that the location of protein antigenic determinants could be predicted by the position of amino acid sequences with the highest local hydrophilicity[16]. Although this analysis may be applicable to sequential determinants, problems may arise for determinants consisting of residues which are adjacent topographically rather than in linear amino acid sequence[15]. Moreover, protein structure outside the discrete determinant sequences also contributes to antigenic specificity, when considering data on closely related proteins which show antigenic disparity despite a complete sequence homology within the regions of described antigenic determinants[17,18]. It has previously been emphasized that evolutionary replacement or deletion of a few amino acids in homologous proteins has the potential of changing the conformation and consequently also the presentation of distantly situated antigenic sites[19,20].

191

Immunochemical analysis using polyclonal antisera has been the
source of extensive information on antigenic structure of immunoglobulins, myoglobin, hemoglobin, lysozyme, albumin and other proteins
[21-24]. In these studies, the structural allocation of antigenic
determinants has been based on the specificity of binding of antibodies towards chemically modified molecules and particularly to
enzymatically derived peptide fragments. This approach however is
complicated by the fact that polyclonal sera may contain antibodies
produced in response to antigenic fragments which had been degraded
by macrophages as well as those directed against the native molecule.
The specificity of sequential determinants on fragmented antigen may
fully correspond to those expressed by the intact molecule. However,
new fragment-specific determinants when adequately presented by
antigen processing cells to helper or suppressor subsets of T cells
are probably of mandatory importance for the regulation of the immune
respons[3,25-28]. On the other hand, when antibodies are considered
as structural probes for native biologically active macro-molecules,
such as hormones, enzymes or toxins, the heterogeneity of antibody
specificities in polyclonal sera represents a serious limitation to
structure/function studies. Hence, monoclonal antibodies (MAB) represent an advanced analytical tool, particularly suitable for a systematic study of conformational determinants and their topographical
relationship to biologically active protein moieties. Furthermore,
they also carry the potential for the immunological manipulation of
the respective biological functions. In the following paragraphs,
the results of partial analysis of antigenic structure of two related
human somatotropic/lactogenic hormones will be discussed.

ANALYSIS OF THE ANTIGENIC STRUCTURE OF hGH AND RELATED HORMONES
WITH POLYCLONAL ANTISERA

Human growth (hGH) and another two structurally and functionally
homologous molecules, ie. chorionic somatomammotropin (hCS) and
prolactin (PRL) are proteins built from a single polypeptide chain
of about 200 amino acids[29]. The primary structure of hGH shares 86%
amino acids in common with hCS but only 16% homology with hPRL.
Accordingly, polyclonal antisera showed cross-reactivity between
hGH and hCS[30] but no shared determinants with hPRL. Antisera raised
against hGH and hCS cross-react significantly with primate hormones
but only marginally with non-primate species[31-33]. Sera from individual patients under hGH therapy manifested various patterns of reactivity to native or reduced/alkylated hGH[34] and distinct antibody
populations binding either human or pig GH had been reported[35].
Analysis of chemically modified derivatives and proteolytic fragments
of hGH and hCS suggested that antigenicity, as well as the biological
and receptor binding properties, required the intact tertiary structure[34]. Study of mixed hGH-hCS recombinant molecules using cross-
absorbed monospecific antisera showed that the specific antigenic
determinants of either of the two hormones resided within the amino-

terminal part of the polypeptide chain[36]. Synthetic peptides of
various chain size have also been examined, but even the most active
peptide corresponding to sequence 98 - 128 competed with the binding
of anti-hGH antibodies only at 10^3 times higher concentration than
native hGH[37]. Recently, an antiserum raised against the subtilisin
digested hGH was reported to bind with the two chain digest and with
its larger 15K dalton chain, but not with the native molecule of
hGH[38]. The latter results are of particular interest towards further
elucidation of the possible biological role of proteolytically cleaved
hormonal fragments.

PROPERTIES OF MONOCLONAL ANTIBODIES AND FAILURE TO RELATE THEIR SPECIFICITY TO SEQUENTIAL DETERMINANTS

Eight cloned hybridoma cell lines were established by fusion
of NS1 myeloma cells with spleen cells from immunized BALB/c mice[39-41].
Two monoclonal antibodies (QH68 and NA27) were specific for hGH only,
one antibody (NA71) showed also weak binding (by displacement but
not in direct RIA) of hCS, four antibodies (NA39, EB01, EB02, EB03)
reacted equally with hGH or hCS and one antibody (EB04) reacted
specifically with hCS only. All antibodies were of the IgG1 isotype
and in contrast to polyclonal murine antisera, neither of the MABs
was precipitating when tested by double-diffusion either alone or in
mixture. None of the MABs reacted with either human prolactin or
non-primate growth hormones. Various affinity constants for binding
to hGH were determined; antibody NA71 was of the highest affinity
whereas NA27 formed immune complexes of very poor stability[41]. Radio
immunoassay (RIA) analysis using radiolabelled hGH showed that equal
counts were bound in the excess of any of the eight MABs (about 60%
of counts added) when added either alone or in mixture. This result
suggested that the antigenic determinants detected by any of these
MABs are expressed by the same molecules.

Most preparations of hGH display various degrees of charge and
size heterogeneity[38]. We have examined the binding of MABs with the
20K dalton variant of hGH which lacks 15 amino acid residues at the
N-terminal of the polypeptide chain and comprises about 15% content
in pituitary preparations of hGH. In contrast with the original
studies using rabbit polyclonal antisera in which the 20K variant
was reported to carry only about 30% of antigenicity, we could not
find a significant difference between the RIA displacing activities
of 20K and standard hGH[41]. In another study the antigenic potency
of dimeric, trimeric and polymeric fractions were compared with the
monomers prepared from therapeutic grade hGH[42]. Here, the results
showed that antibody NA71 was much less reactive with the dimeric
and trimeric fractions than with the monomer, presumably because of
selective masking of the corresponding antigenic determinant. Four
MABs (QA68, NA27, NA71 and NA39) were used also for the analysis of
a tripeptidyl amidopeptidase fragment of hGH, deleted of the first

33 amino acid residues[43] and for the synthetic 44–128 peptide[37].
However, selective expression or deletion of the MAB-directed determinants could not be demonstrated in either of these preparations[41].
In view of this failure to assign the specificity of MABs to sequential determinants, alternative experiments were pursued in an attempt
to demonstrate the individual combining site specificity of MABs
with native hormone molecules.

DEFINITION OF NATIVE HORMONE EPITOPES ON THE BASIS OF ANTIBODY-ANTIBODY COMPETITION ASSAYS

The initial binding assays of MABs with hGH and hCS indicated
the existence of distinct antigenic determinants within the hormone-
specific and shared portions of these molecules[39]. Further demonstration of specific epitopes expressed by native hormones was
provided by two antibody competition assays: (1) the binding of
^{125}I-labelled MABs to hormone coated polystyrene plates was inhibited
by unlabelled MABs of homologous combining site specificity; (2) incubation of MABs with ^{125}I-hormone inhibited the binding to polystyrene plates which had been coated with MABs of homologous specificity[41,45]. Although various antibodies may be expected to bind to
a determinant by different affinity constants, the results with a
pair of relatively high or low affinity MABs (QA68 and NA27) indicated
that this factor did not interfere with the basic qualifications of
combining site specificities. Taking the results together, as presented schematically in Figure 1, specific epitopes were identified
on hGH (QA68/NA27) and hCS (EB4), one marginally cross reactive site
on hGH (NA71) and two distinct determinants (NA39/EB2 and EB1/EB3)
in the structural moiety shared by hGH and hCS[41,45].

DEMONSTRATION OF CROSS-INHIBITION BETWEEN ANTIBODIES BINDING TO NON-OVERLAPPING DETERMINANTS

The results of the "plate antibody competition test" for eight
MABs are summarized in Table 1. It is apparent that in two instances,
using antibodies specific for non-overlapping epitopes, ie. NA27
versus EB2/NA39 and NA71 versus EB1/EB3, an unexpected partial crossinhibition was observed. This was demonstrable only at 10–100 times
higher cross-inhibitory MAB concentrations using reciprocal antibody
combinations. Since the determinants recognized by EB2/NA39 and
EB1/EB3 are localized within structural moieties shared equally by
hGH and hCS, whereas the determinant bound by NA27 is specific for
hGH and that bound by NA71 is only marginally cross-reactive with
hCS, it appears that the mechanism of the observed competitive inhibitions could not be attributed to direct blocking of antigenic
determinants.

Steric hindrance would seem to be the simplest explanation,
but this appears unlikely when considering that the effect was

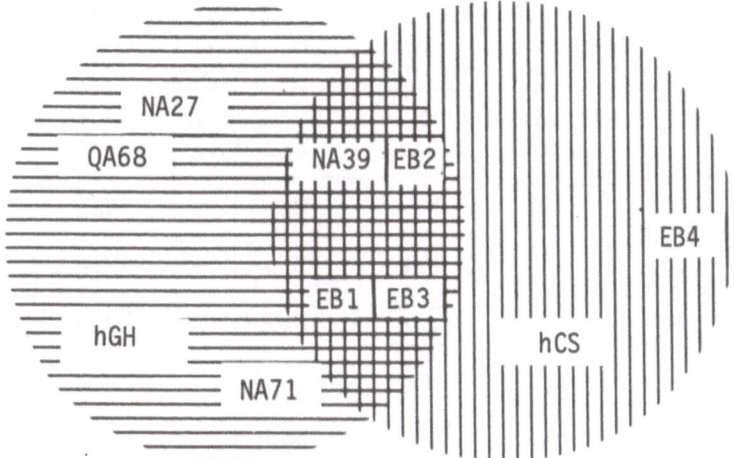

Fig. 1. Schematic diagram indicating the topography of antigenic
         determinants of human growth hormone (hGH) and chorionic
         somatomammotropin (hCS).
         Data are based on the analysis antibody-antibody competitive
         inhibition of antigen binding by radioimmunoassay[41,45].
         The symbols in the diagram correspond to individual mono-
         clonal antibodies.

demonstrable only to NA27 binding, without concomitant inhibition
of antibody QA68 which has an overlapping specificity with NA27.
Therefore "heterologous" inhibitions are tentatively attributed to
conformational changes in the MAB-bound hGH molecule. It is conceivable that this protein hormone with an antigenic structure determined by conformational rather than sequential determinants is
particularly prone to "allosteric" changes when bound in immune
complexes. This interpretation is supported also by the fact that
the heterologous cross-inhibitions were observed only with soluble
immune complexes but not by an alternative assay, in which the
antigen was absorbed to the surface of polystyrene plates[41,45].

     It has been known for a long time that antigen-antibody inter-
actions result in changed Fc-fragment-mediated reactions, but only
recently was it possible to demonstrate directly the conformational
changes in antibody molecules. In particular, the accessibility of
J-chain determinants was taken as a function of Fc conformation and
used to assay the allosteric effects of antigen binding[46]. Probably
the most explicit example of a conformational effect produced by
antibody binding is represented by the activation of defective
β-galactosidase enzymes from mutant strains of Escherichia coli[47].
Subsequent studies showed also the immunological dominance of
enzyme activating, ie. conformational, over the sequential determinants[48]. It was reported recently that the binding of two MABs
of distinct specificity was inhibited by a third antibody and also

Table 1. Competitive inhibition of MAB-complexed hormone
binding to solid-phase bound monoclonal antibodies

| Solid phase bound MABs | MABs mixed with ^{125}I-hormone in soluble phase | | | | | |
	QA68	NA27	NA71	EB2/NA39	EB1/EB3	EB4
QA68	1.2*	2.0	Neg	Neg	Neg	Neg
NA27	0.6	0.7	Neg	2.5	Neg	Neg
NA71	Neg	Neg	1.0	Neg	2.9	Neg
EB2/NA39	Neg	3.0	Neg	2.1	Neg	Neg
EB1/EB3	Neg	Neg	2.9	Neg	2.1	Neg
EB4	Neg	Neg	Neg	Neg	Neg	1.0

Serial dilutions of monoclonal antibody ascites (MAB) were mixed
with ^{125}I-hGH (except for EB4 when ^{125}I-hCS was used), incubated on
wells of polystyrene microtiter plates which had been precoated with
each of the tested MABs and plate-bound radioactivities were determined.

Figures framed in solid lines represent effective steric blocking
by MABs of homologous specificity, whereas figures framed in broken
lines correspond to a pair of MABs where conformational changes of
the antigen in a soluble MAB-complex masked the binding of a heterologous antigenic determinant to the solid phase MAB.

*The figures represent $\log_{10}$ ABT$_{50}$ (ABT$_{50}$ = reciprocal titre giving
50% binding of added ^{125}I-labelled hormone) values of soluble-phase
antibody concentrations giving 90% inhibition of ^{125}I-hormone binding to MAB-coated micrototer plates. These ID$_{90}$ values were inter-
polated from graphs expressing specific antigen binding in the
presence of serially diluted (10^0-10^4 ABT$_{50}$) MABs. Neg = lack of
significant inhibition at 10^4 ABT$_{50}$ MAB dose.

For further details see references 41 and 45.

reciprocally, binding of the third MAB was inhibited by the first
pair of MABs[50]. The authors concluded that the observed crossinhibition could be attributed to antibody-mediated conformational
or steric changes. The results on hGH-binding monoclonal antibodies
correlate the interpretation of experiments on β-galactosidase.
The competitive inhibition assays by which monoclonal antibodies
can determine the topographical relationship of conformational antigenic determinants may qualify also as a valuable tool for further
elucidation of the biologically active sites of hormones and enzymes.

REFERENCES

1. N.A. Mitchison, The carrier effect in the secondary response to hapten-protein conjugates. V. Use of antilymphocyte serum to deplete animals of helper cells, Eur.J.Immunol. 1:68 (1971).

2. G. Senyk, E.B. Williams, D.E. Nitecki, and J.W. Goodman, The functional dissection of an antigen molecule specificity of humoral and cellular immune responses to glucagon, J.Exp.Med. 133:1294 (1971).

3. E.E. Sercarz and D.W. Metzger, Epitope-specific and idiotype-specific cellular interactions in a model protein antigen system, Springer Seminars in Immunopathol. 3:145 (1980).

4. J.A. Berzofsky, L.K. Richman, and D.J. Killion, Distinct H-2 linked Ir genes control both antibody and T cell responses to different determinants on the same antigen, myoglobin, Proc. Natl.Acad.Sci.USA 76:4046 (1979).

5. K. Keck, Ir gene control of carrier recognition. III. Co-operative recognition of two or more carrier determinants on insulin of different species, Eur.J.Immunol. 7:811 (1977).

6. M.J. Crumpton, Protein antigens: the molecular basis of antigenicity and immunogenicity, in:"The Antigens II" (ed. M. Sela), Academic Press, New York, p.1. (1974).

7. F. Loor, On the existence of heterospecific antibodies in sera from rabbits immunized against tobacco mosaic virus determinants, Immunology 21:557 (1971).

8. J. Ivanyi, Recall of antibody synthesis to the primary antigen following successive immunization with heterologous albumins. A two-cell theory of the original antigenic sin, Eur.J.Immunol. 2:354 (1972).

9. I.J. East, P.E.E. Todd, and S.J. Leach, Original antigenic sin: experiments with a defined antigen, Mol.Immunol. 17:1539 (1980).

10. B. Cinader, Sepcificity and inheritance of antibody response: a possible steering mechanism, Nature 188:619 (1960).

11. J. Ivanyi and V. Valentova, The immunological significance of taxonomic origin or protein antigen in chickens, Folia Biologica (Praha) 12:36 (1966).

12. M. Reichlin, Amino acid substitution and the antigenicity of globular proteins, Adv.Immunol. 20:71 (1975).

13. R. Jemmerson and E. Margoliash, Specificity of the antibody response of rabbits to a self-antigen, Nature 282:468 (1979).

14. S. Twining, C.S. David, and M.Z. Atassi, Genetic control of the immune response to myoglobin. IV. Mouse antibodies in outbred and congenic strains against sperm whale myoglobin recognize the same antigenic sites that are recognized by antibodies raised in other species, Mol.Immunol. 18:447 (1981).

15. M.Z. Atassi, Precise determination of protein antigenic structures has unravelled the molecular immune recognition of proteins and provided a prototype for synthetic mimicking of other protein binding sites, Mol.Cell.Biochem. 32:21 (1980).

16. T.P. Hopp and K.R. Woods, Prediction of protein antigenic determinants from amino acid sequences, Proc.Natl.Acad.Sci.USA 78: 3824 (1981).

17. J.G.R. Hurrell, J.A. Smith, P.E. Todd, and S.J. Leach, Cross-reactivity between mammalian myoglobins: linear vs spatial antigenic determinants, Immunochemistry 14:283 (1977).

18. I.M. Ibrahimi, E.M. Prager, T.J. White, and A.C. Wilson, Amino acid sequence of California quail lysozyme. Effect of evolutionary substitutions on the antigenic structure of lysozyme, Biochemistry 18:2736 (1979).

19. A.F.S. Habeeb, Influence of conformation on immunochemical properties of proteins, in:"Immunochemistry of Proteins I" (ed. M.Z. Atassi), Plenum Press, p.163 (1977).

20. T.J. White, I.M. Ibrahimi, and A.C. Wilson, Evolutionary substitutions and the antigenic structure of globular proteins, Nature 274:92 (1978).

21. M. Reichlin, Localizing antigenic determinants in human hemoglobin with mutants: molecular correlates of immunological tolerance, J.Molec.Biol. 64:485 (1972).

22. M.Z. Atassi, The complete antigenic structure of myoglobin: approaches and conclusions for antigenic structures of proteins, in:"Immunochemistry of Proteins II" (ed. M.Z. Atassi), Plenum Press, New York, p.77 (1977).

23. M.Z. Atassi and A.F.S.A. Habeeb, The antigenic structure of hen egg white lysozyme: a model for disulfide-containing proteins, in:"Immunochemistry of Proteins II" (ed. M.Z. Atassi), Plenum Press, New York, p.177 (1977).

24. S. Sakata, R.G. Reed, T. Peters (Jr.) and M.Z. Atassi, Immune recognition of serum albumin. XII Evidence for time-dependent immunochemical cross-reactivity of subdomains 3, 6 and 9 of bovine serum albumin by quantitative immuno-adsorbent titration studies, Mol.Immunol. 18:553 (1981).

25. A.S. Rosenthal, M.A. Barcinsk, and J.T. Blake, Determinant selection is a macrophage dependent immune response gene function. Nature 267:156 (1977).

26. L.R. Richman, W. Strober, and J.A. Berzofsky, Genetic control of the immune response to myoglobin. III Determinant-specific, two Ir gene phenotype is regulated by the genotype of reconstituting Kupffer cells, J.Immunol. 124:619 (1980).

27. I.R. Cohen and J. Talmon, H-2 genetic control of the response of T lymphocytes to insulins. Priming of non-responder mice by forbidden variants of specific antigenic determinants, Eur. J.Immunol. 10:284 (1980).

28. P. Erb, Analysis of the in vitro immune response to insulin. I Primary induction of insulin-specific T helper cells and characterization of the genetic control of the helper cell response to bovine and porcine insulin, Immunology 40:385 (1980).

29. M. Wallis, The chemistry of pituitary growth hormone, prolactin, and related hormones, and its relationship to biological activity. in:"Chemistry and Biochemistry of Amino Acids, Peptides and Proteins 5" (ed. B. Weinstein), p.213 (1978).

30. J.B. Josimovich and J.A. MacLaren, Presence in the human pla-
     centa and term serum of a highly lactogenic substance immu-
     nologically related to pituitary growth hormone, Endocrinology
     71:209 (1962).
31. T. Hayashida, Immunochemical and biological studies with antisera
     to pituitary growth hormones, in:"Hormonal Proteins and Peptides
     3" (ed. C.H. Li), Academic Press, p.41 (1975).
32. A. Tashjian, L. Levine, I.E. Wilhelm, and M.L. Parker, Rabbit,
     guinea pig, rat and human antibodies to human growth hormones:
     immunological reactions with human and non-human primate growth
     hormones, Endocrinology 79:615 (1966).
33. J.P. Gusdon, D.H. Leake, A.H. van Dyke, and W. Atkins, Immu-
     nochemical comparison of human placental lactogen and placental
     proteins from other species, Amer.J.Obstet.Gynec. 107:441 (1970).
34. M.L. Aubert, T.A. Bewley, M.M. Grumbach, and S.L. Kaplan,
     Studies on the relation of molecular conformation of human
     growth hormone and human chorionic somatomammotropin to activity,
     in:"Advance in Human Growth Hormone Research" (ed. S. Raiti),
     Dhew Publ. no.(NIH), p.435 (1974).
35. G. Murphy and E.E. McGarry, Antibodies to porcine growth hormone
     induced by treatment with human growth hormone, J.Clin.Endocrinol
     Metab. 32:641 (1971).
36. S. Burstein, M.M. Grumbach, S.L. Kaplan, and C.H. Li, Immuno-
     reactivity and receptor binding of mixed recombinants of human
     growth hormone and chorionic somatomammotropin, Proc.Natl.Acad.
     Sci.USA 75:5391 (1978).
37. C. Pena, E. Poskus, and A.C. Paladini, A relevant antigenic
     site in human growth hormone localised in sequence 98-120, Mol.
     Immunol. 17:1487 (1980).
38. U.D. Lewis, R.N.P. Singh, G.F. Tutwiler, M.B. Sigel, E.F.
     Vanderlaan, and W.P. Vanderlaan, Human growth hormone - a com-
     plex of proteins, Recent Prog.Horm.Res. 36:477 (1980).
39. J. Ivanyi and P. Davies, Monoclonal antibodies against human
     growth hormone, Mol.Immunol. 17:287 (1980).
40. J. Ivanyi and P. Davies, Monoclonal antibodies to human prolactin
     and chorionic somatomammotropin, in:"Proteins and related sub-
     jects 29 Protides of Biological Fluids" (ed. H. Peeters), in
     press (1981).
41. J. Ivanyi, Analysis of monoclonal antibodies to human growth
     hormone and related proteins, in:"Monoclonal Hybridoma Anti-
     bodies: Techniques and Applications" (ed. J.G.R. Hurrell),
     C.R.C. Press, Inc. in press (1982).
42. T.K. Surowy, M. Daviels, A. Stockell Hartree, and A. Fosten,
     An investigation of the heterogeneity of clinical grade human
     growth hormone, J.Endocrinol. 87:54 (1980).
43. T.W. Doebber, A.R. Divor, and S. Ellis, Identification of a
     tripeptidyl aminopeptidase in the anterior pituitary gland:
     Effect on the chemical and biological properties of rat and
     bovine growth hormones, Endocrinology 103:1794 (1978).

44. J. Ivanyi, Study of antigenic structure and inhibition of activity of human growth hormone and chorionic somatomammotropin by monoclonal antibodies, (submitted for publication).

45. J. Schlessinger, I.Z. Steinberg, D. Givol, J. Hochman, and I Pecht, Antigen-induced conformational changes in antibodies and their Fab fragments studied by circular polarization of fluorescence, Proc.Natl.Acad.Sci.USA 72:2775 (1975).

46. J.C. Brown and M.E. Koshland, Evidence for a long-range conformational change induced by antigen binding to IgM antibody, Proc.Natl.Acad.Sci.USA 74:5682 (1977).

47. M.B. Rotman and F. Celada, Antibody-mediated activation of a defective β-D-galactosidase extracted from an Escherichia coli mutant, Proc.Natl.Acad.Sci.USA 60:660 (1968).

48. F. Celada, J.A. Fowler, and I. Zabin, Probes of β-galactosidase structure with antibodies. Reaction of anti-peptide antibodies against native enzyme, Biochemistry 17:5156 (1978).

49. R.S. Accolla, R. Cin, E. Montesoro, and E. Celada, Antibody-mediated activation of genetically defective Escherichia coli β-galactosidases by monoclonal antibodies produced by somatic cell hybrids, Proc.Natl.Acad.Sci.USA 78:2478 (1981).

AN IMMUNOCHEMICAL STUDY OF THE CHANGES IN CONFORMATION UNDERGONE BY

TWO DOMAINS OF THE TRYPTOPHAN-SYNTHETASE β_2 SUBUNIT UPON ASSOCIATION

Michel E. Goldberg

Institut Pasteur
Unité de Biochimie des Régulations Cellulaires
28 rue du Docteur Roux, 75015 Paris, France

INTRODUCTION

Specific antibodies raised against pure proteins have for long been used as tools to investigate the three dimensional structure of proteins. The structural features which the use of specific antibodies allow to approach are quite varied and can reach very different levels of complexity. For instance a first approach to the comprehension of the three dimensional structure of complex, globular proteins was the demonstration that large fragments isolated from a polypeptide chain can sometimes preserve enough conformational analogy with the initial protein to be recognized by specific anti- bodies directed against the whole native protein. This was shown both with peptides obtained by an in vitro controlled proteolysis of the protein[1] or with fragments synthesized (and hence folded) in vivo by mutants carrying a partial deletion or a chain-termination mutation within the structural gene of the protein[2].

Another type of immunochemical study related to the three dimensional structure of globular protein consists in an "exploration" of the protein surface with specific antibodies to determine which regions of the molecule are accessible to the immunoglobulin (ie. exposed on the outside of the protein). A recent successful example of that approach is the work of Zakin et al.[3] on the organization of the subunits and domains in e.coli aspartokinase-homoserine dehydro- genase I.

The third problem, of general interest in the field of protein conformation, for which immunochemistry has turned out to be quite useful is to detect differences or analogies between distinct con- formational states of a given protein. This approach has been most successful in recent studies on the mechanisms of protein folding to elucidate the pathway of formation of the native structure[4,5,6] as well as in investigations on the allosteric transconformation of a regulatory enzyme[3].

Yet, in the vast majority of these studies, the experiments could give only an "all or none" answer to the questions asked, essentially because a quantitative analysis of the antibody-protein interaction was precluded by the heterogeneity in specificity and affinity of the antisera used. As a result of this situation, the habit has been taken to interpret the results of immunochemical studies on proteins in terms of "absence" or "presence" of a given antigenic determinant and, consequently, in terms of a "denatured" or "native" conformation of that region of the protein which carries this antigenic determinant. This type of interpretation is well illustrated by the beautiful work of Furie et al.[4] in which the immunochemical reactivities of protein fragments either isolated or in the native protein are used to estimate the equilibrium constant between the denatured and native conformations of the isolated fragments. It should be pointed out that this view of the "immuno-reactive" and "non-immunoreactive" states of a protein as corresponding to its native and denatured states fits well with the still widely accepted "two-state" model[7] which, in a first approximation, accounts for the denaturation-renaturation equilibrium of several proteins.

However, in the same way that recent progresses made in the field of protein folding have shown that folding intermediates exist thus rendering the two state model insufficient, the development of more precise and quantitative immunochemical methods make it possible to expect a more refined understanding of the structural features which distinguish immunochemically reactive and unreactive conformations of a protein.

The work reported in this communication shows how a semi-quantitative analysis of a protein-antibody interaction has allowed to distinguish two, very distinct yet well organized, conformations of protein fragments, none of which corresponds to the denatured state.

Functional and chemical properties of the F_1 and F_2 fragments

The protein used in this study is the β_2 subunit of E.coli tryptophan-synthetase, a protein made of two identical β chains (M.W. = 44,000 for each chain). In the presence of its coenzyme, pyridoxal-P, β_2 catalyzes the condensation of indole and L-serine into L-tryptophan. When, in the absence of indole, β_2 is incubated with pyridoxal-P and L-serine, a fluorescent ternary complex, the "aqua-complex", is formed and can serve as a convenient probe of the functional state of the enzyme[8].

Through a mild treatment with trypsin, β_2 can be converted into "nicked-β_2", a modified form of the protein which has nearly the same molecular weight as β_2 and which, though inactive in the tryptophan-synthetase reaction, is able to form the aqua complex in the presence of the coenzyme and serine[9]. Through various criteria[9], including a very sensitive comparison of their immunochemical reactivity[10], β_2 and nicked-β_2 seem to have extremely similar conformations.

It has been shown that the proteolysis results in the nicking of each β-chain into one N-terminal fragment, F_1 (M.W. = 29,000), and one C-terminal fragment F_2 (M.W. = 12,000)[11]. The F_1 and F_2 fragments, which remain associated within nicked-β_2, can be dissociated and isolated by gel filtration in the presence of urea[12]. After separation, they can be "renatured" by dialysing out the urea and, upon mixing again the isolated fragments, they reassociate into a structure which, by all criteria including the ability to form the aqua-complex with pyridoxal-P and serine, appears identical to the β_2 protein[12]. Hence the F_1 and F_2 fragments, which originally were parts of the same polypeptide chain, now behave like two distinct subunits of an oligomeric protein.

Lastly, it should be pointed out that none of the isolated fragments seems able to bind the ligands (pyridoxal-P, indole or serine) of nicked-β_2[13]. It therefore appeared particularly interesting to find out the reasons why the functional properties of nicked-β_2 arise only when the fragments associate. Is it because each isolated fragment fails to renature in the absence of its partner, or because all the binding sites lie at the interface between two domains?

Physical-chemical characterization of the conformations of the isolated fragments

It first was shown that, from the determination of their frictional ratios f/f_o, the two isolated fragments definitely appear as globular, and not denatured, proteins[12]. The F_1 fragment was shown to contain a tryptophan residue and two cystein residues which are shielded from the solvent[12,13] thus confirming the existence of a globular structure. Furthermore, the circular dichroism spectra in the far U.V. of the two isolated fragments are characteristic of proteins rather rich in secondary structure since it could be estimated that F_1 contains about 15% α-helix and 35% β-structure, while F_2 contains about 15% α-helix and 20% β-structure[12].

That these ordered, globular, structures of the isolated fragments are stable and well defined has been confirmed by their relatively high stability and by the very cooperative aspect of the denaturation curves observed both for the thermal inactivation and for the guanidine induced unfolding of the fragments[14,15].

Finally, it was observed that the circular dichroism spectra of the fragments do not change when F_1 and F_2 associate to one another to regenerate the native nicked-β_2 protein, thus showing that no significant modification of their secondary structures occurs upon association[12].

It was therefore concluded, on the basis of all these observations, that the F_1 and F_2 fragments are both able to spontaneously, and independently of each other, acquire the same conformation as that which they have in the native β_2 subunit.

Immunochemical analysis of the conformation of the F_1 and F_2 fragments

In an attempt to more precisely ascertain the conformational analogy between the isolated fragments and the corresponding domains within the native protein, we undertook a comparison of the immunoreactivities of the isolated and associated fragments towards antisera of rabbits hyper-immunized with the native β_2 protein or with the isolated F_1 fragment.

The first conclusion which could easily be reached was that β_2 and the F_1 fragment exhibit a marked cross-reactivity, as judged by complement fixation measurements made with anti-native β_2 or anti-isolated F_1 antisera[10]. However, the F_2 fragment failed to give any reaction with anti-β_2 antisera, at least as judged by complement fixation. Furthermore, it was observed that the total amount of complement fixed by the isolated F_1 fragment was far from accounting for all the immunoreactivity of the intact or nicked β_2 proteins (Figure 1). Hence, it appeared that, in spite of the fact that the isolated fragments are well organized, stable, nearly native structures, their conformations do not allow the expression of most of their antigenic determinants unless the fragments are associated within the nicked protein.

This unexpected result prompted us to undertake a more detailed and quantitative analysis of the antigen-antibody interaction. This was done by direct binding experiments performed with an immunoadsorbent prepared with IgG molecules purified from anti-β_2 antisera. Two important observations were made[10]:

a)   while complement fixation had failed to detect a reaction of F_2 with anti-β_2 antibodies, these direct binding experiments revealed that the majority (about 2/3) of the specific antibody molecules of the antisera used, were indeed directed against antigenic determinants carried by the isolated F_2 fragment;

b)   the average affinities of the heterogeneous antibody population for the F_1 and F_2 fragments were about 10 and 30 fold lower, respectively, than for β_2 (or nicked β_2).

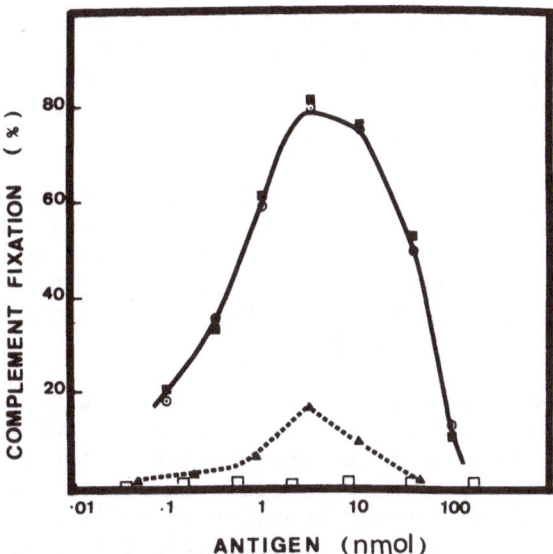

Fig. 1.   <u>Immunological reactivity of β_2, nicked-β_2 and the F_1</u>
<u>and F_2 fragments</u>
Complement fixation was performed as described by Zakin
et al.[10] with an antiserum directed against the native
enzyme.   The antiserum dilution was 1:1000

■ — ■ — ■ β_2

○ — ○ — ○   nicked β_2

▲ — ▲ — ▲ F_1 fragment

□ — □ — □ F_2 fragment

These findings were interpreted as a strong evidence that, upon
associating to each other, the F_1 and F_2 fragments undergo a conformational change which, though too subtle to be detected by conventional physical-chemical methods, is nonetheless important in the
emergence of the immunochemical and functional (see above) properties
of the whole protein.

Comparison of the kinetics of association and isomerisation of
the fragments

The interpretation just above was made according to the assumption that the antigenic determinants, as well as the substrate binding sites, which appear when the fragments associate to one another
are not all located at the interface between a F_1 and a F_2 domain,

and made by the mere juxtaposition of atoms belonging to the two
fragments. This assumption has been verified at least for the active
site, by experiments in which the kinetics of association of the F_1
and F_2 fragments have been compared to the kinetics of reappearance
of the "active site" upon mixing F_1 and F_2.

This has been done in the following way: the isolated, pseudo-
native, F_1 and F_2 fragments have been mixed, in the presence of
pyridoxal-P and L-serine, and the formation of the native functional
nicked β_2 protein has been monitored by recording the fluorescence
of the ternary "aqua-complex" formed between pyridoxal-P, L-serine
and nicked β_2 (Figure 2, curve B).

For the direct determination of the kinetics of association of
the fragments, two fluorescent derivatives of F_1 and F_2 have been
prepared, with the hope that the association of the labeled fragments
might bring the fluorescent groups close enough to each other to
allow for a fluorescence energy transfer signal[16]. The F_1 fragment
has been specifically modified on one particular cysteinyl-residue,
by treatment with N-iodoacetyl-N'-(5-sulfo-1-naphtyl)-ethylene-
diamine. This treatment introduced one fluorescent dansyl group
per F_1 fragment. Similarly, by reacting the F_2 fragment at neutral
pH with fluorescein-isothiocyanate, about 0.2 moles of fluorescein
per mole of F_2 was covalently linked, presumably to the α-amino-
group of the N-terminal residue of F_2. The preparation of these
fluorescent fragments, as well as the experiments which will now be
reported are described in a manuscript by Zetina and Goldberg which
has just been submitted for publication.

It was first verified, by analytical ultracentrifugation, that
these chemical modifications do not interfere with the association
of the labeled fragments. Then, the fluorescence emission spectra
of the modified fragments were recorded, either separately or after
mixing the labeled F_1 and F_2 fragments in stoichiometric amounts.
The wavelength chosen for the excitation (350nm) corresponded to the
absorption maximum of the dansyl-group carried by F_1. It was observed that, after mixing the fragments, the fluorescence intensity
at about 520nm (the emission maximum of the fluorescein group carried
by F_2) increased by more than 60%. This increase could be unambigu-
ously interpreted as an energy transfer from the dansyl- to the
fluorescein-group, occuring when F_1 and F_2 are kept in close proximity
within the associated nicked-protein.

Using this fluorescence energy transfer signal to test the
association of F_1 and F_2, it was then possible to follow directly
the kinetics of reassociation of the labeled F_1 and F_2 fragments
(Figure 2, A). The very striking result obtained is that the reassociation of the fragments by far precedes the appearance of the
functional properties of the nicked protein. A detailed analysis
of the protein concentration dependance of the kinetics of association

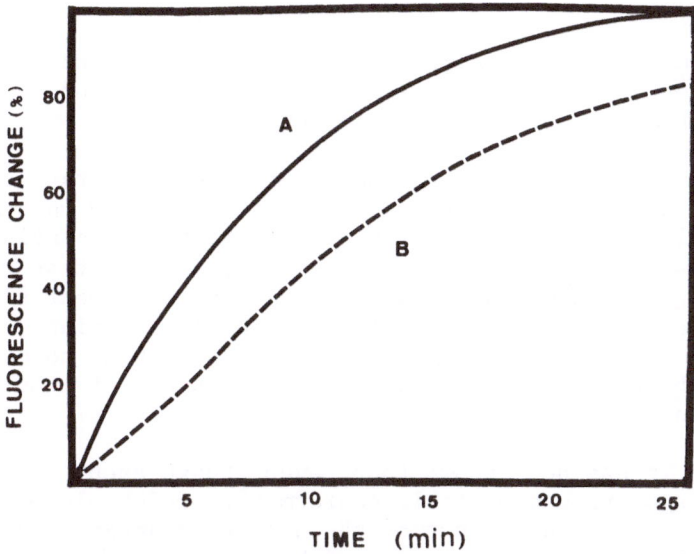

Fig. 2. Kinetics of reassociation and renaturation of the F_1 and F_2 fragments within nicked-

A/ Association of the labeled fragments: dansyl-F_1 (3μM) and fluoresceinated-F_2 (3.2μM) were mixed at 12°C in 100mM potassium phosphate pH 7.8, 2mM β-mercaptoethanol, and 2mM EDTA. The increase in fluorescence ($\lambda_{excitation}$= 350nm; $\lambda_{emission}$= 520nm) was recorded as a function of time (solid line).

B/ Appearance of the "aqua-complex": F_1 (3μM) and F_2 (3μM) were mixed at 12°C in the same buffer as above, supplemented with 0.05mM pyridoxal-P and 100mM L-serine. The fluorescence of the aqua-complex ($\lambda_{excitation}$= 440nm; $\lambda_{emission}$= 500nm) was recorded as a function of time (broken line).

and of formation of the aqua-complex has allowed to conclude that the renaturation of nicked β_2 from the isolated, pseudo-native, fragments proceeds through the following pathway: first, F_1 and F_2 associate rapidly to one another; second, two (F_1F_2) monomers associate to form an "inactive" dimer $(F_1F_2)_2$; finally, this inactive dimer undergoes a slow isomerization to yield the functional nicked β_2 protein.

Whether this isomerization, which is the rate limiting step in the complete renaturation of nicked β_2, is also involved in the recovery of the antigenic determinants of the native protein still remains an open question.

CONCLUSION

The results reported above lead to several important conclusions pertaining to the relations between the conformation of a protein and the "immunochemical signals" produced by that protein.

First, it should be pointed out that, even qualitatively, the "perception" of the signal by antibody molecules very much depends on the experimental method used to detect the protein-antibody inter-action. This has been demonstrated above by showing that the isolated F_2 fragment appeared as "unreactive" by complement fixation while it appeared as "highly reactive" by direct binding experiments. It therefore seems very important, whenever possible, to describe the protein-antibody system in terms of affinities and of number of binding sites.

Second, it has been shown that, even though each of the isolated F_1 and F_2 fragments has a stable conformation which closely resembles that of the corresponding part of the native immunogen (ie. the whole β_2 protein) they react only poorly with specific-anti-β_2 sera. This is specially true of the F_2 fragment which, by complement fixation experiments, failed to give any detectable cross-reaction with β_2. This observation clearly illustrates that a protein may fail to give rise to the "native" immunochemical signals even if its conformation is only slightly different from the native state, and in no way resembles the denatured state.

In this respect, the immunochemical reactivity of a protein appears as an extremely sensitive probe to study minor conformational rearrangements which the classical physical-chemical methods are unable to detect, but which can be crucial for the biological activity and the regulation of the protein molecule.

Although, for reasons mentioned earlier, this tool has presently given essentially quatitative information, it is very likely that the advent of homogeneous monospecific antibodies will lead to a much more precise and quantitative understanding of such essential questions as:

a) how is the protein conformation modified by the binding of specific antibodies?
b) can such "conforming" effects of specific antibody molecules be of general use in modulating the biological function of proteins?
c) to what extent are the functionally defined and immunochemically defined "native" states identical?
d) how cooperative are conformational changes within a protein molecule?

It is hoped that studies presently carried out in our laboratory with monoclonal antibodies directed against native β_2 will bring answers to several of these questions.

REFERENCES

1. C. Lapresle, Etude de la degradation de la serum albumine humaine par un extrait de rate de lapin. II/ Mise en evidence de trois groupements spécifiques differents dans le motif antigenique de l'albumine humaine et de trois anticorps correspondants dans le serum de lapin anti-albumine humaine, Annales de l'Institut Pasteur, 89:654-665 (1955).

2. F. Celada, A. Ullmann, and J. Monod, An immunological study of complementary fragments of β-galactosidase, Biochemistry 13: 5543-5547 (1974).

3. A. Dautry-Varsat and M.M. Zakin, Private communication and unpublished results cited by G.N. Cohen, Limited proteolysis of Escherichia coli aspartokinase I-homoserine dehydrogenase I, in:"Limited proteolysis in Microorganisms", G.N. Cohen and H. Holzer, eds., DHEW Publication nº (NIH) 79-1591, Washington D.C., pp. 209-211.

4. B. Furie, A.N. Schechter, D.H. Sachs, and C.B. Anfinsen, An immunological approach to the conformational equilibrium of Staphylococcal Nuclease, J.Mol.Biol. 92:497-506 (1975).

5. T.E. Creighton, E. Calef, and R. Arnon, Immunochemical analysis of the conformational properties of intermediates trapped in the folding and unfolding of bovine pancreatic trypsin inhibitor, J.Mol.Biol. 123:129-147 (1978).

6. C. Ghelis, M. Tempete-Gaillourdet, and J. Yon, The folding of pancreatic elastase : independent domain refolding and interdomain interaction, Biochem.Biophys.Res.Commun. 84:31-36 (1978).

7. C. Tanford, Protein Denaturation, Adv.Protein Chem. 23:121-282 (1968).

8. M.E. Goldberg, S. York, and L. Stryer, Fluorescence studies of substrate and subunit interactions of the β₂ protein of Escherichia coli tryptophan-synthetase, Biochemistry 7:3662-3667 (1968).

9. A. Högberg-Raibaud and M.E. Goldberg, Preparation and characterization of a modified form of the β₂ subunit of Escherichia coli tryptophan-synthetase suitable for investigating protein folding, Proc.Natl.Acad.Sci.USA, 74:442-446 (1977).

10. M.M. Zakin, G. Boulot, and M.E. Goldberg, Immunochemical study of the β-chain of Escherichia coli tryptophan-synthetase and its proteolytic fragments, Eur.J.Immunol. 10:16-21 (1980).

11. I.P. Crawford, M. Decastel, and M.E. Goldberg, Assignment of the ends of the β-chain of Escherichia coli tryptophan-synthetase to the F₁ and F₂ domains, Biochem.Biophys.Res.Commun. 86:309-316 (1978).

12. A. Högberg-Raibaud and M.E. Goldberg, Isolation and characterization of independently folding regions of the β-chain of Escherichia coli tryptophan-synthetase, Biochemistry 16:4014-4020 (1977).

13. M.E. Goldberg and A. Högberg-Raibaud, Conformational and ligand
     binding properties of the isolated domains from the β_2-subunit
     of Escherichia coli tryptophan-synthetase investigated by the
     reactivity of their cysteines, J.Biol.Chem. 254:7752-7757 (1979).
14. C.R. Zetina and M.E. Goldberg, A comparative study of the
     thermal inactivation of the isolated and associated domains
     within the β_2 subunit of Escherichia coli tryptophan-synthetase:
     evidence for strong interdomain interactions, J.Biol.Chem. 255:
     4381-4385 (1980).
15. C.R. Zetina and M.E. Goldberg, Reversible unfolding of the β_2
     subunit of Escherichia coli tryptophan-synthetase and its pro-
     teolytic fragments, J.Mol.Biol. 137:401-414 (1980).
16. L. Stryer, Fluorescence energy transfer as a spectroscopic
     ruler, Ann.Rev.Biochem. 47:819-846 (1978).

VI: THE CHANGES IN PROTEIN ANTIGEN CONFORMATION INDUCED

BY ANTIBODY

INTRODUCTION

If two or more different kinetically accessible conformations
are compatible with the primary structure of a protein, they can be
envisaged to be in a dynamic equilibrium no matter how much this
equilibrium is displaced in favor of one of them. It is conceivable
that the transition from one to another conformation may be marked
by the disappearance, the new appearance and/or the modification of
conformational antigenic determinants. In each of these cases the
presence of specific antibody is expected not to be neutral, but
rather to influence the conformation by increasing the stability of
one form or inducing/selecting the change to the other form. Therefore, the requirement for an antibody-induced "change in conformation"
is for the antibody molecule to have a higher binding affinity
directed against the less frequently presented form.

There are numerous examples in the literature of these types
of antibody-mediated effect (see, for an early review[1]), starting
with the classical examples by Crumpton[2] where interaction with
anti-apomyoglobin causes metmyoglobin to dissociate and release the
heme and by Sela[3] where antibodies directed towards synthetic amino
acid sequences in helical conformation cause amorphous fragments to
assume a helical shape.

Possibly the most dramatic examples in terms of functional
effects are those involving antibodies elicited against a structurally
"normal" species, which then are confronted with genetically defective
molecules. Here the outcome may be the "cure" of the genetic damage
at the product level, and this may be perceived as new appearance of,
eg., enzyme activity.

211

This kind of phenomenon is illustrated by three articles, all focusing on activation of β-galactosidase, an enzyme for which Zabin et al. (paper 22) have already given background information. By making use of the monoclonal antibody technology each investigation seeks a deeper understanding of the underlying mechanisms. Accolla et al. establish that a single antibody may activate one or more genetically distinct mutants and that the relevant determinants in the mutant are crossreacting but not identical with those in the wild type. The resulting change in affinity in the binding of the activating antibody is found to be 3.8kcalories/mole, a thermodynamic gradient within the range of noncovalent protein bonds. Duncan finds that activation can be the result of cooperative action between two monoclonal antibodies acting on the same β-galactosidase molecule. This finding has a great interest as it can be generalized to other systems where hybridomas are used, and explain previously puzzling results.

Strom et al. assess the binding capacity of the same antibody on mutant molecules in their active and nonreactive forms; as predicted, the antibody discriminates, since the active tetramers are bound more readily, but also inactive molecules can be bound and are subsequently recruited into activity. Thus, two kinetic pathways are followed to achieve the same thermodynamic result.

REFERENCES

1.   F. Celada and R. Strom, Quarterly Rev.Biophys. 5:395 (1972).
2.   M.J. Crumpton, Biochem.J. 100:223 (1966).
3.   M. Sela, Adv.Immunol. 5:29 (1966).

ANTIBODY-MEDIATED ACTIVATION OF GENETICALLY DEFECTIVE ESCHERICHIA COLI GALACTOSIDASES BY MONOCLONAL ANTIBODIES

Roberto S. Accolla, Renata Cina',
Elisabetta Montesoro and Franco Celada

Ludwig Institute for Cancer Research, Lausanne Branch
1066 Epalinges, Switzerland
and Cattedra di Immunologia, Università di Genova
Viale Benedetto XV, 10 16132 Genova, Italy

The effects of antibody binding on protein conformation can be dramatic, and some examples have been known for several years. A great deal of information has been gathered recently from Escherichia coli β-galactosidase, a tetrameric enzyme of M_r 465,000 composed of four identical subunits. By using wild-type enzyme and a battery of mutant products first as immunogens and then as antigens in a series of serological reactions, a number of "conformational effects" have been discovered in this system[1]. These include (a) protection from thermal denaturation; (b) activation of defective enzyme; (c) facilitation of complementation between deletion products; and (d) inactivation of native enzyme. Apparently, with the exception of the first one, in which the antigen and the immunogen used to raise antibody are identical and the immune interaction stabilizes the conformation of the antigen molecule, all other cases involve a change in the tertiary or quaternary conformation of the enzyme. Such a mechanism may require antibodies directed toward conformation-dependent determinants which are capable of interacting with these determinants even when the molecular conformation is modified, or unstable.

The precise mechanism is not yet clear. Following one view, the effect of the binding antibody is to favor, for thermodynamic reasons, the assimilation of the crossreacting determinant to one present in the immunogen. This, in turn, brings about change or stabilization of the antigen conformation to make it similar to the immunogen molecule. Otherwise, it has been hypothesized that antibody shifts the equilibrium between inactive and active molecules by freezing the latter in the wild-type conformation (see Strom et al., this volume). From the dose-response curves it seems that the

213

conformational change or stabilization is accomplished via a single interaction per active center and that this interaction is monovalent. Formal demonstration of the simplicity of the first steps of the conformative action has been obtained by isoelectro-focusing[2] and by the splenic technique[3]. These studies have also uncovered a surprisingly high frequency of mutant-activating antibodies (1 in 4) and of heat-stabilizing antibodies (1 in 5) over the total repertoire of β-galactosidase-specific precursor cells. This finding can be explained by the immunogenic dominance of conformational over sequential determinants as demonstrated in an investigation of the antigenicity of peptides from the enzyme[4]. Recently, some of the conforming antibody-enzyme reactions were attributed to a change in quaternary structure[5].

These findings have increased the interest of conformation dependent antigenic determinants as a model to monitor the phenomenon of subunit polymerization in proteins. Therefore it appeared urgent to dissect the mechanisms involved in mediated conformational changes in β-galactosidase and to this goal it was imperative to use monoclonal antibodies. Hybridomas capable of activating β-galactosidase have been produced independently by Frackelton and Rotman[6] and ourselves[7].

We derived six monoclonal antibodies against β-galactosidase from two separate somatic cell fusions and we studied both their specificity and the relation between binding native β-galactosidase and activating a mutant product.

Table 1 shows the results obtained when three activating hybridomas were confronted with a series of Z-gene products affected by different point mutations. Each hybridoma was able to activate several but not all of the mutant enzymes.

Table 2 shows the quantitative relationship of Ag-binding and activation. In antigen excess, one monoclonal antibody shows similar enzyme binding and mutant-activating capacity. Characteristically, the former reaction has a 200-fold high equilibrium constant, indicating that each antibody molecule is able to bind wild type-β-galactosidase and to activate the mutant, but that the determinant involved is not identical, although it is crossreactive. Taken together these data provide evidence that the enzyme-activation reaction is a single-hit event in which one antibody site favors the correct conformation of one active center of the enzyme. Because each "activating" hybridoma is able to activate several but not all point mutant enzymes tested, it appears that the correction of the genetic defect is produced by binding key sites of the protein three-dimensional structure rather than the sites affected by the mutation.

Table 1. Capacity of different hybridomas to activate
β-galactosidase mutants

Hybridomas*	Activation, -fold[+]				
	627	645	6101	959	918
ZL1-1b	1.2	1.8	1.9	1.2	1.8
ZL2-1b	1.1	1.7	4.6	2.1	1.1
ZL2-2	1.1	1.2	15.4	6.9	6.1
S62[++]	3.1	8.4	4.6	7.8	2.2

*Hybridomas culture fluids were used at concentrations
such that they had the same binding capacity for wild-type
β-galactosidase.
[+]Values are expressed relative to activity found when
normal rabbit serum diluted 1:5 was used. Mutants 627,645,
959 and 918 were described by Messer and Melchers (see ref.1).
[++]Rabbit anti-β-galactosidase antiserum diluted to give the
same binding capacity with wild type-β-galactosidase as the
hybridomas culture fluids.

Table 2. Maximal binding capacity and affinity of AL2-2
monoclonal antibody for wild-type β-galactosidase
and mutant 6101

Antibody*	β-Galactosidase		6101	
	Bound, nmol[+]	Ka,M[++]	Activated, nmol[+]	Ka,M
T3	25.5	ND	1.2	3×10^7
S62	15.6	ND	1.4	3×10^7
ZL2-2	0.0466	96×10^8	0.0685	5.2×10^6

* T3 and S62 are rabbit anti-wild-type β-galactosidase antisera.
+ Values are expressed as nmol of enzyme bound or activated
by 1ml of undiluted antiserum or spent medium.
++ ND, not determinable. Whole antisera contain antibodies
against a number of different sites; thus, it is impossible
to measure binding affinity.

Acknowledgement

This work was supported in part by a grant from The Italian
Research Council (CNR), Progetti bilaterali Italy-USA.

REFERENCES

1. F. Celada and R. Strom, Antibody-induced conformation changes,
    Quart.Rev.of Biophysics. 5:(3)395-425 (1972).
2. G. Kohler, Frequency of precursor cells against the enzyme
    β-galactosidase. An estimate of the BALB/c strain antibody
    repertoire, Eur.J.Immunol. 6:340-346 (1976).
3. R.S. Accolla and F. Celada, Immune responses against the
    β-galactosidase enzyme of E.coli at precursor cell level. I.
    Analysis of the secondary repertoire in BALB/c mice, Eur.J.
    Immunol. 8:688-696 (1978).
4. F. Celada, A.V. Fowler, and I. Zabin, Probes of β-galactosidase
    structure with antibodies. Reaction of anti-peptide antibodies
    against native enzyme, Biochemistry 17:5156-5160 (1978).
5. F. Conway De Macario, J. Ellis, R. Guzman, and M.B. Rotman,
    Antibody mediated activation of a defective β-galactosidase:
    Divμ form of the activable mutant enzyme, PNAS, USA. 75:720-724
    (1978).
6. A.R. Frackelton and M.B. Rotman, Functional diversity of anti-
    bodies elicited by bacterial β-D. galactosidase; monoclonal
    activating, inactivating, protecting and null antibodies to
    normal enzyme, J.Biol.Chem. 255:5286 (1980).
7. R.S. Accolla, R. Cina, E. Montesoro, and F. Celada, Antibody-
    mediated activation of genetically defective E.coli β-galac-
    tosidases by monoclonal antibodies induced by somatic cell
    hybrids, PNAS.USA. 78:2478 (1981).

SYNERGISTIC ACTIVATION BY MONOCLONAL ANTIBODIES OF

β-GALACTOSIDASE FROM GENETICALLY DEFECTIVE E. COLI

R.J.S. Duncan

The Wellcome Research Laboratories
Beckenham
Kent, England

The β-galactosidase (EC 3.2.1.23) proteins synthesised by certain mutant forms of Escherichia coli have little or no ability to catalyze the hydrolyses of galactosides unless activated by treatment with antibodies directed against the enzyme from a strain of E.coli not mutated at the same genetic locus[1,2]. This activation is usually regarded as being mediated by conformational changes in the β-galactosidase protein[1,3] sometimes with the additional assumption that the enzyme protein is constrained to a single active conformation by the antibodies. The results presented here imply that the activation is not a simple correction of a conformational defect, that many distinct active states of the enzyme exist, and that the kinetic mechanism followed during catalysis differs depending on the particular active state conferred by the antibodies.

MATERIALS AND METHODS

β-Galactosidase

The various forms of β-galactosidase used in this study were purified to virtual homogeneity in protein content by methods related to those of Craven et al.[4] from strains of E.coli grown aerobically at 37°C[5]. The purified proteins were stored as precipitates in 50% saturated ammonium sulphate and were stable for many months. In the incubation times of the experiments described in this report the hydrolyses of 2-nitrophenylgalactoside by unactivated forms of the β-galactosidases used were unmeasurably low; moreover the enzymes were stable, with or without activation, under the conditions of the assays and all reaction rates (with the exception of those measured in pre-steady state experiments) were linear for a least 1 hour.

Monoclonal antibodies

    Antibodies directed against normal β-galactosidase were raised
in BALB/c mice and in HO rats. Animals were immunized with purified
β-galactosidase and the splenic lymphocytes from the mice were fused
with P3/NS1-Ag4-1 (NS1) myeloma cells; those from the rats were fused
with either NS1- or 210 RCY 3-Ag 1.2.3 (Y123)-myeloma cells[6]. Colonies of cells which produced antibodies capable of activating the
mutant forms of β-galactosidase were cloned by limiting dilutions
and the isolated, stable monoclonal lines were cultured either in
vitro or in vivo as ascites tumours in syngeneic animals. Hybridoma
BG18 is a rat x mouse hybrid and was grown only in vitro. Antibodies
were harvested from the culture medium or ascitic fluid by precipitation with ammonium or sodium sulphate and purified by chromatography
on DEAE cellulose. The immunoglobulin isotype of each hybridoma line
was determined with isotype specific antisera from Miles Labs. Ltd.,
Slough, U.K. by double immunodiffusion in agar.

    A sample of purified activating antibody (25-60µg) was radio-
iodinated by the iodogen technique[7].

Assay conditions

    All assays and incubations were conducted at 37°C. β-galacto-
sidase was assayed with kinetically saturating concentrations of
2-nitrophenyl (oNPG)- or 4-nitrophenyl (pNPG)-β-D-galactoside in
appropriate buffer systems as described by Roth and Rotman[8] but on
occasion supplemented with 2.5 molal methanol. In order to assay
the mutant forms of the enzyme it was necessary to activate them by
incubation with antibodies raised against the normal enzyme. A
variety of enzyme-and antibody- concentrations were required for the
present study but the usual procedure was to incubate for 20 mins
the enzyme solution and antibody, both dissolved in 50mN-phosphate
buffer pH 7.6 containing 0.5mg normal rabbit globulin and 10 mol
dithiothreitol per ml, and then to initiate the enzymic reaction by
the addition of sufficient substrate solution to dilute the incubation
mixture at least 10-fold.

    The β-galactosidase protein used in this study was, except where
noted, isolated from E.coli lac⁻$_{aba}$[13]. The equilibrium constants
and stoicheiometry of the overall reaction between enzyme and activating antibody BG79 were estimated both kinetically and by a comparative binding technique as described in the legends to Tables and
Figures. Rate constants of the activation by BG79 in the presence
or absence of BG81 were measured by kinetic study of the approach to
a steady state following mixing of the enzyme and antibody.

    Experimental values obtained from equilibrium procedures were
fitted to theoretical equations by methods similar to those of

Duggleby[9]; median values were compared by the Mann-Whitney U-test.
Kinetic constants were obtained from pre-steady state equations of
fitting to theoretical equations using a non-linear least squares
regression method.

RESULTS AND DISCUSSION

Activating antibodies

    Four stable hybridoma lines secreting antibodies which activate
mutant 13 galactosidase were established and their hybrid nature and
immunoglobulin isotype are described in Table 1. Each of the antibodies activate all of the four different purified mutant β-galactosidases used in this study but to different extents, as can be seen
for β-galactosidase from mutant 13 from Table 2. An implication of
the activation of a single protein to different extents by an excess
of different monoclonal antibodies is that a number of active states
must exist with different catalytic properties. This conclusion
could not be reached from experiments with classical antisera because
of the possibility of differing amounts of activation blocking antibodies[2]. Further evidence for the existence of several distinct
active states is also presented in Table 2 where it is shown that
BG79 in combination with another antibody, activates the enzyme
synergistically - hence one active state may be converted to a yet
more active state. This synergistic activation must be due to both
types of antibody interacting with a single molecule of galactosidase
rather than independent activation of two or more different populations of enzyme molecules, for two reasons. Firstly, the combined
activation is synergistic, being greater than the sum of the activations by the individual antibodies (Table 2). Secondly, the enzyme
activated by BG79 plus a second antibody is further stimulated by
methanol (Table 2) whereas when the antibodies are used singly
methanol generally causes inhibition. These activations were true
activations and not merely the result of protection against some
form of enzymic instability as the enzyme is stable under the conditions of assay and as may be seen in Figure 1 the enzyme preactivated by BG81 is activated to a new linear steady state by BG79
(compare lines 1 and 3 of Figure 1). The property of synergistic
activation is restricted to BG79 among our antibodies, no combinations
of antibodies not including BG79 show synergism.

    A second unique property of BG79 which will be fully described
elsewhere is that of inhibition of the enzyme from unmutated E.coli
11,12.

    The stimulation by methanol of the synergistically activated
enzyme is similar to that seen with the normal enzyme from unmutated
E.coli[10] in that it occurs when 2- but not 4-nitrophenylgalactoside

Table 1. Monoclonal hybridoma lines

Designation	Origin	Ig Isotope
BG18	Ho–rat x NSI	rat IgG_1
BG19	Ho–rat x Y123	rat IgG_1
BG79	BALB/c–mouse x NSI	mouse IgG_{2a}
BG81	BALB/c–mouse x NSI	mouse IgG_1

Table 2. Activation of β-galactosidase-13

Antibodies used	No methanol	2.5 molal methanol
BG 18	12 ± 0.3	17 ± 0.3
BG 18 + BG 79	42 ± 1	93 ± 4
BG 19	1 ± 0.1	0.9 ± 0.01
BG 19 + BG 79	11 ± 0.2	17 ± 0.3
BG 81	0.86 ± 0.04	0.2 ± 0.004
BG 81 + BG 79	28 ± 1	46 ± 2
BG 79	7.6 ± 0.4	4.6 ± 0.2
No antibody	< 0.01	< 0.01

The enzyme was incubated for 1 hour with an excess of the
antibodies indicated then 2-nitrophenylgalactoside (8mM,
0.5ml final volume) was added along with, where appropriate,
methanol. The reactions were monitored spectrophotometrically
at 420nm.

is used as the substrate. It is thought that increases in the
catalytic rate of β-galactosidase caused by methanol occur when de-
galactosylation of the enzyme is to some extent rate limiting, but
not otherwise[10]. On this basis the present results show that not
only are different catalytic states available to the enzyme, but
that the detailed reaction pathways of these states may themselves
differ.

Rate constants for activation

Activation of β-galactosidase proteins by classical antisera
is known to have a unimolecular rate limiting step under some conditions of enzyme and antibody concentration[3]. Analysis of curves
such as those in Figure 1 but at a series of antibody and enzyme

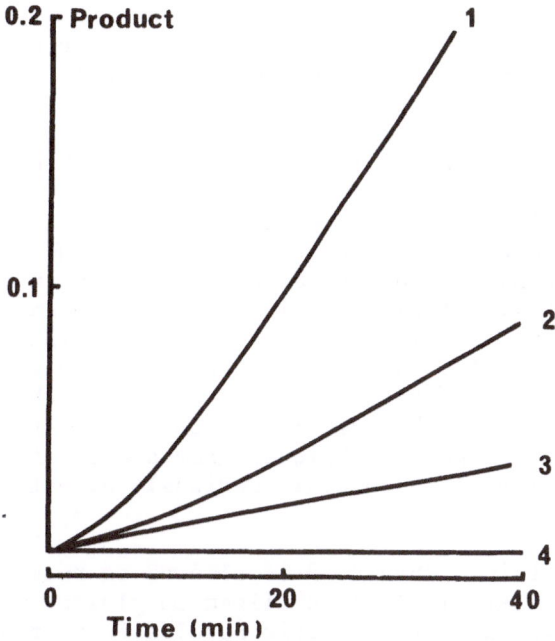

Fig. 1. Pre-steady state of activation by monoclonal antibodies.

1.5 pmol of β-galactosidase-13 was preincubated for 2 hours with tissue culture medium (2,4) or BG81 in the medium (1,3). At time, t, = 0 a solution of 2-nitrophenylgalactoside containing the medium (3,4) or BG79 in the medium (1,2) was added, and the absorbance changes were as shown. Rate constants for the activation by BG79 were obtained by standard analytical techniques from similar curves obtained over a range of BG79 and enzyme concentrations.

concentrations is fully consistent with activation by monoclonal BG79 having a rate limiting unimolecular step with a rate constant of 0.17 ± 0.01^{-1} (19 separate curves analyzed) giving a half-time for activation of about 4.2 mins. If the enzyme has been preactivated by incubation with BG81 in large excess (more than 10 mol BG81 per mol of enzyme) the rate constant for subsequent synergistic activation by BG79 is 0.12 ± 0.01 min^{-1} (10 curves) a value that is significantly less ($p < 0.008$) than that for activation by BG79 alone - ie. this reaction takes longer to occur.

Binding equilibria

The activation of β-galactosidase strain 13 as a function of the concentration of BG79 in the presence or absence of an excess

of BG81 is shown in Figure 2. The binding constant expressing the
overall equilibrium of the reaction, including contributions from
the enzyme-antibody interaction, any conformational equilibria, and
changes occurring as a result of measuring the catalytic activity
of the enzyme, is virtually unchanged by the presence of BG81 (5;
3.1 - 5.9µl with BG79 alone; 3.1; 1.6 - 4.0µl with BG81 plus BG79;
median estimates, 99% confidence limits; 0.05 > p > 0.02), although
the maximum catalytic rate is increased as in Table 2. The method
of Dixon[13] allows determination of the number of binding sites on
an enzyme for a ligand, in this case the activating antibody, to
which it binds tightly. Applying this technique to experiments
similar to those shown in Figure 2 shows that the binding stoicheiometry of BG79 is not altered by the presence of BG81. The values
in this section are presented in terms of µl rather than moles because it is unlikely that the solution of monoclonal BG79 prepared
by classical chromatographic techniques from ascitic fluid is free
from protein contaminants which would seriously effect the calculation
of molarity.

The results obtained when BG79 is allowed to react with β-galactosidase from mutant strain 13 immobilized in plastic wells are
shown in Table 3. As with the kinetic results it is found that the
presence of BG81 does not affect the binding of BG79 - neither in
terms of binding constants nor in binding stoicheiometry. Uniod-
inated BG79 does, however, compete with the labelled BG79 as expected.

The results from these binding experiments imply that preacti-
vation of the β-galactosidase protein by BG81 does not greatly affect
the site to which BG79 binds, for if this were to happen then some
change in the binding constant for BG79 would be seen. At a more
speculative level the results may be interpreted as showing that
combination with an activating antibody does not constrain the mutant
forms of β-galactosidase into a form widely different from that which
occurs predominantly in solution or, as before, changes in the binding
constants should ensue. Obviously the binding sites for BG79 and
BG81 are spatially separated otherwise steric interference would
occur.

SUMMARY

Activation of β-galactosidase proteins by monoclonal activating
antibodies may be interpreted as showing that many activated states
of the enzyme exist, which differ in both catalytic efficiency and
in kinetic mechanisms. Moreover, any changes constrained upon the
protein during activation are not transmitted to distant immunological epitopes, and the rate of activation mediated by binding to
one epitope is only marginally affected by binding at a distant
epitope.

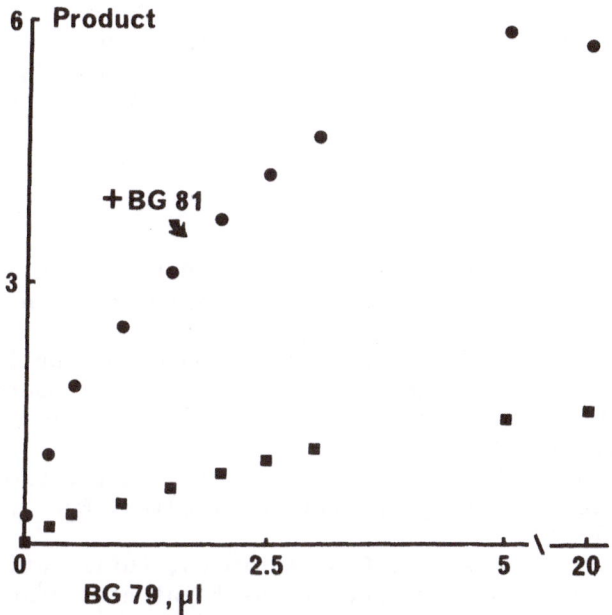

Fig. 2. Kinetic binding studies with BG79.

0.5 pmol of β-galactosidase-13 was incubated with varying concentrations of BG79 in the presence or absence of 10 pmol of BG81. After 2 hours 2 nitrophenylgalactoside (8mM, 0.5ml final) was added and the reaction monitored spectrophotometrically.

Table 3. Constants for the equilibrium between BG79 and β-galactosidase-13

	K_{App}	Max. Binding
BG 81 Absent	8.48 (5.3 − 12.6)	0.48 (0.3 − 0.7)
BG 81 Present	10.2 (4.9 − 14.9)	0.57 (0.4 − 0.9)

Values tabulated are μg of radio-iodinated protein/well.

β-Galactosidase was immobilized in the wells of polyvinyl-chloride microtitre plates and the constants were obtained by standard solid-phase radio-immunoassay techniques in the presence or absence of excess BG81.

I thank Dr. W. Messer of the Max Planck Institute for Molecular Genetics, Berlin, who generously supplied samples of his mutant E.coli; Dr. P.D. Weston for productive discussions throughout this work and Dr. Jane Hewitt who performed the solid phase binding experiments.

REFERENCES

1. F. Celada, J. Ellis, K. Bodlund, and B. Rotman, Antibody mediated activation of a defective β-D-galactosidase, J.Exp.Med. 134:751 (1971).
2. W. Messer and F. Melchers, The activation of mutant β-galactosidase by specific antibodies, in:"The Lactose Operon", J.R. Beckwith and D. Zipser, eds., pp 305-315, Cold Spring Harbor Press. (1970).
3. F. Melchers and W. Messer, The mechanism of activation of mutant β-galactosidase by specific antibodies, Eur.J.Biochem. 35:380 (1973).
4. G.R. Craven, E. Steers, and C.B. Anfinsen, Purification, composition, and molecular weight of the β-galactosidase of E.coli K12, J.Biol.Chem. 240:2468 (1965).
5. L.A. Herzenberg, Studies on the induction of β-galactosidase in a cryptic strain of E.coli. Biochim.Biophys.Acta. 31:525 (1959).
6. J. Hewitt and E. Liew, Antigen-specific suppressor factors produced by T cell hybridomas for delayed-type hypersensitivity, Eur.J.Immunol. 9:572 (1979).
7. P.J. Fraker and J.C. Speck, Protein and cell-membrane iodinations with a sparingly soluble chloroamide, 1,3,4,6-tetrachloro-3a,6a diphenylglycoluril, Biochem.Biophys.Res.Commun. 80:849 (1978).
8. R.A. Roth and B. Rotman, Inactivation of normal β-D-galactosidase by antibodies to defective forms of the enzyme, J.Biol.Chem. 250:7759 (1975).
9. R. Duggleby, A non-linear regression program for small computers, Anal.Biochem. 110:9 (1981).
10. O.M. Viratelle, J.P. Tenu, J. Garnier, and J. Yon, A preliminary study of the nucleophilic competition in β-galactosidase catalyzed reactions, Biochem.Biophys.Res.Commun. 37:1036 (1969).
11. J. Hewitt, R.J.S. Duncan, and P.D. Weston, Hybridoma antibody mediated activation of β-galactosidases from genetically defective E.coli, in:"Protides of the Biological Fluids," M. Peeters, ed., vol 29 in the press. (1981).
12. P.D. Weston and R.J.S. Duncan. Unpublished observations.
13. M. Dixon, The graphical determination of K_m and K_i, Biochem.J. 129:197 (1972).

ANTIBODIES AS PROBES OF PROTEIN STRUCTURE[†]

Irving Zabin, Audrèe V. Fowler
and Franco Celada*

Department of Biological Chemistry
School of Medicine and Molecular Biology Institute
University of California, Los Angeles and
*Cattedra di Immunologia, Università di Genova, Italy

β-Galactosidase of Escherichia coli is a favorite protein for
many different lines of biological enquiry. It is an inducible
enzyme, it is very easy to measure, it can be readily isolated in
pure form, it is an active immunogen, and many "mutant" forms of
the protein, produced by strains with LacZ mutations, are known.
In this paper, immunological studies involving β-galactosidase are
presented. The primary aim has been to use the tools of immunology
to investigate protein structure; another purpose is to use the tools
and techniques available from β-galactosidase biology and chemistry
to help answer basic questions in immunology.

β-Galactosidase is a tetramer containing four identical polypeptide chains. The amino acid sequence has been determined; each
chain contains 1021 amino acids[1]. The sequence was worked out for
the most part by conventional methods of protein chemistry. Peptides
were produced by cleavage of the whole protein with cyanogen bromide,
trypsin, and other reagents; they were isolated, sequenced, and
placed in correct order by inspection of overlapping peptides.

During the course of this work, antibodies were prepared against
a number of cyanogen bromide and tryptic peptides. Peptides were
homogenized in complete Freund's adjuvant without the use of any

[†]Supported in part by U.S. Public Health Service Grant AI 04181
and National Science Foundation Grant 78-1-9974 and a grant of the
Italian Research Council (CNR), Progetti Bilateriali to F.C.
(United States - Italy Co-Science Program, Division of International
Programs, National Science Foundation)

carrier protein and were injected into rabbits. Antibodies were
produced from 18 cyanogen bromide and tryptic peptides ranging in
size from 15 to 96 amino acid residues and representing more than
three-fourths of the polypeptide chain. They were relatively active
antibodies and there was no proportionate relationship between the
size of the peptide and the titer of the resulting antibody[2]
Table 1).

The first use of some of the anti-peptide antibodies was in
primary structure determination. To place peptides in correct order
in a polypeptide chain, and to ensure that none are missing, overlapping peptides from a second digest must be isolated and sequenced.
Sometimes, such overlapping peptides are elusive and are exceedingly
difficult to obtain from protein digests. It occurred to us that an
antibody prepared against one peptide could be expected to bind
another peptide that contains part of the same sequence. Thus, an
anti-peptide antibody can serve as a probe to monitor the purification of an overlapping peptide. An assay procedure was worked
out for this purpose[3]. It is based on inhibition of binding of a
labeled peptide to its antibody. The inhibitory substance is the
overlapping peptide. For the last stages of the β-galactosidase
sequence determination, cyanogen bromide peptides 21 and 22 were
ordered with the help of this method[3].

When the primary structure of proteins is carried out by a
combination of protein and DNA sequencing techniques, ordering of
peptides is of course considerably simplified. Peptides need only
be aligned against the DNA chain. However, anti-peptide antibodies
have been used in several other ways, for example in the study of
proteins produced by mutant strains. Anti-peptide antibodies have
been used to help determine the structure of proteins produced by
deletion mutants[4] and by gene fusions[5]. The method is the same but
instead of searching for an over-lapping peptide, a peptide containing an alteration caused by a mutation is detected by the inhibition assay.

Hybrid β-galactosidase enzymes have been produced by fusions
of a variety of genes to LacZ, the gene specifying β-galactosidase.
Such enzymes contain a "foreign" amino acid sequence in place of
the amino terminal 20 or 25 amino acids of β-galactosidase. The
hybrid proteins retain β-galactosidase activity so that the enzyme
assay can serve as a marker for the substitute sequence. Genes that
have been fused to LacZ include those coding for lac repressor[5]
alkaline phosphatase[6], the maltose binding protein[7], and the lambda
receptor protein[8] of E.coli. The sequences substituted are amino
terminal fragments of as many as 355 [5] and as few as 5 amino acid
residues[6]. Some of these fusions have been prepared to study the
requirements for transport of proteins to various cellular compart-
ments. Leader or signal sequences that are removed from mature
periplasmic and outer membrane proteins are not cleaved from the

Table 1. Anti-Peptide Antibodies

	Peptide		Antibody
designation	size (amino acid residues)	position in β-galactosidase (residue no.)	binding capacity (nMol of antigen per mL of serum)
CNBr2	90	3-92	1.3
T8	81	60-140	1.0
CNBr3	95	93-187	2.3
CNBr4	15	188-202	0.4
T16	20	211-230	4.0
T28-30	36	351-386	8.0
CNBr10	41	378-418	1.0
CNBr14	59	442-500	0.45
CNBr15	40	501-540	2.2
CNBr16	61	541-601	1.7
CNBr18	90	654-743	5.9
CNBr19	23	744-766	1.4
CNBr20	96	767-862	11.9
CNBr21	61	863-923	3.5
CNBr22	43	924-966	6.7
CNBr23	23	967-989	0.6
CNBr24	32	990-1021	5.0

hybrids. Their structures have been determined by protein sequencing methods with the use of anti-CNBr2 to locate peptides containing parts of the foreign sequence.

Another use of the anti-peptide antibodies was to probe the topology of the native protein. The test was carried out by incubation of β-galactosidase with each (rabbit) antibody, precipitation with a second (goat anti-rabbit) antibody, and measurement of enzyme activity in the precipitate and in the supernatant solution. Of eighteen different anti-peptide antibodies, six bound to β-galactosidase[2] (Figure 1). These six antibodies were prepared from peptides derived mostly from the central part of the polypeptide chain. We interpret the results to indicate that sequences in each of the six peptides are exposed on the surface of the enzyme.

For any globular, soluble protein, amino acid residues on the outside of the structure are part of secondary structures such as α-helical segments, β-sheets, turns, or of random lengths of polypeptide chain. Residues distant from each other in the linear sequence may be brought close together in the tertiary structure by

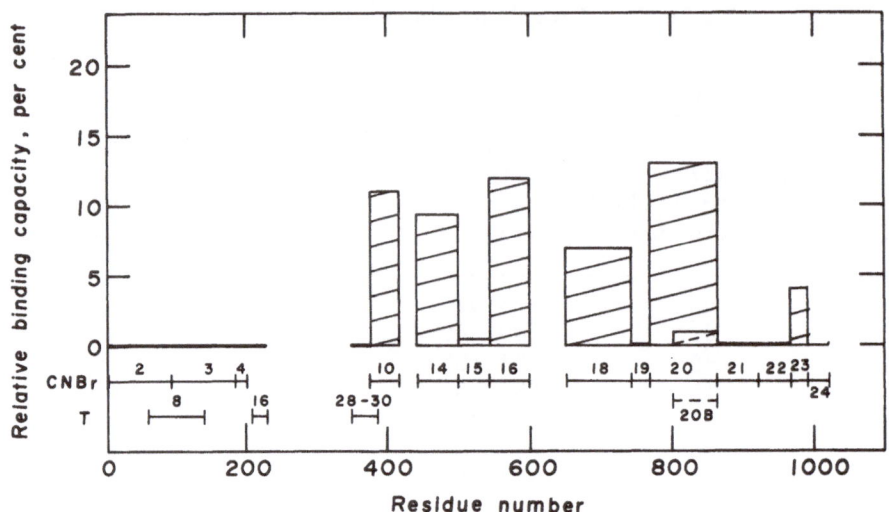

Fig.1.   Relative binding of β-galactosidase by anti-peptide anti-
         bodies. The figure shows a linear map of the polypeptide
         chain of β-galactosidase and the cyanogen bromide (CNBr)
         and tryptic (T) peptides which were used to prepare anti-
         bodies. Solid lines and hatched segments represent the
         binding of antibodies to β-galactosidase as compared with
         the homologous peptide. In an earlier paper[2] anti-CNBr24
         was reported to have a high relative binding to β-galac-
         tosidase. Subsequent experiments indicated that this was
         an error.

the overall folding of the polypeptide chain. For oligomeric proteins, some residues on the surface of subunits become buried within
the multimer. It is difficult to know whether isolated peptides
such as the typical cyanogen bromide peptides of β-galactosidase
have any structure other than random. Examination of the 90-residue
peptide CNBr2 by circular dichroism revealed no secondary structure[9].
Very likely most or all of the peptides have random shapes, although
some secondary and even tertiary conformations could be present for
short times.

     In any event binding of an anti-peptide antibody to the native
protein must indicate that the corresponding sequence makes up a
"sequential" rather than a "conformational" antigenic determinant.
Clearly, conformational determinants are derived from different
parts of the polypeptide and hence cannot be present in relatively
short peptides. This is not to imply that sequential determinants
are not three-dimensional; of course they must be if they are to
fit the antibody combining site.

For the anti-peptide antibodies that did not react with β-galactosidase the negative result can be explained in a number of ways. The corresponding sequence may in fact be buried within the molecule. Or it may be folded in the protein but not in the peptide and antibodies made against the free peptide do not recognize the new shape of that segment. For at least one anti-peptide antibody, the former explanation must be true. Anti-CNBr2 was prepared against CNBr2, residues 3-92. It does not bind to β-galactosidase. However it does bind to a defective β-galactosidase produced by deletion mutant strain M15 or E.coli[10]. The M15 protein lacks amino acid residues 11 through 41 of the wild-type chain and is a dimer rather than a tetramer[11]. Almost certainly anti-CNBr2 binds to M15 protein because all or part of regions 3-10 and/or 42-92 are exposed in the dimer. From other results it is known that anti-CNBr2 does not bind to the segment 3-10 [12], and antigenic determinants for this antibody must be in the segment 42-92.

An experiment with the proteolytic enzyme, trypsin, supports the conclusion that the segment 3-92 (or part of it) is buried within the native enzyme. CNBr2, derived from residues 3-92, is able to complement (restore enzyme activity to) the M15 dimer. This has been termed α-complementation. The α-complemented enzyme, like β-galactosidase itself, is a tetramer[13]. The biological activity of CNBr2 was used as a test to determine whether this part of the polypeptide is exposed or buried within β-galactosidase. Treatment of CNBr2 directly with trypsin results in loss of its ability to complement. When β-galactosidase unfolded with urea was hydrolysed with trypsin, active CNBr2 could not be obtained following treatment with CNBr. However, native β-galactosidase treated with the two reagents under identical conditions yielded fully active CNBr2 [10]. This result indicates that the CNBr2 segment able to complement is unavailable to trypsin in the native enzyme, and confirms the results obtained from studies with anti-CNBr2.

Complementation reactions may be models for the study of protein-protein interactions. Anti-peptide antibodies have been used to clarify some aspects of α-complementation. Anti-CNBr2 interferes with the formation of complemented enzyme from CNBr2 and M15 protein. However once formed, complemented enzyme is not inhibited or dissociated by anti-CNBr2 [10].

α-Complemented enzyme formed from CNBr2 (residues 3-92) and M15 protein, lacking wild-type residues 11-41, contains overlapping or redundant sequences in each protomer. These are the segments 3-10 and 42-92. It is also possible to complement M15 protein with a peptide containing residues 3-41 that is derived from CNBr2 by cleavage with a protease from Staphylococcal aureus. Complemented enzyme from peptide 3-41 and M15 protein no longer contains the 42-92 segment overlap. The power and specificity of antipeptide antibodies as structural probes is illustrated by the fact that anti-

CNBr2 binds α-complemented enzyme formed from CNBr2 and M15 protein, but not that formed from peptide 3-41 and M15 protein[12].

Do antibodies prepared using whole, native proteins as immunogens recognize fragments? The answer seems to depend on the nature, ie. the size and shape, of the fragment. Anti-β-galactosidase binds many prematurely terminated polypeptide chains produced by LacZ nonsense mutants[14], and binds to the omega fragment, the carboxyl terminal third of β-galactosidase[15]. But these fragments have conformations that are similar to that of the whole protein. When CNBr and tryptic peptides from β-galactosidase were tested for binding to anti-β-galactosidase, the results were not surprising. Anti-β-galactosidase did not bind to most peptides. However, it did bind several peptides, CNBr16, CNBr18, and CNBr24 but at titers of less than two per cent that of the homologous antigen. This indicates that anti-β-galactosidase is directed primarily against conformational determinants.

M15 protein binds strongly to anti-β-galactosidase. It has a less stable structure than β-galactosidase as indicated by an increased susceptibility to denaturation by urea or heat[16]. Thus it has properties intermediate between those of a compact, folded protein and the mostly unfolded random structures of free peptides. What kind of antibodies could be produced with M15 protein as immunogen? The answer to this question was surprising. Anti-M15 protein binds most peptides. It also binds both β-galactosidase and M15 protein as expected but the sum of the titers towards the group of peptides is ten times greater than towards M15 protein (F. Celada, D. Centis, J.K. Welply, A.V. Fowler, and I. Zabin, unpublished data).

The most reasonable explanation for these results is that M15 protein has been unfolded and/or degraded in large part at the site of immunogen processing. Most of the antibodies made are therefore directed against sequential determinants. Sufficient M15 protein remains, however, to elicit antibodies towards the conformational determinants in dimeric protein and in β-galactosidase. We conclude that the stability of the protein immunogen must be an important factor in governing the nature of the immune response.

REFERENCES

1. A.V. Fowler and I. Zabin, Amino acid sequence of β-galactosidase XI. Peptide ordering procedures and the complete sequence, J.Biol.Chem. 253:5521-5525 (1978).
2. F. Celada, A.V. Fowler, and I. Zabin, Probes of β-galactosidase structure with antibodies. Reaction of anti-peptide antibodies against native-enzyme, Biochemistry 17:5156-5160 (1978).

3. A.J. Brake, F. Celada, A.V. Fowler, and I. Zabin, An immuno-
    chemical aid to sequence determination of proteins, Analyt.
    Biochem. 80:108-115 (1977).
4. J.K. Welply, A.V. Fowler, J.R. Beckwith, and I. Zabin, Positions
    of early nonsense and deletion mutations in lacZ. J.Bact. 142:
    732-734 (1980).
5. A.J. Brake, A.V. Fowler, I. Zabin, J. Kania, and B. Müller-Hill,
    β-Galactosidase chimeras: Primary structure of a lac repressor-
    β-galactosidase protein, Proc.Natl.Acad.Sci.USA 75:4824-4827
    (1978).
6. A. Sarthy, A.V. Fowler, I Zabin, and J. Beckwith, Use of gene
    fusions to determine a partial signal sequence of alkaline
    phosphatase, J.Bact. 139:932-939 (1979).
7. H. Bedouelle, P.J. Bassford Jr., A.V. Fowler, I Zabin, and J.
    Beckwith, Mutations which alter the function of the signal
    sequence of the maltose binding protein of Escherichia coli.
    Nature 285:78-81 (1980).
8. F. Moreno, A.V. Fowler, M. Hall, T.J. Silhavy, I. Zabin, and
    M. Schwartz, A signal sequence is not sufficient to lead β-
    galactosidase out of the cyto-plasm, Nature 286:356-359 (1980).
9. J.K. Welply, A.V. Fowler, and I. Zabin, β-Galactosidase α-com-
    plementation. Effect of single amino acid substitutions, J.
    Biol.Chem. 256:6811-6816 (1981).
10. F. Celada and I. Zabin, A dimer-dimer binding region in β-ga-
    lactosidase, Biochemistry 18:404-407 (1979).
11. K.E. Langley, M.R. Villarejo, A.V. Fowler, P.J. Zamenhof, and
    I. Zabin, Molecular basis of β-galactosidase α-complementation,
    Proc.Natl.Acad.Sci.USA 72:1254-1257 (1975).
12. J.K. Welply, A.V. Fowler, and I. Zabin, β-Galactosidase α-com-
    plementation. Overlapping sequences, J.Biol.Chem. 256:6804-6810
    (1981).
13. K.E. Langley and I. Zabin, β-Galactosidase α-complementation.
    Properties of the complemented enzyme and mechanism of the
    complementation reaction, Biochemistry 15:4866-4875 (1976).
14. A.V. Fowler and I. Zabin, β-Galactosidase, immunological studies
    of nonsense, missense and deletion mutants, J.Mol.Biol. 33:35-
    47 (1968).
15. F. Celada, A. Ullmann, and J. Monod, An immunological study of
    complementary fragments of β-galactosidase, Biochemistry 13:
    5543-5547 (1974).
16. H. Jörnvall, A.V. Fowler, and I. Zabin, Probe of β-galactosidase
    structure with iodoacetate. Differential reactivity of Thiol
    Groups in wild type and mutant forms of β-galactosidase, Bio-
    chemistry 17:5160-5164 (1978).

IS THERE A PREFERENTIAL PATHWAY FOR

ANTIBODY-MEDIATED ENZYME ACTIVATION?

Roberto Strom[1], Jasna Radojkovic[2],
Paola Minale[2] and Franco Celada[2]

[1]Istituto di Chimica Biologica, Università di Roma
[2]Cattedra di Immunologia
Università di Genova

INTRODUCTION

The phenomenon of activation of a β-galactosidase mutant (AMEF, a product of strain W6101), described by Rotman and Celada[1] has been studied from a kinetic angle by Celada et al.[2], Celada, Strom and Bodlund[3], Celada, Radojkovic and Strom[4]. It was shown that a) the gain of enzymatic activity after the contact with anti-β-gal antiserum is a relatively slow process (Figure 1) and b) the pseudo first order rate constant is asymptotically independent of either antibody (Fab) or AMEF concentration (Figure 2).

These data postulated the existence of a monomolecular step in the rearrangement of the enzyme, but did not indicate whether it preceded or followed the interaction with monovalent antibody.

In the first alternative the active and inactive states of the mutant enzyme molecules would be visualized in equilibrium with each other, and the effect of the added antibody would be to selectively bind the active form. The consequent stabilization would cause a shift of the equilibrium toward the active conformation (scheme 1,a).

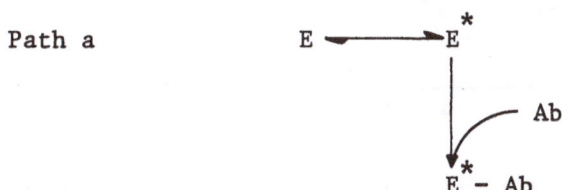

Supported in part by the Italian Research Council (CNR), Progetti bilaterali Italy-USA, and by the Italian Ministry of Education (Progetto di interesse nazionale Immunnologia di base).

The second case could require binding of antibodies also to inactive molecules, followed by the monomolecular process of activation (scheme, 1,b).

Path <u>b</u>

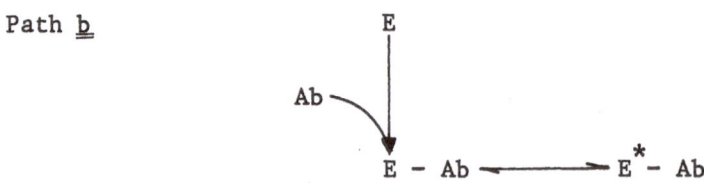

Both pathways were compatible with the results obtained by using rabbit or mouse conventional antisera (directed against wild type enzyme) as the activating agent. However, with these reagents it could not be readily rested whether also non-active AMEF molecules were capable of binding activating antibody, since in the antisera the latter species constitutes a minority in comparison of the antibodies - directed to a number of determinants irrelevant to the activation event - that would bind to both active or nonactive enzyme molecules.

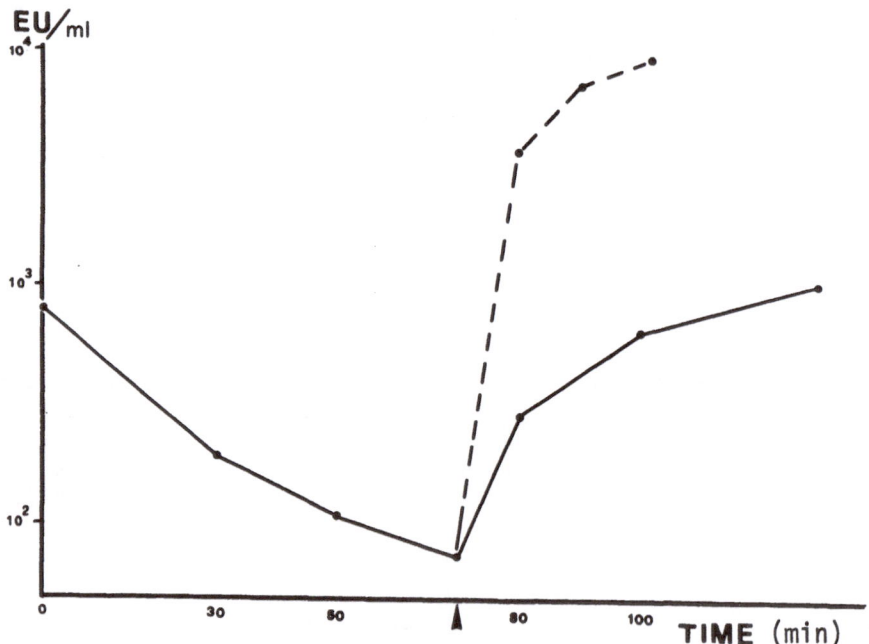

Fig. 1. Decay and rise of the enzymatic activity of an AMEF preparation. At time 0 the mutant enzyme is equilibrated to 37°C. At time 70 min, anti-β-galactosidase antibody is admixed. (●——● monoclonal antibody; ●---● Rabbit polyclonal antibody).

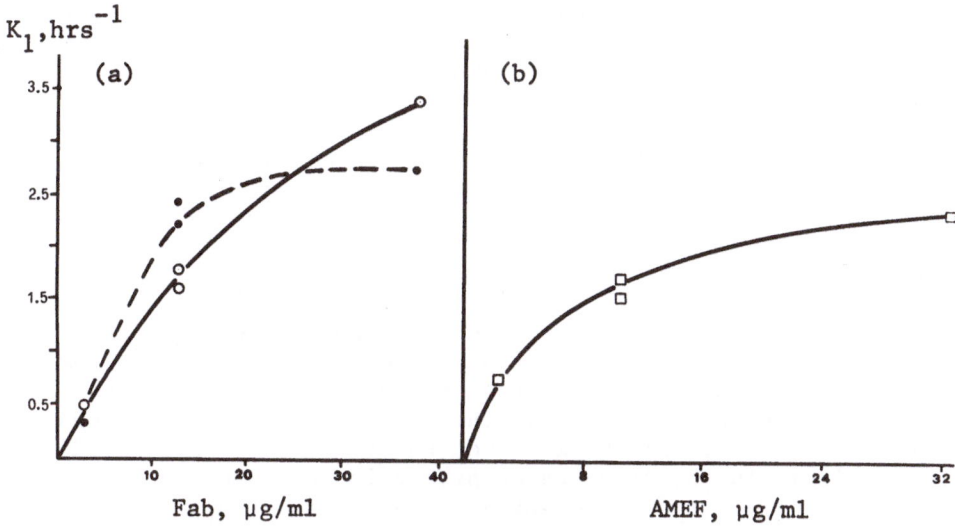

Fig. 2. a) Dependence of the pseudo-first order rate of activation K_1 (expressed as hrs^{-1}) on the concentration of anti-β-gal Fab. AMEF preparation used contained 10.8 (O—O) or 32.4 (●—●)μg protein /ml (final concentration).
b) Dependence of K_1 on AMEF concentration. Anti β-gal Fab final concentration was 12.7μg/ml. The whole experiment was performed at 25°C.

A further aspect of this antibody-mediated activation process involves not only the conformation of single AMEF polypeptide chains, but also their association into tetramers, since the non-activated defective enzyme exists largely in a dimeric form[5].

From the enzymologist's point of view, the activating antibody appears to behave as a positive effector acting essentially on the catalytic step rather than on the substrate-binding ones (as indicated by hyperbolic shape of the velocity vs. substrate curves both in the defective and in the activated enzyme). Pathway a would then be interpreted as falling in the Monod-Wyman-Changeux "concerted" model[6], while pathway b would be more in the line of the Koshland-Nemethy-Filmer "sequential" mechanism[7]. The existence, in the defective enzyme, of a dimer-tetramer equilibrium, shifted to the right by the binding of antibody, would add a further "polysteric" dimension[8].

From the immunologist's point of view the definition of the activation mechanism bears directly on the concept of specificity since a crossreacting determinant is discriminated from the homologous one by a lower binding constant. Flexibility of the antigen could in this case be considered a limitation to the specificity of

the antibody. On the other hand, pathway a would explain the activation by binding only "homologous" conformations, with a specificity high enough to distinguish two shapes of the same genetic
product.

Since there is no basic distinction between the antigen and
antibody partners of the interaction and either one can undergo a
conformational transition, other consequences bear on the concept of
idiotype and consequently on the concept of idiotype network regulation[9].

There is proof that in vivo each idiotype can serve as auto-
immunogen and elicit an anti-idiotype response. There is also ample
evidence that the interaction of the antigen with the paratope may
affect the antigenic properties of the idiotype. If this happens by
means of a conformational change, then the same idiotype, depending
on whether it is in the "free paratope" or in the "bound paratope"
situation may elicit two different anti-idiotypes. In turn, by the
principle of linked transitions , each of these may have a distinct
effect on the binding properties of the antibody; ie. the anti-
"bound idiotype" antibody should favor the binding of the antigen,
while the anti "free idiotype" antibody should decrease the ab-ag
affinity and thus weaken the bond with the antigen. The overall
result may be a further dimension added to the regulatory network.

RESULTS AND DISCUSSION

Several monoclonal antibodies endowed with the capacity of activating β-galactosidase have recently been described by Frackelton
and Rotman[10], Accolla et al.[11], Duncan et al.[12].

In this article we report a series of experiments, aimed at the
definition of the activation pathways, performed with antibody ZL2-2,
one of the hybridoma products obtained by Accolla et al.[11].

ZL2-2 as described previously has the following characteristics:

(a) At infinite ag concentration 1ml of hybridoma fluid binds
    46pmoles of β-gal, and activates 68pmoles of AMEF. From
    these values it can be assumed that each AMEF molecule which
    is bound will be activated.
(b) There is a substantial (200 fold) difference in affinity
    between the β-gal binding and the AMEF activation reactions,
    the respective K_a's being $9.6 \times 10^8 M^{-1}$ and $5.2 \times 10^6 M^{-1}$.

In the present experiments ZL2-2 was used in a solid phase,
having been covalently bound to Sepharose. This was done with the
aim of obtaining a physical separation of bound from free antigen
molecules and directly measuring the distribution of activity between
the two phases.

This approach had also the advantage of simplifying the system by focusing exclusively on the pathways indicated in scheme 1; a possible cycling of antibody molecules is thus kept to a minimum, as well as antibody-mediated association of dimers into tetramers.

With soluble antibody, these association processes may indeed be of importance; in fact the yield of activation was significantly lower with Sepharos-bound than with soluble antibody, when AMEF preincubated at 37°C was used.

As shown in Table 1 both fresh and 37°C preincubated AMEF are retained to a similar extent when passed through a Sepharose ZL2-2 Enzyme molecules which already possessed activity were however preferentially retained. These results demonstrate beyond doubt the validity of path $\underline{a}$ (Figure 1) in the activation; the binding of inactive molecules can however not be neglected, being rather large in absolute amount (lines P of Table 1). The question was therefore raised whether path $\underline{b}$ could co-exist with path $\underline{a}$. As a test to this hypothesis, two treatments with Sepharose-bound antibody were performed in succession on the same AMEF preparation, as shown in scheme 2.

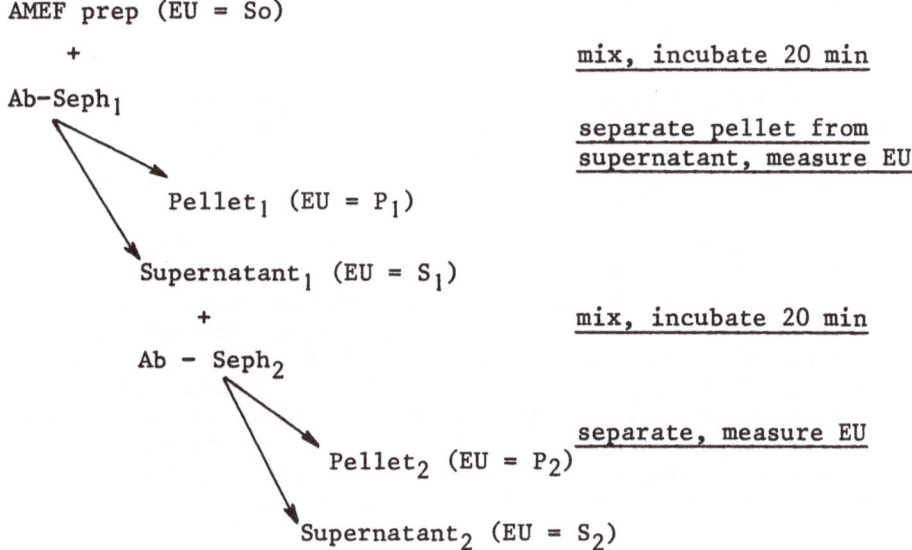

Follow-through of the <u>actual</u> enzyme activity (EU) of an AMEF preparation treated with Sepharose-bound monoclonal antibody <u>in batch</u>.

To evaluate the results, the enzyme activity (EU) associated with the first pellet was compared with the EU lost by the AMEF preparation during the incubation with the first batch of Sepharose-bound ab (ratio $R_1 = P_1/(S_0 - S_1)$). Similarly the activity associated

Table 1.   Activity, Potential (P) and Actual (A) of an
           AMEF preparation passed through a Sepharose-
           bound monoclonal antibody column

		Initial	Eluate	Retained (%)	
Fresh AMEF	P	1200	900	300	(25)
	A	133	17	116	(87)
AMEF incubated at 37°C	P	900	700	200	(22)
	A	28	12	16	(57)

Actual activity (A) of the AMEF preparation was measured
directly by hydrolysis of ONPG (o-nitrophenyl-β-galactoside)
and expressed as $OD_{420nm} \cdot min^{-1}$.

Potential activity (P) was estimated by performing the
determination of the enzyme activity after 2 hrs incubation
at 37°C within optimal amount of hyperimmune rabbit anti-β-
gal serum.

with the second pellet was compared with the loss of activity
suffered by the first supernatant during the incubation with the
second batch of Sepharose-bound ab (ratio $R_2 = P_2/(S_1 - S_2)$).   The
values of three separate experiments are listed in Table 2.

     Had the values of R_1 and of R_2 been equal to unity, this would
have meant that Sepharose-bound ZL2-2 only had the ability of preferentially retaining the AMEF molecules already endowed with some
enzymatic activity(E*).   The antibody-mediated "activating" process
would, then, have consisted solely in a shift to the right of the
$E \rightleftharpoons E^*$ equilibrium, as in path a of scheme 1.   Such a shift, if it
takes place, must be a fairly slow process, since no significant
activity was regained within the time of the experiment, when the
AMEF solution was deprived of the already active enzyme molecules
(column 2 of Table 1; supernatant 1 of scheme 2).

     By contrast, R_2 values significantly higher than unity, such
as found experimentally (Table 2), indicate that Sepharose-linked
ZL2-2, besides binding preferentially pre-existing E* molecules, is
also capable of binding inactive E molecules (as shown on the other
hand by the results illustrated in Table 1) and of activating some
of them.   Moreover, since the values of R_2 are significantly higher
than those of R_1 (which are only slightly greater than unity) it
appears that when the level of preformed activity is low the "re-
cruitment" of an otherwise inactive enzyme appears to prevail over

Table 2. Distribution of actual enzyme activity (as EU) of an AMEF preparation sequentially treated with Sepharose-bound monoclonal anti-β-gal antibody (see scheme 2)

Exp	Initial preparation	First treatment			Second treatment		
	S_0	P_1	S_1	$R_1 = \dfrac{P_1}{S_0 - S_1}$	P_2	S_2	$R_2 = \dfrac{P_2}{S_1 - S_2}$
1	240	420	13	1.84	60	0.2	4.60
2	243	335	19.3	1.49	54	4.4	3.62
3	243	333	28.3	1.54	45	1.1	1.65

a possible increase of the activity of enzyme molecules already endowed with some basal activity.

Therefore, there is indeed antibody-induced activation of inactive enzyme molecules according to b of scheme 1. In fact, since the first step of path a is such a slow process, it seems plausible to hypothesize that the high efficiency of soluble Ab is due to some recycling of the Ab (formation of active E* molecules occurring by dissociation of the E*-Ab complexes generated through path b) rather than directly from a shift of the E ⇌ E* equilibrium of path a. Within this framework, the greater activating capacity of immune antisera as compared to monoclonal antibodies (such as can be seen ie. in Figure 1) could be attributed to a higher efficiency in this recycling process resulting from the lower degree of diffusional freedom existing in a multiliganded complex.

The demonstration of the existence of the kinetic path b prompts a further question: considering that all existing evidence is consistent with the requirement of the tetrameric structure in order to express enzymatic activity, how does the antibody help the achievement of this structure? Does it act indirectly, by introducing a conformational change in the single polypeptide chain, which may in turn favor the polymerization to tetramer, or directly, functioning as some kind of cross-linking agent? At present there does not seem to be a definite answer to this question, but the following facts may be more consistent with the first possibility: a) the basal activity is not increased more than twofold by immobilized monoclonal antibody (experiments of Table 2); this is very modest, since it is known that AMEF enzymatic activity can in optimal conditions the presence of soluble antibody rise up to 500-fold over the initial baseline[1,3]: also allowing for the small fraction of inactive molecules actually bound (17% according to Table 1) the expected rate would be at least 85-fold. b) the Sepharose-bound antibody is quite unable to promote activation of the 37°C-treated AMEF, a condition in which the dimer-tetramer equilibrium, which in fresh AMEF has its mid-point at 20µg protein/ml[5], is shifted towards the dimeric species, and, c) the existing kinetic evidence[4] has never shown a second-order dependence of the activation rate upon the concentration of AMEF.

Because of these hints we feel that the effect of Sepharose-bound antibody can be visualized as favoring the conversion of AMEF protomers into potentially active structures, ready to polymerize. The actual tetramerization however, may be difficult to occur under these experimental conditions. This may limit the actual activation achieved since the appearance of the enzymatic activity requires the existence, the completion or the stabilization of tetramers.

REFERENCES

1. M.B. Rotman and F. Celada, Antibody-mediated activation of a defective β-D-galactosidase extraction from an Escherichia coli mutant, Proc.Nat.Acad.Sci.U.S.A. 60:660 (1968).
2. F. Celada, J. Ellis, K. Bodlund, and M.B. Rotman, Antibody-mediated activation of a defective β-D-galactosidase. II. Immunological relationship between the normal and the defective enzyme, J.Exp.Med. 134:751 (1981).
3. F. Celada, R. Strom, and K. Bodlund, Antibody mediated activation of a defective β-galactosidase (AMEF). Characteristics of binding and activation processes, in: "The Lactose Operon," J.R. Beckwith and D. Zipser eds., p.291 Cold Spring Harbor Laboratory. (1970).
4. F. Celada, J. Radojkovic, and R. Strom, Antibody mediated activation of β-galactosidase mutants and complementary fragments, J.de Chimie Physique 71:1007 (1974).
5. E. Conway de Macario, J. Ellis, R. Guzman, and M.B. Rotman, Antibody mediated activation of a defective β-D-galactosidase: dimeric form of the activable mutant enzyme, Proc.Natl.Acad. Sci.U.S.A. 75:720 (1978).
6. J. Monod, J. Wyman, and J.P. Changeux, On the nature of allosteric transitions: a plausible model, J.Molec.Biol. 12:88 (1965).
7. D.E. Koshland, G. Nemethy, and D. Filmer, Comparison of experimental binding data in proteins containing subunits, Biochemistry N.Y. 5:365 (1966).
8. A. Colosimo, A. Brunori, and J. Wyman, Polysteric linkage, J.Molec.Biol. 100:47 (1976).

9. N.K. Jerne, Towards a network theory of the immune system, Ann.Immunol.(Inst.Pasteur) 125c:353 (1974).

10. A.R. Frackelton and M.B. Rotman, Functional diversity of antibodies elicited by bacterial β-D-galactosidase; monoclonal activating, inactivating, protecting and null antibodies to normal enzyme, J.Biol.Chem. 255:5286 (1980).

11. R.S. Accolla, R. Cinà, E. Montesoro, and F. Celada, Antibody-mediated activation of genetically defective E.coli β-galactosidases by monoclonal antibodies produced by somatic cell hybrids, Proc.Natl.Acad.Sci.U.S.A. 78:2478 (1981).

12. J. Duncan, these proceedings.

VII: THE RECOGNITION BY T CELLS OF PROTEIN ANTIGENS

GENERAL INTRODUCTION

It was realized quite early[1], that contrary to the situation in the B cell universe in which reduced and native proteins are completely non-cross-reactive, denatured protein antigens cross-react very well with native proteins in delayed-hypersensitivity responses. This has been a consistent finding subsequently in many T cell systems using a variety of antigens. It appeared that T cells were restricted therefore to "seeing" continuous or sequential antigenic determinants and in most systems, it could be concluded that reduction and fragmentation of antigen were prerequisites to activation.

In 1973 came the papers of Rosenthal and Shevach[2] showing that activation of proliferative T cells required presentation of antigen in conjunction with self-MHC. The experiments of Zinkernagel and Doherty[3] complemented this work in demonstrating that cytotoxic T cells also recognized their specific target antigens in conjunction with MHC structures. Actually, a major paradigm of immune recognition was hereby overturned: whereas earlier it had been assumed that the system responded to "foreign" antigens, it was now clear that an element of self-recognition was inherent in T cell activation.

MHC restriction experiments dominated the remainder of the decade, setting forth a puzzle which seemed to defy final analysis, in the absence of structural data on the T cell receptor: does the T cell see antigen in the context of MHC restriction elements with a single receptor or are there separate receptors for "nominal antigen" and self-structures? The subject matter of the following ten papers is quite homogeneous, representing an effort to define what it really is that the T cell recognizes. There has been a growing realization that the problem of T cell specificity reflects the union of three distinct elements in a ternary complex: the determinant on the antigen, a "restriction element" on MHC molecules, and the T cell receptor itself. The division of papers into three subgroups utilizes this trinity in forming somewhat arbitrary and overlapping categories but in fact the focus of these reports shifts from (A) the antigen to (B) MHC to (C) T cell receptor.

REFERENCES

1. P.G. Gell and B. Benacerraf, Delayed hypersensitivity to simple protein antigens, Adv.Immunol. 1:319 (1961).
2. A.S. Rosenthal and E.M. Shevach, Function of macrophages in antigen recognition by guinea pig T lymphocytes. I. Requirement of histocompatible macrophages and lymphocytes, J.Exp.Med. 138: 1194 (1974).
3. R.M. Zinkernagel and P.C. Doherty, MHC-restricted cytotoxic T cells: Studies on the biological role of polymorphic major transplanation antigens during T-cell restriction specificity, function and responsiveness, Adv.Immunol. 27:52 (1979).

VII A.   WHICH ARE THE ANTIGENIC DETERMINANTS

THAT T CELLS RECOGNIZE?

In this first group, the authors have focused on the nominal antigen and have sought to define epitopes which are utilized by the T cell. The clear distinction is made between those areas of the peptide molecule interacting with the T cell receptor and those areas which react with the MHC product on the antigen-presenting cell.

In the work of David Thomas and his colleagues, small peptide antigen systems were examined: the 14-residue human fibrinopeptide B, and the octapeptide hormone angiotensin II. The critical role of a single amino acid residue (arg-14 in the fibrinopeptide system) seemed to be related to macrophage Ia antigen expression, and in this work the effort is made to define both of the sites predicted to exist. From these results it appears that octapeptides are the smallest immunogenic peptide entities, presumably bearing both required sites.

In the paper by David Benjamin et al., the large protein, bovine serum albumin, is depicted as a multi-domain molecule. The domains subdivide into two antigenic regions each of which contains determinants recognized by T cells, B cells and Ir gene products (or MHC molecules). Their conclusion is that although the Ir-recognition sites may be conserved among different domains, T cell and B cell determinants are not only unique on each domain but different from each other.

Why are some domains of a protein immunodominant for the T cell? In studies with lysozymes and cytochrome Cs, this could be attributed to a limited number of MHC recognition sites shared on a series of species variants, so that non-crossreactive T cells

245

raised against two of these variants recognize the identical part
of the molecule. Berkower and his colleagues have reported a similar
finding with myoglobin, and have also suggested the likelihood that
a strong immunogenicity for T lymphocytes may be inherent in particular amino acid residues.

The possibility arises in the report by Corradin et al. that T
cells may indeed be able to bind directly to a small peptide of apocytochrome c in the absence of MHC involvement. These data remind
us that binding studies of T lymphocytes are perhaps better probes
of recognitive potential than activation studies.

Many surprises remain, but it appears that these and other
studies on defined antigens will soon yield a detailed picture of
the boundaries of T cell recognition.

T LYMPHOCYTE RECOGNITION OF SMALL PEPTIDE ANTIGENS: FORMATION OF NEOANTIGENIC DETERMINANTS AND CLONAL FREQUENCY RESULTING IN IR GENE CONTROL

David W. Thomas, Michael D. Hoffman
Kun-hwa Hsieh and George D. Wilner

Department of Microbiology and Immunology
The University of Michigan Medical School
Ann Arbor, Michigan 48109, and
The Department of Pathology and Laboratory Medicine
The Jewish Hospital of St. Louis
St. Louis, Missouri 63110

Efforts to understand the nature of T lymphocyte recognition of exogenous antigens have been complicated by the findings that antigen must be associated with stimulator cells, which in some cases are macrophages[1-3]. Moreover, it is clear that helper or proliferating T cells recognize exogenous antigenic determinants only in association with stimulator cell Ia antigens, and that Ia antigens are responsible for Ir gene functions[4-8]. Hypotheses to explain these observations are that either exogenous antigen is physically complexed with Ia antigens and recognized by a single T cell receptor, or that Ia and exogenous antigens are maintained separately on the stimulator cell surface and interact with two distinct T cell receptors. The approach we have taken to distinguish between those possibilities utilizes a series of closely related small synthetic peptide antigens. Our intent was that by precisely defining the antigenic determinants involved in the recognition process, it may be possible to detect those areas of the peptide which interact with clonally distributed T cell receptors, and those areas, if any, that potentially interact with macrophages. Thus by understanding how an exogenous antigen is being "seen" may allow better definition of those cellular structures with which the antigen interacts. Using this approach, two peptide antigen models were developed: the first is based on the 14 residue peptide human fibrinopeptide B(hFPB, Bβ1-14), and the second is based on the octapeptide hormone angiotensin II (AII). In this article we will summarize some of the results obtained in both of these systems and discuss their implications for T cell recognition and Ir gene control.

247

In initial studies strain 2 and strain 13 guinea pigs were immunized with hFPB or various sequential hFPB homologues and responsiveness determined by <u>in vitro</u> T cell proliferation, as shown in Table 1. It is clear from these results that T cell responses to native hFPB are under apparent Ir gene control and that strain 2 are responders and strain 13 are nonresponders. However, by removing the carboxyl terminal Arg^{14}, as in $B\beta 1$-13, the pattern of responsiveness is reversed and strain 13 are now responders and strain 2 nonresponders. Further experiments using immune $(2 \times 13)F_1$, T cells stimulated by antigen-pulsed parental macrophages showed that only strain 2 macrophages presented $B\beta 1$-14, and only strain 13 macrophages presented $B\beta 1$-13 [9]. Therefore, the critical role that Arg^{14} plays in Ir gene control is intimately associated with macrophage Ia antigen expression. The other important information shown in Table 1 is that the smallest immunogenic and antigenic peptide is the octapeptide $B\beta 7$-14, which was additionally found to contain the major antigenic determinant involved in T cell responsiveness[10].

To more carefully define the role Arg^{14}, as well as other peptide residues, serve in T cell recognition of hFPB a series of hFPB analogues were synthesized which differed from the parent molecule by only a single residue. Each of these analogues was tested for immunogenicity and antigenicity for guinea pig T cell responses, as shown in Table 2. Results obtained with these analogues show that single residue substitutions produce distinct effects on strain 2 T cell responses: 1) substitutions that have no effect (position 5); 2) substitutions that destroy or alter immunogenicity (positions 11 and 7-8); and 3) substitutions that create a unique antigenic specificity (positions 14 and 10). The fact that substitutions for Phe^{10} and Arg^{14} dramatically alter the antigenic specificity of T cell responses makes it likely that these two residues make a major contribution to the overall specificity of T cell recognition. The implication of these results is that Phe^{10} and Arg^{14} may serve as contact residues for clonally distributed T cell receptors. The primary structural features of these residues that are responsible for the fine specificity of the T cell clonal response seem to be the side chains. Thus, subtle changes in the para position of the Phe^{10} and the ϵ-amino of Arg^{14} each create unique antigenic species for T cells. Since it is unlikely that these substitutions alter the conformation of these small peptides, the specificity is probably directed against side chains rather than specific for a change in overall configuration.

Unlike those substitutions in positions ^{10} and ^{14} which alter antigenic specificity, substitutions for Phe^{11} result in analogues that are neither antigenic nor immunogenic. This is particularly curious since those same substitutions for Phe^{10} were made for Phe^{11}. One explanation for this result may be that the -OH and $-NO_2$ substitutions add bulk that cannot be accomodated within the antigen combining site. Alternatively, Phe^{11} may serve as a "fixed" residue that contacts more conservative regions of a T cell or macrophage

Table 1. Relative Antigenicity and Immunogenicity of hFPB and hFPB homologues in strain 2 and strain 13 guinea pigs

Antigen		T Cell Proliferation Guinea Pig	
		Strain 2	Strain 13
hFPB, Bβ1-14	PCA^1-Gly^2-Val^3-Asn^4-Asp^5-Asn^6-Glu^7-Glu^8-Gly^9-Phe^{10}-Phe^{11}-Ser^{12}-Ala^{13}-Arg^{14}-OH	+++	-
Bβ3-14	H_2N-Val—————	++	-
Bβ5-14	H_2N-Asp—————	++	-
Bβ7-14	H_2N-Glu—————	++	-
Bβ9-14	H_2N-Gly—————	-	-
Bβ1-13	—————Ala-OH	-	+++

Table 2. Effect of single residue substitutions on strain 2 guinea pig T cell responses to hFPB

Analogue	Effect of Residue Substitution on T Cell Responses
Bβ5-14 H_2N-Asp^5-Asn^6-Glu^7-Glu^8-Gly^9-Phe^{10}-Phe^{11}-Ser^{12}-Ala^{13}-Arg^{14}-OH	
——————— Lys	Each analogue is a unique antigen and primes non-cross-reactive T cell clones.
——————— Lys (Tfa)	
——— Tyr ———	These analogues are neither antigenic nor immunogenic.
——— Phe(4-NO$_2$) ———	
——— Tyr ———	Each analogue is a unique antigen and primes non-cross-reactive T cell clones.
——— Phe(4-NO$_2$) ———	
——— Gln-Gln ———	Immunogenicity is altered, but T cell specificity is not affected.
Asn ———	Analogue behaves the same as Bβ5-14.

receptor. In this light it is of interest that the analogue containing Tyr[11] instead of Phe[11] was somewhat immunogenic in strain 13 guinea pigs (responders to the copolymer of Glu and Tyr), whereas all other analogues were nonimmunogenic[11].

The results presented above indicate that the antigenic specificity of T cell responses is for amino acid side chains and suggest that conformation may have little role in T cell recognition. Indeed, other studies using denatured proteins have shown that T cells recognize the same determinants in both denatured and undenatured antigen[12-14]. These results again suggest that T cells recognize mainly the primary sequence of protein antigens and that conformation is less important. One explanation for these results is that macrophages degrade larger proteins to small peptide fragments that represent the moiety recognized by T cells, and that the same fragments are produced from denatured or undenatured proteins. If this is correct, then one would hypothesize that inverting the residue sequence of a small peptide antigen would have no effect on its antigenicity, since the same residues are present in the same spatial order and the side chains are identical. To test this prediction, an inverted Bβ7-14 peptide sequence was synthesized and tested for antigenic identity with Bβ7-14, as shown in Table 3. It is clear from these results that inverting the sequence of Bβ7-14 results in a peptide that is neither immunogenic nor antigenic for strain 2 T cells. Therefore, not only are the amino acid side chains critical for T cell responses, their directed ordering is essential. One explanation for this result is that the conformation is altered by inverting the primary sequence and that T cells recognize overall shape as well as side chains. However, since the overall charge and chemical properties of these peptides are identical, it is less likely that an altered conformation is responsible for this difference, particularly since Bβ1-14 shows only random conformation[15]. Therefore, we favor the interpretation that T cells exhibit polarity in the "reading" of peptide antigens. The simplest explanation for this is that the peptide is "fixed" with reference to self and that the T cell receptor "reads" from the fixed self association points from one end of the peptide. This would be consistent with formation of a neo-antigenic determinant created by the interaction of exogenous antigens with "self" moieties in some fashion. This association may be either a covalent or noncovalent attachment of peptide antigens to a "self carrier", which may or may not be the Ia antigens themselves.

If the interpretation that exogenous peptides become associated with self for presentation to T cells is correct, it is likely that for Bβ7-14 this occurs via the carboxyl terminus. The reasoning for this is that: 1) altering or deleting residues from the amino terminus of hFPB has no effect on the antigenicity or fine specificity for T cell responses to Bβ7-14; and 2) altering or deleting the carboxyl terminal residue Arg[14] has profound effects on both the

Table 3.   T cell responses of strain 2 guinea pigs to Bβ7-14 and inverted Bβ7-14

		T Cell Proliferative Response	
Immunizing Antigen		Bβ7-14	Inverted Bβ7-14
Bβ7-14	H$_2$N-Glu^7-Glu^8-Gly^9-Phe^{10}-Phe^{11}-Ser^{12}-Ala^{13}-Arg^{14}-OH	+++	—
Inverted Bβ7-14	H$_2$N-Arg -Ala -Ser -Phe -Phe -Gly -Glu -Glu -OH	—	—

specificity of T cell responses and on Ir gene control. If this is
the case, it is possible that T cell responsiveness may be manipulated
by further changes at the carboxyl terminus. Accordingly, a Gly
residue was added to the carboxyl end of Bβ7-14. Since this residue
does not contain a major side chain, it should serve primarily as a
spacer such that any attachment to self via the carboxyl end would
be different than for native Bβ7-14. As shown in Table 4, addition
of the carboxyl terminal Gly converts Bβ7-14 into an immunogenic and
antigenic peptide for strain 13 T cell responses. Thus, strain 13
animals are unresponsive to Bβ7-14, but show intermediate T cell
proliferative responses to Bβ7-15 following immunization with that
antigen. In strain 2 guinea pigs Bβ7-15 is antigenically indistinguishable for Bβ7-14. It would therefore seem that for strain
13 T cell recognition, the addition of a single Gly residue to the
carboxyl end of Bβ7-14 changes Ir gene control. According to the
preceeding hypothesis, this may indicate that the spacing of Bβ7-14
relative to "self" is essential to create antigenic determinants
capable of being recognized by T cells. This interpretation suggests
that Ir gene control may function at several levels. At the first
level, it is possible that for some peptide antigens there may be no
association with self macrophage components and consequently, no
immunogenic moiety is created. Secondly, even though an appropriate
immunogenic moiety is formed between exogenous peptides and macrophages, there may be few or no T cell clones with the complementary
recognition structures for those particular antigenic determinants.
An example of this latter possibility was obtained in the angiotensin
system, as shown in Table 5. Strain 2 animals are responders to AII,
but AIII, which lacks the amino terminal Asp, is neither antigenic
nor immunogenic for strain 2 T cell proliferative responses. Strain
13 animals, on the other hand, are unresponsive to AII by T cell
proliferation following immunization with that antigen, but do show
significant proliferation with AII, as well as AIII, after priming
with AIII. The implication of this result is that strain 13 macrophages can create an immunogenic moiety with AII, and that the lack
of T cell proliferation resides with the responder cells in some
manner. This was substantiated in experiments in which $(2 \times 13)F_1$
animals were immunized with AII or AIII and challenged with AII or
AIII pulsed parental macrophages, as shown in Table 6. Again, it
can be seen that T cells from AIII-immune $(2 \times 13)F_1$ animals respond
to AIII-pulsed strain 13, but not strain 2 macrophages. In addition,
AIII-immune T cells also showed significant proliferation with AII-
pulsed strain 13 macrophages. By contrast, AII-pulsed strain 13
macrophages failed to stimulate AII-immune $(2 \times 13)F_1$ T cells. It
is therefore clear that strain 13 macrophages are capable of creating
an immunogenic moiety with AII, but that these antigenic determinants
are evident only after priming T cells with AIII. The lack of responsiveness of AII-immunized strain 13 T cells could be due to either
suppression or a low clonal frequency of AII-specific T cells. Since
we have been unable to demonstrate suppression in experiments in
which AII and AIII-immune strain 13 T cells were cultured together

Table 4. Guinea pig T cell responses to Bβ7-14 and Bβ7-15

Immunogen			Relative T Cell Proliferation Antigen in Culture	
		Guinea Pig	Bβ7-14	Bβ7-15
Bβ7-14	NH_2-Glu^7-Glu^8-Gly^9-Phe^{10}-Phe^{11}-Ser^{12}-Ala^{13}-Arg^{14}-OH	Strain 13	−	−
Bβ7-15	——————— Gly^{15}-OH		−	++
Bβ7-14		Strain 2	+++	+++
Bβ7-15			+++	+++

Table 5. Responsiveness of strain 2 and strain 13 guinea pigs to AII and AIII

Immunogen		Guinea Pig	Relative T Cell Proliferation Antigen in Culture	
			AII	AIII
AII	H₂N-Asp -Arg -Val -Tyr -Ile -His -Pro -Phe -OH	Strain 2	+++	−
AIII	H₂N-Arg ———————————————————		−	−
AII		Strain 13	−	−
AIII			++	+++

Table 6. Response of $(2 \times 13)F_1$ guinea pigs to AII and AIII in association with parental macrophages

$(2 \times 13)F_1$ T Cells Immunized with	Macrophages Pulsed with Antigen	Relative T Cell Proliferation Antigen-Pulsed Guinea Pig Macrophages	
		Strain 2	Strain 13
AII	AII	+++	-
	AIII	-	-
AIII	AII	-	++
	AIII	-	+++

with AII and AIII, we favor the interpretation that in strain 13 guinea pigs there are simply too few AII-specific T cells to be detected by in vitro proliferation following immunization with AII.

In conclusion, the use of small peptide antigens to probe the immune system has yielded several insights into the nature of T cell recognition of macrophage-associated antigens. First, it has been shown that T cells seem to exhibit polarity in the reading of exogenous peptide antigenic determinants. The simplest explanation for this observation is that foreign determinants are read in conjunction with some self moieties as a neo-antigenic determinant. Accordingly, it is likely that antigenic peptides are covalently or noncovalently associated with self, which for Bβ7-14 probably involves the carboxyl terminus. The self moiety with which the antigen becomes associated is at present unknown, but could be macrophage Ia antigens themselves as proposed by Determinant Selection theories[16,17]. Secondly, these results suggest several mechanisms for Ir gene control of T cell responses. As discussed above, one possibility is that for certain peptides there may be no appropriate association with "self" moieties and therefore no immunogenic complex is formed. Secondly, for some peptide antigens Ir gene control is due to the functional absence of the corresponding T cell clones. This may be due to either the lack of expression of the appropriate antigen-specific T cell receptors, a low clonal frequency of those cells, or from a cellular regulatory mechanism. Thus, the use of peptide antigens has provided several new insights into the nature of T cell recognition and should allow further definition of how the immunogenic complex is formed and of those cellular structures that specifically interact with exogenous antigens.

ACKNOWLEDGMENTS

Supported by U.S. Public Health Service Grants AI-14226 from the NIAID and HL-14147 from the NHLBI, NIH. David W. Thomas is recipient of Research Career Development Award AI-00352 from the NIAID, NIH.

REFERENCES

1. J.A. Waldron, R.G. Horn, and A.S. Rosenthal, Antigen induced proliferation of guinea pig lymphocytes in vitro: obligatory role of macrophages in the recognition of antigen by immune T-lymphocytes, J.Immunol. 111:58 (1973).
2. R.C. Seeger and J.J. Oppenheim, Synergistic interaction of macrophages and lymphocytes in antigen-induced transformation of lymphocytes, J.Exp.Med. 132:44 (1970).
3. D.L. Rosenstreich and A.S. Rosenthal, Peritoneal exudate lymphocyte. II. In vitro lymphocyte proliferation induced by brief exposure to antigen, J.Immunol. 110:934 (1973).

4. E.M. Shevach and A.S. Rosenthal, Function of macrophages in antigen recognition by guinea pig T lymphocytes. II. Role of the macrophage in the regulation of genetic control of the immune response, J.Exp.Med. 138:1213 (1973).

5. E.M. Shevach, The function of macrophages in antigen recognition by guinea pig lymphocytes. III. Genetic analysis of the antigens mediating macrophage-T lymphocyte interaction, J.Immunol. 116:1482 (1976).

6. R.H. Schwartz, A. Yano, and W.E. Paul, Interaction between antigen-presenting cells and primed T lymphocytes: an assessment of Ir gene expression in the antigen-presenting cell. Immunol.Rev. 40:153 (1978).

7. P. Marrack and J.W. Kappler, The role of H-2-linked genes in helper T-cell function. III. Expression of immune response genes for trinitrophenyl conjugates of poly-L (Tyr, Glu)-poly-D,L-Ala-poly-L-Lys in B cells and macrophages, J.Exp.Med. 147:1596 (1978).

8. A. Singer, C. Cowing, K.S. Hathcock, H.B. Dickler, and R.J. Hodes, Cellular and genetic control of antibody responses in vitro. III. Immune response gene regulation of accessory cell function, J.Exp.Med. 146:1611 (1978).

9. D.W. Thomas, S.K. Meltz and G.D. Wilner, Nature of T lymphocyte recognition of macrophage-associated antigens. II. Macrophage determination of guinea pig T cell responses to human fibrinopeptide B. J.Immunol. 123:1299 (1979).

10. D.W. Thomas, J.L. Schauster, and G.D. Wilner, Nature of T lymphocyte recognition of macrophage-associated antigens. III. Definition of the antigenic regions of human fibrinopeptide B involved in guinea pig T cell responses, Cellular Immunol. 58:215 (1981).

11. D.W. Thomas, K-h, Hsieh, J.L. Schauster, M.S. Mudd, and G.D. Wilner, Nature of T lymphocyte recognition of macrophage-associated antigens. V. Contribution of individual peptide residues of human fibrinopeptide B to T lymphocyte responses, J. Exp. Med. 152:620 (1980)

12. R.W. Chesnut, R.O. Endres, and H.M. Grey, Antigen recognition by T cells and B cells: recognition of cross-reactivity between native and denatured forms of globular antigens, Clin.Immunol. Immunopathol. 15:397 (1980).

13. V. Schirrmacher and H. Wigzell, Immune responses against native and chemically modified albumins in mice. I. Analysis of non-thymus-processed (B) and thymus processed (T) cell responses against methylated bovine serum albumin, J.Exp.Med. 136:1616 (1972).

14. K. Ishizaka, H. Okudaira, and T.P. King, Immunogenic properties of modified antigen E. II. Ability of urea-denatured antigen and polypeptide chain to prime T cells specific for antigen E, J.Immunol. 114:110 (1975).

15. R.M. Huseby, Conformational structure of the fibrinopeptides released during fibrinogen to fibrin conversion, Physiol.Chem. Phys. 5:1 (1973).

16. A.S. Rosenthal, Determinant selection and macrophage function in genetic control of the immune response, Immunol.Rev. 40:136 (1978).

17. B. Benacerraf, A hypothesis to relate the specificity of T lymphocytes and the activity of I region-specific Ir genes in macrophages and B lymphocytes, J.Immunol. 120:1809 (1978).

THE ANTIGENIC STRUCTURE OF BOVINE SERUM ALBUMIN:*

T-CELL, B-CELL, AND Ia DETERMINANTS

David C. Benjamin, Lise A. Daigle**
and Richard L. Riley***

Department of Microbiology, School of Medicine
University of Virginia, Charlottesville
Virginia 22908, USA

Bovine serum albumin (BSA) is a single polypeptide chain of 582 amino acids. It is composed of three structural and functional domains which are phylogenetically related[1]. Each domain is an independent folding unit[2] and current evidence suggests that during synthesis the formation of the disulfide bonds proceeds linearly from the amino to the carboxy terminus as the protein emerges from the polyribosome and is secreted[3]. Native surface structure, however, appears to form in the opposite direction[4], ie., from the carboxy to the amino terminus, indicating that the three dimensional structure of each domain begins to assemble only after synthesis of that domain is complete and then by nucleation around a structure residing within the carboxy terminal double disulfide loop of each domain[4,5].

Although considerable homology, in general structure, can be seen between domains, the amino acid sequence and function of each domain has diverged considerably[1]. For example, there is, at most, 25% sequence identity between any two domains and no more than 10% between all three domains. The majority of the identity between all three domains (about 12/19 conserved residues) centers around the disulfide bonds which are highly conserved throughout evolution[1]. Indeed 18/19 of the residues shared between domains are also shared with human albumin.

---

* Supported by NCI grant CA24431 and NIH grant GM 28241
** Supported by training grant 5-T32-CAO9109 from U.S.P.H.S.
***Supported by training grant 5-T32-CAO9109 from U.S.P.H.S. Present address: Scripps Clinic and Research Foundation, La Jolla, California.

Bovine albumin has been a long-standing tool for the immunologist and immunochemist. Yet, until recently little has been known about its chemical and physical structure, not to mention its detailed antigenic structure and the control of immune responses to this protein. From the immunological point of view each domain of albumin can be separated into two antigenic regions for a total of six antigenic regions on the entire molecule[6,7] (Figure 1). Each antigenic region contains one or more B-cell determinants, one or more T-cell determinants, and at least one "determinant" seen by the product of an Ir gene[6,8-12]. Each of these classes of "determinants" will be briefly discussed (in reverse order) with emphasis on the B-cell determinants on serum albumin.

The Ia "Determinant(s)"

Murine immune responses to serum albumin are under the control of Ir genes mapping within the H-2 complex[8-12]. B10.BR (H-2^k), B10. D2 (H-2^d), B10.A (H-2^a), B10.A (4R) (H-2^{h4}), and B10.GD (H-2^{g2}) are responders whereas B10 (H-2^b) and B10.A (5R) (H-2^{i5}) are low responders when immunized with either intact BSA or with fragments representing the first or third domains of BSA. This is true for antibody responses, for T-cell proliferative responses, and for helper T-cell responses (Table 1)[8-12].

Classical genetic mapping (Table 1) and anti-Ia (monoclonal) antibody blocking (Table 2) studies demonstrate that two separate Ir genes control the immune responses to BSA in mice. The genes map to the I-A and I-E/C subregions and have been termed the Ir-BSA$_1$ and Ir-BSA$_2$ genes respectively[10]. The specificity of the monoclonal anti-Ia antibodies used have been determined by their ability to block the T-cell proliferative responses to the terpolymers GAT (Table 2 and reference 13) and GLØ[13]. Other experiments[12] (Table 3) demonstrate the same Ir region multigenic control of T-cell and B-cell responses to immunization with either domain 1 (fragment CNBr1-183, Figure 1) or domain 3 (fragment T377-582, Figure 1).

It is interesting to note that although the E$_\alpha^k$E$_\beta^k$ product is functionally capable of presenting BSA to T-cells in high-responder B10.BR mice, the E$_\alpha^k$E$_\beta^b$ product in B10.A(5R) low responder mice cannot do so (at least with the same efficiency). Although this observation is not new, it would suggest that the E$_\beta$ chain determines the "specificity" of this Ia molecule. The 14.4.4s monoclonal anti-Ia (anti-IE/C^{k,d}) did inhibit the majority of the response of these B10.A(5R) low responder mice (Table 2).

Smaller fragments representing subdomains of domain 1 and 3 (ie., fragment N1 - residues 1-87 and fragment T115-184 of domain 1; and fragment P504-582 of domain 3) are immunogenic in responder mice

BOVINE SERUM ALBUMIN

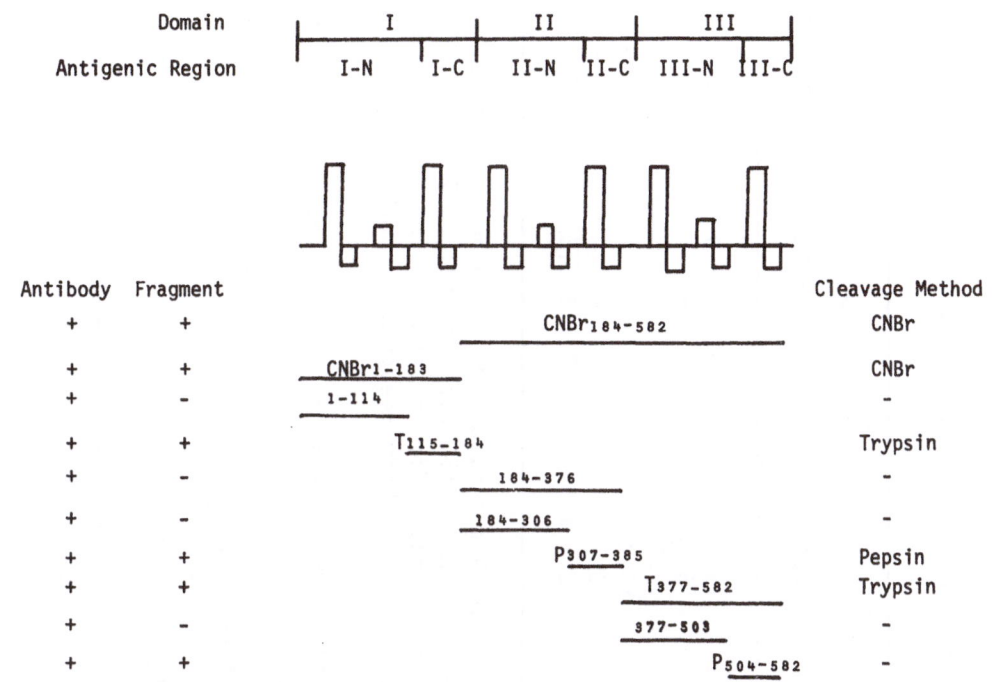

Fig. 1. Schematic representation of the domain structure of albumin,
the position of the double disulfide loops, and the fragments of the molecule that have been isolated and used in
our studies. Included in the figure are the cleavage methods
used to obtain the fragments plus the regions to which antibody populations have been isolated from an antiserum produced in rabbits upon immunization to whole BSA. Also included are the six antigenic regions which these antibody
subpopulations define.
Adapted from: J.M. Teale and D.C. Benjamin, Antibody as
immunological probe for studying the refolding of serum
albumin: Refolding within each domain, J.Biol.Chem. 252:
4521 (1977).

(data not shown). This demonstrates that each antigenic region
(Figure 1) most probably has Ia "determinants" as well as T-cell and
B-cell determinants.

Detailed analyses of the domain specificity of anti-BSA produced
by high responder B10.BR and low responder B10 mice showed a specificity shift in the high responder strain (Table 4)[11]. Domain 3

Table 1. Genetic control of immune responses to bovine serum albumin*

Strain	MHC Alleles	T-Cell Proliferation**			nmoles Antigen Bound/ml***	
		ΔCPM	SI	BSA(BSA)	DNP(DNP-BSA)	DNP(DNP-KLH)
B10.BR	kkkkkkkk	17,000	9.5	13.3 ± 3.0	16.6 ± 4.0	140 ± 20
B10.D2	ddddddd	10,100	6.9	6.0 ± 0.8	-	-
B10.A	kkkkkddd	12,800	13.8	9.2 ± 2.0	-	-
B10.A(4R)	kkbbbbbb	13,400	13.2	15.2 ± 2.4	-	-
B10.A(5R)	bbbkkddd	2,900	2.4	0.7 ± 0.3	-	-
B10	bbbbbbbb	3,000	2.5	2.1 ± 0.4	0.7 ± 0.3	130 ± 20

*Adapted from data presented in: 1) R.L. Riley, R.N. Germain, and D.C. Benjamin. Genetic Control and Specificity of Immune Responses to Bovine Albumin. I. MHC Gene Control of T-cell Proliferative Responses, and 2) L.A. Daigle and D.C. Benjamin. Genetic Control and Specificity of Immune Responses to Bovine Albumin. II. MHC Gene Control of Antibody Responses. Both submitted for publication.

**Mice were primed in the hind footpads with BSA in CFA, sacrificed 7 days later, and their popliteal lymph node cells challenged in vitro with either BSA, PPD, no antigen, or an irrelevant antigen. No significant differences were seen in the responses to PPD. Response in the absence of antigen or with an irrelevant antigen ranged from 1000-2500 cpm. SI = stimulation index relative to the no antigen control.

***Mice were immunized with 100µg BSA, DNP-BSA, or DNP-KLH three times. Immunogen is indicated in parentheses. Anti-DNP was measured in a modified Farr assay using ^{3}H-DNP-lysine. Anti-BSA and anti-KLH (not shown) were measured using ^{125}I-labeled antigen in a double antibody assay. No significant differences were seen in the anti-KLH responses of B10 and B10.BR.

Table 2. Blocking of BSA-specific T-cell proliferative responses with monoclonal anti-Ia antibodies[a]

Strain	Antigen	%SN[§]	Monoclonal Antibody Used[¶]				
			MKD/6(IA^d)	11.5.2(IA^k)	14.4.4S(IE/Ck,d)	Mix[†]	Control[¨]
B10.BR	BSA	10	87(102)	30(35)*	54(64)*	–	85(100)
B10.D2	BSA	10	37(44)*	86(101)	52(61)*	21(25)*	85(100)
B10.A(4R)	BSA	10	83	30*	71	–	–
B10.A(5R)	BSA	10	79(106)	–	38(51)*	–	74(100)
B10.GD	BSA	10	20(33)*	–	50(83)	–	60(100)
BALB/C	GAT$_{10}$	1	60*	–	95	–	–
		4	34*	–	75	–	–
		10	28*	–	88	–	–

[a] A portion of this data has been adapted from: R.L. Riley, R.N. Germain, and D.C. Benjamin. Genetic Control of Immune Responses to Bovine Albumin. I. MHC Gene Control of T-Cell Proliferative Responses. Submitted for publication. Values are expressed as the % of the total BSA induced response in the absence of any inhibiting antibody. Numbers in parentheses indicate the % of the total response corrected for the control hybridoma.

[¶] Hybridoma used is as indicated with its specificity given in parentheses.

[§] %SN = percent of the total culture volume which has been added as hybridoma supernatant. The 11.5.2 hybridoma was added as ascites fluid at a final concentration of 1% of the culture volume.

[†] A mixture of the two positive hybridomas was added: each at a final concentration of 4%.

[¨] Various controls were used including supernatants of hybridomas producing irrelevant antibody and supernatant from clone 11.4.1 (Salk Cell Distribution Center) which has a H-2Kk specificity.

[*] P values range from 0.02 to 0.001. For the mixture of monoclonals with the B10.D2 response, the mixture is significantly different from either anti-Ia alone with P = 0.02.

Table 3. Genetic control of immune responses to the domains of BSA

Strain	Immunogen**	T-Cell Proliferative Responses*				Antibody
		ΔCPM	MKD/6(IA^d)	14.4.4S(IE/C^{k,d})	Control	nmoles Domain Bound/ml
B10.BR	Domain 1	42,500 ± 3,700	n.d.	n.d.	n.d.	5.8
B10		2,200 ± 600	n.d.	n.d.	n.d.	0.1
B10.A(4R)		n.d.	n.d.	n.d.	n.d.	4.7
B10.D2		39,000 ± 1,400	47	36	94	n.d.
B10.BR	Domain 3	41,000 ± 3,200	n.d.	n.d.	n.d.	3.0
B10		6,800 ± 700	n.d.	n.d.	n.d.	0.4
B10.A(4R)		n.d.	n.d.	n.d.	n.d.	1.5
B10.D2		16,800 ± 2,500	49	54	95	n.d.

*T-cell proliferative responses were measured as described in Table 1. Blocking of Domain-specific responses was carried out with the indicated monoclonal anti-Ia antibodies – the specificity of each hybridoma is given in parentheses – results are expressed as % of the control responses in the absence of any monoclonal antibody.

**Domain 1 was obtained by CNBr cleavage of BSA whereas Domain 3 was obtained following trypsin digestion of BSA. For T-cell responses mice were immunized with 5μg in the hind footpads and sacrificed 7 days later for in vitro proliferative responses. For antibody responses mice were injected with 33μg in CFA i.p. and again two weeks later with 33μg in saline i.p. Serum was taken 7 days later. Antibody was determined using 125I labeled Domain in a modified Farr assay.

determinants are dominant in the primary response. During subsequent
responses the relative proportion of antibody specific for domains
1 and 2 increases such that by the tertiary response the amount of
the total anti-BSA that is specific for each domain is approximately
the same. This shift of specificity is reminiscent of, yet different
from, that seen in the response of low responder strains of mice to
staphylococcal nuclease[14] which has been interpreted as an Ir gene
controlled determinant selection similar to that proposed by
Berzofsky et al.[15] for myoglobin. The quantitative difference in
the total anti-staphylococcal nuclease, between high and low responders, disappears by the tertiary response[14]. In contrast the
specificity shift in the response of B10.BR mice to BSA (Table 4)
occurs while the quantitative difference is maintained. Interestly
the specificity of the antibody produced in a tertiary anti-BSA response by low-responder mice also shows the dominance of domain 3
determinants (Table 4).

     Immune responses to structurally complex antigens are usually
not under apparent Ir gene control. How, then, are we to account
for such control in mice to the multi-determinant antigen bovine
serum albumin? Previous studies[15] by others suggest that immune
responses to each antigenic determinant (or group of determinants)
on a complex antigen are regulated by a different Ir gene. This hypothesis would require the existence of a relatively large repertoire
or Ir gene products. Indeed the ability of Ir genes to discriminate
between related synthetic antigens[16] is consistent with this view.
However, only a few Ir gene products (Ia molecules) have been identified[17] and it is difficult to envisage how so few Ia molecules (Ir
genes) could modulate immune responses to such a wide range of diverse antigen molecules. It is possible that many Ir genes (and
thus Ia molecules) do in fact exist but remain undiscovered or that
combinatorial associations among Ia polypeptide chains (as so often
proposed for immunoglobulin heavy and light chains) generate the
large number of specificities required to mediate the Ir gene function[17]. Alternatively, as suggested by Benacerraf[18], if the specificity of the Ia molecules was directed toward some structure, ie.,
a tripeptide sequence, that occurs relatively commonly in proteins,
then the Ir gene repertoire would be substantially increased over
the number of known Ia molecules. However, the demonstration of
antigen binding by individual receptor sites on Ia molecules have
proven difficult. Therefore the nature of the interaction between
antigens and Ia molecules and their recognition by T-cells remains
obscure and certainly controversial.

     The domains of albumin do possess different antigenic determi-
nants; yet the ability of the same monoclonal antibody to inhibit
proliferative responses to both domains (Table 3) suggests that the
Ia molecules involved in regulating the responses to both domains
(eg., the I-A encoded molecules) share a common antigenic determinant
(detected by the monoclonal anti-Ia) and indeed may be one and the

Table 4. Domain specificity of antibody responses to BSA* in high and low responder mice

| Strain | Response | % Total Anti-BSA** | | | Total Antibody |
		Domain 1	Domain 2	Domain 3	nmoles Antigen Bound/ml
B10.BR	1	20	9	71	1.39
	2	33	12	55	5.68
	3	33	28	39	11.80
B10	3	7	15	78	0.59

*Adapted from: L.A. Daigle and D.C. Benjamin. Genetic Control and Specificity of Immune Responses to Bovine Albumin. II. MHC Gene Control of Antibody Responses. Submitted for publication.

**Assays were carried out by a modified Farr technique using ^{125}I labeled fragments CNBr1-183 (Domain 1), CNBr184-582 (Domain 2+3), and T377-582 (Domain 3). Assays using CNBr184-582 to determine Domain 2 specific antibody were carried out in a 750X excess of non-labelled Domain 3. Preliminary studies demonstrated this amount to be more than sufficient to inhibit any anti-domain 3 antibodies from binding to CNBr184-582.

same molecule. We would like to suggest that although each domain
of BSA is antigenically distinct, they are still structurally homologous to the extent that certain structures shared by the domains are
recognized by, or associate with, the same Ia molecule. These would
be simple structures, eg., di- or tripeptides[18], a hydrophobic cleft,
or a stretch of hydrophilic residues, that are common to each of the
domains (and indeed common to many other apparently unrelated protein molecules). In other words, although the B-cell and T-cell
antigenic determinants on each domain are distinct, the Ia "determinants" would be "crossreacting". If the B-cell and T-cell antigenic determinants (although distinct) and the Ia "determinants"
(although similar) are located in similar positions on each domain,
then the orientation of the antigenic determinants on each domain
relative to the high and low responder Ia "determinants" would be
similar (Figure 2). As a result the orientation of the domains,
either alone or as part of the intact BSA molecule, on the surface
of the antigen presenting cell would be similar and a similar pattern
of Ir gene control to the determinants on each domain would be observed. Thus BSA would appear less complex than it really is. This
would imply that, if the I-A and I-E/C molecules recognize different
Ia "determinants" on BSA, determinant selection must result[15]. A
schematic representation of such a model is shown in Figure 2 for a
single domain.

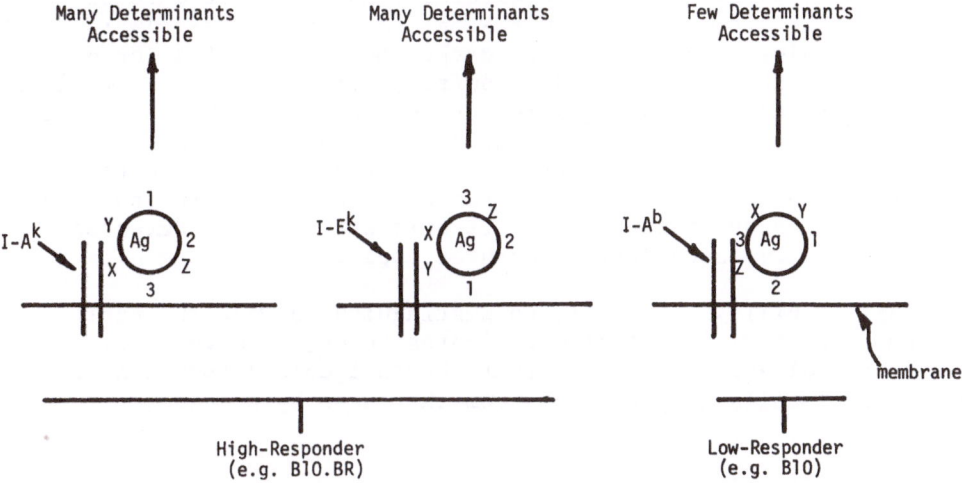

Fig. 2.   Schematic representation of the interaction between an Ia
          molecule and an Ia "determinant" on a domain of albumin.
          The Ia "determinants" X, Y, and Z would be present on each
          domain. The B-cell and T-cell determinants (eg., 1, 2,
          and 3) are distinct and not shared between domains. The
          Ia "determinants" are, of course, not immunogenic for B-
          cells or T-cells.

The T-cell Determinant

Using fragments of albumin representing domains or subdomains
we have recently demonstrated that each domain of albumin contains
one or more antigenic determinants recognized by T-cells (Table 3;
ref.[8,9,12]). We have also demonstrated that each of these T-cell
determinants is unique to a given domain (Figure 3), ie., there is
little or no crossreactivity at the T-cell level between any two
antigenic determinants on BSA. This, of course, agrees with all
our results on the antibody responses to BSA and its constituent
domains (see below, and ref.[6,11,12,19,20]). In many other experiments carried out over a number of years we have, as have others,
shown that T-cell determinants on BSA are different from B-cell
determinants in that antibody appears to require the three-dimensional native structure on the antigen for interaction to occur.
In contrast, T-cell proliferative responses can be either induced
and/or recalled equally well using native or denatured albumin or
fragments of albumin.

A major criticism of studies along this line on albumin is
that the experiments have been conducted using totally reduced and
carboxymethylated albumin and that the bulky carboxymethyl group
added to each cysteine residue had structural effects that would
not be present if such bulky groups were absent. That this is not
the case is readily seen if one carefully examines the data on the
kinetics of in vitro folding of BSA or its domains[2,4,5,21,22,23].
In these studies reduced, but not carboxymethylated, BSA (or domains
of BSA) is diluted into refolding buffer, aliquots taken at various
times thereafter, and the kinetics of appearance of native secondary
structure, antigenic structure, and domain function assessed. The
studies from all three laboratories readily show that no native
antigenic structure exists in the reduced albumin or its isolated
reduced domains and that native antigenic structure is detected
only after native secondary structures appear.

This inability of T-cells to distinguish between the native
and denatured forms of albumin indicates to us, as have similar experiments indicated to others, that the antigenic determinant recognized by T-cells is different from that seen by B-cells.

The B-cell Determinants

Just how antigenically complex is albumin? As stated above,
from the T-cell point of view our evidence indicates a minimum of
six distinct antigenic determinants. At this time we can say little
more concerning the BSA-specific T-cell repertoire. However, regarding the B-cell determinants we have much more to say.

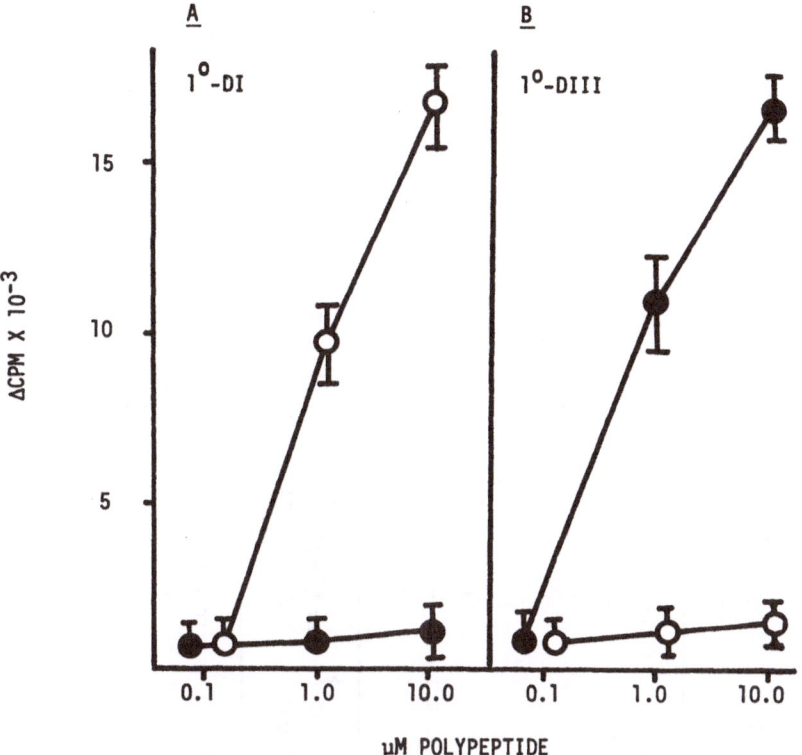

Fig. 3. Domain Specificity of T-cell Proliferative Responses.
Responder mice were primed with either CNBr1-183 (Domain 1)
or T377-582 (Domain 3) in the hind footpads in CFA. Seven
(7) days later the popliteal lymph nodes were removed and
the cells placed in culture with either fragment as
challenge antigen in varying doses. Panel A - Domain 1
primed lymph nodes; Panel B - Domain 3 primed lymphnodes.
○——○ Domain 1 in vitro challenge. ●——● Domain 3 in vitro
challenge.

Recent studies in our laboratory have left no doubt in our
mind that BSA is a multi-determinant antigen on which each and every
antigenic determinant is distinct. This is true whether the responding animal is a rabbit (Table 5; ref.[6,19,20]) or a mouse (Table 3;
Table 4; Figure 4; and ref.[12]). Table 5 presents a summary of the
data we have accumulated over the years with hyperimmune rabbit
antisera. The rabbit antiserum was fractionated into subpopulations
of antibody, each specific for one of the subdomains of BSA. This
fractionation was accomplished using a number of immunoadsorbents
constructed using the fragments[6] shown in Figure 1. Overall, greater
than 90% of the total anti-BSA was recovered with essentially full
activity - thus there was no loss of a significant portion of the

Table 5. Specificity of antibody directed to the subdomains of bovine serum albumin. For table legend see next page.

Inhibitor*	Anti-BSA** (N1-582)	Anti-DI** (N1-183)	Anti-DI-N** (N1-114)	Anti-DI-C** (N115-183)
BSA	100(1.1)***	100(1.0)	100(1.1)	100(1.0)
CNBr1-183	32(12.0)	100(4.5)	98(5.4)	97(5.0)
T377-582	44(2.4)	0	0	0
CNBr184-582	72(6.0)	0	0	0
CNBr1-183 + CNBr184-582	100(8.9)	–	–	–
T115-184	15(45.0)	–	4	81(90.0)
P307-385	–	–	–	–
P504-582	–	–	–	–

Inhibitor*	Anti-DII** (N184-376)	Anti-DII-N** (N184-306)	Anti-DII-C** (N307-376)	Anti-DIII (N377-582)	Anti-DIII-N** (N377-504)	Anti-DIII-C** (N504-582)
BSA	100(1.1)	100(1.0)	100(1.0)	100(1.1)	100(1.2)	100(1.1)
CNBr1-183	6	0	0	7	–	–
T377-582	3	0	0	100(2.4)	100(2.9)	98(4.7)
CNBr184-582	96(7.0)	95(5.7)	98(7.2)	100(5.0)	–	–
T115-184	–	–	–	–	–	0
P307-385	–	7	98(84.0)	–	–	0
P504-582	–	–	–	–	6	90(21.0)

Table 5. Legend

*Peptides used in the inhibition of the various subpopulations of antibody reacting with ^{125}I-BSA in a modified Farr assay. Numbers represent the position of the peptide in the sequence of BSA.

**Antibody solution used in the inhibition assay. Numbers in parentheses represent the amino acid position number in the sequence of albumin against which the antibody is specific. See text for method of isolation.

***Maximum inhibition used at the highest concentration of inhibitor used, ie., a 1000X molar excess over the ^{125}I-BSA in the assay. Numbers in parentheses represent the molar ratio, of inhibitor to ^{125}-I-BSA, at which 50% maximum inhibition was observed.

anti-BSA. It should be emphasized that all the subpopulations (those which did and those which did not adhere to the immunoadsorbents) were used in these studies. Inhibition of the reaction of anti-native-albumin with ^{125}-I-BSA by fragments representing one or more domains of albumin was never complete even at a 1000-fold molar excess of the inhibiting fragment. The maximum inhibition observed for the regions representing domains 1, 2, and 3 was 32%, 28%, and 44% respectively indicating that the antigenic determinants were relatively evenly distributed over the surface of the BSA molecule. This, of course, assumes an equal response to each determinant. The fact that no fragment would give complete inhibition of the BSA-anti-BSA reaction suggests that each fragment possessed some, but not all, of the antigenic determinants of albumin. Indeed the maximum extents of inhibition by two non-related fragments are additive - a result which could only occur if the two fragments were antigenically distinct. Any subpopulation of antibody was specific for a given region in that its reaction with ^{125}-I-albumin was inhibited only by intact albumin or by fragments containing that region of albumin. Whereas the reaction between albumin and anti-albumin was only partially inhibited by any given fragment, the reaction between albumin and an antibody to any given region was completely inhibited by a fragment containing that region but not at all by fragments chemically lacking that region. These latter results can only be interpreted to mean that: a) albumin and the corresponding fragment bear the same antigenic determinants with reference to a given subpopulation of antibody; and b) these antigenic determinants are not found elsewhere on the albumin molecule. Overall these results clearly demonstrate the existence of six different antigenic regions on albumin (two per domain) with a minimum of one B-cell determinant per antigenic region.

Although we have not pursued studies using hyperimmune murine antisera to BSA, the results shown in Table 4 clearly demonstrate that the domains of BSA are antigenically distinct using murine antisera.

Recently we have established more than 200 hybridoma cell lines
producing antibody to BSA. We have, to date, cloned approximately
40 of these and have determined the domain specificity of each
(Figure 4) by solid phase radioimmunoassay (plate coating) and by
the Farr assay. All domain 1 specific clones reacted with Fragment
CNBr1-183 but not with CNBr184-582 (domains 2 + 3) or with T377-582
(domain 3). Those specific for domain 3 reacted with both CNBr184-
582 and T377-582 whereas those specific for domain 2 reacted only
with CNBr184-582. Each monoclonal antibody reacts with only one
domain, ie. they are specific for domains 1, 2, or 3 with no exceptions. The frequency of clones producing antibody for a given domain
is approximately equal - reflecting the relative proportion of antibody in whole antisera specific for each domain (Table 4; ref[6]).
Furthermore we have determined the subdomain specificity for many
of the domain 1 and domain 3 specific clones. This was accomplished
using fragments T115-184 and P505-582 representing the carboxy terminal double disulfide loop of domains 1 and 3 respectively (loops
3 and 9 of the intact BSA, ref[1]). In addition, the crossreactivity
of most of the clones was established using a panel of BSA and five
crossreacting serum albumins. Our initial interest resided with
monoclonals to domains 1 and 3 - therefore we have not assessed the
crossreactivity of all our anti-domain 2 monoclonals. The cross-
reactivity patterns and the domain or subdomain specificities are
shown in Figure 4.

First of all, the use of these monoclonal antibodies confirms
the presence of six antigenic regions as previously suggested[6]. In
addition, they demonstrate that each antigenic region bears more
than one distinct antigenic determinant as determined by the cross-
reactivity patterns. The presence of distinct antigenic determinants
in the III-C antigenic region has been confirmed by blocking studies
(data not shown). Overall, the results demonstrate six distinct
antigenic determinants on domain 1 (three in each subdomain), three
distinct determinants on domain 2, and seven distinct determinants
on domain 3 (three in the III-N subdomain and four in the III-C subdomain). Other determinants most probably exist, eg. we have concentrated on domains 1 and 3 and all the clones specific for domain
2 have not been characterized. In addition, mouse anti-BSA cross-
reacts with horse serum albumin and since our hybridomas have been
raised in media containing horse serum we would not detect this antigenic determinant on BSA.

The hybridoma cell lines to be cloned and used were chosen at
random. This, in combination with the fact that the frequency of
hybridomas specific for each domain parallels the proportion of
antibody in hyperimmune antisera specific for each domain (Table 4,
ref.[6,12]) suggests that we are seeing the full BSA-specific B-cell
potential in the strain of mouse used to construct the hybridomas.
This brings us to a total of 17 distinct antigenic determinants on
bovine serum albumin. Only a few of the monoclonal antibodies have

CROSSREACTIVITY PATTERNS AND SPECIFICITY OF
ANTI-BSA MONOCLONAL ANTIBODIES

Domain Specificity	Pattern	Subdomain Specificity
Domain I		I-N
		I-C
Domain II		
Domain III		III-N
		III-C

B S R P G H

Fig. 4. B = bovine albumin; S = sheep albumin; R = rabbit albumin;
P = pig albumin; G = guinea pig albumin; H = human albumin.

been tested for reactivity with reduced albumin or reduced fragment.
None react to any significant extent - again demonstrating the specificity of antibody for native three-dimensional structures.

All these results are in direct contrast to those described by
Atassi and his colleagues[30,35]. They suggest that BSA contains only
two determinants as detected by late antisera, that these two determinants are present on the carboxy-terminal double disulfide loop
of a domain, and that these two determinants are shared between domains, eg. two determinants repeated three times on the entire BSA
molecule. These results are in disagreement with our result and
essentially all other studies on the antigenic structure of serum
albumins (see below). Indeed they are at variance with their own
results: a) they have shown that sera taken early after immunization
with BSA do indeed detect multiple distinct antigenic determinants[35]
in that reaction of these early antibodies with BSA is never completely inhibited by any one fragment - thus early antisera would

not seem to contain much antibody to these repeating sites. b) Although the reaction of late antisera (day 398)[35] with BSA is inhibited almost entirely (80-100%) by fragments representing single
domains, using immunoadsorbent techniques a maximum of 50% of the
anti-BSA reactive antibody reacts with the fragment. This discrepancy is not easily explained. c) In crossreactivity studies[30] they
show that this same late antisera - putatively directed to two repeating determinants - show varying degrees of crossreactivity with
a series of heterologous albumins. Since he used a method which
detects both high and low affinity antibody, the results can only
be interpreted as demonstrating 5 distinct antigenic determinants
on BSA, four of which are shared with one or more crossreacting
albumins. Thus, much of the data they report is inconsistent with
their own hypothesis.

It would appear at first glance that data from two other laboratories support (at least in part) this repeating determinant hypothesis : a) Peters et al.[7] have also suggested six antigenic regions on BSA defined by distinct antigenic determinants. They have
also suggested that some antibodies in their antisera see repeating
determinants. Their data, however, is internally inconsistent as
follows: antibody prepared to fragment P307-582 is partially inhibited in its reaction with BSA by fragments CNBr1-183 and P1-306
at molar excess of 50. Thus it would seem that the amino terminal
and carboxy terminal protions of BSA share determinants. It is
interesting however that fragment P1-385 does not inhibit the reaction although it would be expected to bear all the determinants
present on the previous two fragments. In addition their studies
showed that the reaction of antibody to fragment P1-306 is not significantly inhibited by fragments from the carboxy-terminal half of
BSA. b) Porter[25] showed that only a portion of the antibody, taken
from any one of a number of rabbits immunized with BSA, reacted with
a 12,000 dalton chymotryptic fragment of BSA. Sera taken from one
rabbit, after an additional 6 injections of BSA, contained antibody
most of which reacted with the fragment. These results only suggest
that a portion of the determinants detected earlier have become
immunodominant. Thus, these results are consistent with a repeating
determinant model but certainly do not prove it.

Thus albumin would seem to be a multi-determinant protein antigen for which each antigenic determinant is unique. This concept
is not new with us, but is supported by a vast amount of literature
spanning almost three decades beginning with the pioneering work on
human albumin by Claude Lapresle[24]. In these early studies, Lapresle
demonstrated that rabbit spleen cathepsins would degrade human albumin into three antigenically distinct fragments. Porter and his
group[25,26] have shown that fragments from BSA and from HSA bear only
a portion of the total antigenic determinants present on the intact
molecule. Indeed in their earlier study[25] they demonstrate four

distinct antigenic determinants on BSA and state that when other crossreacting antigens were used "... further subdivisions could be made and it seems that all antisera contain a complex mixture of antibodies directed against different combining sites...". In 1961 Weigle[27] showed that BSA could be cleaved into four distinct fragments with pepsin each of which was antigenically distinct and each of which shared an antigenic determinant with one or more heterologous albumins – thus again four distinct determinants. In the same study[27] Weigle conducted an extensive series of crossreactions using 11 different mammalian albumins and both precipitation and inhibition of precipitation methods. He found that absorption with a given albumin (except for sheep albumin) removed some but not all the anti-BSA that crossreacted with that and other albumins. Therefore his results indicate many (more than four) distinct antigenic determinants on BSA. Similar results have been obtained by others[19, 20,28-30]. Indeed, Timpl et al.[28] have definitively demonstrated 13 different antigenic determinants on BSA. Other studies on antibody synthesis by single cells[31] and on specific acquired immunological tolerance to albumins[32,33] or to fragments of albumin[34] support this multi-distinct determinant nature of BSA.

What then is the nature of this native structure recognized by B-cells on BSA. Reduced and carboxymethylated BSA and fragments of BSA do contain significant secondary structure[2] as determined by circular dichroism measurements. Yet the lack of reactivity of conventional anti-BSA and the monoclonal antibodies with reduced albumin would suggest that these secondary structures are not antigenic determinants. Most probably each B-cell determinant consists of side chains of amino acids from relatively distant positions in the linear sequence which have been brought together by the specific folding mechanism of BSA and stabilized by numerous covalent and non-covalent forces. Studies are currently underway in our laboratory to antigenically characterize intermediates in this folding process.

In summary the structural entities on BSA recognized by T-cells and by B-cells are distinct from each other and each probably different in turn from that structure recognized by the Ir gene product.

ACKNOWLEDGMENTS

The authors would like to thank Dr. Ronald Germain for helpful discussion and for certain of the monoclonal anti-Ia antibodies used in these studies. They would also like to thank Dr. Jay Berzofsky for certain of the congenic lines and Drs. Judy Teale and David Katz for the B10.A(4R) breeding pairs. Last, but not least, the excellent technical assistance of Mr. Daniel Hershey and Ms. Kristi DeCoursey is gratefully appreciated.

REFERENCES

1. J.R. Brown, Structural origins of mammalian albumin, Fed.Proc. 35:2141 (1976).
2. J.M. Teale and D.C. Benjamin, Antibody as an immunological probe for studying the refolding of bovine serum albumin. II. Evidence for the independent folding of the domains of the molecule, J.Biol.Chem. 251:4609 (1976).
3. T. Peters Jr., Study of the intracellular site of formation of disulfide bonds of rat albumin in vivo. Fed.Proc. 39:1869 (1980).
4. J.M. Teale and D.C. Benjamin, Antibody as an immunological probe for studying the refolding of bovine serum albumin: Refolding within each domain, J.Biol.Chem. 252:4521 (1977).
5. L.G. Chavez Jr., and D.C. Benjamin, Antibody as an immunological probe for studying the refolding of bovine serum albumin. An immunochemical approach to the identification of possible nucleation sites, J.Biol.Chem. 253:8081 (1978).
6. D.C. Benjamin and J.M. Teale, The antigenic structure of bovine serum albumin. Evidence for multiple, different, domain-specific antigenic determinants, J.Biol.Chem. 253:8087 (1978).
7. T. Peters Jr., R.C. Feldhoff, and R.G. Reed, Immunochemical studies of fragments of bovine serum albumin, J.Biol.Chem. 252: 8464 (1977).
8. R.L. Riley and D.C. Benjamin, Genetic control and specificity of T-lymphocyte proliferative responses to bovine serum albumin, in: "Immunobiology of the Major Histocompatability Complex," Proceedings of the 7th International Convocation on Immunology, 1981. ed., M. Zaleski, (1980).
9. R.L. Riley and D.C. Benjamin, MHC control and specificity of T-Lymphocyte proliferative responses to bovine serum albumin, Fed.Proc. 40:999 (1981).
10. R.L. Riley, R.N. Germain, and D.C. Benjamin, Genetic control and specificity of immune responses, submitted for publication.
11. L.A. Daigle and D.C. Benjamin, Genetic control and specificity of immune responses to bovine albumin. II. MHC gene control of antibody responses, submitted for publication.
12. R.L. Riley, L.A. Daigle, and D.C. Benjamin, Genetic control and specificity of immune responses to bovine serum albumin. III. MHC gene control of T-cell proliferative responses to distinct antigenic determinants on BSA domains, in preparation.
13. J.T. Nepon, B. Benacerraf, and R.N. Germain, Analysis of Ir gene function using monoclonal antibodies: Independent regulation of the GAT and GLPhe T-cell responses by I-A and I-E subregion products on a single accessory cell population, J.Immunol. 127:31 (1981).
14. D.H. Sachs, J.A. Berzofsky, D.S. Pisetsky, and R.H. Schwartz, Genetic control of the immune response to staphylococcal nuclease, Springer Seminars in Immunopathol. 1:51 (1978).
15. J.A. Berzofsky, L.K. Richman, and D.J. Killion, Distinct H-2 linked Ir genes control both the antibody and T-cell responses

to different determinants on the same antigen, myoglobin, Proc.Nat.Acad.Sci.U.S.A. 76:4046 (1979).

16. M.E. Dorf, Complementation of H-2 linked genes controlling immune responsiveness, Springer Seminars in Immunopathology 1:171 (1978).

17. H.O. McDevitt, The role of H-2 I region genes in regulation of the immune response, J.Immunogenetics 8:287 (1981).

18. B. Benacerraf, A hypothesis to relate the specificity of T lymphocytes and the activity of I-region specific Ir genes in macrophages and B-lymphocytes, J.Immunol. 120:1809 (1978).

19. D.C. Benjamin and W.O. Weigle, The termination of immunological unresponsiveness to bovine serum albumin in rabbits. III. Structural and serological relationships among various serum albumins and their cyanogen bromide fragments, Immunochem. 8: 1087 (1971).

20. D.C. Benjamin, Ph.D. Thesis. University of California at San Diego. (1969).

21. J.M. Teale and D.C. Benjamin, Antibody as an immunological probe for studying the refolding of bovine serum albumin. I. The catalysis of reoxidation of reduced bovine serum albumin by glutathione and a disulfide interchange enzyme, J.Biol.Chem. 251:4603 (1976).

22. K.O. Johanson, D.B. Wetlaufer, R.G. Reed, and T. Peters Jr., Refolding of bovine serum albumin and its proteolytic fragments. Regain of disulfide bonds, secondary structure, and ligand-binding ability, J.Biol.Chem. 256:445 (1981).

23. D.B. Wetlaufer, Folding of protein fragments, in: "Advances in Protein Chemistry".

24. C. Lapresle, Etude de la degradation de la serumalbumine humaine par un extrait de rate de lapin. II. Mise en evidence de trois groupements specifiques differents dans le motif antigenique de l'albumine humaine et de trois anticorps correspondants dans le serum de lapin anti-albumine humaine, Ann.Inst.Pasteur 89: 654 (1955).

25. R.R. Porter, The isolation and properties of a fragment of bovine serum albumin which retains the ability to combine with rabbit antiserum, Biochem.J. 66:677 (1957).

26. E.M. Press and R.R. Porter, Isolation and characterization of fragments of human serum albumin containing some of the antigenic sites of the whole molecule, Biochem.J. 83:172 (1962).

27. W.O. Weigle, Immunochemical properties of the cross-reactions between anti-BSA and heterologous albumins, J.Immunol. 87:599 (1961).

28. R. Timpl, H. Furthmayr, and I. Wolff, Antigenic determinant analysis of bovine serum albumin, Int.Arch.Allergy. 32:318 (1967).

29. T. Kamiyama, Immunological crossreactions and species-specificities of cyanogen bromide fragments of bovine, goat, and sheep serum albumins, Immunochem. 14:91 (1977).

30. S. Sakata and M.Z. Atassi, Immunochemistry of serum albumin. VI. A dynamic approach to the immunochemical crossreactions of proteins using serum albumins from various species as models, Biochem.Biophys.Acta. 576:322 (1979).

31. D.C. Benjamin and W.O. Weigle, Frequency of single spleen cells from hyperimmune rabbits producing antibody of two different specificities, J.Immunol. 105:537 (1970).

32. D.C. Benjamin and W.O. Weigle, The termination of immunological unresponsiveness to BSA in rabbits. I. Quantitative and qualitative response to crossreacting albumins, J.Exp.Med. 132:66 (1970).

33. D.C. Benjamin and W.O. Weigle, The termination of immunological unresponsiveness to BSA in rabbits. II. Response to a subsequent injection of BSA, J.Immunol. 105:1231 (1974).

34. D.C. Benjamin and C.W. Hershey, The termination of immunologic unresponsiveness to the cyanogen bromide fragments of BSA in rabbits, J.Immunol. 113:1593 (1974).

35. S. Sakata and M.Z. Atassi, Immunochemistry of serum albumin. V. A time dependent examination of the antibody response to bovine serum albumin by the activity of its third domain, Immunochem. 16:451 (1979).

ATTEMPTS TO UNDERSTAND THE FUNCTION OF MACROPHAGES AND

ANTIGEN SPECIFIC CELLS IN THEIR MUTUAL INTERACTION

Giampietro Corradin, Serge Carel, Yolande Buchmüller
Patrick Amiguet, Howard D. Engers and Claude Bron

Institut de Biochimie, Université de Lausanne
and the Department of Immunology
Swiss Institute for Experimental Cancer Research
1066 Epalinges, Switzerland

Our main research interest in the last several years has centered on the understanding of the function of macrophages in the interaction with the antigen specific T cell and the fine antigen specificity of the latter. While several studies, including our own, point out that a single amino acid substitution is critical for the activation of antigen specific T cells[1-5], no data are available to determine whether or not processing and presentation by macrophages or recognition of antigen specific T cells is the critical step in this phenomenon. The dilemma will remain unresolved as long as one is unable to determine the contribution of each single cell type to the activation process.

We attempted to dissect the system by studying macrophage processing and/or presentation and direct antigen binding on cloned T cells.

Role of macrophages in processing and presentation

While it is easy to understand that in some cases specificity of T cell activation resides at the recognition level, ie., the antigen specific T cells, as it was shown for antigen-antibody interactions, a more complex explanation is needed to interpret specificity at the level of macrophages. The notion that in most cases activation is independent of the three-dimensional structure of a protein being used for in vivo priming or in vitro activation of primed T cells is best explained by the presence in the macrophage of an enzymatic system which is operative prior to presentation of antigen to specific T cells[6,7]. Therefore, if such an enzymatic system is present in a macrophage, different proteins should give rise to different peptides. Thus, in the case of closely related proteins, in which T cell activation is dependent on the presence of certain amino acid residues[1,5], the specificity step might reside not only at the level of recognition by the T cell but also at the level of macrophages by generation of different sets of immunogenic peptides.

Another corollary that derives from the hypothesis that protein degradation is a necessary step prior to activation of T cells, is that protein conformation should be, in certain cases, crucial for T cell activation. In fact, protein degradation has been long recognized to be a function of protein conformation[8]. For proteins which do not become denatured at pH 4.5 in the lysosomes where enzymatic hydrolysis is likely to occur, degradation of native and denatured proteins would result in the formation of different peptides and as a consequence, different T cell clones are likely to be activated. To test this typothesis, several studies were performed to demonstrate that in several strains of mice, in vitro and in vivo acitvation of cytochrome c specific T cell was dependent on the three-dimensional structure of the protein[9,10,11]. This analysis was carried out not only at the population level but also at the clonal level as summarized in Table 1. Furthermore, it was shown that in the case of denatured apocytochrome c primed BDF$_1$ mice, T cells were stimulated by heme peptide 1-65 only in the presence of C57BL/6 and BDF$_1$ macrophages, but not in the presence of DBA macrophages. This lack of activation is not due to the inability of DBA macrophages to present the heme peptide 1-65 since native cytochrome c primed BDF$_1$ T cells were stimulated by this peptide in the presence of DBA macrophages[11]. It seemed to us that these results suggested that C57BL/6 and DBA macrophages have a different capacity to process the heme peptide 1-65 and that some distinct non overlapping immunogenic peptides are generated by these two macrophages. This would imply that these two strains of mice have different lysosomal proteases which would confer some restriction as to which peptides are generated and capable of inducing an immune response, prior to any other interaction of macrophages with antigen specific T cells. However, no biochemical evidence is available to support this notion at the present time.

It seems to us that a biochemical approach to studying processing and presentation would be possible only if a pure population of macrophages capable of presenting and processing antigen were available in large quantity. As shown in Table 2, we found that the macrophage line IC-21 of C57BL/6 origin[12] is capable of supporting a proliferation of short or long term antigen specific T cells. This antigen specific proliferation is inhibited by monoclonal anti-Ia^{b,s} antibodies[13] suggesting that the IC-21 macrophage line is the cell responsible for processing and presentation of antigen to T cell. This finding should enable us to carry our biochemical experiments aimed at the understanding of the mechanism by which macrophages process and present antigens to T cells.

Table 1.   Proliferative response of apocytochrome c
           specific T cells[*]

Antigen[**]	CPM ± SE	
	Lymph node cells	Clone 2-42
Apocytochrome c	52.1 ± 5.2	35.3 ± 1.8
Apo peptide 1-65	65.5 ± 4.5	38.6 ± 2.3
Heme peptide 1-65	0.5 ± 0.2	54.2 ± 3.2
Cytochrome c	0.5 ± 0.1	3.2 ± 1.1
Control	2.1 ± 0.3	5.2 ± 0.5

[*]Cells and clone 2-42 were obtained from beef apocytochrome
c primed BALB/C mice.

[**]At optimal antigen concentration.

Table 2.   Proliferative response of ovalbumin specific T
           cells in the presence of IC-21 macrophages[*]

Cultures[**]	CPM ± SE
T cells + OVA	0.6 ± 0.4
T cells + IC-21	0.3 ± 0.1
T cells + IC-21 + OVA	15.9 ± 0.6
T cells + IC-21 + OVA + anti Ia^{b,s}[***]	2.5 ± 1.8

[*]Ovalbumin specific T cells were obtained from BDF$_1$ primed
lymph node cells which were propagated in vitro for 5 days
followed by a percoll purification and one day in vitro
culture in the presence of a mixed lymphocyte culture
supernatant as the source of interleukin 2.

[**]10^4 T cell blasts and 20,000 mitomycin c treated IC-21
macrophages were used.

[***]Monoclonal anti-Ia^{b,s} antibody was used at a 1/200
dilution.

Antigen binding by cloned T cells and hybrids

    Antigen specific T cell clones and T cell hybrids were derived
in order to study the interaction between antigen and its receptor(s)

on T cells. This was based on the assumption that direct antigen
binding could be measured without preincubation of antigen with
macrophages if the preparation used for binding was the smallest
fragment still capable of inducing specific proliferation. In these
cases, the processed peptide might not differ dramatically in its
length and conformation from the original material.

Two types of cells were extensively studied: the first one was
a cloned T cell (clone 1-16) derived from in vitro stimulated lymph
node cells obtained from apocytochrome c primed BALB/c mice and the
second one (clone IIC-3) resulted from the fusion of BW-5147 thymoma
and long term cultured T cells, specific for the cytochrome c peptide
66-80 derivatized with a dinitroaminophenyl group (DNAP; ∿ 1 DNAP
group/peptide) which was used to prime SJL mice.

In the first case, we were able to show that clone 1-16 was
stimulated by beef and horse apocytochrome c, apo peptide 1-38 and
apo peptides 1-38 and 11-25 in which the cysteine residues 14 and 17
were derivatized with a 2-nitrophenylsulfenyl group (NPS)[10,13]. The
region recognized by this clone seems to be located around amino
acid residue 13-14 (Lys-Cys[14]). When directed binding was determined
by using labeled apocytochrome c, apo 1-38, NPS 11-25 and 1-38 peptides, no binding was observed. It should be pointed out, however,
that all the antigen preparations tested for binding were modified
at the cysteine residues 14 and 17 either by the introduction of the
NPS group or by formation of a disulfide bridge between the two
cysteine residues. If the integrity of the cysteine residues is
important for activation of clone 1-16 and interaction of processed
peptide and binding T cell receptors, peptides carrying the free
cysteine residues should then be employed.

On the other hand, evidence for specific binding was obtained
when hybrid cells derived from the fusion of BW 5147 thymoma and
long term T cell blasts specific for DNAP 66-80 peptide were analysed[15]. As indicated in Table 3, this binding is specific since it
was only inhibited by preincubation of unlabeled homologous antigen
but not by other DNAP peptides.

To our surprise, the only peptide other than the original antigen which exhibited a positive binding was NPS 11-25. This peptide
does not have any sequence homology with peptide 66-80 and as expected does not bind in its unmodified form. On the other hand, the
introduction of the NPS group on the cysteines 14 and 17 seemed to
result in the expression of antigenic determinants similar to those
present in the peptide DNAP 66-80. In particular it seemed to us
that structural homology could be found between the sequences Lys-
Cys-NPS and Lys-Tyr-DNAP as shown in Figure 1.

To test this hypothesis, peptide 66-80 was modified at the
lysine residues by the reversible reagent citraconic anhydride and

Table 3. Binding of ^{125}I-DNAP 66-80 to T cell hybrids[*]

Cold antigen added	CPM			
	II C3	B II A2	B II B4	B W5147[**]
———————	7400	420	410	235
DNAP 66-80	720	-	-	-
DNAP 81-104	7450	515	540	260

[*] Cells were incubated with constant amount of labelled antigen; a 200 fold excess of cold antigen was also added prior to addition of labeled antigen.

[**] II C3, B II A2 and B II B4 were clones derived from the fusion of antigen specific T cells and the BW-5147 thymoma.

66-80:

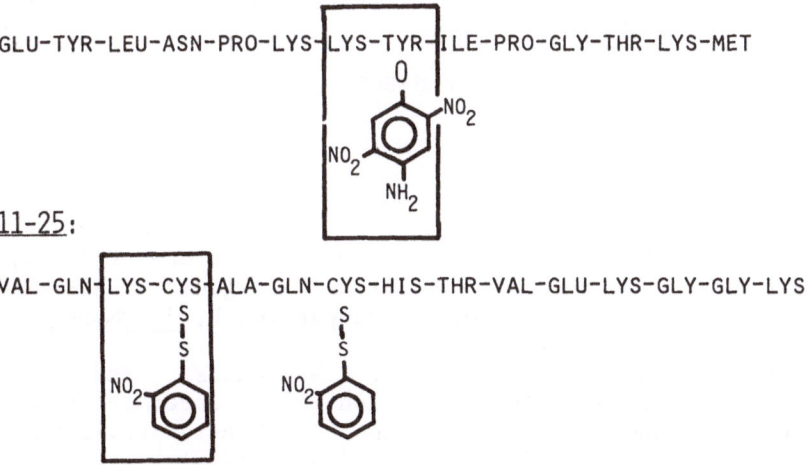

11-25:

Fig. 1. Amino acid sequence of beef cytochrome c peptides 11-25 and 60-80 to which NPS and DNAP groups were attached respectively.

then modified by the DNAP group, which now would be attached only to a tyrosine residue. Lysine residues were then deblocked. As shown in Table 4, while the citraconic DNAP 66-80 peptide is not capable of inhibiting the binding of labeled DNAP 66-80, removal of the citraconic groups resulted in complete inhibition of the binding.

Table 4. Effect of citraconic derivatization on the
          inhibition of binding of DNAP 66-20 by II C3
          hybrid cells*

Cold antigen added	CPM
——	5650
DNAP 66-80	645
Citraconic DNAP 66-80	5550
Deblocked citraconic DNAP 66-80	1210

*Cells were incubated with a constant amount of labeled
antigen; a 200 fold excess of cold antigen was added
prior to addition of labeled antigen.

This would strongly argue that the segment seen by the hybrid is
Lys-Tyr-DNAP and that this modified dipeptide is necessary for the
binding observed. However, a more stringent analysis is needed to
confirm this hypothesis.

This work was supported by the Swiss National Science Foundation,
Grant   3.643-0.80.

REFERENCES

1. D.W. Thomas, S.K. Meltz, and G.D. Wilner, Nature of T lymphocite
    recognition of macrophages-associated antigens. I. Response of
    guinea pig T cells to human fibronopeptide B, J.Immunol. 123:
    759 (1979).
2. L.Y. Rosenwasser, M.A. Barcinski, R.H. Schwartz, and A.S.
    Rosenthal, Immune response gene control of determinant selection.
    II. Genetic control of the murine T lymphocyte proliferative
    response to insulin, J.Immunol. 123:471 (1979).
3. G. Corradin and J.M. Chiller, Lymphocyte specificity to protein
    antigens. II. Fine specificity of T cell activation with cyto-
    chrome c and derived peptides as antigenic probes, J.Exp.Med.
    149:436 (1979).
4. R.M. Maizels, J.A. Clarke, M.A. Harvey, A. Miller, and E.E.
    Sercarz, Epitope specificity of T cell proliferative response
    to lysozyme: proliferative T cells react predominantly to
    different determinants from those recognized by B cells, Eur.
    J.Immunol. 10:509 (1980).
5. M.L. Wolff and M. Reichlin, Antigenic specificity of T cell
    receptor for cytochrome c on guinea pig lymphocytes, Immuno-
    chemistry, 15:289 (1978).

6. B. Benacerraf, A hypothesis to relate the specificity of T lymphocytes and the activity of I region-specific Ir genes in macrophages and lymphocytes, J.Immunol. 120:1809 (1978).

7. G. Corradin, R.W. Chesnut and H.M. Grey, Antigen degradation by macrophages as an obligatory step in the presentation of antigen to T lymphocytes, Ric.Clin.Lab. 9:311 (1979).

8. C. Tanford, Protein denaturation, in: "Advances in Protein Chemistry," C.B. Anfinsen, M.L. Anson, J.T. Edsall, and F.M. Richards, eds., Academic Press, New York, 23:122 (1968).

9. G. Corradin, J.M. Chiller, H.D. Engers, C. Bron, and Y. Buchmüller, Lymphocyte specificity to Protein Antigens. IV. In vivo and in vitro activation of cytochrome c specific T cell is dependent on protein conformation, Amer.J.Reprod.Immunol. In Press. (1981).

10. G. Corradin, R.H. Zubler, and H.M. Engers, Clonal analysis of the BALB/c T-cell response to apo beef cytochrome c. J.Immunol. In Press. (1981).

11. Y. Buchmüller and G. Corradin, Lymphocyte specificity to protein antigen. V. Conformational Dependence of activation of cytochrome c specific T cells. Submitted for Publication. (1982).

12. J. Mauel and V. Defendi, Infection and transformation of mouse peritoneal macrophages by simian virus 40, J.Exp.Med. 134:336 (1971).

13. Y. Buchmüller and G. Corradin, Antigen presentation in vitro by a murine macrophage line. 9th International Congress of Ret. End.Soc., Davos, Switzerland. (1982).

14. G. Corradin, M. Juillerat, C. Vita, and H.M. Engers, Proliferation analysis of apocytochrome c specific T-cell clone. Submitted for publication.

15. S. Carel, C. Bron, and G. Corradin, Antigen binding analysis of a T-cell hybrid. Submitted for publication.

A POSSIBLE IMMUNODOMINANT DOMAIN ON

MYOGLOBIN RECOGNIZED BY T LYMPHOCYTES

Ira Berkower, Gail K. Buckenmeyer
Frank R.N. Gurd and Jay A. Berzofsky

NCI, NIH, Bethesda, Maryland and Indiana University
Bloomington
Indiana

SUMMARY

Sperm whale and equine myoglobin, which share 87% of their sequences, do not cross-stimulate proliferation of T lymphocytes, although they do crossreact extensively with antibodies made against either myoglobin, both for antisera and for several monoclonal antibodies. This failure to cross-stimulate T lymphocytes occurs in strain B10.S despite the fact that these mice are genetic high responders to both myoglobins.

Mixing experiments using splenic adherent cells from nonimmune animals and syngeneic T lymphocytes immune to sperm whale or equine myoglobin showed that the same population of antigen presenting cells can present either myoglobin. In addition, we found that both sperm whale and equine-specific proliferation could be blocked by monoclonal anti-I-A^s antibodies. Thus, the failure of these myoglobins to cross-stimulate T lymphocytes was not due to association of each myoglobin with different Ia antigens on the surface of antigen presenting cells.

In order to map the antigenic determinants recognized by myoglobin-specific T lymphocytes, we immunized mice with sperm whale or equine myoglobin and measured T lymphocyte proliferation in response to a series of species variants. All stimulatory myoglobins for sperm whale immune T lymphocytes share Glu 109, while all myoglobins which differ at this position are nonstimulatory. Conversely, all stimulatory myoglobins for equine immune T lymphocytes share

Asp 109, while all myoglobins which have substitutions at this site
are nonstimulatory. The failure of pig myoglobin, which has Glu 109,
to stimulate sperm whale immune T lymphocytes suggests that His 113
is also part of this determinant. Glu 109 is also one of the critical residues for T cell recognition of sperm whale myoglobin by the
congenic strain B10.D2. It is striking that T cells are so sensitive
to the relatively conservative substitution of Asp for Glu. Why
should this particular residue be critical for several different
strains of mice and different species of myoglobins? The data suggest
that the fine specificity for Glu 109 vs Asp 109 is that of the T
cell. However, the apparent immunodominance of this antigenic site
on the surface of myoglobin could be due to its peculiar immunogenicity for T cells, or could reflect binding in a preferred orientation
(or site specific processing) by the antigen presenting cell.

INTRODUCTION

     We have studied the fine specificity of T lymphocytes immune to
either sperm whale or equine myoglobin. Myoglobin is a small protein
antigen capable of acting as its own antigenic carrier, and therefore it should have one or more antigenic determinants recognized
by T lymphocytes. This paper presents evidence identifying an antigenic site on myoglobin which is recognized by T cells and which explains the cross-reactive pattern with other myoglobins. In addition,
the immune response to myoglobin shows genetic control by H-2 linked
genes. Furthermore, antibody blocking studies suggest that the T
cell response to this antigenic determinant depends on expression
of genes in the I-A subregion in H-2^S mice. Finally, using data
presented elsewhere in this symposium[1], we have compared the spatial
relationship between antigenic determinants recognized by B cells
and T cells, when both cellular responses are controlled by genes
located in the same I subregion.

METHODS

     We have followed the T cell proliferative assay method of
Corradin et al.[10]. B10.S mice were immunized subcutaneously in the
tail with 100 micrograms myoglobin in complete Freund's adjuvant.
Eight days later, draining inguinal and periaortic lymph nodes were
removed. Single cell suspensions were passed through a small nylon
wool column (.04g per mouse used) and resuspended in RPMI 1640 plus
HEPES, penicillin, streptomycin, 2mM L-glutamine, and 10% heat inactivated fetal calf serum. Four hundred thousand cells were added
to 0.2ml microtiter wells in the presence of various concentrations
of each myoglobin tested. After 4 days at 37° in 5% CO_2, 1 micro-
curie of ^{3}H-TdR was added to each well. Cultures were harvested
the following day and radioactivity measured with a scintillation
counter.

RESULTS AND DISCUSSION

We began by making a survey of sperm whale and equine myoglobin responses in different B10 congenic strains (Table 1). These strains clearly show genetic control of the T cell proliferative response to sperm whale and equine myoglobin as described previously[3,4,5]. Since the strains differ only at the $\underline{H-2}$ locus, the high or low response gene must be inherited along with the $\underline{H-2}$ region. However, sperm whale and equine myoglobin do not cross stimulate proliferating T lymphocytes. It is surprising that the population of T lymphocytes immune to sperm whale myoglobin does not respond to equine myoglobin and vice versa, since the $\underline{H-2}^s$ strain is a high responder to both myoglobins. In addition, antiserum prepared against sperm whale myoglobin cross reacts extensively with equine myoglobin, as do half of our hybridoma antibodies to sperm whale myoglobin. (Ref. 6, and J.A. Berzofsky, G.K. Buckenmeyer, and G. Hicks, submitted for publication.)

We examined the cellular and biochemical basis for the failure of the two myoglobins, which share 87% of their primary sequence, to stimulate any subsets of antigen-specific T lymphocytes in common. In order to determine which cell type carried the antigen specificity, we cultured nonimmune splenic adherent cells with purified T lymphocytes immune to sperm whale or equine myoglobin (Figure 1). Under these conditions, T lymphocyte proliferation was totally dependent on added adherent cells plus antigen. The population of adherent cells was capable of presenting either myoglobin to appropriately primed T lymphocytes. However, neither population of myoglobin immune T lymphocytes could be stimulated by the myoglobin not used for immunization. Therefore, the observed specificity is that of the T cell, not the antigen presenting cell, although the antigen presenting cell may still influence what the T cell sees.

Current views of antigen recognition suggest that T lymphocytes see antigen not alone, but rather in association with Ia antigens on the surface of the antigen presenting cell[7,8]. Therefore, we asked whether the failure to cross stimulate was due to presenting each myoglobin with a different Ia antigen. Normal splenic adherent cells were pretreated with normal mouse serum, alloantiserum to Ia^s, or monoclonal anti-I-A antibodies specific for $\underline{I-A}^s$ or $\underline{I-A}^b$. These cells were washed and used as antigen presenting cells for a population of purified T lymphocytes, specific for sperm whale or equine myoglobin (Figure 2). For sperm whale immune T lymphocytes, allo-antiserum pretreatment (1:80) blocked proliferation almost completely, while normal mouse serum (1:80) had no effect. Pretreatment with monoclonal anti-I-A^s (1:10) blocked proliferation nearly as well, while a control supernatant (1:10) from the parent cell line had no effect. In a separate experiment, monoclonal anti-I-A^b (1:10) had no effect, and the monoclonal anti-I-A^s did not suppress the response of B10.D2 T lymphocytes to sperm whale myoglobin. For equine immune

Table 1. Proliferative response of B10 congenic T cells
          immune to equine or sperm whale myoglobin

| | | | Stimulating Myoglobin | |
Strain	H-2	Immune To	Equine	Sperm Whale
B10	b	Equine	−	−
		Sperm Whale	−	−
B10.D2	d	Equine	+	+
		Sperm Whale	−	+
B10.S	s	Equine	+	−
		Sperm Whale	−	+

Cross reactive stimulation represents greater than half the
stimulation by the immunizing myoglobin, while nonstimulatory
myoglobins gave less than one quarter as great proliferation.

T lymphocytes, anti-Ia^s alloantiserum and monoclonal anti-I-A^s antibodies inhibited proliferation to the same extent. Therefore, both
myoglobins are presented in association with the same Ia determinant.

From the above studies, we conclude that the observed specificity is due to the T cell antigen receptor. We suspected that T
lymphocytes primed to sperm whale myoglobin recognize one major
determinant, while T cells primed to equine myoglobin recognize a
different determinant. Therefore, we attempted to map the major
antigenic determinant of each myoglobin, by immunizing mice with
sperm whale or equine myoglobin and testing T lymphocyte proliferation
in response to a series of species variants of myoglobin.

The proliferative response of B10.S T lymphocytes immune to
equine myoglobin was tested against a variety of mammalian myoglobins
(Figure 3). As might be expected, beef and dog myoglobin cross
stimulate, while several whale myoglobins (including sperm whale)
do not. However, some whale myoglobins do stimulate, sometimes to
a greater extent and at lower concentrations than does equine myoglobin. We hoped that by comparing amino acid sequences we would
be able to find the evolutionary substitutions which occurred in the
antigenic site (or sites). These substitutions probably occurred
sometime during whale evolution, since some whale myoglobins cross
stimulate and others do not. Similar comparative studies with
insulin[8], cytochrome c [2,9,10], and lysozyme[11] have identified critical amino acid residues of the T cell determinants for these antigens.

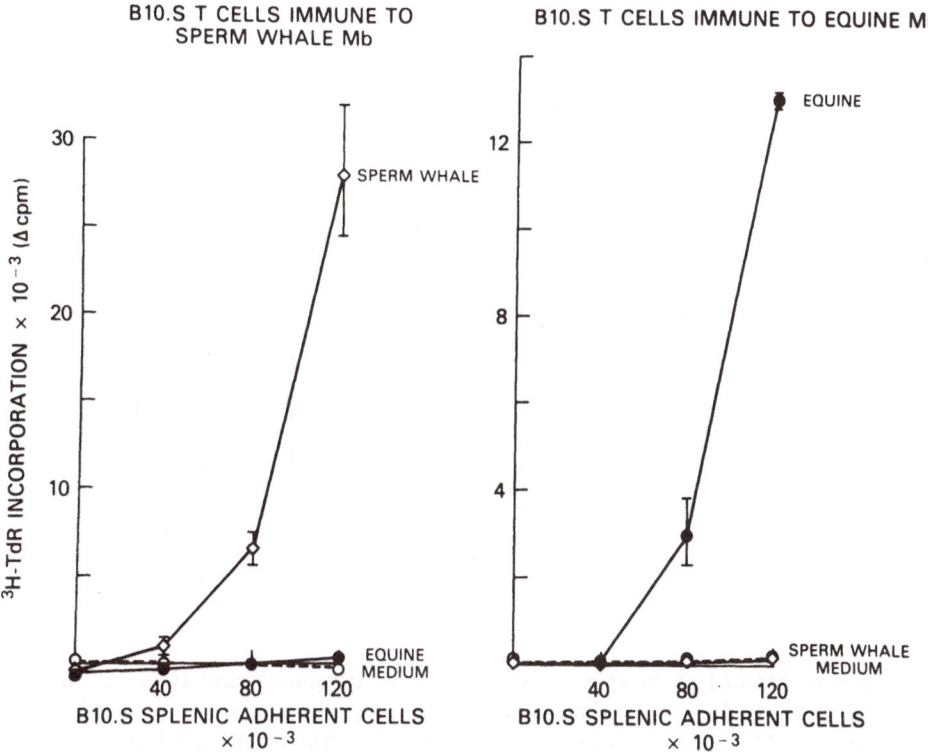

Fig 1. Proliferative response of B10.S T lymphocytes immune to
sperm whale or equine myoglobin. T lymphocytes from four
sperm whale immune or four equine immune animals were purified by passage through two serial nylon wool columns and
cultured at 2×10^5 per well in the presence of graded
numbers of B10.S splenic adherent cells and either medium
alone or each myoglobin at 2.5 micromolar. Control cultures
without myoglobin or adherent cells gave 504cpm for sperm
whale immune and 46cpm for equine immune T lymphocytes.
Error bars are SEM of triplicate cultures.

In analyzing the sequence differences between stimulatory and
nonstimulatory myoglobins, we assume that substitutions which correspond to contact residues within the antigenic site will greatly
alter antigenicity, while substitutions located far from the antigenic site will have little effect on overall structure or antigenicity. The twenty amino acid differences[12,13] between equine
and sperm whale myoglobin (and the substitutions in other myoglobins
at these 20 positions) were examined (Table 2). Sei whale and minke
whale myoglobin cross stimulate and have only one residue in common
with equine myoglobin which is not shared with the nonstimulatory
myoglobins, namely Asp 109. Of the myoglobins tested, all stimulatory myoglobins share Asp 109, while all myoglobins which have

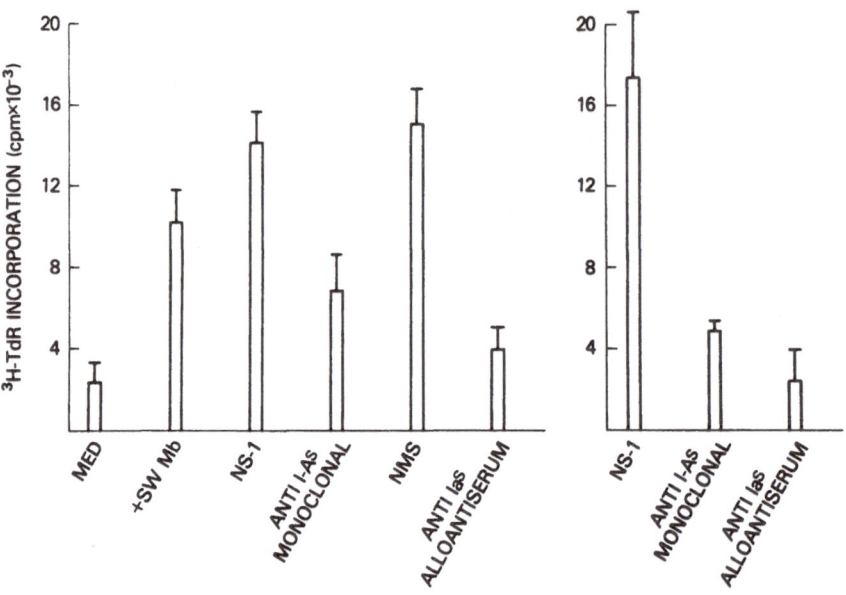

Fig 2.   Proliferative response of B10.S T lymphocytes immune to
         sperm whale or equine myoglobin after pretreating antigen
         presenting cells with anti-Ia^S alloantiserum or monoclonal
         anti-I-A^S antibody from culture supernatants in the absence
         of complement. T lymphocytes from five sperm whale immune
         or four equine immune B10.S mice were purified by passage
         through two nylon wool columns (0.6g followed by 0.3g nylon
         wool) and cultured at 2 x 10^5 per well in the presence of
         sperm whale or equine myoglobin at 2.5 micromolar and 8 x
         10^4 B10.S splenic adherent cells. The latter has been pre-
         incubated overnight with anti-Ia as described in Materials
         and Methods. Control cultures without myoglobin or splenic
         adherent cells gave 371cpm for sperm whale immune and 33cpm
         for equine immune T lymphocytes. P values by a 2-tailed
         student's test were: for sperm whale myoglobin: NS-1 vs
         anti I-A^S, P<.05, NMS vs anti Ia^S, P<.005; for equine myo-
         globin: NS-1 vs anti I-A^S, P<.005.

substitutions at this site are nonstimulatory. No other single
residue can explain the observed pattern of cross stimulation. We
conclude that Asp 109 is part of an antigenic determinant recognized
by a large enough proportion of T lymphocytes that they appear to
dominate the entire response to equine myoglobin.

    In a similar way, we studied the T cell determinant of sperm
whale myoglobin, by immunizing B10.S mice with sperm whale myoglobin

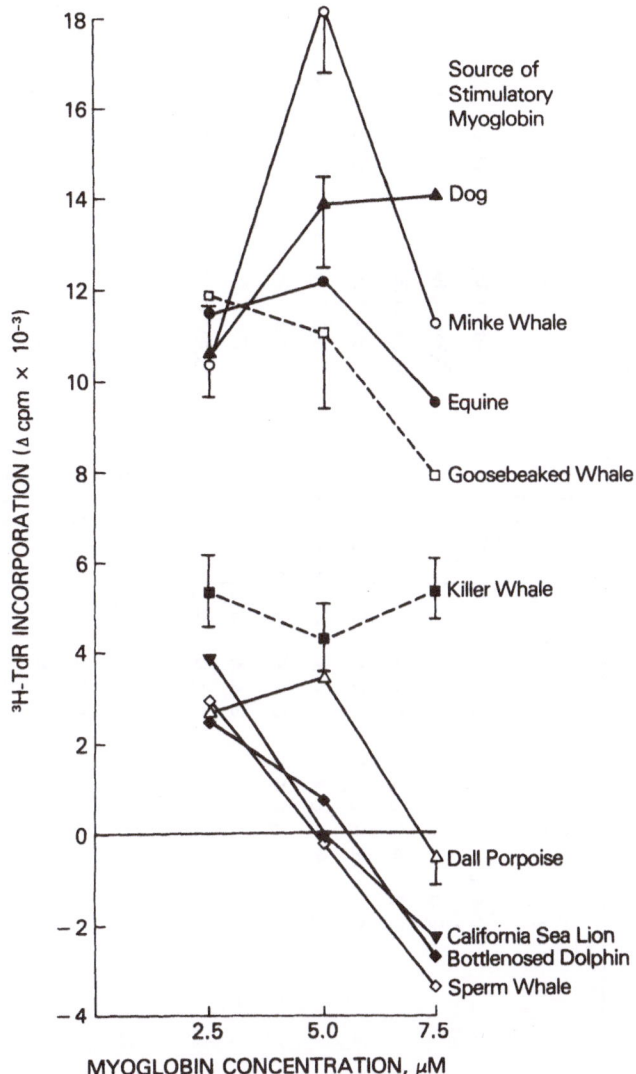

Fig. 3. Proliferative response of B10.S T lymphocytes immune to
equine myoglobin. Unpurified lymphocytes from 3 animals
were cultured at 4 x 10[5] per well in the presence of various
concentrations of each myoglobin. Control cultures without
myoglobin gave 4,862cpm.

and testing T cell proliferation to a series of myoglobin variants
(Table 2). Harbor seal and California sea lion myoglobin stimulate
and have all the substitutions of minke whale, sei whale, goosebeaked
whale and finback whale, which do not stimulate, except for Glu 109.
Of the myoglobins tested, all stimulatory myoglobins for sperm whale

Table 2. Proliferation of B10.S T Cells. For table legend see next page.

Residue numbers (read vertically):

```
                     1 1 2 2 2 3 3 3 4 6 6 7 1 1 1 1 1 1
                     4 9 2 5 1 7 8 4 5 5 6 4 0 1 2 3 4 5
                                             9 8 2 2 0 1
```

STIMULATING MYOGLOBIN	RESIDUE NUMBER (sequence)	IMMUNE TO SPERM WHALE MYOGLOBIN Δcpm (SEM)	EQUINE MYOGLOBIN Δcpm (SEM)
1. SPERM WHALE	V E L H A V D I K S R V T A [E] R D N K Y	18,593 (1.34)	3,500 (1.12)
2. KILLER WHALE	G D N G L G K N [E] F	16,745 (1.15)	2,568 (1.18)
3. RABBIT	G D N G L E V H T K N K S N F	14,351 (1.29)	–
4. DALL PORPOISE	G N G L V G K N G [E] F	7,654 (1.04)	1,759 (1.11)
5. HARBOR SEAL	G D N G L E V K N G [K] E K N F	9,592 (1.32)	3,132 (1.33)
6. CALIFORNIA SEA LION	G D N G L E V G K K G K K N F	6,869 (1.24)	683 (1.24)
7. SEI WHALE	D N G K N G D F	1,205 (1.60)	20,927 (1.21)
8. MINKE WHALE	D N G K N G D [E] F	3,694 (1.16)	13,604 (1.25)
9. GOOSEBEAKED WHALE	G L E G K H G D T F	–	10,373 (1.15)
10. FINBACK WHALE	D N G K N G D F	–	–
11. PIG	G D N G E V G K N G K S N F	–448 (1.24)	–
12. DOG	G D I N G L E V N K N G D K K N F	–	6,863 (1.09)
13. EQUINE	G D Q N G I E V T G K T V G D K N T N F	–72 (1.82)	11,584 (1.07)
MEDIUM CONTROL		611	1,164

Table 2. Legend
In the sperm whale experiment, lymph node cells from four
B10.S mice, immunized with 100 micrograms sperm whale
myoglobin, were passed through an 0.2g nylon wool column,
and 4 x 10⁵ cells were cultured in the presence of sperm
whale or a variety of other myoglobins.
Results obtained at the optimal concentration of 2 micromolar are shown.
In the equine myoglobin experiment, lymph node cells from
three equine myoglobin-immune B10.S mice were purified by
passage through an 0.15g nylon wool column, and 4 x 10⁵
cells were incubated with various concentrations of each
myoglobin. Results obtained at the optimal concentration
of 2 micromolar are shown. The complete table of sequence
differences between equine and sperm whale myoglobin is
given, as well as the amino acid differences from the sperm
whale sequence, at these residues only, for the other myoglobins tested.

immune T lymphocytes share Glu 109, while all myoglobins which differ
at this position are nonstimulatory. Rabbit myoglobin cross stimulates, despite numerous amino acid substitutions outside the antigenic determinant.

    Pig myoglobin is the only one which shares Glu 109 but does not
stimulate. This finding may be due to a second substitution within
the determinant, so we compared the sequence differences between
sperm whale and pig myoglobin (Table 3), as well as all the other
stimulatory myoglobins having Glu 109. Three substitutions Lys 87,
His 113, and Ile 142 are unique to pig myoglobin and might account
for the failure to stimulate. Of these three, only His 113 is close
to Glu 109 on the surface of myoglobin in its native structure[14],
suggesting that His 113 is also part of this determinant.

    Changing to another high responder haplotype, H-2d, we found
similar results. B10.D2 mice were immunized to sperm whale myoglobin,
and T lymphocyte proliferation was tested against a variety of myoglobins (data not shown). Once again, killer whale and harbor seal
myoglobin cross stimulate, while equine and dog myoglobin do not.
The only difference from the response of H-2s mice is that California
sea lion myoglobin does not cross stimulate in H-2d mice. Comparing
sequences, we find that Glu 109 is a critical part of the antigenic
determinant recognized by H-2d T lymphocytes. Only California sea
lion and pig myoglobin have Glu 109 but fail to stimulate. Comparing
the differences between sperm whale and pig myoglobin (Table 3) reveals only one substitution, His 116 to Gln, which can explain the
failure of pig and California sea lion myoglobin to stimulate. However, His 113 may also be a part of this site, since we have no myoglobin variant with a change at residue 113 without a change at
residue 116 as well.

Table 3.

STIMULATING MYOGLOBIN	RESIDUE NUMBER
	1 4 1 1 2 2 3 4 5 5 6 7 8 8 1 1 1 1 1 1 1 1
	2 5 7 8 5 5 1 3 6 4 6 7 0 1 1 1 3 4 4 5
	1 3 6 8 2 0 2 1
1. Sperm Whale	V E H A D I S R T A V A L K I H H R N K I Y
2. Killer Whale	G D N G — — G K — — N — — — — — — — — — — F
3. Rabbit	G D N G E V T K S — N — I — V — — K S N — F
4. Dall Porpoise	G — N G — V G K — — N G — — — — — — — — — F
5. Harbor Seal	G D N G E V — K S D N G — — — — — K K N — F
6. California Sea Lion	G D N G E V G K S D K G — — — — Q K K N — F
7. Pig	G D N G E V G K S D N G V T V Q Q K S N M F

The complete table of sequence differences between sperm whale
and pig myoglobin are given, as well as the amino acid differences from the sperm whale sequence, at these residues only,
for the other myoglobins sharing Glu 109.

We conclude that residue 109 is part of a critical determinant
for both H-2^s and H-2^d high response haplotypes and for a number of
different myoglobins. The data suggest that the specificity for
Glu 109 vs Asp 109 is due to the T lymphocyte. The failure of sperm
whale and equine myoglobins to cross stimulate is due to the fact
that both myoglobins are recognized at residue 109, but they have
different amino acids substituted at this position. The apparent
immunodominance of this site could be due to its high immunogenicity
for T lymphocytes; or it could be due to binding to the antigen presenting cell in a preferred orientation, or processing by the antigen presenting cell to preferentially present a certain part of the
antigen. Alternatively, it remains a formal possibility that the
specificity resides in the macrophage, which might process myoglobins
with Glu 109 differently from those with Asp 109, but we think this
explanation less likely.

Recent work from our laboratory has produced and characterized
6 monoclonal antibodies to sperm whale myoglobin in H-2^s animals[1,6],
[15,16]. Two of these bind equine myoglobin with high affinity. The
antigenic determinants of all six are clearly not Glu 109, since the
binding affinities to a series of myoglobins are not affected by
substitutions at residue 109. Three antigenic determinants recognized by monoclonal antibodies to myoglobin have been mapped un-
ambiguously to the left side of native myoglobin (Figure 4) and consist of residues Ala 144, Lys 145, Glu 83 (Clone 1); Glu 4, Lys 79
(Clone 3.4); and Lys 140 (Clone 5). In contrast, the T cell antigenic determinant Glu 109, His 113 is located on the back surface

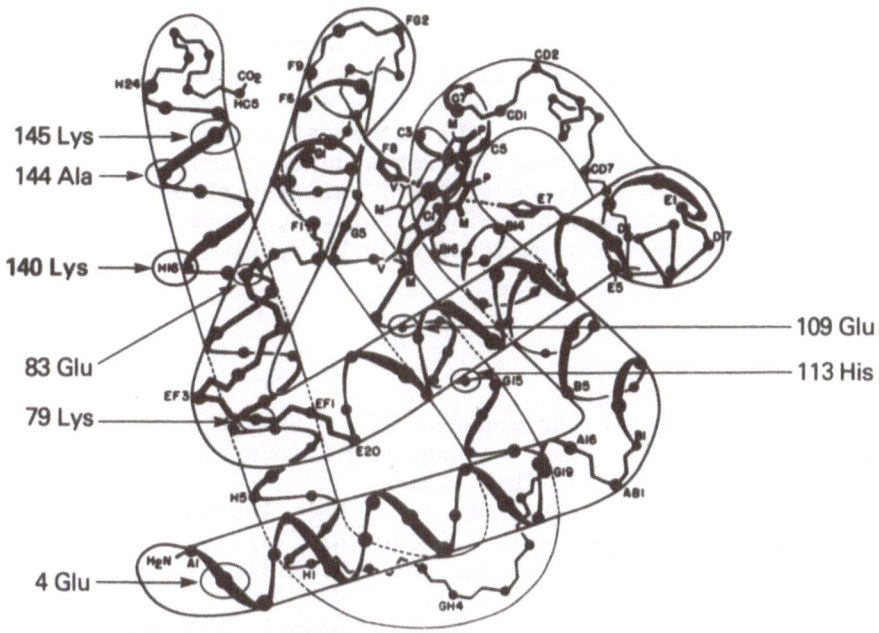

Fig. 4. Schematic representation of the native 3 dimensional
structure of myoglobin showing the binding sites for
monoclonal hybridoma antibodies[1] and for T lymphocytes.
The binding site for hybridoma clone 1 includes Ala 144,
Lys 145, and Glu 83; for hybridoma clone 3.4 includes
Glu 4 and Lys 79; for hybridoma clone 5 includes Lys
140, as shown on the left of the figure. T lymphocyte
recognition of sperm whale myoglobin depends on Glu 109
and His 113, as shown on the right of the figure.
Drawing modified from R.E. Dickerson, The Proteins
(2nd ed.), Vol.II, H. Neurath, ed. Acad. Press, (1964).
(with permission of publisher).

of native myoglobin and is indicated on the right side of Figure 4.
Additional data[17,18] plus the anti-I-A^s blocking data presented here,
suggest that both T cell and antibody sites are under control of the
same immune response gene, which maps to I-A^s. Perhaps the juxtaposition of T cell and B cell determinants is the structural basis
for genetic control of antibody specificity for fragments of myoglobin. Antigen bridging between a single T helper cell site and
nearby B cell sites, would result in selective expansion of many B
cell clones specific for sites within range of the T cell site.
Genetic control of response to the T cell site would regulate the
expression of a number of B cell clones specific for nonoverlapping
sites nearby[4]. Whether cloned myoglobin specific T cell lines will
provide site specific help to antibody producing B cells is under
investigation.

Finally, we should point out that residues 109, 113, and 116 are together on the surface of myoglobin because they are separated, respectively, by one turn of the alpha helix. Therefore, although the data do not directly address the role of tertiary structure in T cell activation, they suggest that at least secondary structure (the alpha helix) is important. Other studies are in progress to determine whether, in addition to the T cells which we know will recognize non-native fragments of the antigen, there exist T cell populations which can be stimulated only by myoglobin in its native tertiary conformation.

REFERENCES

1. J.A. Berzofsky, G.K. Buckenmeyer, G. Hicks, D.J. Killion, I. Berkower, Y. Kohno, M.A. Flanagan, M.R. Busch, R.J. Feldmann, J. Minna, F.R.N. Gurd, this volume.

2. G. Corradin and J.M. Chiller, J.Exp.Med. 149:436-447 (1979).

3. J.A. Berzofsky, J.Immunol. 120:360-369 (1978).

4. J.A. Berzofsky, L.K. Richman, and D.J. Killion, Proc.Natl.Acad. Sci.U.S.A. 46:4046-4050 (1979).

5. J.A. Berzofsky and L.K. Richman, J.Immunol. 126:1898-1904 (1981).

6. J.A. Berzofsky, G. Hicks, J.Fedorko, and J. Minna, J.Biol.Chem. 255:11188-11191 (1980).

7. J.A. Berzofsky, in: "Biological Regulation and Development," Vol.II. R.F. Goldberger ed. New York, Chap.10, pp:467-594 (1980).

8. A.S. Rosenthal, Immunol.Rev. 40:136;152 (1978).

9. A.M. Solinger, M.E. Ultee, E. Margoliash, and R.H. Schwartz, J.Exp.Med. 150:830-848 (1979).

10. G. Corradin, H.M. Etlinger, and J.M. Chiller, J.Immunol. 119: 1048-1053 (1977).

11. R.M. Maizels, J.A. Clarke, M.A. Harvey, A. Miller, and E.E. Sercarz, Eur.J.Imm. 10:509-515 (1980).

12. A.B. Edmundson, Nature 205:883-886 (1965).

13. M. Dautrevaux, Y. Boulanger, K. Han, and G. Biserte, Eur.J. Biochem. 11:267-277 (1969).

14. T. Takano, J.Mol.Biol. 110:537-568 (1977).

15. Y. Kohno, I. Berkower, J. Minna, and J.A. Berzofsky, submitted for publication.

16. J.A. Berzofsky, G.K. Buckenmeyer, G. Hicks, F.R.N. Gurd, R.J. Feldmann, and J. Minna, J.Biol.Chem. in press, (1981).

17. K. Ozato, J.K. Lunney, M. El-Gamil, and D.H. Sachs, J.Immunol. 125:940-945 (1980).

18. Y. Kohno and J.A. Berzofsky, submitted for publication.

VIIB. WHAT IS THE ROLE OF MHC AND OTHER GENES IN THE ANTIGEN-

PRESENTING CELL?

The genetic failure to respond to an antigen is often the
result of a gene or genes mapping within the major histocompatibility
complex of the species. Such <u>Ir</u> genes control the response to T-
dependent antigens, and for protein antigens are found in the I subregions of the complex. The critical question has been the nature
of the Ir gene product. Thus, does a failure to respond reflect
the lack of an appropriate I-region structural molecule that can interact with certain determinants on the antigen molecule? Or is
such a failure representative of the lack of suitable T cell receptors available for combination with antigen in the context of MHC
structures?

In the report by Rosenthal and his colleagues, a line of evidence is presented that supports the expression of Ir genes at the
antigen-presenting cell level. The B6.C-H-2^{bml2} mouse with a mutation in an I-A molecule resulting in a loss of a serological Ia
specificity (Ia. W39), fails to respond to certain proteins such
as insulin, towards which the parent can respond. All the evidence
in this and another mutant (CBA/N x B6)F1 mouse which also does not
express Ia. W39, supports the identity between this missing specificity and the Ir gene product. This is the first case in which a
structural mutation in an Ia molecule has a defined effect on the
response to a particular antigen.

In continuing this theme, Alex Miller considers the complex
point of whether the existence of a suitable Ia presentation structure for a particular epitope on an antigen such as lysozyme, ensures
that it will always be utilized. A model is presented in which the
utilization of the low resolution Ia system for highly focussed presentation is explained by making use of the T cell's specificity in

stabilizing the presenting complex. Interesting aspects of the
model are its use of Ia molecules not only in an enzymatic capacity
for antigen processing, but also later playing a key role in T cell
activation.

In paper 28, a non-MHC gene (Xid) was implicated in controlling
the expression of an Ia^b specificity. Rotman and Depasquale, in
probing the ability of mice to respond to beta-galactosidase in its
native, soluble form (without adjuvants) have discovered a most intricate example that may be very similar. An Ir gene (Ir-Z3) outside of H-2 is postulated to control expression of an Ia gene encoded
in the I-E subregion, which is usually not expressed in H-2^s mice
lacking the complementing Ir-Z3 allele.

Taylor and Morrison's manuscript attempts to explain adjuvanticity in broad biological perspective, examining several recognition
systems with low-specificity, not involving immunoglobulin. The
type of adjuvant used often sends the antibody response in separate
directions. The authors suggest that differing localization owing
to the type of adjuvant may determine the type of T cell response,
via utilization of specialized antigen-presenting cells in each
locale. Interestingly, in experiments designed to probe intrinsic
adjuvanticity, not only the hydrophobic fluorescein but also glucosamine appeared to have this property.

ANTIGEN PRESENTING FUNCTION OF MACROPHAGES

C. Shirley Lin[1], Ted. H. Hansen[1], Lanny T. Rosenwasser[2],
Brigitte T. Huber[2] and Alan S. Rosenthal[1]

[1]Department of Immunology, Merch Sharp & Dohme Research
Laboratories, Rathway, New Jersey 07065
[2]The Allergy Division, Department of Medicine & Department
of Pathology, Cancer Research Center
Tufts University School of Medicine, Boston
Massachusetts 02111

INTRODUCTION

The immune response to many natural and synthetic protein
antigens has been shown to be controlled by genes (Ir) linked to
the major histocompatibility complex (MHC). Animals possessing Ir
genes can elicit a strong immune response against a particular antigen. Animals lacking the specific Ir gene are deficient in either
humoral or cellular immunity. Ir genes were shown to control only
T cell mediated responses to T-dependent antigens[1]. Antibody production was affected secondarily as a result of this impaired T
cell – B cell interaction. Responses to T-independent antigens which
activate B lymphoctes directly are not under Ir gene control.

Two hypothesis have been suggested to account for Ir gene controlled restrictions. One hypothesis states that Ir gene defects
reflect the absence of T cell clones bearing receptors recognizing
a product between the foreign antigens and an MHC antigen[2]. The
second hypothesis proposes that Ir gene products on the cell surface
of macrophages select and interact with certain determinants on the
protein antigen which then are recognized by the T lymphocytes[3].
Using the T-cell proliferation assay, previous studies in both guinea
pigs[4] and mice[5] have provided strong evidence that immune response
genes operate at the level of the antigen-presenting cell.

A novel approach for further studies of Ir gene function has
been made possible by using mutant mice. This manuscript will review

recent findings in Ir gene expression utilizing an I-A gene mutant, B6.C-H-2^{bm12} and Xid gene mutant, (CBA/Nx B6)F$_1$ male mice.

MATERIALS AND METHODS

Animals: B10.A, B10.RIII, B10.G, C57B1/6 (B6), (B6xDBA/2)F$_1$ were purchased from the Jackson Laboratory, Bar Harbor, Maine. (CBA/N x B6)F$_1$ and (B6xCBA/N)F$_1$ mice were bred in the animal facilities at Tufts University School of Medicine, Boston, Mass. B6.C-H-2^{bm12}, (B10.AxB6.C-H-2^{bm12})F$_1$, (B10.RIIIxB6.C-H-2^{bm12})F$_1$, (B10.GX B6.C-H-2^{bm12})F$_1$ and (B10.A(4R)xB6.C-H-2^{bm12})F$_1$ were bred in the animal facilities at Merck Sharp & Dohme Research Laboratories, Rahway, New Jersey, from breeding stock originally obtained from Roger Melvold (North-western University).

Antigen. Beef insulin was obtained from Novo Research Institute, Denmark. It contained less than 1ppm of proinsulin contamination. Multichain poly (L-Tyrosine: L-Glutamic acid) poly DL-Alanine: poly-L-Lysine, (TG)AL; Lot No.MC9; average M.W. 280,000 was purchased from Miles (Kankakee, IL). Five times crystallized ovalbumin (OVA) was purchased from Miles. Type II collagen was obtained from Sigma (St. Louis, MO) and was prepared as described previously[6]. Purified protein derivative of tuberculin (PPD) was obtained from Connaught Medical Research Laboratory, Toronto, Canada.

Immunization. A mixture comprising beef insulin, (TG)AL, collagen and OVA antigens was emulsified in complete Freund's adjuvant (CFA) (0.5mg/ml killed Mycobacterium tuberculosis, H37Ra, Difco Laboratories, Inc., Detroit, MI). Mice were immunized by injection into the hind foot pads and subcutaneously into the base of the tail with a total of 0.2ml emulsion containing 50μg of each antigen.

Cell collection and separation. The T-cell proliferation assay described previously[7] was followed with some modification. Inguinal, periaortic and popliteal lymph nodes were dissected 11 to 14 days after immunization. The nodes were removed asceptically and single cell suspensions obtained by pressing between two glass slides. Following two washings, the lymph node cell suspension was passed through two consecutive nylon wool adherence columns (nylon wool, Fenwall Laboratories, Morton Grove, IL) to deplete antigen presenting cells. The enriched T-cell populations obtained were suspended in RPMI 1640 (Grand Island Biological Co., Grand Island, New York) supplemented with fresh L-glutamine (0.3mg/ml), penicillin (100U/ml), gentamicin (10μg/ml), 2-mercaptoethanol (5 x 10^{-5}M) and 10% heat inactivated fetal calf serum (Grand Island Biological Co., Grand Island, NY).

Cell culture and assay of DNA synthesis. Aliquots of 0.2ml of (1 to 1.5 x 10^5) T-cell suspension were pipetted into each flat or round bottomed well of microtiter plates (Flow Laboratories). Slpeen cells from normal, nonimmunized animals were irradiated (2000 rads) and added to the cell cultures as the source of antigen presenting cells. Ten microliters of antigens at appropriate dilutions were added to the wells. Each experimental point was set up in triplicate. The microtiter plates were incubated for 96 hours at $37°C$ in a humidified atmosphere of 5% CO_2 and 95% air. Sixteen to twelve hours before harvesting, $1\mu Ci$ of tritiated thymidine (6.7Ci/mM, New England Nuclear Co., Boston, Massachusetts) was added to each well. Cells were harvested onto glass fiber filters by the use of a microharvester.

Tritiated thymidine incorporation was then determined by liquid scintillation spectrometry and the results were reported as ΔCPM (counts per minute above control).

Antigen pulsed antigen presenting cells. Antigen presenting adherent accessory cells (APC) were obtained from two sources: spleens were aseptically removed from the animal donors, single-cell suspensions were made, and the spleen cells were allowed to adhere to sterile glass petri dishes for 2h at $37°C$. Simultaneously, the peritoneal cavities of splenic donor mice were lavaged with heparinised Hanks Balanced Salt Solution (HBSS) (10U heparin/ml), and these peritoneal washout cells (PC) were also allowed to adhere for 2h. Then, the splpeen-adherent cells and adherent PC cells were collected as described previously, pooled and treated with 40μg/ml mitomycin C (Sigma Chemical Co.) for 60min at $37°C$. These cells were then washed vigorously five times to remove excess mitomycin C and then used as accessory cells for T cell proliferation or as targets for antigen pulsing.

Blocking experiments. Anti I-A^b alloantiserum, (AxB10.A)F$_1$ anti-B10.A(5T) was obtained from the Research Resource Branch, National Institute of Allergy and Infectious Diseases, and its anti-K^b activity was absorbed with El-4 tumor cells. Anti-Ia.W39 was prepared by immunizing (N x B$_6$)F$_1$ male mice with B6 spleen cells, as previously described[8]. Normal mouse sera (NMS) from nonimmune male (N x B6)F$_1$ mice was used as a control. To assess the blocking effect of antisera, 1% NMS, conventional anti-I-A^b, or anti-Ia.W39 were added to the start of culture and were present continuously in the cultures.

RESULTS AND DISCUSSION

An Aβ Gene Mutation Resulted in a Selective Loss of Beef Insulin Response

The H-2^bm12 haplotype was first detected as a gain-loss mutation in skin graft studies using (B6 x Balb/C)F$_1$ mice. Breeding studies

Table 1. Comparison of I-A Controlled Proliferative Responses
Between B6.H-2bm12 and B6 Mice

| Strain | (^{3}H) Thymidine Incorporation, ΔCPM | | |
	Beef Insulin 100μg/ml	(TG)AL 100μg/ml	Collagen 100μg/ml
B6.C-H-2bm12	6,363	70,803	27,830
B6	42,224	68,675	36,861

demonstrated that the mutation occurs in the H-2 haplotype of the
B6 mice. The mutant haplotype was made homozygous and coisogenic
with B6, thus establishing the B6.C-H-2bm12 strain. Besides skin
graft rejection[9,10], the H-2bm12 mutation also induces the mixed
lymphocyte response[9], in vitro cytotoxicity[11] and exhibits changes
in the serologically detected Ia specificities[9,12]. Recently, biochemical data suggested that the mutation is a point mutation occuring in the Aβ chain of the AαbAβb Ia molecules[13]. Using the T-cell
proliferation assay, responses to beef insulin, (TG)AL and collagen
were compared between B6.C-H-2bm12 and B6 mice. Table 1 shows that
B6.C-H-2bm12 mice fail to respond to beef insulin, while other I-A
controlled responses were comparable with those of B6 mice. This
data suggests that a point mutation results in a selective loss in
beef insulin response. Previous studies demonstrated that the mutation also resulted in changes in Ia specificities[9,12]. The present study provides the first direct evidence that Ir gene products,
which mediate the immune response, are Ia molecules[14].

The Defect in Ir Gene Function in B6.C-H-2bm12 is at the Antigen Presenting Cell Level

To investigate the nonresponsiveness of B6.C-H-2bm12 mice to
beef insulin, an APC depleted T-cell population from (B6 x B6.C-H-
2bm12)F$_1$ was tested for its proliferative response to beef insulin
in the presence of APC's from F$_1$ or either parental strain. Table 2
shows that (B6. x B6.C-H-2bm12)F$_1$ T cells when challenged with beef
insulin proliferate in vitro and incorporate (^{3}H) thymidine in the
presence of F$_1$ or B6 antigen presenting cells but not in the presence
of APC from B6.C-H-2bm12 mice. When challenged with (TG)AL or collagen, (B6 x B6.C-H-2bm12)F$_1$ T cells proliferate equally well in
the presence of APC from F$_1$, B6 or B6.C-H-2bm12 mice. The data
strongly suggest that Ir gene expression is at the APC level and not
at the T-cell level.

Table 2. Proliferative Response of (B6 x bml2)F_1 T Cells in the
          Presence of Antigen Presenting Cells from F_1 or Parental
          Strains

Irradiated Spleen Cells	(^{3}H) Thymidine Incorporation, CPM		
	Beef Insulin 100μg/ml	(TG)A-L 100μg/ml	Collagen 100μg/ml
(B6 x bml2)F	28,746	43,294	10,654
B6	38,527	39,458	8,213
bm 12	8,935	26,537	8,986

Earlier studies show that H-2^b mice recognize an amino acid
sequence situated in the A8-A9-A10 region of the A chain of beef
insulin[15]. Our present data thus suggest that the B6.C-H-2^{bml2} mutation results in an alteration of the Ia molecules on the cell surface of APC. The mutated Ia molecules fail to recognize the small
defined area on the A chain of beef insulin. This together with
studies in progress defining the amino acid changes in the Aβ polypeptide of B6.C-H-2^{bml2} mice, may provide insight into complementing
sites of interaction between Ia and restricted domains on the antigen.

Complementation in Beef Insulin Response of (I-A^k x I-A^{bml2})F_1 Hybrid

The F_1 hybrids of two nonresponder strains, (B10.A x B6.C-H-
2^{bml2})F_1 mice were shown to be good responders to beef insulin[14].
Recently, we have tested F_1 hybrids between B6.C-H-2^{bml2} and other
nonresponder strains, B10.RIII and B10.G for their responses to beef
insulin. Table 3 shows that the complementation is only observed
in (I-A^k x I-A^{bml2})F_1 mice. Neither (I-A^r x I-A^{bml2})F_1 nor (I-A^q
x I-A^{bml2}) respond to beef insulin. There are several reports of
F_1 hybrids produced from matings of two low responder parental
strains giving responses higher than that of either parental
strain[16,17]. Our data show that hybrid mice (I-A^k x I-A^{bml2}) respond
very well to beef insulin thus demonstrating complementation. Complementation of the functional lesion in B6.C-H-2^{bml2} mice has been
observed in other assays. B6.C-H-2^{bml2} was found not to respond
to H-Y antigen; however, hybrid mice (I-A^k x I-A^{bml2}) have been reported to respond similar to B6 mice against H-Y antigen in skin
graft rejection[18] and in the F_1 anti-parent proliferation assay[19].
Complementation of the immune response in other test systems has

Table 3.   Complementation in Proliferative Responses to Beef
           Insulin

| Strain | (^{3}H) Thymidine Incorporation, ΔCPM | |
	Beef Insulin 100μg/ml	(TG)AL 100μg/ml
(B10.A x bm12)F$_1$	35,199	23,268
(B10.RIII x BM12)F$_1$	2,458	11,207
(B10.G x bm12)F$_1$	4,845	11,196

been attributed to hybrid Ia molecules ie. an alpha and a beta polypeptide from each parent[20]. If this also holds true with the complementation seen in studies with B6.C-H-2^{bm12} mice, our data suggest
that an Ia^k polypeptide hybridizes with an Ia^{bm12} polypeptide to form
an Ia molecule which functions normally in certain immune responses.

The Complementing Ir Genes are Expressed at the APC Level

     Another (I-A^k x I-A^{bm12})F$_1$ hybrid (B10.A(4R) x B6.C-H-2^{bm12})F$_1$,
was also found to be a good responder to beef insulin. Table 4 shows
that in the presence of APC from F$_1$, (B10.A(4R) x B6.C-H-2^{bm12})F$_1$
T cells respond well to beef insulin. APC from either parental
strain are not able to present beef insulin to the F$_1$ T cells. These
data also suggest that complementing hybrid Ia molecules are expressed at the APC level.

     The proliferative response to polypeptide poly (Glu^{55}Lys^{36}Phe^9)n
(GLφ) has been shown to be controlled by two separate immune response
genes. One maps in the I-A subregion, the other in the I-E/C subregion[21]. The complementation in response to GLφ has been observed
in the recombinant (cis-complementation) and in the F$_1$ hybrids
(trans-complementation) of two nonresponder parental strains[16].
Previous studies usning radiation chimeras demonstrated that at least
one cell type had to express both gene products to function[22].
Further studies using the same type of chimera demonstrated that
neither I-A nor I-E/C gene product has to be present in the T cell
to generate a proliferative response to GLφ[23]. These results indicated that the APC was the cell type which expressed both Ir-GLφ
gene products.

Table 4. Response to Beef Insulin of (B10.A(4R) x (B6.C-H-2^{bm12})F$_1$ T Cells in the Presence of Antigen Presenting Cells from F$_1$ or Either Parental Strains

Irradiated Spleen Cells	(H^3) Thymidine Incorporation, $\triangle$CPM Beef Insulin 100µg/ml	Medium Control CPM
F$_1$	26,132	5,186
B6.C-H-2^{bm12}	9,104	1,679
B10.A(4R)	5,683	2,592

Xid Gene Mutant (CBA/N x B6)F$_1$ Mice Also Fail to Response to Beef Insulin

Recently, one of the Ia specificities missing in B6.C-H-2^{bm12} mice was found to be Ia.W39[24] which was a private specificity of I-A^b. Ia.W39 is regulated via the X chromosome-linked Xid gene. The Xid gene regulates the expression of the late-appearing B cell differentiation antigens Lyb-3[25] Lyb-5[26], Lyb-7[27] and Ia.W39. Mutant CBA/N mice also possess the Xid gene defect. Male (CBA/N x B6)F$_1$ lacks Ia.W39 while female (CBA/N x B6)F$_1$ and male (B6 x CBA/N)F$_1$ express Ia.W39 on their cell surface. Table 5 shows that Xid gene defective (CBA/N x B6)F$_1$ male mice do not respond to beef insulin while the other two normal F$_1$ hybrids are good responders to beef insulin[28].

Table 5. Proliferative Response of Immune F$_1$ T Cells

Strain	(^{3}H) Thymidine Incorporation, $\triangle$CPM Beef Insulin 100µg/ml	TNP-OVA 100µg/ml	PPD 10µg/ml
Female (CBA/N x B6)F$_1$	12,280	49,680	6,980
Male (CBA/N x B6)F$_1$	2,300	42,080	4,910
Female (B6 x CBA/N)F$_1$	17,450	ND	7,580

Table 6.   Effect of Anti-Ia Sera on Proliferative Response of
           Immune (CBA/N x B6)F Female T Cells to Beef Insulin

Antisera Added in Culture	(^{3}H) Thymidine Incorporation, ΔCPM	Percent Inhibition
1% NMS	5,156	-
1% Anti-Ia.W39	1,246	76
1% Anti-I-A^b	0	100

Ia.W39 is Associated with Beef Insulin Response

Anti-Ia.W39 antibody was added to the proliferative assay to
determine if the response to beef insulin was inhibited. Table 6
shows that anti-Ia.W39 antibody greatly inhibits the response to
beef insulin in (CBA/N x B6)F$_1$ female mice. The data in Table 5
and Table 6 strongly indicate that Ia.W39 is associated with the
beef insulin response in B6 mice. The findings that B6.C-H-2^{bml2}
mice have lost Ia.W39 and are nonresponders to beef insulin also
strongly support the view that Ia.W39 is the Ia specificity which
recognizes beef insulin in B6 mice. The mechanisms by which I-A^{bml2}
and Xid genes control the beef insulin response may be different.
The I-A^{bml2} gene appears to be a simple mutation in a structural
gene whilst the Xid gene appears to be a regulating gene which influences several immunologic functions[29].

Other studies have shown that alloantisera blockade in the T
cell proliferative response is at the APC level[30,31]. Experiments
using monoclonal antibodies and chimeras provided evidence showing
that the inhibition of proliferative response by anti-Ia reagents
occurs at the level of the APC[32]. The data shown in Table 6 therefore imply that Ia.W39 is expressed on the surface of the APC.

Only Ia.W39-Positive Antigen Presenting Cells Recognize Beef Insulin

In order to investigate the expression of Ia.W39 at the cellular
level, female (CBA/N x B6)F$_1$T cells were cultured with beef insulin-
pulsed APC from (CBA/N x B6)F$_1$ female or nonresponder (CBA/N x B6)F$_1$
male mice. Table 7 demonstrates that Ia.W39 negative macrophages
fail to present beef insulin to female (CBA/N x B6)F$_1$ T cells.
The proliferative response to TNP-OVA were comparable when macrophages from either female or male mice were used. The data further
support the hypothesis that Ir gene products are expressed at the
APC level.

Table 7. Response of Immune (N x B6)F Female T Cells to
Antigen-Pulsed Adherent Presenting Cells

Antigen Pulsed APC	(^{3}H) Thymidine Incorporation, ΔCPM
F_1 female, beef insulin	11,090
F_1 female, TNP-OVA	24,280
F_1 male, beef insulin	1,770
F_1 male, TNP-OVA	19,320

SUMMARY AND CONCLUSION

The B6.C-H-2^{bm12} mutant is an I-A mutation. Recent biochemical
data suggest that it is a point mutation occurring in the structural
gene coding for the Aβ^b chain of the Aα^bAβ^b Ia molecule. The mutation has resulted in gain and loss of serologically detected Ia
specificities. CBA/N mutant mice possess a recessive defect on Xid
gene which is linked to the X-chromosome. The Xid gene is a regulatory gene which controls the expression of the Ia^b specificity,
Ia.W39. Hybrid male (CBA/N x B6)F_1 mice have the Xid gene and Ia.W39
has been found to be synthesized in the cell cytoplasm of these mice,
although it is not expressed on the cell membrane.

This manuscript reviews the studies of Ir gene function by
using B6.C-H-2^{bm12} and male (CBA/N x B6)F_1 mutant mice. The immune
response to several natural and synthetic protein antigens were compared between B6.C-H-2^{bm12} and B6 mice. B6.C-H-2^{bm12} was found not
to respond to beef insulin while other I-A controlled responses such
as responses to (TG)AL and collagen were comparable with the responses in B6 mice. The nonresponsiveness to beef insulin in the
mutant was found to be at the antigen presenting cell level. Complementation in the beef insulin response was observed in the F_1
hybrid of two nonresponder parental strains. The complementing genes
were found at the antigen presenting cell level. B6.C-H-2^{bm12} was
found lacking Ia.W39 which was a private specificity of B6. Another
mutant, (CBA/N x B6)F_1 male was found also lacking Ia.W39 and not
to respond to beef insulin. The expression of Ia.W39 was found at
the antigen presenting cell level. Major conclusions drawn from
our present studies are: (1) that Ia molecules are Ir gene products;
(2) a small defined area of an antigen interacts with a restricted
region on the Ia molecules and (3) the product resulting from the
interaction of the antigen and the Ia molecules is located on the
antigen presenting cell surface.

ACKNOWLEDGEMENTS

The authors would like to thank Mrs. Marilyn M. Serson for her excellent work in preparing this manuscript.

REFERENCES

1. B. Benacerraf and H. O. McDevitt, Science 175:273 (1972).
2. H. von Boehmer, W. Hass and N. K. Jerne, Proc. Natl. Acad. Sci. USA 75:2439 (1978).
3. A. S. Rosenthal, Immunological Reviews 40:136 (1978).
4. E. M. Shevach and A. S. Rosenthal, J. Exp. Med. 138:1213 (1973).
5. R. H. Schwartz, C. S. David, M. E. Dorf and B. Benacerraf, Proc. Natl. Acad. Sci. USA 75:2387 (1978).
6. L. J. Rosenwasser, R. S. Bhatnager and J. D. Stobo, J. Immunol. 124:2854 (1978).
7. L. J. Rosenwasser and A. S. Rosenthal, J. Immunol. 120:1991 (1978).
8. B. T. Huber, Proc. Natl. Acad. Sci. USA 76:3460 (1979).
9. I. F. C. McKenzie, G. M. Morgan, M. S. Sandrin, M. M. Michaelides, R. W. Melvold and H. I. Kohn, J. Exp. Med. 160:1323 (1979).
10. T. H. Hansen, R. W. Melvold, J. S. Arn and D. H. Sachs, Nature (London) 285:340 (1980).
11. L. P. de Waal, C. T. M. Melief and R. W. Melvold, Eur. J. Immunol. 11:258 (1981).
12. G. M. Morgan, I. F. C. McKenzie and R. W. Melvold, Immunogenet. 11:1 (1980).
13. D. J. McKean, R. Melvold and C. David, Immunogenet., in press (1981).
14. C. S. Lin, A. S. Rosenthal, H. C. Passmore and T. H. Hansen, Proc. Natl. Acad. Sci. 78:6406 (1981).
15. L. J. Rosenwasser, M. A. Barcinski, R. H. Schwartz and A. S. Rosenthal, J. Immunol. 123:471 (1979).
16. M. E. Dord, Springer Seminars in Immunopathology 1:171 (1978).
17. A. M. Solinger and R. H. Schwartz, J. Immunol. 124:2485 (1980).
18. M. Michaelides, M. Sandrin, G. Morgan, I. F. C. McKenzie, R. Ashman and R. W. Melvold, J. Exp. Med. 153:464 (1981).
19. C. G. Fathman, M. Kimoto, R. Melvold and C. David, Proc. Natl. Acad. Sci. 78:1853 (1981).
20. E. A. Lerner, L. A. Matis, C. A. Janeway Jr., P. P. Jones, R. H. Schwartz and D. B. Murphy, J. Exp. Med. 152:1085 (1980).
21. R. H. Schwartz, M. E. Dorf, B. Benacerraf and W. E. Paul, J. Exp. Med. 143:897 (1976).
22. R. H. Schwartz, A. Yano, J. H. Stimpfling and W. E. Paul, J. Exp. Med. 149:40 (1979).
23. D. L. Longo and R. H. Schwartz, J. Exp. Med. 151:1452 (1980).
24. B. T. Huber, T. H. Hansen and D. A. Thorley-Lawson, Fed. Proc. 40:4135 (1981).
25. B. T. Huber, R. K. Gershon and H. Cantor, J. Exp. Med. 145:10 (1977).

26. A. Ahmed, I. Scher, S. O. Sharrow, A. H. Smith, W. E. Paul, D. H. Sachs and K. W. Sell, J. Exp. Med. 145:101 (1977).
27. B. Subbarao, D. E. Mosier, A. Ahmed, J. J. Mond, I. Scher and W. E. Paul, J. Exp. Med. 149:495 (1979).
28. L. J. Rosenwasser and B. T. Huber, J. Exp. Med. 153:1113 (1981).
29. T. H. Huber and D. A. Thorley-Lawson, Proc. Natl. Acad Sci., in press (1981).
30. R. H. Schwartz, C. S. David, D. H. Sachs and W. E. Paul, J. Immunol. 117:531 (1976).
31. D. W. Thomas, U. Yamashita and E. M. Shevach, J. Immunol. 119:223 (1977).
32. E. M. Shevach, Springer Sem. Immunopathol. 1:207 (1978).

THE SORTING-OUT PROBLEM IN ANTIGEN PRESENTATION

Alexander Miller

Department of Microbiology
University of California
Los Angeles, CA 90024, USA

It has become increasingly clear in recent years that there exists a severe restriction in antigen presentation to a large class of T cells. Even when so unrestricted an assay as induction of proliferation (as opposed to generation of help) is used, it is generally found that only small regions (and, in many cases, a single region) of protein antigens are presented. Such a result is amply illustrated in this volume in the papers of Lin et al.; Berkower et al.; Thomas et al.; Krzych et al.; Rude et al.; and Corradin et al.

This restricted recognition and, even, absence of recognition, has clearly been linked to the Ia region in the mouse and to a probably similar region in the guinea pig. As yet, no formal proof exists that the restrictions reflect a limitation in antigen-processing per se rather than a limitation in the functional T cell library. However, several lines of evidence make the latter possibility unlikely[1-3]. As we showed several years ago[4], response to human lysozyme occurs in mice of haplotypes k, v and r but not in b, d, q, s, and u mice. High response is also found in B10.A(4R), AQR and ATL mice, but minimal responsiveness occurs in B10.A(3R) and B10.A(5R) mice, mapping the crucial gene for responsiveness to IA. This same pattern of responsiveness had been found to low levels of (H,G)-A-L, bovine γ-globulin, and ovomucoid[8] and, more recently to human fibrino-peptide B[6]. Thus the Ir-gene control of response to five antigens showing no obvious epitypic relationship shows congruence with respect to strain distribution pattern.

In a more recent study from our laboratory[7], it was found that the T cell proliferative response to HEL (see Figure 1) in H-2^d mice was largely restricted to a determinant within the C-terminal peptide, a.a. 106-129. The response to REL was similarly restricted to this

Amino Acid	70	75	80	85	90	95	100	105	110	115	120	125

```
HEL   DGRTPGSRNLCNIPCSALLSSDITASVNCAKKIVSDGNGMNAWVAWRNRCKGTDVQAWIRGCRL
JEL   --------------------------------------------VH----------------N--------
TEL   -------K------------------------------A-G-------------------H--------
REL   -------K---H----------------------------------------KH------NV-------
NEL   ------------------Q----TA---------------------------KH------RV--K----
OEL   ------TK---H-S-----MGA-AP--R---R---------------------KH------ST--KD-K-
```

Fig. 1. Partial sequences of several avian egg-white lysozymes.
These lysozymes are: HEL: hen (chicken); JEL: Japanese
quail; TEL: turkey; REL: ring-necked pheasant; NEL:
Numida (guinea-hen); OEL: Ortalis (chachalaca). Peptide
106-129 is prepared by cyanogen bromide cleavage at methionine 105.

region but no cross-reactivity was observed. However, the HEL-primed
response was recalled by JEL and TEL, while the REL-primed response
was recalled by NEL and OEL. It can be seen from Figure 1 that HEL,
JEL and TEL uniquely share asparagine-113 and arginine-114 while REL,
NEL and OEL share lysine-113 and histidine-114. This result is consistent with the idea that the antigen-presenting mechanism restricts
T cell recognition to a favored determinant region of homologous
molecules. Depending on the amino acid sequences in this region,
totally non-crossreactive sets of T cells may be activated. In support of this idea, are the studies of Hansburg et al.[8] with the carboxy terminal peptide (a.a. 81-104) of pigeon cytochrome c and its
acetimidyl derivative. These two peptides were found to be antigenically distinct with regard to induction of T cell proliferation
in B10.A mice. Nevertheless, parallel patterns of cross-reactivity
were found when a set of native and derivatized peptides of cytochrome
c from several other species were tested. Thus, it appears that the
antigen-presenting mechanism plays a critical focussing role for
specific parts of a given antigen.

     There are two other aspects of the antigen-presenting mechanism
which are relevant to the model considered below. The focussing is
not absolute but hierarchical. To illustrate, when HEL is used to
stimulate T cells of B10.A mice to proliferation, the responding
cells are almost entirely specific for a determinant within a peptide
consisting of a.a. 13 to 105. Little or no recall is obtained with
the carboxy terminal peptide, a.a. 106-129. However, when the latter
peptide is used to prime mice there is a strong response which is
recalled by not only the peptide but equally well by HEL. Similarly,
it has been shown in our laboratory (Krzych et al., this volume) that
several peptides of E. coli beta-galactosidase will generate T helper
cells active in inducing an in vitro anti-hapten response when hapten-
ated-beta-galactosidase is used as an antigen. Nevertheless, the
entire molecule itself (in either native or denatured form) does not

generate helpers of the above peptide specificities but instead, generates helpers of another specificity.

Another aspect of the presenting system is that there is nearly complete cross-reactivity between native and denatured protein, e.g. native HEL and its reduced, carbozymethylated derivative. It is straightforward to postulate an orderly proteolysis of native protein leading to preservation of a favored fragment equivalent to a fragment of denatured protein. However, it is necessary to postulate a special degrading mechanism to explain how denatured protein (which must be sensitive to proteolysis at many points) is reduced to exactly that fragment which is produced from native protein.

The Sorting-Out Paradox

If a relatively precise focussing function is attributed to the presenting mechanism, then a paradoxical situation exists. It appears likely that presentation is associated with molecules apparently coded for within the Ia complex both on the basis of genetic analysis and isolation of antigen-Ia complexes. Analysis of these molecules by isoelectric focussing, SDS-PAGE gels, peptide mapping etc., gives no indication of great complexity though serological analysis does indicate that some heterogeneity is not being picked up by conventional protein analysis. However, there is certainly no indication of heterogeneity approaching that achieved through the V-gene system.

So one is faced with utilization of a relatively low resolution system for highly focussed presentation. It would appear that this would present a huge "sorting-out" problem. That is, on the average one would expect the presenting structure to be binding the "wrong" peptide fragment and therefore not be available for presentation to any given specificity T cell.

Utilization of the T Cell Specificity

A way around the problem of limited specificity is to make use of the much higher order specificity of the target of presentation - the T cell. The crux of the model to be presented is that union of any peptide fragment and presenting structure is extremely transient unless the peptide fragment is recognized by a T cell. This recognition induces an effector function in the T cell receptor which feeds back on the Ia element of the presenting structure and, in turn, induces a stabilization of the presenting complex.

A Model Involving Ia Molecule as Proteases

Although it is possible to construct models similar to the one which follows in which Ia molecules have no direct enzymatic (or

pseudoenzymatic) function but serve to modify peptidases, for simplicity the two functions of peptidase activity and modifiability
will be combined in single molecules.

It will be assumed here that the first steps in presentation
of a native protein antigen are reduction of the antigen to a set
of peptide fragments (thus making the native protein equivalent to
the denatured protein). It is also assumed that these peptide fragments are held on or within the presenting cell by a relatively non-
specific mechanism. (For generation of antibody requiring B-T collaboration, the B cell receptors may perform an important directive
binding function, both shielding part of the molecule and allowing
attack on other parts). The peptide fragments are accessible to Ia
molecules, eg. IA and IE especially, which act as peptidases of limited specificities whose specificities will vary with H-2 haplotype.
For most portions of most fragments, rapid cleavage will occur and
these will be unavailable for T cell binding. However, certain bonds
will be relatively resistant to cleavage and a peptide fragment -Ia
protease complex with somewhat extended lifetime will be created.
The complex formation will impart a favored conformation to the amino
acids near those directly interacting with the Ia molecule. T cells
recognizing this conformation will bind to the peptide. (Thus, the
failure of T cells to bind to free peptide may be attributed to low
concentration in solution of a preferred conformation. T cells directed, in part, against strongly binding haptens might well bind
free antigen in the absence of Ia).

Alternatively, Ia molecules may be viewed as only binding molecules of limited, but far from singular specificity, but with no
hydrolytic function. In this model, peptide fragments with the most
stable association with Ia would both be most protected from enzymatic attack and most likely to react with T cells recognizing the
peptide in its particular bound conformation. However, this version
of the model is less compatible with the occurrence of stable soluble
complexes of (processed) antigen and Ia bearing molecules (see below).

It is reasonable to expect that some mechanism exists which prevents recognition of bare Ia molecules as antigen by T cells. However, this does not preclude recognition of a "combined" epitope
consisting of elements of a bound peptide and amino acid moieties
of Ia which are in or near the binding site. Such combined binding
may explain the recently observed shifts in peptide specificity for
T cell clones when peptide clones are presented by cells of different
haplotype[9].

Alteration of Ia Function

It is an essential element of this model that the interaction

of Ia and peptide fragment is transient in the absence of T cells.
That is, free Ia molecules continually become available either through
hydrolysis or dissociation of bound peptide fragments. After T cell
binding to peptide-Ia complex, it is imagined that a change is induced
in the Ia molecule which leads to stabilization of the peptide-Ia
complex. For example, it might be that Ia molecules function as enzymes only when bound in a particular orientation to presenting cell
membrane elements and interaction with T cell destabilizes this
binding. Such a scenario could lead naturally to the occurrence of
stable soluble peptide-Ia complexes. In any case, Ia molecules would
thus be preempted by binding to a specific peptide fragment and then
only be available to T cell receptors recognizing that fragment.

Alteration of T Cells by Ia

It has been shown that T helper cells can be specific for haptens
and that T cells, in general, can be specific for all manner of molecules. In the present model, although peptide fragments are seen,
they are seen in a restricted conformation and, thus, present an
epitope not different in kind from that present on native proteins.
It follows, then, that T cells are not selected for expansion merely
on the basis of binding to an epitope. It seems necessary to postulate a role for Ia in activation of the T cell. If T cell receptors
are monovalent, it may be that they are aggregated through Ia-
dependent interactions and that such aggregation is necessary as a
proliferation signal.

Breadth of the T Cell Response

The observed restriction in determinant recognition belies the
potential for response at both the T and presenting cell level. In
recent studies from our laboratory (F. Manca, E. Sercarz and
A. Miller, submitted for publication) it has been found that individual T cell clones derived from mouse lymph node cells primed with HEL
are all directed against a single peptide. (a.a. 74-96). However,
crossreactivity studies with other avian lysozymes of only 16 of these
clones subdivided these clones into at least five specificity groups.
This result suggests a degree of heterogeneity not unlike that found
with antibodies. As described above, the limited presentation observed with a single protein may be hierarchical rather than absolute. That is, in the absence of the primary determinant, a second
determinant may be well presented. Furthermore the presentation of
determinants on most proteins by what appear to be a limited number
of Ia elements is consistent with relatively broad recognition but
dominance by a single or, at least, a small number of peptide fragments as described above.

Apologia

It is clear that our knowledge of the elements involved in antigen presentation and in T cell recognition and interaction with presenting cells are still primitive. Nevertheless, there may be some value in a model which attempts to deal with higher restricted presentation and discrimination by a relatively low information system. The Ia complex at present appears to be such a system. This model involved the specificity of the T cell to improve the resolution of the Ia presenting system.

ACKNOWLEDGEMENTS

This work was supported in part by Grants AI-11183 and CA-24442 from The National Institutes of Health. It is a pleasure to thank Mrs. Vicky Godoy for preparation of the manuscript.

REFERENCES

1. A. S. Rosenthal, Determinant selection and macrophage function in genetic control of the immune response, Immunol. Rev. 40: 136-152 (1978).
2. B. Banacerraf, A hypothesis to relate the specificity of T lymphocytes, J. Immunol. 120:1809-1812 (1978).
3. A. Miller, A model implicating altered macrophage function in H-2 linked nonresponsiveness to hen lysozyme, in:"Immunobiology of Proteins and Paptides-I," (Adv. exp. med. biol. 98) (M.A. Atassi and A. B. Stavitsky), p. 131-139, Plenum Press, New York (1978).
4. D. E. Kipp, L. Adorini, S. Hill, A. Miller and E. E. Sercarz, Partial overlap of Ir gene-controlled responses to two proteins of limited relatedness: hen egg-white lysozyme and human lysozyme, Ann. Immunol. 131:311-325 (1980).
5. J. Klein, Genetic control of immune response, in:"Biology of the mouse histocompatibility-2 complex," p.411-448, Springer-Verlag, New York (1975).
6. L. B. Peterson, G. D. Wilner and D. W. Thomas, Murine T lymphocyte proliferative responses to human fibrinopeptide B, Fed. Proc. 39:1127 (1980).
7. M. E. Katz, R. M. Maizels, B. A. Araneo, A. Miller and E. E. Sercarz, The MHC regulates different specificity restrictions in the T cell proliferative response of mice to the lysozymes, Fed. Proc. 39:1127 (1980).
8. D. Hansburg, C. Hannum, J. K. Inman, E. Appella, E. Margoliash and R. H. Schwartz, Parallel cross-reactivity pattern of 2 sets of antigenically distinct cytochrome c peptides: possible evidence for a presentational model of Ir gene function, J. Immunol. 127:1844-1851 (1981).

9. E. Heber-Katz, R. H. Schwartz, L. A. Matis, C. Hannum, T. Fairwell, E. Appella and D. Hansburg, The contribution of antigen-presenting cell MHC-gene products to the specificity of antigen-induced T cell activation, J. Exp. Med. 155:1086 (1982).

GENETIC ANALYSIS OF RESPONSIVENESS TO ADJUVANT-FREE IMMUNOLOGICAL SIGNALS: TRANSFER OF GENES FROM RESPONDER STRAINS (SJL/J AND CE/J) TO A NONRESPONDER BACKGROUND UNCOVERS A REQUIREMENT FOR H-2 HETEROZYGOSITY*

Boris Rotman and Michael J. Depasquale

Division of Biology and Medicine
Brown University, Providence
Rhode Island 02912

In general, soluble protein antigens administered as deaggregated adjuvant-free solutions serve as poor immunologic signals[1-3]. Although this unresponsiveness to soluble proteins has been under investigation over the years, only recent studies with inbred mice (using bacterial β-D-galactosidase as the antigen) have demonstrated that it is under genetic control[4]. It was observed that while the majority (94%) of the inbred strains tested were unresponsive, SJL/J and CE/J mice produced high levels of specific IgG. This immune response (Ir)[1] was shown to depend on two genetic loci, Ir-Z1 and ir-Z2, each segregating (in F_2 and backcrosses) as a single, autosomal gene: Ir-Z1 of SJL/J as dominant and ir-Z2 of CE/J as recessive. Since there is no discernible association between H-2 haplotype (of independent origin) and responsiveness[4], we concluded at that time that the chromosomal location of Ir-Z genes differs from that

---
*Supported by United State Public Health grant CA-24440 from the National Cancer Institute.
Abbreviations used in the paper: Ir, immune response; Z, β-D-galactosidase from Escherichia coli (EC 3.2.1.23).

of well known Ir genes which are either closely linked to the H-2
region or to immunoglobulin structural genes. Based on these prem-
ises, we introduced Ir-Z genes into the nonresponder background of
C57BL/6J by using conventional backcrossing[5] in conjunction with
selective breeding of responders. Our expectation was that this
procedure would yield congenic inbred lines differing in responsiveness to adjuvant-free β-D-galactosidase. Accordingly, responder mice
obtained from F1 crosses between C57BL/6J (a nonresponder) and either
SJL/J or CE/J were successively backcrossed to C57BL/6J. From each
backcross, responder mice were selected for a subsequent backcross
to C57BL/6J. After a series of 12 backcrosses, brother-sister responder mice were subjected to inbreeding for several generations
in order to promote generalized homozygosity.

     Contrary to original expectations, this selective procedure gave
the following results: i) the frequency of responders among intercross
progeny was very low (25% instead of either the 75% expected for a
dominant gene or the 100% for a recessive); furthermore, the frequency
was the same as that observed among backcross progeny; ii) responders
carried the H-2 allele of the responder donor strain (ie. H-2^s or
H-2^k); iii) responders were H-2 heterozygotes.

     The stringency of these data allowed us to construct a genetic
model with predictive value in which responsiveness to Z requires
allelic complementation at each of two epistatic genes. One of the
genes is closely linked to H-2 (for simplicity, the H-2 haplotype
is used here to designate this gene); the other gene, termed Ir-Z3
(not necessarily related to either Ir-Z1 or ir-Z2), is not linked
to H-2. According to the model, a responder in the B6.SJ line (derived from SJL/J) has the genotype H-2^b/H-2^s Ir-Z3^a/Ir-Z3^b, and in
the B6.CE line (derived from CE/J), H-2^b/H-2^k Ir-Z3^a/Ir-Z3^b. The
model predicts that it should be possible to construct a set of eight
congenic nonresponder inbred strains showing a given complementation
pattern to responsiveness. For B6.SJ derived mice, complementation
should occur in crosses between nonresponder inbred strains of the
following genotype: H-2^b/H-2^b Ir-Z3^a/Ir-Z3^a x H-2^s/H-2^s Ir-Z3^b/Ir-
Z3^b; or in crosses between H-2^b/H-2^b Ir-Z3^b/Ir-Z3^b x H-2^s/H-2^s Ir-Z3^a/
Ir-Z3^a. A similar complementation pattern should be observed among
four congenic complementing nonresponder strains derived from B6.CE
mice. In addition, complementation to responsiveness should be observed in some inter-strain crosses of these strains if the Ir-Z3
alleles of B6.SJ and B6.CE are compatible in terms of complementation.

     Another implication of the model is that an explanation is re-
quired for the responsiveness of inbred H-2 homozygous strains such
as SJL/J. We present a testable hypothesis postulating that SJL/J
has a gene capable of overriding the nonresponder genotype normally
associated with homozygosis at the H-2 and Ir-Z3 loci. Loss of the
overriding gene during construction of the B6.SJ line would explain
the heterozygosity of responders.

MATERIALS AND METHODS

Mice and Testing Procedure

The mouse congenic lines used are listed in Table 1. The B6.SJ
plp[b]/lcy and B6.SJ plp[b]/2cy lines were constructed by Dr. Marianna
Cherry (Research Department, Jackson Laboratories). Mice were immunized at 8-10 weeks of age after a minimum of 7 days in the Brown
University vivarium.

Immunization consisted of a single peritoneal injection of 0.1ml
of Dulbecco's phosphate buffered saline containing 50µg of β-D-
galactosidase (EC 3.2.1.23) purified from Escherichia coli[6] and stored
as described[4]. Fourteen days after immunization, serum was collected
from individual mice and assayed quantitatively for specific antibody
by the activation of a defective β-D-galactosidase as previously indicated[4].

Determination of H-2 Haplotype

Mice were typed by a new method which permits the use of whole
blood for cytotoxicity assays[7]. Briefly, fluorochromatic leukocytes
were prepared by diluting 50µl of heparinized blood into 4ml phosphate-buffered saline containing 30µM fluorescein diacetate. After
a 15 minute incubation at 21 C, the cells were washed and then exposed
to anti-H-2 cytotoxic antibodies in the presence of rabbit complement.
The surviving leukocytes were counted in a flow cytofluorimeter at
the rate of 1500-6000 cells per minute. The following anti-H-2 cytotoxic antisera were obtained from the National Institutes of Health:
D19 (anti-K[s], anti-D[s]), D32 (anti-D[k]), D33 (anti-K[b]), Y1-9-03-15-03
(anti-K[k]), and Y1-9-03-15-02 (anti-K[s]). The antisera were used at
dilutions ranging from 1:225 to 1:2400.

Determination of C3 Complement Protein

Mice were typed for electrophoretic variants of the C3 complement
component by the method of Barry Whitney (personal communication)
using cellulose acetate strips (Helena Laboratories, Beaumont, Texas).
The allele of the fast C3 variant is present in SJL/J mice (C3[b]),
and the allele for the slow in C57BL/6J (C3[a]); hybrids express both
allelic variants[8,9].

RESULTS

Genetic crosses and selective procedures used for construction
of four congenic lines differing in responsiveness to adjuvant-free Z,
H-2 haplotype and C3 electrophoretic variants are outlined in Table 1.

Table 1. Construction of Congenic Lines*

Congenic Line	Background Strain	Donor Strain	Selected Trait
B6.SJ-Ir-Z	C57BL/6J	SJL/J	responsiveness
B6.SJ plp^b/1Cy**	C57BL/6J	SJL/J	H-2^{s}Tla^{a}C3^b
B6.SJ plp^b/2Cy**	C57BL/6J	SJL/J	H-2^{b}Tla^{b}C3^b
B6.CE-Ir1Z	C57BL/6J	CE/J	responsiveness

*The backcross system of producing congenic lines was used[5]. Back-
crosses involved matings of C57BL/6J females to responder males
which were individually selected from the previous backcross
progeny. C57BL/6J is a non-responder H-2^{b}Tla^{b}C3^a; SJL/J is a
responder H-2^{s}Tla^{a}C3^b. Other construction details are given in
the text.
**This line was constructed by Dr. Marianna Cherry; the plp (plasma
protein) gene product appears to be identical to the C3 complement
protein (B. Whitney, personal communication).

For the Ir-Z lines, C57BL/6J mice (nonresponders) were crossed to
either SJL/J or CE/J. Responder males selected from the backcross
progeny were used for further backcrosses to C57BL/6J females. This
procedure was designed to avoid transferring maternal anti-galacto-
sidase antibody to newborns (we observed significant antibody transfer through the milk). Since SJL/J carries the dominant Ir-Z1 gene[4]
construction of the congenic line involved backcrossing of responders
for 12 successive generations followed by 6 generations of intercrosses between responders. For the line carrying the recessive
ir-Z2 gene of CE/J, the backcross progeny were intercrossed in order
to obtain responders. Mice obtained by these procedures were termed
B6.SJ-Ir-Z and B6.CE-Ir-Z when derived from SJL/J or CE/J, respectively.

Two significant observations were made during production of the
B6.SJ-Ir-Z and B6.CE-Ir-Z lines. First, the frequency of responders
among progeny of the 6th backcross generation of the B6.SJ-Ir-Z line
dropped to 25%, clearly below the 50% expected for a dominant gene.
Second, the originally recessive responder trait of the B6.CE-Ir-Z
line appeared to segregate as a dominant trait after the fifth backcross. Therefore, the construction of the B6.CE-Ir-Z line proceeded
as for the B6.SJ-Ir-Z line, namely, intercrosses were omitted.

After the 11th generation of backcrosses, mice from the B6.SJ-Ir-
Z and B6.CE-Ir-Z lines were typed for H-2 haplotype and allelic

Table 2. Segregation of H-2 and Responsiveness in Intercrosses between Congenic Responders

A. Intercrosses Between B6.SJ Responders

Class	Responders	Nonresponders	Total
$H-2^b/H-2^b$	0	61	61
$H-2^s/H-2^s$	1*	75	76
$H-2^b/H-2^s$	68	78	146
Total	69	214	283

B. Intercrosses Between B6.CE Responders

$H-2^b/H-2^b$	0	40	40
$H-2^k/H-2^k$	0	12	12
$H-2^b/H-2^k$	39	43	82
Total	39	95	134

Congenic responder mice obtained after 12 backcrosses
(see text) were intercrossed and the progeny were H-2
typed and tested for responsiveness. Brother-sister
crosses between responders were repeated for six consecutive generations. In B6.SJ matings, the parental
mice were $H-2^b/H-2^s$ responders; in B6.CE matings, they
were $H-2^b/H-2^k$ responders. The data represent the
progeny of about 40 crosses some of which involved the
same pair of mice. Each mouse was immunized and tested
as indicated in the text. A mouse was classified as a
responder if its antiserum titer was more than 380 activating units per ml[4]. The average titer of the responders was 870 units with a SEM of 60 units. The non-
responder average was 102 with a SEM of 11 units. The
average value of normal serum (50 determinations) was
117 units with a SEM of 4. The titers were corrected
for the normal serum value.
*A progeny test indicated that this mouse is a non-
responder.

variants of the C3 complement protein. The results demonstrated
that the H-2 haplotype of the Ir-Z donor strain had been inherited,
namely, B6.SJ-Ir-Z responders were $H-2^b/H-2^s$ and B6.CE-Ir-Z responders
$H-2^b/H-2^k$. Since our breeding procedure did not involve direct selection of H-2 genes, the probability of obtaining these results by
random gene assortment is very low (less than 3×10^{-7} for the two
lines together).

Responder mice obtained from the 11th and 12th backcross were
intercrossed in order to produce homozygosity at the differential
loci. However, in contrast to the expectation, practically all of
the responders (from six successive brother-sister crosses) were
$H-2^{b/s}$ heterozygotes (Table 2). The one exceptional homozygous $H-2^s$
responder gave nonresponder progeny indicating that responsiveness
in that mouse was not inheritable.

Further genetic analysis showed that in each generation, whether
backcross or intercross, the frequency of responders among H-2 heterozygotes was about 50% (Tables 2 and 3), indicating that heterozygosity at an MHC gene(s) is necessary but not sufficient for responsiveness. In both B6.SJ and B6.CE lines, the segregation frequency
of the H-2 markers was normal (ie. it agreed with that expected for
a single codominant locus) except for that of $H-2^{k/k}$ which fell significantly below the expected values (Table 2). Also, sex ratios
were normal, and litter size was comparable to that of C57BL/6J.

Table 3. Segregation of Responsiveness in Backcrosses
          of Congenic Responders

Congenic Line	Responders	Nonresponders	Total
B6.SJ	58(28%)	149(72%)	207
B6.CE	57(24%)	178(76%)	235

Responder mice obtained from the 9th-10th backcross
generation (during construction of congenic responder
lines) were backcrossed again to C57BL/6J and the
progeny was tested as indicated in Table 2. The average responder titer was 771 (SEM=41) and 687 (SEM=43)
units per ml for B6.SJ and B6.CE, respectively. The
nonresponder average titers were 68 (SEM=10) and 77
(SEM=9) for B6.SJ and B6.CE, respectively.

Table 4. Response of B6.SJ plp^b Congenic Mice

Congenic Line	No. of Mice	Titer±SEM
B6.SJ plp^b/1Cy	27	50±16
B6.SJ plp^b/2Cy	8	27±13
F$_1$(B6.SJ plp^b/1Cy X B6.SJ plp^b/2Cy)	18	126±29
F$_1$(B6.SJ plp^b/1Cy X B6.SJ plp^b/2Cy)	9	234±81

The region of chromosome 17 showing heterozygosity was characterized by testing mice of the B6.SJ line for alleles of C3, Tla,
and Qa-1. We found that heterozygosity does not extend beyond the
C3 locus, since B6.SJ responders were homozygous for C3^a, the allele
of C57BL/6J. The C3 gene is located 11.4 units to the right of the
MHC[8],[9]. In contrast, tests for Tla and Qa-1 alleles (kindly performed by Dr. L. Flaherty) showed presence of SJL/J and B6 alleles
of these genes in the responders, indicating that a crossover between
Qa-1 and C3 must have occurred during construction of the B6.SJ
line.

Additional information was obtained from analysis of two congenic inbred strains, B6.SJ plp^b/1Cy and B6.SJ plp^b/2Cy, (constructed
by Dr. M. Cherry) in which the H-2^s and C3^b alleles of SJL/J are
present in the C57BL/6J background. We found these congenic strains
and their F$_1$ progeny to be nonresponders (Table 4). Thus, these
observations agree with the results given above showing that H-2
heterozygotes are not responders per se, but require additional
gene(s) for responsiveness.

In Table 5 we have compared our results with data generated
by a genetic model postulating allelic complementation at each of
two epistatic genes (see above). As shown by the probability analysis, there is excellent agreement between observed and predicted
results. Further support for the model was obtained from two independent experiments. In crosses between B6.SJ-Ir-Z responders and
B6.CE-Ir-Z responders, we observed that all the responder progeny
were heterozygotes, and that individual responsiveness was not associated with any particular H-2 haplotype combination, ie. the
responders were b/k, b/s or k/s (Table 6). Moreover, in contrast

Table 5. Significance Test of the Genetic Model

Data from	H-2 Class	Observed		Observed		χ^2/p
		Responder	Nonresponder	Responder	Nonresponder	
Table 2-A	homozygous	0	137	0	137	0.0/1.0
	heterozygous	68	78	73	73	0.7/0.4
Table 2-B	homozygous	0	52	0	52	0.0/1.0
	heterozygous	39	43	41	41	0.2/0.7
Table 3	n.d.	58	149	51.75	155.25	1.0/0.3
	n.d.	57	178	58.75	176.25	0.1/0.8
Table 4	homozygous	0	5	0	5	0.0/1.0
	heterozygous	20	9	21.75	7.25	0.6/0.4

The predicted values were calculated according to the genetic model outlined in the text.

Table 6.   Segregation of H-2 and Responsiveness in Interline
           Crosses between B6.SJ Responders and B6.CE Responders

Class	Responders	Nonresponders	Total
H-2 heterozygous*	20	9	29
$H-2^b/H-2^b$ homozygous	0	5	5

Parental mice were H-2 heterozygous responders from each congenic
line. The B6.SJ were $N_{12}F_{16}$ and the B6.CE were $N_{11}F_{14}$. Responsiveness was determined as indicated in Table 2.
*Three types of heterozygotes were observed: $H-2^b/H-2^k$, $H-2^b/H-2^s$,
and $H-2^k/H-2^s$. The number of responders in each class was 5, 4,
and 11, respectively; the number of nonresponders 3, 1, and 5,
respectively.

to the crosses within a line, these interline crosses yielded 59%
responders (20/34). These results are predicted by the model under
the assumption that the Ir-Z3 alleles are similar in both congenic
lines or that there is complementation between the alleles. In the
other experiment, we tested the prediction that complementation
to responsiveness should occur in crosses between $H-2^b$ and $H-2^s$ homo-
zygotes in which complementation of Ir-Z3 alleles is possible.
Accordingly, nonresponder progeny from intercrosses between B6.SJ
responders were crossed inter se. There were six crosses involving
two $H-2^s/H-2^s$ females, two $H-2^b/H-2^b$ males, and one $H-2^s/H-2^s$ male.
As shown in Table 7, the progeny of $H-2^b/H-2^b$ x $H-2^s/H-2^s$ crosses
indicated complementation since there were 30 responders among 53
mice.

DISCUSSION

     We have analyzed the genetics underlying immunologic responsive-
ness to adjuvant-free bacterial β-D-galactosidase by tranferring
genes from each of two different responder mouse strains (SJL/J and
CE/J) into a nonresponder background (C57BL/6J). This approach, in
contrast to previous conventional genetic analyses[4], gave two unexpected results: (i) involvement of H-2 linked Ir genes; (ii) H-2
heterozygosity is required but not sufficient for responsiveness.
It should be noted that transferring the $H-2^s-C3^b$ chromosome segment
of SJL/J into C57BL/6J without selection for responsiveness yields
a nonresponder congenic line (B6.SJ plp^b/1Cy). Furthermore, $H-2^{b/s}$
heterozygotes resulting from crosses between B6.SJ plp^b/1Cy and
B6.SJ plp^b/2Cy were nonresponders. These observations led us to

Table 7. Complementation between B6.SJ Nonresponders

Male haplotype	Progeny from A		Progeny from B		Progeny from C		Total	
	R	NR	R	NR	R	NR	R	NR
H-2^b/H-2^b	3	4	4	5	6	5	13	14
H-2^b/H-2^b	8	1	5	5	4	3	17	9
H-2^s/H-2^s	0	8	1	12	1	6	2	26

Three H-2^s/H-2^s females and three males (H-2 indicated) were crossed <u>inter se</u>. The number of responders (R) and nonresponders (NR) obtained from each cross is <u>indicated</u>.

formulate the genetic model outlined above in which responsiveness
is controlled by two epistatic genes (one of them linked to H-2)
each consisting of complementing alleles.

Two predictions of the model were tested, namely, that complementation to responsiveness should be observed in crosses between
nonresponder B6.SJ congenic mice of different H-2 haplotype, and
that crosses between B6.SJ and B6.CE responders should yield 50%
responders instead of the 25% observed in intraline crosses. In
both tests, the results agreed with the predictions. In addition,
the results of the latter experiment indicated that the Ir-Z3 alleles
in the B6.SJ and B6.CE lines can complement each other, and that
the heterozygosity of responders is not limited to $H-2^{b/s}$ and $H-2^{b/k}$,
but $H-2^{k/s}$ mice are also capable of responding.

Implicit in our model is the condition that crossovers can not
generate a responder homozygous at either the H-2 or Ir-Z3 locus,
therefore, one has to explain how the model would apply to inbred
responders such as SJL/J and CE/J. We propose as a working hypothesis that some cells in these mice are phenotypically H-2 heterozygotes. This possibility is not farfetched given that spontaneous
reticulum cell sarcomas of SJL/J exhibit H-2 heterozygosity as shown
by the presence of "alien" $H-2^b$ and $H-2^d$ determinants[10,11]. Thus,
it is conceivable that a subset of SJL/J cells resemble sarcomas in
their expressing alien H-2 antigens; these cells would be under control of an SJL/J gene which was lost during construction of the B6.SJ
congenic line. This hypothesis is currently under test in this laboratory.

With regard to molecular mechanisms, one could hypothesize that
the function of the Ir-Z3 gene is to control expression of I-coded
glycoproteins which according to recent evidence appear to be products of Ir genes[12]. In this respect, the effect of Ir-Z3 would be
similar to that of xid, and X-linked gene which controls concomitant
expression of Ia.W39 and Ir-insulin genes[13]. This is an attractive
possibility for the following reasons: (i) the H-2 linked allelic
complementation observed in these studies could in fact be complementation between I-A and I-E products (A_e and $E\alpha$ chains, respectively)
yielding a functional surface molecule[14] termed E[15] which appears
to be involved in complementation of Ir genes[12,16]; (ii) both C57BL/
6J and SJL/J produce cytoplasmic Ae chain but do not express the surface E molecule because they lack cytoplasmic $E\alpha$[11,12,17,18]. Accordingly, we propose the following working hypothesis. In contrast to
nonresponder inbred strains, SJL/J has a subset of cells expressing
two types of surface E molecules, namely, $A_e^s E_\alpha^b$ and $A_e^b E_\alpha^b$ resulting
from expression of normally repressed $I-A^b$ and $I-E^b$ genes. Recognition of soluble protein antigens is effected by the hybrid molecule
$A_e^s E_\alpha^b$ and not by $A_e^b E_\alpha^b$ molecule. During construction of the B6.SJ
congenic line the gene controlling expression of alien I molecules
was lost and therefore the $A_e^s E_\alpha^b$ molecule required for responsiveness

is produced only in $H-2^b/H-2^s$ heterozygotes. Considering that neither SJL/J nor C57BL/6J express E_α[11,12,17], the role of the Ir-Z3 product (encoded by complementing allesles) would be to promote expression of the I-E gene.

The fact that complementation between nonresponder B6.SJ-Ir-Z mice is observed (Table 7) indicates that it is feasible to construct the set of eight complementing nonresponder congenic inbred strains predicted by our model. These complementing strains would be valuable as tools for cellular and biochemical studies of interactions between H-2 and non-H-2 linked Ir genes.

SUMMARY

Congenic lines of inbred mice differing in responsiveness to adjuvant-free β-D-galactosidase were constructed by a selective transfer of genes from responder strains, SJL/J and CE/J, into a nonresponder C57BL/6J background. A genetic analysis of the resulting congenic responder lines showed that the H-2 region of the responder donor strain had been indirectly selected in the process. This result was unexpected because there is no discernible association between H-2 haplotype of inbred strains and responsiveness to adjuvant-free Z^4. In addition, the analysis yielded the following results:

(i) H-2 heterozygosity is a rigorous requirement for responsiveness;
(ii) only about 50% of the H-2 heterozygotes were responders;
(iii) $H-2^b/H-2^s$ heterozygotes obtained from crosses between two congenic nonresponder lines (constructed without selection for responsiveness) were nonresponders.

To explain these results, we propose a genetic model in which responsiveness requires allelic complementation at each of two epistatic genes. One of the genes is closely linked to H-2 (for simplicity, the H-2 haplotype is used here to designate this gene); the other gene, Ir-Z3, is unlinked to H-2. According to the model, a responder in the B6.SJ-Ir-Z line (derived from SJL/J) has the genotype $H-2^b/H-2^s$ $Ir-Z3^a/Ir-Z3^b$, and in the B6.CE-Ir-Z line (derived from CE/J). $H-2^b/H-2^k$ $Ir-Z3^a/Ir-Z3^b$.

Results of experiments designed to test some predictions of the model indicate good agreement between observed and expected data. The implications of the model in terms of responsiveness of H-2 homozygous inbred strains (eg. SJL/J) and complementation between Ir-Z and H-2 linked Ir genes are discussed.

ACKNOWLEDGEMENTS

The authors thank Dr. Marianna Cherry for helpful discussions and for providing the B6.SJ p1p^b mice which were crucial for this study; Dr. Donald Bailey for helpful discussions; Dr. Lorrain Flaherty for Tla and Qa-1 typing; Dr. Barry Whitney for help with C3 assays; and Paula Bains for technical assistance.

REFERENCES

1. W. O. Weigle, Immunological unresponsiveness, Adv. Immunol. 16:61 (1973).

2. D. W. Dresser and N. A. Mitchison, The mechanism of immunological paralysis, Adv. Immunol. 8:129 (1968).

3. R. T. Smith, Immunological tolerance of nonliving antigens, Adv. Immunol. 1:67 (1961).

4. B. Rotman, Genetic control of immunologic unresponsiveness to adjuvant-free solutions of β-D-galactosidase. I. The Ir-Z1 and ir-Z2 loci in mice, J. Immunol. 120:1460 (1978).

5. J. Klein, "Biology of the Mouse Histocompatibility-2 complex," Springer-Verlag, New York, pp. 31-33 (1975).

6. R. A. Roth and B. Rotman, Inactivation of normal β-D-galactosidase by antibodies to defective forms of the enzymes, J. Biol. Chem. 250:7759 (1975).

7. B. Rotman and M. J. DePasquale, A quantitative cytotoxicity assay using whole blood: Application to H-2 tissue typing, J. Immunol. Meth. (in press).

8. F. Penalava da Silva, G. F. Hoecker, N. K. Day, K. Vienne and P. Rubinstein, Murine complement component 3: genetic variation and linkage to H-2, Proc. Natl. Acad. Sci. 75:963 (1978).

9. S. Natsuume-Sakai, J.-I. Hayakawa and M. Takahashi, Genetic polymorphism of murine C3 controlled by a single co-dominant locus on chromosome 17, J. Immunol. 121:491.

10. D. Fyfe, N. M. Ponzio and J. Finke, Presence of H-2 like specificities on reticulum cell sarcoma cells syngeneic to SJL mice, Pro. Am. Assoc. Cancer Res. 19:154 (1978).

11. S. M. Wilbur and B. Bonavida, Expression of hybrid Ia molecules on the cell surface of reticulum cell sarcomas that are undetectable on host SJL/J lymphocytes, J. Exp. Med. 153:501 (1981).

12. E. A. Lerner, L. A. Matis, C. A. Janeway, Jr., P. P. Jones, R. Schwartz and D. B. Murphy, Monoclonal antibody against an Ir gene product?, J. Exp. Med. 152:1085 (1980).

13. L. J. Rosenwasser and B. T. Huber, The xid gene controls Ia.W39-associated immune response gene function, J. Exp. Med. 153:1113 (1981).

14. P. P. Jones, D. B. Murphy and H. O. McDevitt, Two-gene control of the expression of a murine Ia antigen, J. Exp. Med. 148:925 (1978).

15. J. Klein, A. Juretic, C. N. Baxevanis and Z. A. Nagy, The traditional and a new version of the mouse H-2 complex, Nature 291:455 (1981).

16. M. E. Dorf, J. H. Stimpfling, N. K. Cheung and B. Benacerraf,
    Coupled complementation of Ir genes, in:"Ir genes and Ia
    antigens," H. O. McDevitt, ed., Academic Press Inc., New York
    p. 55 (1978).
17. K. Ozato, J. K. Lunney, M. El-Gamil and D. H. Sachs, Evidence
    for the absence of I-E/C antigen expression on the cell surface
    in mice of the H-2^b or H-2^s haplotypes, J. Immunol. 125:940
    (1980).
18. R. G. Cook, E. Vitetta, J. W. Uhr and J. D. Capra, Structural
    studies on the murine Ia antigens. V. Evidence that the struc-
    tural gene for the I-E/C beta polypeptide is encoded within
    the I-A subregion. J. Exp. Med. 149:981 (1979).

HOW IS ADJUVANTICITY RECOGNISED?

R. B. Taylor and C. A. Morrison

The activities of the immune system are customarily divided into specific functions mediated by immunoglobulin, and a range of other functions which are loosely classed by "non-specific". Many of these latter do however exhibit a degree of specificity. Thus phagocytic cells can recognise and engulf a variety of foreign particles and yet ignore host cells, natural killer cells will lyse certain targets and not others; and, as will be mentioned later, there are grounds for supposing that there is some specificity in the recognition of different adjuvants. The purpose of this paper is to bring together some diverse observations which appear to be moving towards the identification of a system of non-immunoglobulin molecules as being responsible for some low-specificity recognition phenomena, and to focus on the question of whether adjuvanticity is, like antigenicity, recognised by specific receptors.

The necessity for some non-immunoglobulin recognition system is obvious in the invertebrates. These depend basically on phago-cytosis for their defence against invading organisms, and this is often aided by soluble opsonins[1]. These opsonins often have speci-ficity for simple sugars. They thus come within the definition of lectins: small binding-site, usually sugar-specific, non-immuno-globulin structure[2]. Lectins have been found not only in invert-ebrates, but in most groups of vertebrates. In most cases, however, a defense or other function has not been defined. Lectins have also been found in mammals. The best-known exists in the liver on the surface of hepatocytes, and is specific for galactose. It is thought to function in removing from circulation effete glycoproteins, which have lost the terminal sialic acid residue and so come to expose the subterminal galactose residue[3]. This lectin has been purified and characterised as a glycoprotein made up of subunits of 48,000

and 40,000 daltons. Several other lectins have been described, and
some have been implicated in the cellular interactions involved in
differentiation[4]. Although many of these lectins probably play roles
in the internal economy of the body, there is recent evidence that
even in mammals lectins may be involved in the recognition and destruction of foreign material. One line of evidence comes from the
adherence of bacteria to macrophages. Weir and his colleagues found
that this could often be inhibited by simple sugars[5,6]. Specificity
was shown in the fact that any particular sugar was inhibitory only
if it was represented in the bacterial cell wall.

     In addition to simple adherence it is now clear that there are
mechanisms of contact-mediated cytotoxicity which do not depend on
immunoglobulin: for example the ability of NK cells to kill certain
tumour cells, and the ability of macrophages to lyse certain erythrocytes. The specificity of this type of recognition has not been
well defined; but recently it has proved to be possible to inhibit
these reactions with monosaccharides in a rather specific way. Using
a system in which human monocytes (activated by 7-day culture) lysed
red cells of other species, Muchmore and Blaese[7] found that different
saccharides were inhibitory with target red cells of different
species. Likewise tumour cell killing by NK cells was inhibitable
by certain monosaccharides, whereas antibody-dependent cell-mediated
cytotoxicity was not[8]. A feature of the lectin-like receptors both
for binding[6] and lysis[9] is that they are very susceptible to trypsin.

     Little or no data is available so far on the molecular nature
of these lectins. However a number of observations suggest that some
of them may in fact be Ia antigens, or at least be coded in the I-
region. Thus binding of bacteria to mouse macrophages could be inhibited by a broad-spectrum anti-Ia antiserum[10]. The importance of
Ia molecules in antigen-presentation to T cells is well known,
although it is currently debated whether Ia actually binds the antigen. If it does, then the bond must have rather low affinity or it
would have been detected by the co-capping experiments which have
been employed. A binding event may be suggested by the fact that
the initial phase of uptake of antigen for presentation is independent of metabolism[11]. In this case of course the interaction must,
for most protein antigens, involve other than carbohydrate specificities. Nevertheless, protein-carbohydrate interactions seem to
be important in cellular interactions, since the activity of suppressor T cells could be inhibited by a monosaccharide[12] and the
proliferative response of human T cells to diphtheria toxoid was inhibited by various sugars depending on the HLA DRW phenotype of the
donor[7].

     So far we have considered the role of non-immunoglobulin mol-
ecules in recognition without asking what events follow. In invert-
ebrates presumably lectins have to serve the function of self/not-
self discrimination. But once the immunoglobulins had evolved, with

their much greater specificity, it seems likely that the lectins
would be largely redundant in this primary function. However, recognition by immunoglobulin receptors does not necessarily lead to an
immune response. The bi-directional nature of the immune system is
becoming increasingly clear, by which the initial recognition event
can at any time be either amplified to high levels, or else suppressed so profoundly as to result in tolerance to the antigen.
The direction which is actually taken depends on a number of factors.
Of prime importance amongst these is the presence or absence of adjuvant. Besides the "extrinsic" adjuvant which may accompany the antigen, it is necessary to introduce the concept that any antigen can
have more or less of an intrinsic quality, which Dresser has termed
"adjuvanticity"[13]. Without this concept it is very hard to explain
why some proteins such as IgGs and serum albumins tend to induce
tolerance while others such as transferrins and certain alpha-
globulins are highly immunogenic. It seems unlikely that the degree
of foreignness recognised by T and B cells is the deciding factor
since the immunogenicity of an antigen in the presence of extrinsic
adjuvant appears to bear little relation to its immunogenicity without adjuvant. For example chicken IgG, if carefully prepared, is
a good tolerogen on its own for the mouse, but is a very powerful
immunogen if given with adjuvant. On the other hand thyroglobulins,
and even <u>mouse</u> thyroglobulin, readily cause an antibody response in
some strains of mice without extrinsic adjuvant[14].

Intrinsic adjuvanticity thus seems to depend on features of
the antigen which are not directly related to the determinants recognised by T and B cells. Such features have pronounced effects not
only on the quantity, but on the quality of the response. Thus conjugation of protein with aliphatic side-chains tends to enhance
delayed hypersensitivity and suppress antibody production[15,16].

As is well known, long-chain repeating polymers tend to induce
T-independent rather than T-dependent responses. This is often
assumed to be a result solely of multivalence. Several observations
suggest however that the intrinsic nature of the molecule plays an
important role. For example, like other small globular proteins the
monomer fragments of most flagellins behave as T-dependent antigens.
However the monomer prepared form a certain strain of Salmonella
would induce IgM responses independent of T cell help[17]. By contrast
the flagellins of other strains were equally T-dependent whether in
monomer or poly form. Again, there now appear to be three characteristic types of T-independent response given by distinct populations of B cells. The choice between these depends on the nature
of the carrier molecule[18]. Caution is required in using this kind
of interpretation with bacterial antigens since bacteria may bear
special lectins by which they adhere to mammalian cells[19].

It is interesting to note that antigens which induce qualitatively different responses tend to localise in different areas of the

lymphoid tissues. This leads to the intriguing idea that they may
be picked up by specialised antigen presenting cells. Thus, antigens
inducing T-dependent antibody responses localise onto the follicular
dendritic cells of germinal centres[20]; while lipidated proteins go
mainly to the T-dependent areas[16]. Here it is a reasonable guess
that they localise onto the Steinman dendritic cell, which is specialised for induction of cell-mediated immunity[21]. Different patterns
of localisation are seen for T-independent antigens also: neutral
polysaccharides such as Ficoll and dextran go very specifically to
the marginal zone macrophages, while acid polysaccharides are distributed in the red and white pulp[22].

The well-known extrinsic adjuvants also have pronounced qualitative effects on the response. For example complete Freund's adjuvant
enhances IgG and delayed hypersensitivity and suppresses IgE, while
alum enhances IgG and IgE and suppresses delayed hypersensitivity.

The recognition of adjuvanticity could thus be the same thing
as antigen-presentation. An argument against this can be made from
a consideration of the requirements for induction of tolerance in
T cells. If T cells can only recognise antigen together with I-
region products then they should only be made tolerant by this combination. Experimentally, it has been shown that injection of TNP-
conjugated epidermal cells leads to contact sensitivity; but if the
epidermal cells have first been UV-irradiated, then tolerance is
induced[23]. Likewise injection of DNP conjugated to lymphocytes induces tolerance, but this can be converted to immunity by simultaneous injection of concanavalin A[24]. In both of these experiments
it was found that Ia-expression was necessary for induction of either
tolerance or immunity. Thus it appears that adjuvanticity delivers
a signal which is distinct from antigen-presentation but is essential
for a positive response. The general theory of one signal = tolerance, two signals = immunity has been elaborated some years ago[13,25];
but the precise interpretation still eludes us.

Can we find any molecules specialised as receptors for adjuvant?
There is little information so far, but an opportunity was provided
by the finding that C3H mice do not respond to the adjuvant action
of LPA. A rabbit antiserum was prepared which lysed (with complement)
a proportion of B cells of responder, but not non-responder strains[26].
It is very likely that this antiserum recognised a distinct receptor
because (a) it mimicked LPS in being able to stimulate B cells, and
(b) its lytic action (with complement) could be competitively inhibited with LPS or Lipid A[27]. This goes against the commonly-held
notion that adjuvants merely damage membranes in a non-specific way,
and encourages us to look for other specific receptors.

As an approach to the study of adjuvanticity we have begun experiments in which a variety of small molecules suspected of adjuvanticity are covalently attached to a protein antigen. The antigen

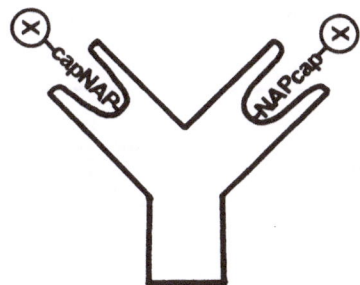

Fig. 1. Method for attaching adjuvant moiety. The photoactivated
affinity-label NAP (4-nitro-2-azido-phenyl-) is coupled
via an aminocaproic acid linking chain to the molecule
under test (X). This is mixed with rabbit anti-NAP antibody, and illuminated so as to form a covalent bond.

is rabbit immunoglobulin, which induces little or no response on
its own and may thus be said to lack adjuvanticity. It is important
for comparative purposes that these molecules are attached to a defined site on the molecule, at a defined molar ratio and without
the denaturing influence of extreme conditions. This was done by
making covalent-bonded monomeric antigen-antibody complexes
(Figure 1)[28]. The photosensitive affinity label 'NAP' (4-nitro-2-
azido-phenyl-) was coupled via an aminocaproic acid linking chain
to the molecule under test ("X") at a molar ratio of not more than
one-to-one. This conjugate was mixed with purified rabbit anti-NAP
antibody and illuminated so as to form a covalent bond between NAP
and the anti-NAP binding site. Mice were injected with 10-20µg of
these preparations, or with uncomplexed rabbit anti-NAP as control,
and then 4-6 weeks later challenged with 10µg rabbit IgG to elicit
the immunological memory which had built up as a result of the first
injection. In most experiments so far the test molecule has been
either bovine ribonuclease (RNase) or ε-fluorescyl-lysine (lys Fl).
(The synthesis of NAPcapRNase, and NAPcaplysFl are described in 28
and 29 respectively). Other test molecules have included some commonly used haptens, and some amino-sugars coupled to NAPcap via the
amino group. The results of a typical experiment with fluorescein
are given in Figure 2, and some other results are summarised in
Table 1. In some experiments both the text complex and control
rabbit anti-NAP were deaggregated by centrifugation at 100,000XG
for 2.5 hours. The top half only of the centrifuge tube was taken
for injection. Deaggregation usually resulted in a small reduction
in the immunogenicity of both test and control, but the differential
was usually maintained. In some experiments, however, no significant
difference was recorded. The source of this variability has not so
far been identified. It could depend on variations in the natural
exposure to adjuvant, eg. by endotoxins from the gut flora.

Table 1. Priming the Response to Rabbit IgG with Various Covalent
          Complexes

Molecule exposed in the complex	Deaggregation	No. of experiments showing:-	
		Sig. priming[b]	No. sig. priming
Fluorescein	−	3	1
	+	2	1
Fluorescein (random coupled[a])	−	1[c]	0
	+	1	0
Ribonuclease	−	4[c]	0
	+	0	1
Arsanyl-hydroxyphenacetyl	−	1	0
Glucosamine	+	2	1
Phosphoryl choline	−	0	1
	+	0	1
Galactosamine	+	0	1
Mannosamine	+	0	1
Bacitracin	+	0	1
None (NAP amino caproic acid only)	−	0[c]	3

[a]Rabbit anti-NAP was reacted with fluorescein isothiocyanate to make
  conjugates of molar ratio approximately 2F1/1 anti-NAP.
[b]$p < 0.05$ by comparison with uncomplexed anti-NAP.
[c]Some of these are based on primary response (as judged from immune
  clearance of the complex) rather than priming.

     Most of the other molecules tested have not had a significant
adjuvant effect, but it is interesting that one of the sugars,
glucosamine, was positive in 2 out of 3 experiments.

     One possible interpretation of these results, (which was the
one initially entertained) is that occupation of the binding site
of the anti-NAP molecule had induced a change in conformation which
caused it to deliver an immunogenic signal. This now seems unlikely

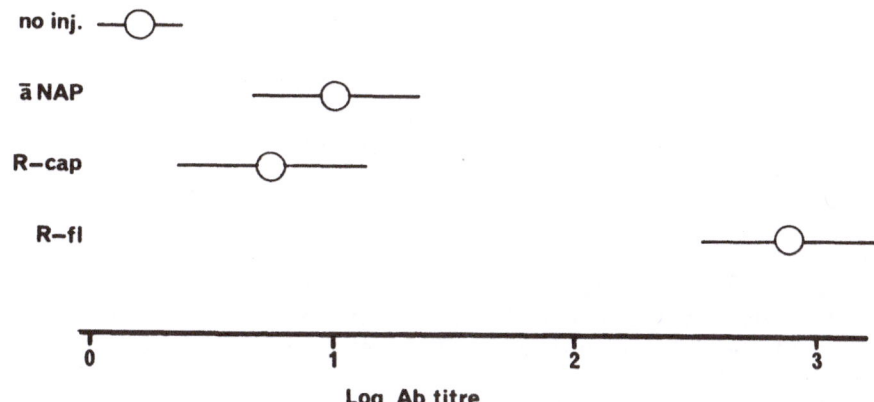

Fig. 2.   "Adjuvant" effect of fluorescein. Mice were injected
with 10μg rabbit anti-NAP, either uncoupled (ā NAP) or
coupled NAP-cap (R-cap) or to NAPcaplysFl (R-Fl). Six
weeks later they were all boosted with 10μg RIgG, and the
serum anti-RIgG titres measured 13 days after the boost.

in view of the negative results obtained with several ligands,
although these must fill the binding site equally as well as fluorescein. We are currently searching for molecules which have stronger
and more consistent effects.

REFERENCES

1. S. W. Hardy, T. C. Fletcher and J. A. Olafsen, Aspects of cellu-
    lar and humoral defence mechanisms in the pacific oyster,
    Crassostrea gigas, in:"Developmental Immunobiology,"
    J. B. Solomon and J. D. Horton, eds., Elsevier, p. 59 (1977).
2. E. R. Gold and P. Balding, Receptor specific proteins, plant
    and animal lectins, Excerpta Medica, Amsterdam (1975).
3. G. Ashwell and A. G. Morrell, Membrane glycoproteins and recog-
    nition phenomena, Trends. Biochem. Sci. 2:76 (1975).
4. D. L. Simpson, D. R. Thorne and H. H. Loh, Lectins: endogenous
    carbohydrate-binding proteins from vertebrate tissues: func-
    tional role in recognition processes, Life Sci. 22:727 (1978).
5. H. M. Ogsmundsdottir and D. M. Weir, The characteristics of
    binding of Corynebacterium parvum to glass-adherent mouse
    peritoneal exudate cells, Clin. Exp. Immunol. 26:334 (1976).
6. D. M. Weir, Surface carbohydrates and lectins in cellular recog-
    nition, Immunology Today 1:45 (1980).
7. A. V. Muchmore and R. M. Blaese, Evidence that monocyte-mediated
    cellular recognition phenomena are mediated by receptors with
    specificity for simple oligosaccharides, in:"Macrophage
    Regulation of Immunity," E. R. Unanue and A. Rosenthal, eds.,
    Academic Press (1980).

8. R. P. Macdermott, L. J. Kienker, M. J. Bertovich and
    A. V. Muchmore, Inhibition of spontaneous but not antibody-
    dependent cell-mediated cytotoxicity by simple sugars: evi-
    dence that endogenous lectins may mediate spontaneous cell-
    mediated cytotoxicity, Immunology 44:143 (1981).
9. R. Kiessling, G. Petranyi, K. Karre, M. Jondal, D. Tracey and
    H. Wigzell, Killer cells: a functional comparison between
    natural, immune T-cell and antibody-dependent in vitro sys-
    tems, J. Exp. Med. 143:772 (1976).
10. D. M. Weir, J. Stewart, E. J. Glass, A. Oliver C. C. Blackwell
    and J. Doyle, in:"Proceedings of International Workshop on
    Macrophage Heterogeneity, O. Forster, ed., Academic Press,
    London (1981).
11. A. S. Rosenthal, J. W. Thomas, J. Schroer, K. Yokomuro,
    J. J. Blake and L. J. Rosenwasser, in:"Immunological Tolerance
    and Macrophage Function, p. 57, P. Baram, ed., Elsevier
    (1979).
12. U. K. Koszinowski and M. Kramer, Selective inhibition of T sup-
    pressor-cell function by a monosaccharide, Nature 289:181
    (1981).
13. D. W. Dress, Specific inhibition of antibody production. II
    Paralysis induced in adult mice by small quantities of protein
    antigen, Immunology 5:378 (1962).
14. M. ElRehewy and Y. M. Kong, Syngeneic thyroglobulin is immuno-
    genic in good responder mice, Eur. J. Immunol. 11_146 (1981).
15. C. R. Parish, Immune response to chemically-modified flagellin.
    I Induction of antibody tolerance to flagellin by aceto-
    acetylated derivatives of the protein, J. Exp. Med. 134:1
    (1971).
16. J. Coon and R. L. Hunter, Selective induction of delayed hyper-
    sensitivity by a lipid-conjugated protein antigen which is
    localised in thymus-dependent lymphoid tissue, J. Immunol.
    110:183 (1973).
17. R. E. Langman, W. D. Armstrong and E. Diener. Antigenic compo-
    sition, not the degree of polymerisation, determines the
    requirement for thymus-derived cells in immune responses to
    Salmonella flagellen antigens, J. Immunol. 113:251 (1974).
18. G. K. Lewis and J. W. Goodman, Carrier-directed anti-hapten re-
    sponses by B-cell subsets, J. Exp. Med. 146:1 (1977).
19. Ciba Foundation Symposium of Adhesion and Microorganism Patho-
    genicity, 1981.
20. G. J. V. Nossal and G. L. Ada, "Antigens, Lymphoid Cells and the
    Immune Response", Academic Press, New York (1971).
21. R. M. Steinman and M. C. Nussenszweig, Dendritic cells: features
    and functions, Immunol. Revs. 53:127 (1981).
22. J. H. Humphrey, Differentiation of function among antigen-pre-
    senting cells, in:"Microenvironments in Haematopoietic and
    Lymphoid Differentiation," Ciba Found. Symp. 84, p. 302
    (1981).
23. D. N. Sauder, K. Tamaki, A. N. Moshell, H. Fujiwara and
    S. I. Katz, Induction of tolerance to topically applied TNCB

using TNP-conjugated ultra-violet light-irradiated epidermal cells, J. Immunol. 127:261 (1981).

24. R. P. Cleaveland and H. N. Claman, T cell signals: tolerance to DNFB is converted to sensitisation by a separate non-specific signal, J. Immunol. 124:474 (1981).

25. P. A. Bretscher and M. Cohn, A theory of self-nonself discrimination, Science 169:1042 (1970).

26. L. Forni and A. Coutinho, An antiserum which recognises lipo-polysaccharide-reactive B cells in the mouse, Eur. J. Immunol. 8:56 (1978).

27. A. Coutinho, L. Forni and T. Watanabe, Genetic and functional characterisation of an antiserum to the lipid A-specific triggering receptor on murine B lymphocytes, Eur. J. Immunol. 8:63 (1978).

28. R. B. Taylor, J. P. Tite and C. Manzo, Immunoregulatory effects of a covalent antigen-antibody complex, Nature 281:488 (1979).

29. J. P. Tite and R. B. Taylor, Immunoregulation by covalent anti-gen-antibody complexes. II. Suppression of a T-cell independent anti-hapten response, Immunology 38:325 (1979).

VII C.   ANALYSIS OF THE TERNARY COMPLEX OF ANTIGEN:

MHC: AND T-CELL RECEPTOR WITH T CELL CLONES

In the analysis of T cell recognition in which at least three components are implicated, it is of undoubted advantage to be able to keep two of them constant at a time. Until recently with the advent of T cell cloning, the heterogeneity of T cell populations made such an analysis impossible. In the next three contributions, the directed question addressed is the existence of "interaction entities" on the antigen-presenting cell, between Ia molecules and antigen.

Papers 31 and 32 describe very similar situations in which receptors on monoclonal T cells combine with MHC element A and epitope-X, or MHC element B and epitope Y. In the case of Rude et al., epitope X̄ and Y were on the same beef insulin molecule, while in the work of Heber-Katz et al., the antigenic specificities were defined by a set of cytochrome variants. Since two ostensibly different antigen determinants can be accommodated by the same T cell receptor, it would appear that the complementary space in the T receptor's active site can be suitably filled by the appropriately chosen MHC element. Thus, each of the two elements recognized contribute to the overall specificity of T cell recognition. It was predicted therefore that recognition of allogeneic MHC + epitope Z should not require any unusual new mechanism.

Lonai studied the binding of soluble Ia-antigen complexes (IAC), similar to the GRF earlier described by Erb and Feldmann[1], as the most direct way to access recognition by T cell hybridoma clones between $H-2^b$ normal cells and an $H-2^k$ lymphoma. Later, selected T clones could be shown to secrete 2 different antigen-specific T helper factors, one restricted by $H-2^b$ and one by $H-2^k$. Since it was doubtful that the BW-5147 lymphoma possessed antigen-binding capacity, the argument was advanced that these findings support a dual recognition model, in which the independent antigen-recognizing receptor of the T cell hybrid associated with a separate MHC-recognizing receptor. This is a result opposite to that reported by Kappler et al.,[2] who fused two T-cell hybridomas of differing restriction specificity and found no reassociation of specificities in the fusoma. It is clear that the use of somatic cell hybridization techniques will eventually solve such current controversies in the study of gene expression in many systems.

REFERENCES

1. P. Erb, M. Feldmann and N. Hogg, Role of macrophages in the generation of T helper cells. IV. Nature of genetically related factor derived from macrophages incubated with soluble antigens, Eur.J.Immunol. 6:365 (1976).
2. J.W. Kappler, B. Skidmore, J. White, and P. Marrack, Antigen-inducible, H-2 restricted, interleukin-2-producing T cell hybridomas: Lack of independent antigen and H-2 recognition, J.Exp.Med. 153:1198 (1981).

SPECIFICITY AND RESTRICTION OF T CELLS

IN A SYSTEM OF COMPLEMENTING Ir GENES[*]

E. Rüde, A.B. Reske-Kunz, and E. Spaeth

Institute for Immunology
Johannes-Gutenberg Universität
D-6500 Mainz, FRG

The phenomenon of Ir gene control of immune responsiveness is intimately related to the fact that antigen-specific stimulation of T cells with the Lyt-1 surface phenotype requires the simultaneous recognition of both external antigen and Ia antigen on the surface of antigen-presenting cells. There is increasing evidence indicating that Ia antigens encoded for by the I-A and I-E/C subregions of the H-2 complex represent the products of Ir genes.

For studies on the MHC influence on antigen recognition by T cells and the dependence of recognition on the structure of the external antigen we have used insulin as model antigen. Its primary and tertiary structures are well established[1]. Furthermore, several species variants with a limited number of amino acid substitutions relative to the mouse insulins are available.

RESULTS AND DISCUSSION

Responsiveness of mice to insulins of various species has been shown to be under Ir gene control by several groups including ourselves. Pertinent results are presented in Table 1. Mice of both $H-2^b$ and $H-2^k$ haplotype are low responders to pig insulin (PI), while F_1 hybrids were found by Keck[2,3] to be capable of mounting an antibody response to this antigen. This response pattern, estab-

---

*This work was supported by a grant of the Deutsche Forschungs-gemeinschaft, SFB 107. Part of this study was carried out by E.S. in partial fulfilment of the requirements for a doctoral thesis at the University of Mainz.

Table 1.   Response Pattern of Mice to Insulins

H-2 Type	Responder status	Insulin	Amino acid substitutions relative to mouse insulin, at position:					
			A4[*]	A8	A9	A10	B3[**]	B30
–	–	mouse	Asp	Thr	Ser	Ile	Lys	Ser
b	low							
k	low	pig(PI)	Glu	Thr	Ser	Ile	Asn	Ala
b x k	high							
b	high							
k	low	beef(BI)	Glu	Ala	Ser	Val	Asn	Ala
b x k	high							
b	low							
k	high	sheep(SI)	Glu	Ala	Gly	Val	Asn	Ala
b x k	high							

[*]A4 refers to the amino acid residue no.4 in the A-polypeptide chain of insulin.

[**]B3 refers to residue no.3 in the B-chain of insulin.

lished by testing for antibody production, was confirmed by using anamnestic T cell proliferation in vitro as an assay system[4,5]. Furthermore, responsiveness of F_1 hybrids to PI was shown to depend on complementation of Ir genes mapping in the I-A^k and K^{b}I-A^b subregions of the two parental strains[5]. PI-primed T cells of (H-2^b x H-2^k) hybrid mice (b x k)F_1) could be restimulated in vitro when PI was presented on syngeneic F_1 macrophages, but no or only very weak stimulation was observed in the presence of parental macrophages (Figure 1). This finding suggests that PI-reactive T cells are restricted for I-A determinants which are unique for F_1-presenting cells. These F_1-unique restriction elements most probably result from an interaction of I-A^b and I-A^k encoded polypeptide chains[6].

Mice of the H-2^b haplotype are low responders to PI but show high responsiveness to beef insulin (BI). Based on the assumption that responsiveness is primarily due to the recognition of "determinants" foreign to the mouse and taking into account the difference in amino acid sequence between PI and BI it was concluded that high responsiveness of H-2^b mice to BI depends on recognition of the A-chain loop including amino acid residues A8 and A10 [7]. According to this reasoning additonal amino acid differences in BI relative to the mouse insulins (A4, B3, B30) may not be relevant for recognition in H-2^b mice.

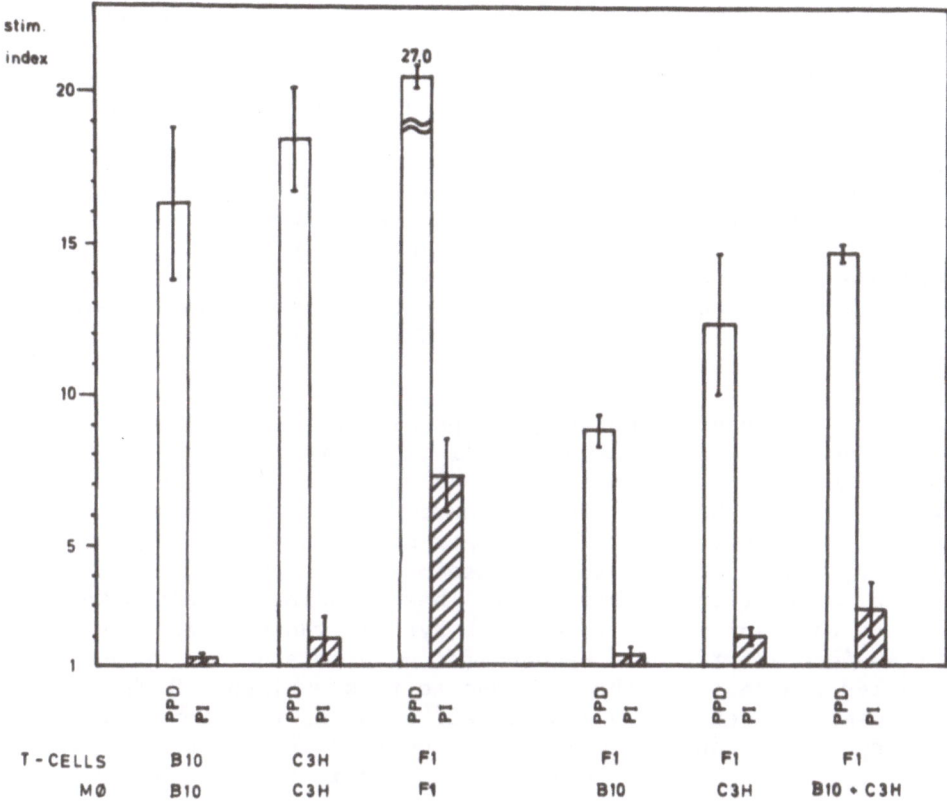

Fig. 1. Proliferative response (stimulation index) of lymph node
         T cells of mice primed with pig insulin (PI) in complete
         Freund's adjuvant. As a source of antigen presenting
         macrophages (Mφ) irradiated spleen cells were added to
         the cultures. The antigens used were purified protein
         derivative of tuberculin (PPD) and PI at a concentration
         of 100µg/ml.

     High responsiveness is generally inherited as a dominant trait
in (high responder x low responder)F_1 hybrids. In the case of
(b x k)F_1 hybrids one could, therefore, expect that T cells primed
to BI are specific for the A-chain loop and are restricted for the
parental H-2^b haplotype. In order to test whether other specificities, possibly in the context of F_1 restriction elements, are also
involved, long term cultured T cell lines were developed from (B10
x B10.BR)F_1 hybrids immunized with BI by repeated restimulation of
lymph node T cells in vitro. Intermittently, T blasts were expanded
with IL-2 containing supernatants. One of the resulting lines was
tested for specificity in more detail. As shown schematically in
Table 2, T cells of this line reacted equally well with BI, PI, SI,

and with the isolated A-chain of BI or PI. No stimulation was observed with the isolated B-chain which is identical for BI and PI.
Thus, the determinant recognized on the insulin molecule is localized
on the A-chain but is distinct from the A-chain loop of BI. It
possibly includes the glutamic acid residue in A4.

Furthermore, restimulation of the T cells with these antigens
was possible only in the presence of $(b \times k)F_1$ but not of either
parental spleen cells or a mixture thereof (Table 2). By using the
appropriate presenting cells of F_1 hybrids derived from H-2 recombinant strains the complementing loci required for the expression
of the F_1-specific Ia-restriction determinant could be mapped to
I-A^k and K^b/I-A^b subregions[8].

These data suggest that high responsiveness of F_1 hybrids is
not only due to the presence of T cell clones recognizing the A-
chain loop of BI in the context of I-A^b restriction determinants,
as is the case in the H-2^b high responder parent, but is also caused
by an additional epitope of the antigen molecule that can be recognized in F_1 hybrids due to the expression and in the context of F_1-
unique Ia-restriction determinants. These findings extend data of
Kimoto and Fathman[9] in the (T,G)-A--L system, who had recently shown
that F_1-unique determinants encoded for by high and low responder
MHC alleles can restrict the response to this antigen. However, the
antigen used in their study did not allow the distinction of whether
the epitope of the antigen molecule recognized in the context of
parental or F_1-unique restriction structures is the same or different.

As already discussed above, PI-specific T cells of $(b \times k)F_1$
hybrids can be restimulated by the homologous antigen only in the
presence of $(b \times k)F_1$-presenting cells. Since H-2^b mice are high
responders to BI, and H-2^b macrophages therefore are capable of presenting this antigen (at least to H-2^b T cells), it was tested
whether PI-specific $(b \times k)F_1$ T cells could perhaps be restimulated
by BI presented on H-2^b macrophages. For carrying out such experiments a long term cultured T cell line was used because normal PI
primed lymph node T cells give relatively low stimulation even with
the homologous antigen (Figure 1). Furthermore, such T cells may
be contaminated with syngeneic F_1 macrophages or dendritic cells
which would complicate interpretation of the results.

The specificity pattern observed upon restimulation of a T cell
line derived from (B10 x B10.BR)F_1 mice immunized with PI is shown
in Table 3. In agreement with previous findings using primed lymph
node T cells PI induced proliferation only in the presence of F_1-
presenting cells. The determinant recognized was present on all
insulins tested including horse insulin (HI). It is located on the
A-chain of both PI and BI and presumably depends on the glutamic
acid residue A4. This reactivity pattern including the restriction
specificity for F_1-unique Ia-determinants coded for in I-A^k and

Table 2. Reactivity pattern of a BI-specific T cell line derived
          from (B10 x B10.BR)F_1 mice towards variant insulins and
          isolated chains

Inducing antigen	Proliferative response (stimulation index) in the presence of presenting cells of			
	(B10xB10.BR)F_1	B10	B10.BR	B10+B10.BR
PPD	1,4	0,8	1,2	0,9
BI	88,7	2,3	1,4	1,3
PI	100,6	1,0	2,3	1,2
SI	103,9	1,8	1,5	1,2
BI-A $(SSO_3^-)_4$ [a]	88,5			
PI-A $(SSO_3^-)_4$ [a]	95,2			
BI-B $(SSO_3^-)_2$ [a]	2,3			

[a] The isolated, S-sulfonated form of the A- or B-chain of beef or
pig insulin.

K^b/I-A^b parental subregions (data not shown) corresponds to the one
established for the BI-specific F_1-T cell line described above
(Table 2). No restimulation was observed with either of the insulins
in the presence of H-2^k presenting cells. When H-2^b presenting cells
were used, strong proliferation was induced by BI and a weak cross-
stimulation was observed with SI and HI. PI was not stimulatory
under these conditions. From this reactivity pattern with H-2^b presenting cells it follows that amino acid residues in the A-chain
loop of BI, mainly A8 and A10, are essential for recognition. However, since the isolated A-chain of BI, in which the disulfide
bridges are opened was not stimulatory it appears that the determinant
recognized was either dependent on the conformation of the intact
loop of BI or that part of the B-chain was also required. Thus,
different determinants of PI and BI are recognized in the context of
F_1 and of H-2^b restriction elements, respectively.

These data suggest that F_1-T cells of this line cross-reactively
recognize PI in the context of F_1 and BI in the context of parental
H-2^b restriction structures. It may be argued that such a reaction
pattern can be due to separate T cells present in this line. Such
an explanation is unlikely because the F_1 hybrids, from which the
line was developed, had been immunized with PI and their lymph node
T cells were repeatedly stimulated with PI and F_1-presenting cells
over a period of several months. At this stage a cross-reaction

Table 3. Reactivity pattern of a PI-specific T cell line derived
          from (B10 x B10.BR)F$_1$ mice towards variant insulins and
          isolated chains

Inducing antigen	Proliferative response (stimulation index) in the presence of presenting cells of		
	(B10xB10.BR)F$_1$	B10	B10.BR
PI	100,1	1,1	1,1
BI	83,6	55,4	1,2
SI	112,6	9,3	1,2
HI	94,6	5,6	0,8
PI-A (SS)$_2$[a]	68,0	1,5	
BI-A (SS)$_2$[a]	45,1	1,1	
BI-A (SSO$_3^-$)$_4$[b]	15,7	1,5	
BI-B (SS)[a]	1,3	0,8	
BI-B (SSO$_3^-$)$_4$[b]	1,1	1,1	

[a] The isolated A- or B-chain of beef or pig insulin after reduction
    and reoxidation of the SH-groups.
[b] The isolated, S-sulfonated form of the A- or B-chain of beef
    insulin.

with BI and H-2^b-presenting cells was detected which was about one
tenth in magnitude as compared with the "homologous" (PI/F$_1$) reaction.
Most probably only a minority of the PI/F$_1$-specific T cell clones
was cross-reactive with BI in the context of H-2^b. Upon further
propagation of the cells by repeated stimulation with BI and H-2^b-
presenting cells the cross-reacting clones were stimulated equally
well by PI/F$_1$ and by BI/H-2^b-presenting cells. This pattern was
maintained when the T cell line was now again restimulated over a
period of four months with PI/F$_1$-presenting cells. One can expect
that any clone exclusively reactive with BI/H-2^b would have been
lost as were the non-cross-reacting PI/F$_1$-specific clones during
selection with BI/H-2^b-presenting cells. Furthermore, several sublines have been grown out of this T cell line. Nine of these showed
the specificity pattern of the original line while one had an additional strong reactivity for PI and H-2^b-presenting cells (low
responder to PI). As shown in Table 3, T cells of the original line
were not significantly stimulated by this antigen combination. The

segregation of such a rare specificity argues for clonality of these sublines and makes it very unlikely that the two major cross-reactive specificities were carried by separate T cells that did not segregate.

A cross-reactivity of T cells between MHC-antigen A + external antigen X_1 and MHC-antigen B + external antigen X_2, as described here, can be most readily explained if an interaction between the restricting MHC-antigen and the external antigen contributes to the overall specificity of T cell recognition. For such an interpretation it is not essential whether a functional unit of MHC and external antigen is recognized by a single or by two different receptors. An important point is that recognition of MHC and external antigen should not be independent: ie. if there are different receptors, they should be functionally linked as suggested by recent hybrid-ization experiments[10]. A specific complex of MHC and external anti-gen may not be stable in isolated form but such an interaction could contribute to the overall stability of a multicomponent complex con-sisting of MHC-antigen, external antigen and one or two T cell re-ceptors[11].

If antigen presentation were not to involve any type of inter-action between MHC and external antigen, the cross-reaction could be interpreted on the basis of two separate receptors which are hetero-clitic. For triggering the cell the combined affinity of the two distinct receptors has to reach a certain threshold value. In the combination PI/H-2^b which is not stimulatory the affinity of the anti-Ia receptor for I-A^b is too low as compared with that for the "homologous" (b x k)F$_1$ restriction element. This lower affinity of the anti-MHC receptor could be compensated if the receptor for the external antigen has a higher affinity for BI than for PI, although the latter was used for immunization. Therefore, the combined af-finity for BI/H-2^b may be sufficient for cross-stimulation. On the basis of the available data no decision can be made between these alternative explanations. Cross-reactions of a similar type have very recently been also reported for cytotoxic T cells[12] and for proliferating T cells specific for cytochrome C (Heber-Katz et al., this volume).

The findings presented in this study concerning the specificity pattern of T cells reactive with insulins indicate that restriction for a particular Ia-antigen is linked with recognition of a distinct determinant or amino acid substitution on the insulin molecule. Thus, I-A^b restriction appears to be associated with recognition of the A-chain loop of BI, whereas F$_1$ restriction was found to be as-sociated with recognition of an A-chain determinant possibly including the A4 residue. This pattern was retained even for the cross-reacting T cell line. It is difficult to explain such associations without assuming a specific interaction between Ia and external antigen. The observation that the presence of "foreign" amino acid residues on an antigen molecule as such is not sufficient to induce a response unless the corresponding Ia-antigens are expressed in the strain tested, is in agreement with such an interpretation. An alternative explanation would require a very intricate coupling of the specificities of separate receptors for MHC and external antigen as has been suggested to be brought about by somatic mutation and positive, Ia-dependent selection of the T cell receptor for external antigen[13].

The existence of T cell clones, both Ia-restricted and K-, D-restricted, which show cross-reactivity between syngeneic MHC + antigen X_1 and semiallogeneic or allogeneic MHC + antigen X_2 is of particular interest with respect to the detection of allo-restricted T cells after depletion of allo-reactivity[14-16]. It seems reasonable to assume that self- and allo-restricted T cells are in fact cross-reactive members of the same pool of cells, as are self-restricted and allo-reactive T cells. Therefore, there would be no need to account for the occurrence of allo-reactive and allo-restricted T cells by special mechanisms.

On the basis of this assumption one could predict that the usual pattern of Ir gene control is changed or breaks down if one tests the reactivity of allo-restricted T cells. This should apply at least to certain haplotype combinations. For instance, T cells of a non-responder strain to PI which are depleted of the respective allo-reactive clones should be stimulated by PI presented on allogeneic, responder macrophages since certain T cell clones specific for self-MHC + antigen X may be cross-reactive with allo-MHC + PI. The reverse combination, ie. allo-restricted T cells of responder haplotype and non-responder macrophages, should only work if no qualitative defect exists with regard to the presentation of the particular antigen by non-responder macrophages.

In several systems this does not appear to be the case. As mentioned above, a T cell subline was isolated which could be stimulated by PI presented on low responder $H-2^b$ macrophages. Furthermore, previous studies indicated that low responder macrophages were capable of presenting non-permissive antigens to helper T cells. This applies to PI in $H-2^b$ mice[5] and to the polypeptide antigens $(H,G)-A--L$ and T_6-A--L in $H-2^b$ and $H-2^k$ mice, respectively[17]. In such systems the allogeneic combinations of high responder T cells and low responder presenting cells should, therefore, allow antigen-specific stimulation due to the presence of some cross-reactive T cell clones. This would mean complete loss of Ir gene control in allo-restricted T cell-macrophage combinations. Direct experimental evidence supporting such an assumption has recently been obtained by in vitro priming of allo-restricted T cells with Ir gene controlled antigens[18].

In the normal syngeneic situation low responsiveness can be most readily explained by the induction of tolerance or the elimination of self-reactive clones. Cross-reactivity of certain of these clones with the external antigen in association with a defined syngeneic restriction element would then result in the inability to respond.

ACKNOWLEDGEMENTS

We like to acknowledge the expert technical assistance of Ms Christiane Bodin and Ms Doris Meier. We thank Drs. H.G. Gattner and D. Brandenbury, Deutsches Wollforschungsinstitut, Aachen, for kindly providing us with samples of the insulin polypeptide chains, and we are grateful to Dr. G. Seipke, Hoechst AG, Frankfurt, for his generous gift of insulins.

REFERENCES

1. T.L. Blundell, G.G. Dodson, D.C. Hodgkin, and D.A. Mercola, Insulin: The structure in the crystal and its reflection in chemistry and biology, Adv.Prot.Chem. 26:279 (1972).
2. K. Keck, Ir gene control of carrier recognition. I. Immunogenicity of bovine insulin derivatives, Eur.J.Immunol. 5:801 (1975).
3. K. Keck, Ir gene control of carrier recognition. III. Cooperative recognition of two or more carrier determinants on insulin of different species, Eur.J.Immunol. 7:811 (1977).
4. L.J. Rosenwasser, M.A. Barcinski, R.H. Schwartz, and A.S. Rosenthal, Immune response gene control of determinant selection. II. Genetic control of the murine T lymphocyte proliferative response to insulin, J.Immunol. 123:471 (1979).
5. E. Spaeth, H. Stötter, A. Reske-Kunz, F. Zimmermann, E. Rüde, and H.J. Hedrich, Function and complementation of Ir genes controlling the responsiveness of mice to insulin, in: "Basic and Clinical Aspects of Immunity to Insulin," K. Keck and P. Erb, eds., Walter de Gruyter, Berlin, New York, in press (1981).
6. C.G. Fathman, Hybrid I region antigens, Transplantation 30:1 (1980).
7. K. Keck, Ir gene control of immunogenicity of insulin and A-chain loop as a carrier determinant, Nature 254:78 (1975).
8. A.B. Reske-Kunz and E. Rüde, Fine specificity of a T cell line reactive to bovine insulin, submitted for publication.
9. M. Kimoto and C.G. Fathman, Antigen-reactive T cell clones. II. Unique homozygous and (high responder x low responder)F_1 hybrid antigen presenting determinants detected using poly(Tyr,Glu)-poly-D,L-Ala--poly-Lys-reactive T cell clones, J.Exp.Med. 153: 375 (1981).
10. J.W. Kappler, B. Skidmore, J. White, and P. Marrack, Antigen-inducible, H-2-restricted, interleukin-2-producing T cell hybridomas. Lack of independent antigen and H-2 recognition, J.Exp.Med. 153:1198 (1981).

11. K. Eichmann, A three site interaction model for antigen speci-
    ficity, MHC-restriction and Ir gene control in lymphocyte
    communication, Immunobiol. 158:145 (1981).
12. Th.R. Hünig and M.J. Bevan, Cloned cytotoxic T-lymphocytes that
    recognize two different conventional antigens in association
    with two different H-2 restriction elements, Immunobiol. 160:45
    (1981).
13. H. von Boehmer, W. Haas, and N.K. Jerne, Major histocompati-
    bility complex-linked immune responsiveness is acquired by
    lymphocytes of low responder mice differentiating in thymus of
    high responder mice, Proc.Natl.Acad.Sci.USA 75:2439 (1978).
14. D.W. Thomas and E.M. Shevach, Nature of the antigenic complex
    recognized by T lymphocytes. Specific sensitization by antigens
    associated with allogeneic macrophages, Proc.Natl.Acad.Sci.USA
    74:2104 (1977).
15. P.C. Doherty and J.R. Bennink, Vaccinia-specific cytotoxic T
    cell responses in the context of H-2 antigens not encountered
    in thymus may reflect aberrant recognition of a virus-H-2
    complex, J.Exp.Med. 149:150 (1979).
16. H. Wagner, C. Hardt, H. Stockinger, K. Pfizenmaier, R. Bartlett,
    and M. Röllinghoff, Impact of thymus on the generation of immuno-
    competence and diversity of antigen-specific MHC-restricted
    cytotoxic T lymphocyte precursors, Immunol.Rev. 58:95 (1981).
17. H. Stötter, A. Imm, M. Meyer-Delius, and E. Rüde, Specificity
    of H-2 linked Ir gene control in mice: Demonstration of T
    helper cells recognizing branched synthetic polypeptides in
    low responder mice, J.Immunol. 127:8 (1981).
18. Z.A. Nagy, N. Ishii, C.N. Baxevanis, and J. Klein, Lack of Ir
    gene control in T cell responses restricted by allogeneic MHC
    molecules. Behring Institute Mitteilungen 70, in press, (1982).

EVIDENCE FOR Ia - ANTIGEN INTERACTION

IN T CELL ACTIVATION

Ellen Heber-Katz, Daniel Hansburg
and Ronald H. Schwartz

The C-terminal cyanogen bromide cleavage fragment of tobacco horn worm moth cytochrome c (moth 81-103) is immunogenic in both B10.A and B10.A(5R) mice, whereas the C-terminal fragment of pigeon cytochrome c (pigeon 81-104) is only immunogenic in the B10.A. When either cytochrome c fragment is used to immunize B10.A mice, the proliferative response of T cells from the draining lymph nodes shows a characteristic response profile (A pattern) elicited by a set of stimulatory cytochrome c fragments from different species. Thus, moth and fly 81-103 always stimulate better than pigeon 81-104, which in turn stimulates better than tuna 81-103. In contrast, T cells from B10.A(5R) mice immunized with moth 81-103 show a different response profile (5R pattern) when tested with the same set of cytochrome c fragments. In this case, moth and fly 81-103 stimulate well while pigeon and tuna fragments stimulate poorly if at all.

To determine whether these patterns were characteristic of single T cells, hybridoma clones were generated by fusing in vitro stimulated cytochrome c primed T cells with the AKR thymoma BW-5147 and selecting in HAT medium for those clones which produced T-cell growth factor (TCGF) when stimulated with antigen and syngeneic spleen cells. TCGF production was assayed on small numbers of thymocytes stimulated with suboptimal amounts of Con A. Three of 3 clones from B10.A mice immunized with pigeon 81-104 showed the A pattern when stimulated with various cytochrome c fragments. Thus, the clones appeared to be a representative sample of the whole population and their similar specificity suggested a limited T cell repertoire for the antigen.

The most interesting aspect of the clones' specificity was discovered when they were examined for MHC-restriction. Although the

B10.A clones did not respond to pigeon 81-104 when tested with B10.A
(5R) antigen-presenting cells (APCs), they did respond to moth and
fly 81-103. In fact, the pattern was similar to that seen with whole
populations of B10.A(5R) T cells immunized to moth 81-103. This
surprising result suggested that the T cell populations in the two
strains were similar and that their apparent specificity differences
resulted mainly from difference in their APCs.

To test this hypothesis B10.A(5R) mice were immunized with a
benzyl tyrosine derivative of moth 81-103 and T cell clones derived
by cell fusion to BW-5147. Only one stable clone was isolated that
produced TCGF when stimulated with native moth 81-103; however, its
response pattern was similar to that of whole B10.A(5R) lymph node
T cells, i.e., it responded to fly 81-103 and moth 81-103 but not
at all to the pigeon or tuna fragments. Amazingly, when this clone
was tested with B10.A APCs, it now responded to both pigeon 81-104
and tune 81-103! Its responses to moth and fly 81-103 were also
better than those achieved with B10.A(5R) APCs, although not as
heteroclitic as usually observed in the A pattern of B10.A T cells
primed to pigeon 81-104. These results eliminated the possibility
that the presentation by allogeneic APCs was due to weak cross-
reactions and suggested instead that the B10.A and B10.A(5R) T cell
repertoires for these cytochrome c antigens are very limited. Subsequent experiments using normal $\bar{T}$ cell populations have confirmed
the existence of this particular degeneracy in the bulk T cell repertoire of the B10.A, and B10.A(5R) mice.

Finally, a comparison of the responses of the B10.A(5R) clone
to a variety of native and chemically derivatized C-terminal fragments in the presence of either B10.A or B10.A(5R) APCs showed that
the particular pattern observed was indeed determined by the APC.
This observation demonstrates, for the first time, objective evidence
that the APC Ia molecule and the nominal antigen interact during
the stimulation of a T lymphocyte proliferative response.

SOMATIC T CELL HYBRIDS IN THE ANALYSIS OF

H-2 RESTRICTION AND ANTIGEN RECOGNITION

Peter Lonai

Department of Chemical Immunology
The Weizmann Institute of Science
Rehovot, Israel

In this synopsis I will discuss a few recent experiments from our laboratory, which studied helper factors of T cell hybridomas. I will also analyze certain old and new approaches, which hitherto were little used in the study of immunological signals, but may help to solve some of these problems. Our recent results in this field have been published, or are now in press; therefore the data will not be described here in detail.

Differences between antigen recognition by B and T cells have been studied since the discovery of T cells as the second main class of immunocytes. This interest was amplified by two more recent discoveries: the control of T cell specificity by H-2 linked Ir genes[1], and the H-2 restriction of antigen recognition by T cells[2-4]. I have attempted here to concentrate the problems raised by these discoveries into two groups of questions:

1) The nature of the T cell receptor: Which genes control it: Igh-1, H-2, or both? Are there separate receptors for self H-2 and antigen, or are both recognized by one receptor?

2) Structure of the immunogen recognized by T cells: Are antigen and self H-2 presented as separate moieties, or as a complex between the antigen and H-2 gene products? If such complexes exist, what is the molecular nature of their association?

The objects of these questions are the receptor and its ligand. Our studies started by seeking appropriate systems to investigate antigen binding by H-2 restricted T cells. For this end we made efforts to obtain at least crude preparations of the ligand of T cells, and to establish T cell clones, which bear homogenous receptors. The first problem was investigated by studying H-2

restricted ligands and immunogens released by macrophages preincubated with antigen, while to analyze the second, we prepared T cell
hybridoma clones as a homogeneous receptor source.

It has been reported by Erb and Feldmann that antigen-fed macrophages release a factor, which can replace macrophages in the in vitro
induction of helper cells (Ref.5, see also in this volume). This
factor, GRF, was found to contain Class II (Ia products of the H-2
complex, together with the antigen. We have in a sense rediscovered
GRF in studying the binding of protein antigens to mouse T cells.
Our findings on the mechanism of antigen binding emphasize the importance of two macrophage derived factors, a nonspecific, non H-2
restricted lymphokine, and an Ia associated antigen containing factor
(IAC), which appears to be similar to GRF.

We had found that T cells of the Lyt-1[+],2,3[-] phenotype bind
various protein antigens optimally after preincubation with supernatants of macrophage cultures[6]. Later it was found that the material
responsible for this effect is present in semipurified preparations
of the lymphokine IL-1 (Laf) isolated from culture fluids of LPS-
treated, macrophage-like tumor cells of the line P388D1. The antigen
binding stimulatory activity of IL-1 is connected with an increase
in plasma membrane viscosity of antigen binding T cells[7]. We think
that IL-1 is a physiological regulator of antigen binding by Lyt-1[+]
T cells, and that this regulation is causally connected with lymphokine induced membrane changes[6,7].

IL-1 treated normal peripheral T cells or cloned helper hybridoma cells bind antigen after it has been released by macrophages,
preincubated with the antigen. This "processed antigen", or IAC
binds to T cells at physiological temperature, and the binding is
restricted to identity in the A subregion of H-2 between the pro-
cessor and the binding cells. IAC contains Ia antigenic determinants,
as shown by affinity chromatography with anti-Ia immunosorbents[8,9].

IAC is immunogenic in vivo[8], and also in vitro, as demonstrated
by antigen dependent T cell proliferation assays. This immunogen-
icity, when low, limiting, antigen doses are used is H-2 restricted[8,10]. Radioactive antigen suicide assays have shown that cells which
bind IAC are helper T cells[8]. These findings were supported also
by the H-2 restricted IAC-antigen binding characteristics of cloned
helper hybridoma cells[9].

Experiments with our present crude IAC preparations gave little
information on the molecular structure of a possible antigen-Ia
complex. Some of our functional experiments however gave certain
insights to the problem. We have compared antigen binding inhibition by cold native antigen, and by homologous IAC-antigen. Both
normal[11] and hybridoma[9] cells were completely inhibited by a ten
fold excess of cold homologous IAC, whereas native antigen did not

inhibit antigen binding even at several thousand fold excess. This
suggested that the T cell receptor has higher affinity for processed
than non-processed forms of the same antigen. Moreover, because
allogeneic IAC is not bound[8], and heterologous antigen processed by
syngeneic macrophages does not compete[9,11], the experiments also
suggest that if the receptor and the ligand are each made up from
two products, these have to be in close connection. Hence, the data
support either the altered self hypothesis, or one version of the
dual recognition hypothesis, which predicts that the anti-self and
anti-foreign antigen parts of the receptor should be subsites on one
receptor molecule, or separate sites on two closely connected receptors[11].

        The information gained from the antigen binding experiments was
used in the development of functional T cell hybridoma clones. This
was most helpful for rapid functional screening of fusion hybrid
lines. More than five hundred lines were prepared by fusing C57BL/6
($H-2^b$) or B10.BR($H-2^k$) splenic T cells immunized with NP-chicken
gamma globulin (CGG) with the AKR lymphoma BW-5147 ($H-2^k$). These
were screened for binding (C3H x B6)F1 ($H-2^{k/b}$)-processed NP-CGG.
Twelve lines were found to be positive. These positive lines then
were tested for the production of CGG specific helper factors, and
indeed were found to be functional, and CGG specific[9]. The positive
lines were cloned by limiting dilution.

        The helper factors of these hybridoma clones are carrier (CGG)
specific and stimulate hapten specific antibody production in spleen
cells immunized with heterologous carrier coupled with the NP or
NIP hapten in the presence of NP-, or NIP-CGG in vitro, suggesting
that they activate B cells. The factors contain antigenic determinants of Ig H chain variable region (V_H) "framework" determinants[9,
12], as detected by a rabbit antibody against the V_H fragment of the
M315 myeloma protein[13]. Affinity chromatography with monoclonal
antibodies revealed that they contain in addition also Ia determinants controlled by the I-A, and possibly also by the I-E subregion
of H-2 [9,12]. The most interesting feature of these monoclonal T
cell replacing factors is that their effect is H-2 (I-A) restricted,
i.e., they stimulate only those B cell sources which carry similar
I-A loci to the genotype of the hybridoma cells[12]. Because of this,
and because of the similarities between the antigen binding and
helper factor activity of the same clones[9,10], it appears that these
factors may represent an essential part of H-2 restricted helper T
cell receptors.

        The H-2 heterologous hybridoma clones, which derived from fusing
$H-2^b$ normal T cells specific to NP-CG with $H-2^k$ lymphoma cells,
made it possible to investigate the number of genes involved in the
control of H-2 restricted helper factors. Our reasoning was as
follows: The BW-5147 lymphoma has no known specificity, and does
not bind NIP-CGG specific helper factors[9,10,12]. Therefore, it is

highly unlikely that the parental lymphoma cells could have donated
an active carrier specific locus to the hybridoma cells. Hence, the
clones could have only one functional carrier specific locus of B6
origin. If there are independent self H-2 specific receptors controlled by independent "anti-self" loci, than it can be assumed that
both parental cells could have donated such a locus. Moreover,
because the recognition of self in H-2 restricted T cells is directed to their own H-2 gene products, the hybridoma clones deriving
from H-2 heterologous fusions could have inherited two anti-self
alleles of different specificity (see Table 1). It follows that
H-2 heterologous hybridoma clones could produce two helper factors,
one specific to CGG and H-2^b, and the other for CGG and H-2^k. According to the above reasoning the CGG+H-2^k specific factor should be
the product of two independent specificity defining loci, that is
of a B6 derived carrier specific locus, and an AKR-derived (BW-5147)
anti-self locus (Table 1).

The validity of this speculation was tested by three different
experiments[14]. The culture supernatants of most H-2 heterologous
clones contained helper factors which could stimulate both H-2^b and
H-2^k type B cell sources, but they were inactive with H-2^q or f B
cell sources. Additional experiments have demonstrated that this
"double producer" activity is due to the production of two helper
factors by one hybridoma clone. First, the helper factors were adsorbed on spleen cells of H-2^b of H-2^k genotype. It was shown previously that the adsorption of helper factors by normal spleen cells
is H-2 (I-A) restricted[12]. From the culture fluid of H-2 heterologous hybridomas, H-2^k spleen cells removed the activity which
stimulates H-2^k B cell sources, and the H-2^b restricted activity
remained. The opposite was observed when H-2^b spleen cells were
used for adsorption. In the third experiment the helper factors
were fractionated on I-A^b or I-A^k specific monoclonal antibody affinity columns. I-A^b specific columns bound the H-2^b restricted,
and passed the H-2^k restricted activity, whereas I-A^k specific columns
bound the H-2^k restricted, but passed the H-2^b restricted helper
activity. The adsorbed helper factors could be eluted from the affinity columns with 0.1N NH$_4$OH. Similar observations were made using
this method on four additional clones and subclones, two of which
derived from an independent cell fusion experiment[14].

It appears therefore that H-2 restricted helper factors, of the
kind secreted by our hybridomas, are controlled by two separate loci,
one of which controls carrier-specificity, and the other controls
the specificity to self Ia. In present immunogenetic terms these
results support the validity of dual recognition theories over the
altered self hypotheses for the case of H-2 restricted helper factors.

Preliminary experiments suggest that the relevance of these
findings may be extended to the cellular antigen binding receptor
of the same clones. We have found with one of the H-2 heterologous

Table 1. Possible expression of anti self and anti carrier receptor loci in T hybridomas deriving from fusing C57BL/6 anti NP-CGG T cells with the BW-5147 lymphoma

| | Receptor loci | | |
	Anti Carrier	Anti Self	Phenotype (specificity)
C57BL/6 ($H-2^b$)	+	+	Anti CGG + anti I-A^b
BW-5147 ($H-2^k$)	–	+	Anti CGG + anti I-A^k

clones, in antigen binding studies that it binds both $H-2^b$- and $H-2^k$ IAC-CGG, but not $H-2^f$-IAC-CGG[10].

These results suggest that the dual specificity of our $H-2$ restricted hybridoma clones may be regulated by two separate genes. They do not indicate however what is the nature of these genes. In these[9,10,12,14], as well as in other studies[15-17] we have demonstrated that T cell receptors, or factors, contain antigenic determinants which crossreact with framework determinants of Ig V_H. These, similarly to crossreactive idiotypic determinants of T cells[18], were found to be linked to the H chain allotype[19]. It is reasonable to assume therefore that at least a part of the receptor should be controlled by loci linked to Igh-1 on the 12th chromosome of the mouse[20]. The identity of the second of the two loci is known even less well. In our experiments helper factors with different anti-self specificity were separated by anti-Ia affinity columns. This finding does not allow us to exclude $H-2$ gene products in the definition of the $H-2$ restricted specificity of T cells. Supportive evidence for this alternative derives from data in the literature suggesting that the expression of T cell idiotypes may be controlled partially by $H-2$ loci[21]. The present data however do not distinguish between the alternatives whether T cells may have two V_H-like and an $H-2$-like specificity defining element.

The interpretation of our experiments derives from genetic evaluation of the data. There are however results in the literature which may be in conflict with our interpretation. Kappler et al. reported experiments with $H-2$ restricted double hybridoma clones, which derived from fusing two hybridomas of different anti-antigen and anti-self $H-2$ specificities. None of twenty such clones was found to express cross-complementing specificities between the parental genotypes[22]. They have tested the specificity of the clones by measuring the antigen specific induction of TCGF production; thus, there are differences in the methodologies used by the two laboratories. It is also possible that heterologous and homologous products of two double genomes do not pair with equal frequency. More importantly we do not know with what frequency one may expect the expression of two antiself specificities in a somatic T cell hybrid even in the case of the hybridomas studied by us[10,14]. Hence, the clarification of this possible controversy awaits further investigation.

In my mind the importance of such studies is as much in the approach, as in the theoretical conclusions. Use of somatic cell hybrids in immunology became extensive following the finding by Köhler and Milstein that hybridization of normal lymphocytes with tumor cells can lead to their "immortalization"[23]. The main impact of this discovery, for most workers in the field, was to produce lymphocytes and their products in large quantity and high purity. Less emphasis was given to the original purpose of the somatic hybrid technology, to the study of complementary gene expression and gene segregation in heterokaryons. Combined with recombinant DNA techniques, the usefulness of somatic hybrid technology can be enhanced several fold. The combined approach allows the quick and simple detection of known alleles by DNA hybridization, their localization by somatic genetic techniques, as well as the definition of unknown traits as structurally defined alleles, by DNA transfer and by the detection of their phenotypic expression[24]. These approaches require changing the tools to solve some of the important problems of immunological recognition.

REFERENCES

1. H.O. McDevitt, B.D. Deak, D.C. Shreffler, J. Klein, J.H. Stimpfling, and G.D. Snell, Genetic control of the immune response. Mapping of the Ir-1 locus, J.Exp.Med. 135:1259 (1972).
2. R.M. Zinkernagel and P.C. Doherty, Immunological surveillance against altered self components by sensitized T lymphocytes in lymphocytic choriomeningitis, Nature 251:547 (1974).
3. G.M. Shearer, Cell mediated cytotoxicity to TNP-modified syngeneic lymphocytes, Eur.J.Immunol. 4:527 (1974).
4. E.M. Shevach and A.S. Rosenthal, Function of macrophages in antigen recognition by guinea pig T lymphocytes II. J.Exp.Med. 138:1213 (1973).
5. P. Erb and M. Feldmann, The role of macrophages in the generation of T helper cells. III. Influence of macrophage-derived factors in helper cell induction, Eur.J.Immunol. 5:759 (1975).
6. P. Lonai and L. Steinman, Physiological regulation of antigen binding to T cells, role of a soluble macrophage factor and interferon, Proc.Natl.Acad.Sci.USA. 74:5662 (1977).
7. J. Puri, M. Shinitzky, and P. Lonai, Concomitant increase in antigen binding and in T cell membrane lipid viscosity by the lymphocyte activating factor, LAF. J.Immunol. 124:1937 (1980).

8. J. Puri and P. Lonai, Mechanism of antigen binding by T cells. H-2 (I-A) restricted binding of antigen plus Ia by helper cells, Eur.J.Immunol. 10:273 (1980).

9. P.Lonai, J. Puri and G.J. Hämmerling, H-2 restricted antigen binding by a hybridoma clone which produces specific helper factor, Proc.Natl.Acad.Sci.USA. 78:549 (1981).

10. P. Lonai, E. Arman, H.F.J. Savelkoul, V. Friedman, J. Puri, and G.J. Hämmerling, Factors, receptors and their ligands; studies with H-2 restricted helper hybridoma clones, in: "Isolation, characterization and utilization of T lymphocyte clones," C.G. Fathman and F.W. Firch eds., Acad. Press, N.Y., in press. (1981).

11. P. Lonai, L. Steinman, V. Friedman, G. Drizlikh, and J. Puri, Specificity of antigen binding by T cells; competition between soluble and Ia associated antigen, Eur.J.Immunol. 11:382 (1981).

12. P. Lonai, J. Puri, S. Bitton, Y.Ben-Neriah, D. Givol, and G.J. Hämmerling, H-2 restricted helper factor secreted by cloned hybridoma cells, J.Exp.Med. 154:942 (1981).

13. Y. Ben-Neriah, C. Wuilmart, P. Lonai, and D. Givol, Preparation and characterization of anti framework antibodies to the heavy chain variable region (V_H) of mouse immunoglobulins, Eur.J. Immunol. 8:797 (1978).

14. P. Lonai, S. Bitton, H.F.J. Savelkoul, J. Puri, and G.J. Hämmerling, Two separate genes regulate self-Ia and carrier recognition in H-2 restricted helper factors secreted by hybridoma cells, J.Exp.Med. in press. (1981).

15. P. Lonai, Y. Ben-Neriah, L. Steinman, and D. Givol, Selective participation of immunoglobulin V-region and MHC products in antigen binding by T cells, Eur.J.Immunol. 8:827 (1978).

16. K. Eichmann, Y. Ben-Neriah, D. Hetzelberger, C. Polke, D. Givel and P. Lonai, Correlated expression of V_H framework and V_H idiotypic determinants on T helper cells and on functionally undefined T cells binding group A streptococcal carbohydrate, Eur.J.Immunol. 10:105 (1980).

17. J. Puri, Y. Ben-Neriah, D. Givol, and P. Lonai, Antibodies to immunoglobulin heavy chain variable regions protect helper cells from specific suicide by radiolabeled antigen, Eur.J.Immunol. 10:281 (1980).

18. K. Eichmann, Expression and function of idiotypes on lymphocytes, Adv.Immunol. 26:195 (1979).

19. Y. Ben-Neriah, D. Givol, P. Lonai, M.M. Simon, and K. Eichmann, Allotype-linked genetic control of a polymorphic V_H framework determinant on mouse T-helper cell receptors, Nature 285:257 (1980).

20. T. Meo, J. Johnson, C.V. Beechey, S.J. Andrews, J. Peters, and A.G. Searle, Linkage analyses of murine immunoglobulin heavy chain and pre-albumin genes establish their location on chromosome 12 proximal to the T(5;12)31H breakpoint in band 12F1, Proc.Natl.Acad.Sci.USA. 77:550 (1980).

21. J.W. Kappler, B. Skidmore, J. White, and P. Marrack, Antigen
     inducible H-2 restricted, IL-2-producing T cell hybridomas.
     Lack of independent antigen and H-2 recognition, J.Exp.Med.
     153:1198 (1981).
22. P. Krammer and K. Eichmann, T cell receptor idiotypes are con-
     trolled by genes in the heavy chain linkage group and the major
     histocompatibility complex, Nature 270:733 (1977).
23. G. Köhler and C. Milstein, Continuous cultures of fused cells
     secreting antibody of predefined specificity, Nature 256:495
     (1975).
24. F.H. Ruddle, A new era in mammalian gene mapping: somatic cell
     genetics and recombinant DNA methodologies, Nature 294:115
     (1981).

VIII: STRUCTURAL ASPECTS OF T-B AND T-T INTERACTION

INTRODUCTION

Having explored the recognition by B cells of conformational determinants, and the specialized requirements for T cell recognition of antigen in the context of MHC molecules, we approach one of the most complex mechanistic puzzles confronting cellular immunologists. In what way do the various collaborative cell types in the immune system recognize each other and communicate on or off signals?

A few generalized views of this problem can be introduced here which will later be discussed within the individual papers of this Topic. The antigen-bridging view presented by Mitchison et al.[2], p.379, was that T and B cells interacted across a molecular bridge of antigen, so that the T cells recognized one receptor while the B cell recognized another. Presumably, the part of the antigen buried within the B cell receptor is in native conformation, but what is the nature of the protruding remainder of the antigen molecule? Is it fragmented and denatured, operated on by neighboring antigen-processing cells? If not, can the helper T cell recognize the available native determinants on the free surface of the antigen? If so, can a single helper T cell recognize either the native antigen or the fragmented antigen-MHC complex by which they apparently are initially activated? Is the recognition of the MHC entity originally seen by the T cell on the macrophage surface identical to that found on the B cell surface, and do the B cells then process antigen according to the same rules as the macrophages? How do idiotype-recognizing T cells fit into the picture? What is the role of the non-specific lymphokines secreted by the T cells? These questions and others were considered in the seven papers that follow.

369

The first five papers in this Topic examine the anti-bridging problem. Goodman et.al., consider T-B collaboration and present evidence from work with bifunctional molecules in which the epitopes are separated by polyproline spacers, that supports the conclusion that the flexibility or rigidity of the spacer determines responsiveness. The experiments are consistent with an antigen-bridging model, and also are best reconciled with a scheme in which T cells have independent recognition receptors for antigen and MHC. The authors suggest that it is possible that the T cell bound receptor may have a different reactivity (requiring simultaneous MHC recognition) from the helper factor secreted by the cell, which may recognize nominal carrier epitopes.

An alternative model of T-B collaboration was presented by Grey and Chesnut. According to their view, B cells can present antigen to the T cell in an MHC-restricted and specific manner, and they detail evidence with B cell lymphomas and in certain cases, with macrophage depleted spleen cells, to support their model. Although the B lymphoma can act as a processing cell to "present" a variety of antigens, unrelated to the specificity of the B cell receptor for antigen, it is presumed that under physiological circumstances, B cells will concentrate the specific antigen to which they are directed, and display the processed antigen on their surface in connection with Ia antigens, which will serve as the site for T cell recognition.

In the reports of Krzych et. al., and Green and Gershon, the nature of the determinants recognized by the different functional populations of T cells is considered. In the beta-galactosidase model (paper 36) the nature of the bridging entity was postulated to be a small fragment of the antigen possessing a suppressor determinant and a helper determinant. What was crucial to the detection of the suppressor determinant was the necessity that the helper T cell targeted could in turn bridge with a B cell recognizing a neighboring epitope. The frequency of such functional epitopes, even on a molecule as large as beta-galactosidase, was very limited. Green and Gershon set forth views on the newest member of the T cell functional orchestra, the contrasuppressor cell. In the two different cases, the sheep red cell and lysozyme systems, the same antigen will trigger suppressor cells and contrasuppressor cells. In fact, the contrasuppressor cells could be shown in the lysozyme system to recognize a different epitope on the molecule than the helper T cells. The basis for the lack of overlap in the functional T cell repertoire in different subpopulations needs to be clarified, although it is suggested in these latter two papers that antigen-presenting cells are probably responsible.

Antonio Lanzavecchia examined T-B cell interaction in a human system using soluble antigens. T cell lines were MHC restricted

and required the presence of antigen-presenting cells. Two types of
interaction between T cells and B cells were described: one involved
soluble mediators from T cells which promoted maturation of activated
B cells; a second antigen-bridging interaction, occurring in the
presence of high concentrations of antigen, could even take place
non-specifically with B-cell absorbed antigen serving as a target for
T helper cells. At lower concentrations of antigen, specificity
reigned.

In the papers by Jorgensen et. al., and Weaver et. al., the
important problem of V-domain recognition by T cells is examined.
Recently, it has become evident that in the immune network, not only
must anti-idiotypic antibody be considered, but also a role must be
found for the idiotype-recognizing T cell. It is such helper T cells
specific for idiotype which may lead to the existence of predominant
idiotypes in the serum response to an antigen or hapten. Do such T
cells recognize the same structures as anti-idiotypic antibody? In
paper 39, it was found that V-lambda-315-recognizing T helper cells
recognize either the assembled or free form of the antigen, while
the serologically defined idiotypes require the quarternary structure
of assembled domains. H-2 linked Ir genes also could control the
recognition and response to the V domains, suggesting that idiotypic
regulation may further involve this parameter.

Weaver et. al., examine what appears to be a simple system to
study the interplay of epitope-specific and idiotype-specific factors.
Ferredoxin possesses two antigenic determinants, an N-terminal and
C-terminal one. Anti-idiotype directed against an N-specific hybridoma seems to also react with regulatory T cells. Perturbation
of the network was also accomplished by injection of idiotype itself. Interestingly, anti-idiotype also has the effect of amplifying
responses to idiotypically unrelated epitopes on the molecule! It
is evident that intense study of idiotypic network effects on defined protein antigen systems will yield important regulatory insights.

ANTIGEN BRIDGING IN T CELL-B CELL INTERACTION: FACT OR FICTION?

Joel W. Goodman, Danute E. Nitecki
Sherman Fong and Zehra Kaymakcalan

Department of Microbiology and Immunology
University of California
San Francisco, California 94143

Antigen-bridging models of the interaction between T and B lymphocytes rose to prominence in the late 1960s, based principally on two different experimental approaches[1-3]. One involved the "carrier effect", in which physical association between hapten and carrier is required for a secondary anti-hapten antibody response in animals primed separately with hapten and carrier, and the other involved thymus-marrow reconstitution of irradiated animals, demonstrating that hapten and carrier were recognized by different cell types in the anti-hapten antibody response.

The approach we have taken to the general question of mechanisms of cell cooperation has made use of small, structurally defined antigens in which the hapten and carrier components have spatial relationships which are reasonably well understood[4]. These antigens are epitomized by L-tyrosine-p-azobenzenearsonate (RAT), which induces cellular immunity without antibody formation in several animal species and serves as a carrier for haptens[5,6]. Early evidence in this system favoring bridging models stemmed from enhanced anti-hapten antibody responses in guinea pigs immunized with bifunctional antigens in which the hapten and carrier epitopes were separated by a spacer (DNP-spacer-RAT) as compared to when they were not (DNP-RAT)[7], and the requirement for rigid spacers separating the epitopes of symmetrical bifunctional antigens (RAT-spacer-RAT) in cases where the epitopes tend to associate intramolecularly due to charge complementarity[8,9].

The ability to construct synthetic bifunctional antigens with rigid spacers (DNP)-(proline)$_n$-RAT) permitted more rigorous probing of the validity of antigen bridging. Consider the plausibility that T-dependent B cell activation requires contact between the B

373

cell and either a helper T (T_H) cell itself or a soluble product
released by the T_H cell, with antigen serving as the focusing agent
in either case. For this argument, we shall ignore the role of the
macrophage, which has definitively been shown to be essential for
activation of the T_H cell, but for whose role in interaction between
B cells and activated T_H cells there is less evidence. The argument
is supported by abundant evidence favoring genetic restriction of
T-B collaboration in at least some circumstances[10,11], and by the
elaboration of antigen-specific helper factors bearing markers coded
by the major histocompatibility complex[12]. It is independent of the
possible existence of more than one type of T_H cell[13]. If the T cell
or its product recognizes the carrier epitope of a bifunctional antigen which is bound to a B cell by its haptenic epitope, and if contact between the B cell and a "cell interaction structure" on the T
cell or its product is required for B cell activation, then it should
be possible to construct bifunctional antigens with rigid spacers of
dimension sufficient to prohibit cell contact. Rigid spacers are
required for this purpose to prevent flexing of the spacer chain
with consequent indeterminate variance of the distance between the
carrier and haptenic epitopes. If such proved to be the case, the
maximum permissible distance between the epitopes which still allowed
B cell activation could be estimated.

Bifunctional antigens with rigid polyprolyl spacers (DNP-$(PRO)_n$-
RAT), were synthesized by the solid phase technique[14]. The spacers
comprised as many as 40 proline residues. Guinea pig were immunized
in the footpads with varying molar quantities of the antigens emulsified in complete Freund's adjuvant. The antigens used and the
anti-DNP antibody responses mounted by the animals are summarized
in Table 1. The PRO_{10} and PRO_{22} compounds evoked responses, measured
as precipitating antibody, that peaked about 3 weeks after immunization at levels of 150 to 200µg of anti-DNP antibody per milliliter
of serum. These responses were quantitatively and kinetically similar to those obtained using bifunctional antigens with flexible
spacers[7]. The optimal antigen dose of those tested was 1.0µmol in
each instance. This dose of DNP-PRO-RAT did not raise significant
anti-DNP responses, suggesting that a single proline residue is an
ineffective spacer for implementing cell interaction. Of particular
interest, DNP-$(PRO)_{31}$-RAT was negative at all doses and times assayed. Extending the spacer to 40 proline residues also gave negative results, indicating that the absence of a response to DNP-
$(PRO)_{31}$-RAT was not idiosyncratic.

Assessment of the distance between the two epitopes in these
antigens relies on several assumptions concerning the rigidity and
dimension of polyproline chains, which can exist in two alternative
forms[15,16]. Form I is a _cis_ right-handed helix with an axial translation of 1.9Å per residue. This conformation is preferentially
assumed in non-polar solvents. Form II is a _trans_ left-handed helix
with three residues per helical turn and an axial translation of

Table 1. Antibody responses to bifunctional antigens
with polyprolyl spacers

Antigen (40μM)	Spacer* Length Å	Anti-DNP Antibody μg/ml
DNP-PRO-RAT	3	0
DNP-(PRO)$_{10}$-RAT	31	70 ± 26**
DNP-(PRO)$_{22}$-RAT	69	143 ± 14
DNP-(PRO)$_{31}$-RAT	97	0
DNP-(PRO)$_{40}$-RAT	125	0
DNP-(PRO)$_{16}$SAC-(PRO)$_{15}$-RAT	–	153 ± 26***

*   Based on axial translation of 3.12 Å per proline
    residue in poly-L-proline[15,16].
** Arithmetic mean ± S.E.; 3-6 guinea pigs per group.
*** Compound tested in separate experiment with DNP-
    (PRO)$_{22}$-RAT and DNP-(PRO)$_{31}$-RAT. Value normalized
    to response against DNP-(PRO)$_{22}$-RAT.

3.12Å per residue. The trans conformation is assumed in aqueous
(polar) solvents. The form II coordinates were used in calculating
the distances between carrier and haptenic epitopes of the bifunctional antigens (Table 1).

The reliability of the figures is supported by several studies.
Schimmel and Flory[17] showed that the mean square end-to-end distance
of a hexadecamer of L-proline closely approximates the value obtained
from the helix coordinates for form II. For longer chain lengths
of about 50 residues, intrinsic viscosities are in excellent agreement with predicted values[18]. Moreover, in studies more relevant
to the work presented here, in which oligomers of proline were used
to separate donor and acceptor chromophores in energy transfer
measurements, the observed efficiency of energy transfer as a function of distance was also in excellent agreement with predictions[19].
These observations, in aggregate, provide solid support for the
assumption that polyproline chains with 10-50 residues behave as
rigidly extended rods in solution (and, presumably, at the locus of
cell interaction, although this clearly represents a greater extrapolation).

The figures thus derived lead to the conclusion that epitopes separated by 69A (PRO_{22}) are fully operative for cell cooperation, whereas a span of 97A (PRO_{31}) obviates delivery of the helper signal to the B cell. The maximum permissible separation could, of course, be more closely estimated by fabricating antigens with intermediate spacers, but the principle has been established and the additional synthetic work required would not justify the information to be gained. It might be argued that the long polyprolyl spacer antigens which do not induce anti-DNP responses induce suppression instead. This possibility is unlikely inasmuch as all animals, regardless of the antigen they received, developed comparable arsonate-specific delayed skin reactivity when tested with ABA-bovine serum albumin conjugates[14]. Some animals developed small quantities of anti-poly-prolyl antibody, but there were no significant differences between the groups. Thus, animals immunized with 1.0μmol of the $(PRO)_{10}$, $(PRO)_{22}$ and $(PRO)_{31}$ compounds made 23±23, 31±23 and 16±14μg of anti-polyproline antibody per milliliter of serum, respectively[14].

The model can be tested further by constructing an antigen with a non-permissive polyproline spacer in which a single proline residue in the middle of the chain is replaced by a different group, permitting a bend or kink in the spacer. If the model is correct, such an antigen might be expected to induce anti-DNP antibody responses. Accordingly, the bifunctional antigen $DNP-(PRO)_{16}-SAC-(PRO)_{15}-RAT$ was synthesized with a 6-aminocaproyl (SAC) group in the middle of the $(PRO)_{31}$ spacer chain. Guinea pigs immunized with this compound responded with anti-DNP antibody levels fully comparable to those mounted against the bifunctionals with $(PRO)_{10}$ and $(PRO)_{22}$ spacers (Table 1). Thus, the introduction of a flexible residue in the middle of a non-permissive polyproline spacer restored the antibody response. This result is very difficult to explain without invoking some type of antigen bridging mechanism in B cell activation.

The present findings are clearly consistent with both the carrier effect[2,3] and the MHC restriction of T helper cell interaction with B cells[10,11], alluded to earlier. These phenomena are most readily explicable by postulating the focusing of T help on B cells by antigen. Whether T help can recognize unmodified antigen sufficiently well to be focused in this way, as indicated by some reports[12], is still problematic, but the responses to rigidly spaced bifunctional antigens in the present study are difficult to reconcile with a single T cell receptor site which recognizes a complex of nominal antigen and a cell surface structure ("altered self"). They are compatible with a model in which functional T help bears a receptor for nominal antigen and a cell interaction structure coded by the MHC[12], through which the triggering signal is delivered to the B cell.

A scenario that satisfies most of the observations about T_H cells is one in which the receptor on the surface of the cell and the soluble helper factor secreted by the cell do not have identical reactivities. Thus, helper T cells do not "see" nominal antigen sufficiently well to be enriched by affinity chromatography, "suicided" or activated. They are activated in an MHC-restricted fashion by antigen associated with antigen-presenting cells. Once activated, they release soluble antigen-specific helper factors which find antigen-binding B cells by virtue of their ability to recognize nominal carrier epitopes. Only one helper factor has been studied in any detail thus far[12]. This "TGAL"-specific factor binds nominal antigen and bears H-2-coded determinants, features which are required to explain antigen bridging and MHC-restriction in T-B cell interaction. It remains to be seen if additional evidence supports this model.

REFERENCES

1. P.A. Bretscher and M. Cohn, Minimal model for the mechanism of antibody induction and paralysis by antigen, Nature 220:444 (1968).

2. N.A. Mitchison, K. Rajewsky and R.B. Taylor, Cooperation of antigenic determinants and of cells in the induction of antibodies, in: "Prague Sym. on Developmental Aspects of Antibody Formation and Structure," J. Sterzl and I. Riha, eds., Academic Press, New York (1970).

3. K. Rajewsky, V. Schirrmacher, S. Nase, and N.K. Jerne, The requirement of more than one antigenic determinant for immunogenicity, J.Exp.Med. 129:1131 (1969).

4. J.W. Goodman, Antigenic determinants and antibody combining sites, in: "The Antigens," Vol.III, M. Sela, ed., Academic Press, New York (1975).

5. S.S. Alkan, D.E. Nitecki, and J.W. Goodman, Antigen recognition the immune response. The capacity of L-tyosine-p-azobenzenearsonate to serve as a carrier for a macro-molecular hapten, J. Immunol. 107:353 (1971).

6. J.W. Goodman, S. Fong, G.K. Lewis, R. Kamin, D.E. Nitecki, and G. Derbalian, Antigen structure and lymphocyte activation, Immunol.Rev. 39:36 (1978).

7. S.S. Alkan, E.B. Williams, D.E. Nitecki, and J.W. Goodman, Antigen recognition and the immune response. Humoral and cellular responses small monofunctional and bifunctional antigen molecules, J.Exp.Med. 135:1228 (1972).

8. M.E. Bush, S.S. Alkan, D.E. Nitecki, and J.W. Goodman, Antigen recognition and the immune response. Self-help with symmetrical bifunctional antigen molecules, J.Exp.Med. 136:1478 (1972).

9. J.W. Goodman, C.J. Bellone, D. Hanes, and D.E. Nitecki, Antigen structural requirements for lymphocyte triggering and cell cooperation, in: "Progress in Immunology II," Vol.2, L. Brent and J. Holbrow, eds., North Holland Publishing Co., (1974).

10. J. Andersson, M.H. Schreier, and F. Melchers, T-cell dependent
     B-cell stimulation is H-2 restricted and antigen dependent only
     at the resting B cell level, Proc.Nat.Acad.Sci.USA. 77:1612
     (1980).

11. B. Jones and C.A. Janeway Jr., Cooperative interaction of B
     lymphocytes with antigen-specific helper T lymphocytes is MHC
     restricted, Nature 292:547 (1981).

12. R.N. Apte, I. Lowy, R. De Baetselier, and E. Mozes, Establish-
     ment and characterization of continuous helper T cell lines
     specific to poly(L Tyr, L Glu)-poly(DL Ala)-poly(L Lys), J.
     Immunol. 127:25 (1981).

13. C.A. Janeway Jr., R.A. Murgita, F.I. Weinbaum, R. Asofsky, and
     H.W. Wigzell, Evidence for an immunoglobulin-dependent antigen-
     specific helper T cell, Proc.Nat.Acad.Sci.USA. 74:4582 (1977).

14. S. Fong, D.E. Nitecki, R.M. Cook, and J.W. Goodman, Spatial
     requirements between haptenic and carrier determinants for
     T-dependent antibody responses, J.Exp.Med. 148:817 (1978).

15. V.D. Gupta, R.D. Singh, A.M. Dwivedi, Vibrational spectra and
     dispersion curves of poly-L-proline II chain, Biopolymers 12:
     1377 (1973).

16. P.M. Cowan and S. McGavin, Structure of poly-L-proline, Nature
     176:501 (1955).

17. P.R. Schimmel and P.J. Flory, Conformational energy and con-
     figurational statistics of poly-L-proline, Proc.Nat.Acad.Sci.
     USA. 58:52 (1967).

18. I.Z. Steinberg, W.F. Harrington, A. Berger, M. Sela, and E.
     Katchalski, The configurational changes of poly-L-proline in
     solution, J.Am.Chem.Soc. 82:5263 (1960).

19. L. Stryer and R.P. Haugland, Energy transfer: A spectroscopic
     ruler, Proc.Nat.Acad.Sci.USA. 58:719 (1967).

B CELLS AS MHC RESTRICTED ANTIGEN PRESENTING CELLS:

A MODEL FOR T-B INTERACTION[*]

Howard M. Grey and Robert W. Chesnut

Department of Medicine, National Jewish Hospital and
Research Center/National Asthma Center and
Departments of Pathology and Medicine
University of Colorado Health Sciences Center
Denver, CO

There is considerable evidence to indicate that antigen-specific
T helper cells and B cells recognize the same antigen in quite
different ways. For instance, native and denatured forms of the same
protein are highly cross-reactive antigens as recognized by helper
and proliferating T cells, whereas they are completely noncross-
reactive at the B cell level[1-3]. Furthermore, although B cells can
readily be shown to bind conventional soluble antigens[4], helper T
cells cannot[5,6]; however, after antigen processing by syngeneic Ia-
positive macrophages, antigen-dependent binding of helper T cells
to macrophages has been demonstrated[7,8]. Data such as these have
led to the postulation that B cells recognize intact, unprocessed
antigen, whereas helper T cells recognize processed antigen presented
in the context of macrophage Ia antigens. This postulation presents
some difficulties for how T-B collaboration works, especially if
antigen is important in bridging the 2 cells types as has been proposed in the hapten-carrier system first described by Mitchison[9].
A mechanism of T-B collaboration via antigen bridging has been postulated[10] which involves the processing of antigen by antigen-
specific B cells and subsequent presentation of the antigen, in the
context of the B cell IA, to antigen-specific helper T cells.

For the past few years we have been conducting experiments to
test this hypothesis. We reasoned that if B cells could serve as
antigen presenting cells it was likely that they did so in a clonally

---
[*]This work supported in part by USPHS grants AI-09758, AI-17510 and
CA-21825.

restricted manner by means of the interaction between antigen and antigen specific Ig receptors. Because of the difficulty in obtaining sufficient numbers of purified antigen specific B cells to test their capacity to present antigen we decided to use rabbit IgG as the antigen and rabbit anti-mouse Ig (RAMIG) as a reagent that would polyclonally react with all B cells via their Ig receptors. Assuming that the intracellular fate of RAMIG is similar to that of a protein antigen in antigen specific B cells then the capacity of mouse B cells to process and present rabbit IgG in the form of RAMIG would be a test of the general concept that B cells can act as antigen presenting cells. Macrophage depleted spleen cells (MDSC) were used as a source of B cells to present the rabbit IgG to T cells sensitized with normal rabbit IgG (NRGG). For comparative purposes, NRGG, an antigen that would be expected to bind to only a very few RGG-specific B cells was used. Normal spleen adherent cells were also used as a source of accessory cells. The data shown in Table 1 indicate that when macrophages were pulsed with NRGG or RAMIG nearly equivalent T cell proliferation was observed in response to either antigen. In contrast to this when MDSC were pulsed with the same two antigen preparations RAMIG consistently stimulated a significant response whereas NRGG failed to elicit T cell proliferation. The extent of T cell proliferation observed with RAMIG pulsed MDSC varied from 20 to 100% of that obtained when antigen pulsed spleen macrophages were used as presenting cells.

A variety of experiments have been carried out utilizing this system in an attempt to rule out the possibility that contaminating macrophages might have been responsible for the results[11]. The fact that NRGG pulsed B cells did not induce proliferation indicated that insufficient macrophages were present in the MDSC population to account for the proliferative response by a mechanism involving the direct uptake of antigen by such contaminating cells. That the T cells also present in the MDSC population were not involved in antigen presentation was demonstrated by the fact that treatment of the MDSC with anti-Thy-1 and complement did not decrease the capacity of RAMIG pulsed cells to stimulate a proliferative response. The inability of MDSC to present another antigen, ovalbumin in a similar T cell proliferative response strengthens the conclusion that macrophages were not present to a significant extent in the MDSC population. The possibility that the immune complexes formed between RAMIG and B cell Ig created a "super antigen" that could be taken up very efficiently by a small number of contaminating macrophages also seemed unlikely. The evidence against this possibility was 1) heat aggregated NRGG did not stimulate a proliferative response when MDSC was induced. Furthermore the presentation of RAMIG by MDSC was shown to be genetically restricted in that T cells incorporated 10 to 30 fold more tritiated thymidine when presented antigen by syngeneic compared with allogeneic RAMIG pulsed MDSC.

Table 1. Presentation of Rabbit Anti-Mouse Immunoglobulin to Rabbit
          IgG-Primed T Cells by Macrophage and Macrophage-Depleted
          Spleen Cells (MDSC)

Antigen Presenting Cells	Antigen[a]	^{3}H-thymidine Incorporation $(E-C;\ cpm \times 10^{-3})$[b] in Expt. No.			
		1	2	3	4
Macrophage	NRGG	124	234	121	142
Macrophage	RAMIG	99	186	103	144
MDSC	NRGG	0	0	4	0.5
MDSC	RAMIG	169	68	148	32

[a]MDSC were pulsed with 1mg RAMIG or NRGG/2 x 10^6 cells for 1 hr at
37°C; macrophages were pulsed with 250!g/well of RAMIG or NRGG for
4hrs at 37°C.

[b]^{3}H-thymidine incorporation expressed as experimental(E) cpm minus
unstimulated control (C) cpm.

        Although all of the available data indicated that in this test
system B cells were in fact capable of presenting antigen to T cells,
we felt that until we could more formally rule out the possibility
of macrophage contamination the issue would still be uncertain. To
this end we have more recently utilized B lymphoma lines and antigen
specific T cell hybridomas to further test the antigen presenting
capacity of B cells[12]. A collection of BALB/c B cell tumors previously shown to be Ia positive were obtained from Dr. Richard Asofsky.
These cloned tumor cell lines were used to present antigen to T cell
hybridomas that were constructed by Drs. Kappler and Marrack and
shown to be both antigen specific and MHC restricted[13]. The stimulation of these hybridomas in an antigen specific, MHC restricted
manner could be detected by their elaboration of interleukin-2 (IL-2)
in the culture medium. The IL-2 was assayed by its capacity to
maintain the viability of a T cell line (HT-2) whose viability is
absolutely dependent upon an exogeneous source of IL-2 being added
to its culture medium[14]. Table 2 shows some representative results
from these experiments and clearly indicates that several of the B
cell lines could present antigen to several T cell hybridomas. As
was the case with normal spleen filler cells, the capacity of the
B cell lines to present antigen was MHC restricted as evidenced by
the BALB/c lymphoma cells ability to present antigen only to T cell
hybridomas that were restricted to responding to antigens in the
context of Ia^d. In addition to a histocompatible accessory B cell
the activation of T hybridomas required specific antigen and antigen
alone in the absence of accessory B cells or accessory cells alone

Table 2. Antigen presentation by B cell tumor lines

		T cell hybridoma IL-2 Production (units/ml)		
Tumor cell line	H-2 type	AO-40.10	AODK 1.16	AODH 7.1
A20-1.11	d	0	640	> 640
L10A 6.2	d	0	80	> 640
K46R	d	0	40	> 640
BAL 17.7.2	d	0	269	320
Antigen/MHC Restriction		OVA/H-2^k	KLH/H-2^d	HGG/H-2^d

did not lead to T hybridoma IL-2 production. These data formally demonstrate the capacity of B cells to present antigen, and extended our previous work with normal B cells[11].

There was, however, one striking discrepancy between the data obtained with the B cell tumor lines and the T cell hybridomas and the previous data we obtained with RAMIG pulsed normal B cells and NRGG primed T cells. In the latter situation conventional antigens such as normal rabbit gamma globulin and ovalbumin which did not get taken up by the Ig receptors on B cells as did RAMIG, were not presented in an effective manner by B cells to the primed T cells. In the tumor situation, the B cell lines could present a number of antigens to T cell hybridomas. This presentation was undoubtedly not via the Ig receptors on these B cell lines since the different antigens used are known not to share any antigenic determinants with one another and there is no evidence that the membrance Ig of the B cell lines studied was specific for any of the antigens used. Since the two systems studied were quite different (that is, one used T cell hybridomas and IL-2 formation, and the other used normal T cells and a proliferation response), we decided to determine whether this inconsistency in results persisted in a more controlled experiment. To that end, normal T cells primed to KLH were tested for their capacity to produce IL-2 when stimulated by antigen presented by normal B cells, B lymphomas or spleen adherent cells. Table 3 illustrates the results we have obtained. Whereas both the B cell lymphoma and the spleen adherent cells presented antigen very efficiently to the KLH primed T cells, normal B cells lacked this capacity.

Table 3. Comparison of Macrophages, B cells and B Lymphomas in Stimulation of KLH Primed T Cells[a]

Accessory Cells	Antigen (1 mg/ml)	IL-2 (Units/ml)
Spleen Adherent Cells	−	0
	+	80
B	−	0
	+	0
A20-1.11	−	0
	+	320
BAL 17.7.2	−	0
	+	160

[a]T cells were obtained from the lymph nodes of KLH primed BALB/c mice and depleted of macrophages[11].

There are at least two possible explanations for why B cells did not work in this system whereas they did work in the RAMIG system. It is possible that cofactors elaborated by macrophages and B cell lymphomas necessary for induction of IL-2 are not made by resting B cells but are made by RAMIG stimulated B cells. This possibility is currently being tested in our laboratory. Alternatively, the capacity of resting B cells to take up conventional antigen via mechanisms other than their antigen specific Ig receptors may be deficient relative to that of B lymphomas and macrophages. We have tested this latter possibility by studying the capacity of these different populations of accessory cells to take up antigen. There are two possible mechanisms by which antigen can be taken up by these cells: fluid phase pinocytosis, and physical adsorption to the cell surface. Fluid phase pinocytosis was evaluated by measuring the uptake of horseradish peroxidase (HRP) a study made possible because an insignificant amount of this protein is bound to the surface of these cells[15]. The physical adsorption of antigen to the surface of cells was assayed by measuring the uptake of radiolabelled antigen after a 1 hr incubation in the cold. The data shown in Table 4 indicates that significant differences are observed in both the adsorptive uptake as well as the fluid phase pinocytic rate of the three cell types studied. Macrophages were the most active cells in both fluid phase pinocytosis as well as binding of iodinated KLH to its surface. The B cell lymphoma was approximately 10-50 fold lower in both of these assays and normal B cells were still another order

Table 4.   Adsorptive Binding of Antigen and Fluid Phase Pinocytic
           Activity of B Lymphomas, Normal B Cells and Macrophages

	I*KLH uptake (ng/10^7 cells)	I*RAMIG uptake (ng/10^7 cells)	HRP uptake ($1H_2O/10^7$ cells/hr)
Normal B[a]	0.26	148	0.0014
BAL lymphoma	4.49	117	0.012
Macrophages[b]	57.72	216	0.58

[a]Normal B cells were isolated from the spleens of BALB/c mice and
depleted of macrophages by passage over G-10 columns and depleted
of T cells by anti-Thy 1 and C' treatment.

[b]Macrophages in the form of peritoneal exudate cells were obtained
by peritoneal lavage of BDF_1 mice 3 days following injection with
protease peptone.

of magnitude lower than the B lymphoma. In striking contrast to
these results were the results obtained when iodinated RAMIG was
studied. All three preparations bound this reagent very efficiently,
the two B cell populations being comparable to the macrophages by
virtue of the specific binding of the RAMIG preparation to their
membrane immunoglobulin. These data are therefore consistent with
the hypothesis that the inability of normal B cells to present
conventional antigen to sensitized T cells is due, at least in part
to their relative inefficiency in taking up antigen by adsorption
or by fluid phase pinocytosis.

In conclusion we have presented data with both normal T and B
cells as well as cloned tumor cell lines of B and T cell lineage
that indicate that under certain circumstances B cells can present
antigen to T cells in a highly efficient manner. As is the case
with the more classic accessory cell, the macrophage, this presentation is both antigen specific and MHC restricted. Differences between the capacity of spleen adherent cells, B cell lymphomas and
normal B cells to present antigen have also been documented. It is
suggested that unlike B lymphomas and macrophages, normal B cells
may be restricted in their capacity to present antigen by their
relative inefficiency in binding antigen via nonspecific means so
that under physiologic circumstances, sufficient antigen uptake by
B cells occurs only when relatively large amounts of antigen can be
taken up by these cells via their antigen specific Ig receptors.
As illustrated in Figure 1 this situation provides a possible
mechanism by which antigen specific T-B interaction might occur.
Antigen is first bound by the Ig receptors on antigen specific B

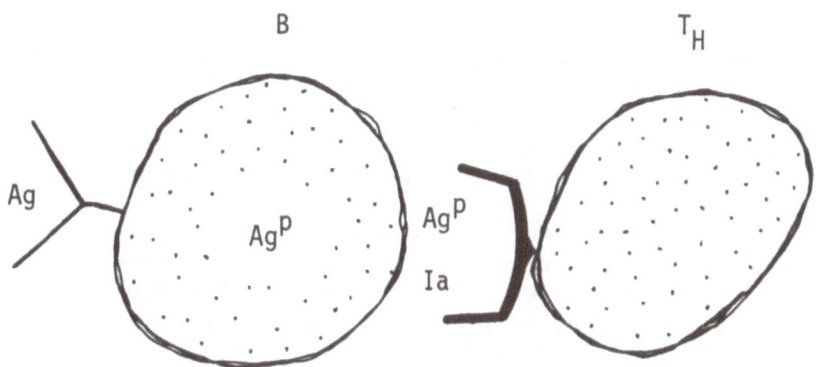

Fig. 1. Model of T-B interaction based on the capacity of B cells
to process and present antigen to T cells.

cells. This is followed by endocytosis of the immune complexes and
"processing" of the antigen (Ag^P). Following the processing event
the antigen is reexpressed on the B cell plasma membrane in association with the B cell Ia antigens. This processed form of the antigen
plus Ia can then be recognized by antigen specific helper T cells
and provides a mechanism for the binding of antigen specific T and
B cells to one another to allow for a variety of signals to be passed
between the two cells.

REFERENCES

1. R. W. Chesnut, R. O. Endres, and H. M. Grey. Antigen recognition
   by T cells and B cells: recognition of cross-reactivity between native and denatured forms of globular antigens. Clin.
   Immunol. Immunopathol. 15:397. (1980).
2. V. Schirrmacher and H. Wigzell. Immune responses against native
   and chemically modified albumins in mice. I. Analysis of
   non-thymus-processing (B) and thyumus-processed (T) cell
   responses against methylated bovine serum albumin. J. Exp.
   Med. 136:1616. (1972)
3. K. Ishizaka, H. Okudaira, and T. P. King. Immunogenic properties
   of modified antigen E. II. Ability of unrea-denatured antigen
   and polypeptide chain to prime T cells specific for antigen
   E. J. Immunol. 114:110. (1975).
4. H. Wigzell, K. G. Sundquist and T. O. Yoshida. Separation of
   cells according to surface antigens by the use of antibody-
   coated columns. Fractionation of cells carrying immunoglobulins and blood group antigen. Scand.J. Immunol. 1:75. (1972).

5. A. Basten, J. F. A. P. Miller and R. Abraham. Relationship between Fc receptors, antigen-binding sites on T and B cells, and H-2 complex-associated determinants. J. Exp.Med. 141:547. (1975).

6. J. P. Lamelin, B. Lisowska-Bernstein, A. Matter, J. E. Ryser and P. Vassalli. Mouse thymus-independent and thymus-derived lymphoid cells. J. Exp. Med. 136:984. (1972).

7. P. E. Lipsky and A. S. Rosenthal. Macrophage-lymphocyte interaction. II. Antigen-mediated physical interactions between immune guinea pig lymph node lymphocytes and syngeneic macrophages. J. Exp. Med. 141:138. (1975).

8. J. E. Swierkosz, K. Rock, P. Marrack and J. W. Kappler. The role of H-2-linked genes in helper T-cell function. II. Isolation on antigen-pulsed macrophages of two separate populations of F_1 helper T cells each specific for antigen and one set of parental H-2 products. J. Exp. Med. 147:554 (1978).

9. N. A. Mitchison. The carrier effect in the secondary response to hapten protein conjugates. II. Cell cooperation. Eur. J. Immunol. 1:18. (1971).

10. B. Benacerraf. A hypothesis to relate the specificity of T lymphocytes and the activity of I region-specific Ir genes in macrophages and B lymphocytes. J. Immunol. 120:1809 (1978).

11. R. Chesnut and H. Grey. Studies on the capacity of B cells to serve as antigen presenting cells. J. Immunol. 126:1075. (1981).

12. E. Walker, N. Warner, R. Chesnut, J. Kappler and P. Marrack. Antigen specific, I-region restricted interactions in vitro between tumor cell lines and T cell hybridomas. J. Exp. Med. Submitted. (1981).

13. J. Kappler, B. Skidmore, J. White and P. Marrack. Antigen-inducible, H-2 restricted, interleukin-2-producing T cell hybridomas. Lack of independent antigen and H-2 recognition. J. Exp. Med. 153:1198. (1981).

14. J. Watson. Continuous proliferation of murine antigen specific helper T lymphocytes in culture. J. Exp. Med. 150:1510. (1979).

15. R. Steinman and Z. Cohn. The interaction of soluble horseradish perozidase with mouse peritoneal macrophages in vitro. J. Cell Biol. 55:186. (1971).

A VIEW FROM THE BRIDGE: ANTIGENIC DETERMINANTS IN IMMUNOREGULATION

Douglas R. Green[1] and Richard K. Gershon[2]

[1]Departments of Pathology and Biology, Yale University
School of Medicine, 310 Cedar Street, New Haven, CT 06510
[2]Laboratory of Cellular Immunology, Howard Hughes Medical
Institute at Yale University School of Medicine
310 Cedar Street, New Haven, CT 06510

Although Louis Pasteur was not the first person to apply the principles of immunity, he was the first immunologist since he emphasized the role of antigen to immune responses (though he misinterpreted this role, it was nevertheless a landmark discovery). Since that time, immunologists have worked to dissect the immune response from both ends, that is, to ascertain the various components of the interacting forces of the immune system and to discover the roles that each portion of the antigen have in eliciting these effects.

Landsteiner[1] proposed specific regions of an antigen, defining hapten moieties as those entities against which a response is directed, but which would be incapable of inducing a response on their own. The carrier, then, was the immunogenic moiety which allows the response against the hapten to occur.

After Ovary and Benacerraf[2] demonstrated the carrier effect, the idea developed that cells directed against different portions of an antigen cooperate in producing a maximal immune response[3-7]. Thus, T helper cells preferentially recognize carrier determinants while B cells direct their attention toward the hapten. The rules governing these interactions changed, however, when it was found that haptens could act as carriers and carriers as haptens[8,9] suggesting that it was the antigen <u>bridge</u> between the hapten and the carrier that served to bring together the interacting elements of T cell help and B cell activation to produce an immune response.

Since that time, the antigen bridge has become an extremely useful working hypothesis for a mechanism of immunoregulatory cell recognition. By such a mechanism not only may helper T cells recognize B cells, but, in turn, suppressor T cells may recognize helper T cells (sometimes by taking the "B cell determinant", to bridge to the helper cell).

Thus, the idea that different subsets of lymphocytes may react with different epitopes on a given molecule has been promulgated. We might ask then whether the concept of identifying different epitopes by employing distinct subsets can be inverted, that is "can lymphocyte subsets be distinguished on the basis of the specific epitope they recognize and by their requirement for bridging of this epitope with additional determinants in order to achieve their regulatory effect?"

Recently, we described a novel regulatory T cell activity which functions to block suppression (contrasuppression)[11]. Helper T cells can serve as a target of both suppression[12] and contrasuppression, the latter acting to render their targets relatively resistant to subsequent suppression. While both helper T cells and contrasuppressor cells are Ly-1 cells, they can be readily differentiated by virtue of their regulatory activity and additional cell surface markers. Unlike helper cells, contrasuppressor effector cells are I-J[+] and adhere to the V. villosa lectin[13].

As we will discuss, these cells can also be distinguished by virtue of their preference for recognizing different antigenic determinants. Suppressor and contrasuppressor cells recognize an epitope distinct from that of helper cells, thus allowing the possibility for interaction via an antigen bridge.

Evidence to support these ideas came from studies in two antigen systems, sheep red blood cells (SRBC) and hen egg lysozyme (HEL). The latter studies were conducted in collaboration with Eli Sercarz, Alex Miller, and their colleagues at UCLA.

The logic of compartmentalizing antigenic fragments into sets of determinants which are recognized preferentially by lymphocyte subsets has been applied to the sheep red blood cell (SRBC antigen system). Evidence has accumulated which suggests that a major suppressor cell determinant exists on the glycophorin molecule. This evidence comes from several sources:

1.   An SRBC specific suppressor cell clone produces a factor which is capable of inactivating helper T cells for the SRBC response. This factor binds specifically to sheep glycophorin[14].

2.   During the course of hyperimmunization, significant amounts of T suppressor factor appears in the serum. The T derivation of the molecule can be verified by an anti-T- "isotype" serum (it bears no immunoglobulin markers). This molecule binds glycophorin and bears

a dominant idiotype which has subsequently been shown to be present on the above clonal product[15].

3.   Immunization _in vitro_ with nanogram amounts of purified glycophorin leads to the production of active suppressor cells that can suppress the entire _in vitro_ antibody response to SRBC[16].

The reason why suppressor T cells should express this limited repertoire is unlikely to be due to a paucity of available receptors, since a very large number of antigens are capable of activating suppression. The _potential_ repertoire is likely to be there, but determinant selection is limiting.

It is therefore very unlikely to be an accident that the molecule bearing the predominant suppressor determinant is also capable of inducing contrasuppressor cells. Education of T cells _in vitro_ with amounts of glycophorin which are slightly supraoptimal for generation of suppressor cells yields contrasuppressor cells capable of interfering with the activity of the suppressor cells generated with smaller doses[16].

On the other hand, the glycophorin molecule is incapable of inducing helper T cells for the SRBC response, under the conditions so far studied.

Thus, both suppressor and contrasuppressor cells seem to be able to recognize the same molecule on the surface of the sheep red blood cell (glycophorin), whereas the helper T cell sees some unrelated "carrier" molecule (the "helper" determinant).

A similar situation has been observed in the HEL system. In non-responder animals, suppressor T cells preferentially recognize a determinant (N-C) which is not recognized by helper T cells[17]. Destruction of this suppressor determinant by removal of the amino terminal residues produces a molecule (AP-HEL) which primes for helper cell activity in non-responders[18].

Experimental evidence suggests that hyperimmunization produces potent contrasuppressor cells which are capable of blocking suppression of helper T cells[19]. Using such a protocol, responder mice were repeatedly immunized with the whole molecule (HEL), the suppressor determinant (N-C), or the molecule lacking the suppressor determinant (AP-HEL). Helper T cells were produced by a single injection of HEL in complete Freund's adjuvant, then purified by collection of the non-adherent spleen cell fraction from goat anti-mouse immunoglobulin plates. Non-specific suppressor T cells were produced by _in vitro_ culture of spleen cells without antigen, as described elsewhere[20]. Using a modification of the intermediate culture system[12], the helper cells were placed in the lower chamber of a periscopic double Marbrook chamber (Bellco). Suppressor cells

plus or minus Ig⁻ cells from the hyperimmunized animals were placed
in the upper chamber, separated from the helpers by a membrane (0.1
micron pore). All cultures received HEL coupled sheep red blood
cells. After 48 hours all helper cells were recovered and tested for
activity by addition to B cells (anti-Thy-1 plus complement treated
spleen) plus HEL-SRBC. After five days, cultures were assayed for
anti-HEL plaque forming cells.

The suppressor cells were capable of inactivating the helper
cells for the response. T cells from animals hyperimmunized with the
whole molecule (HEL) or the suppressor fragment (N-C) were capable of
blocking this suppression. Animals similarly immunized to the fragment lacking the suppressor determinant (AP-HEL) completely lacked
this contrasuppressive activity.

Helper cells are generated in such animals in response to HEL
or AP-HEL. Contrasuppressor cells seem to be generated in response
to HEL or N-C. On the basis of preferred antigenic determinant,
then, these functional T cell populations can be distinguished. More
importantly, contrasuppressor cells and suppressor cells seem to show
preference for the same region of the antigen.

Thus, in two different antigen systems, the same phenomenon has
appeared, suggesting that suppressor and contrasuppressor determinants tend to be more closely related to one another than they are
to helper cell determinants.

On the basis of these findings alone, we might invoke antigen
bridging and predict that both the suppressor cell and the contrasuppressor cell act upon the helper T cell as their mutual target. That
this has independently been shown to be the case[13] greatly strengthens this contention.

A few predictions arise from these considerations. Many Ir gene
effects are attributed to the activation of suppressor T cells in the
non-responder animals[17,21]. Responder animals apparently fail to
produce such suppressor cells[22]. We might predict, instead, that
this "failure" of some suppressor determinants to induce suppressor
cells in responder haplotype strains (such as the case of the N-C
fragment of HEL) is due to concurrent activation of the contrasuppressor circuit. Removal of these contrasuppressors (or disruption
of the contrasuppressor circuit[23]) would then reveal this latent
suppressor activity.

Thus, unresponsiveness may sometimes be due to a failure to
dominantly activate contrasuppressor cells in non-responder animals.
The observation that contrasuppressor cells can overcome an Ir gene
defect in A strain animals (Durum, Madri et al., to be published)
supports this notion.

Further support comes from the finding that antigen presenting cell subsets have a role in determining the dominance of suppression versus contrasuppression[24]. We might expect that Ir genes operating at the level of the antigen presenting cell could produce such complex regulating effects leading to responsiveness or non-responsiveness.

There may be other factors which contribute to the state of balance of suppression and contrasuppression. While both may recognize closely related antigenic determinants, it is possible that the antigen receptors differ qualitatively. Like that of the helper cell, the receptor of the contrasuppressor effector cell is difficult to examine, but studies have been conducted on a factor which induces contrasuppression[23]. This factor was found to be significantly more cross-reactive in binding to antigen than a concurrently activated suppressor factor. This agrees very well with the observation that cross-reactive antigens are effective in breaking immune tolerance[25], [26]. Thus, slight modification in the structure of a suppressor determinant may preferentially activate contrasuppression.

Another contributing factor which is not clearly understood is antibody. Evidence has accumulated to suggest that B cells and antibody may be extremely important in the induction of contrasuppression: B cell deficient mice do not generate contrasuppressor cells under conditions that are optimal for their generation in normal mice[27]; T cells induced by anti-H-2 antibodies block the appearance of GVH induced suppression[28]; neonatal injection of anti-allotype antibodies may, in addition to generating allotype suppressor cells, generate T cells which are capable of interfering with these allotype suppressor cells[29]. Thus, production of antibody and the formation of antigen-antibody complexes may have enhancing effects on immune responsiveness[30] attributable, perhaps, to preferential activation of contrasuppression over suppression. This dichotomy would fit nicely with the observation that IgM antibodies tend to be enhancing antibody production (ie. are contrasuppressive) while IgG antibodies, particularly those of high affinity tend to inhibit antibody production (ie. are suppressive).

We are left with the question of why specific epitopes might be preferentially recognized by some subpopulations and not others. Perhaps some determinants have a predilection for activating suppressor T cells by a mechanism that has co-evolved with suppressor cell recognition mechanisms and have been selected for by virtue of their ability to prevent autoimmunity. Contrasuppressor cells must then function to allow responses against antigens which sufficiently resemble self, to activate the suppressor mechanism. The fact that suppressor cells arise before contrasuppressor cells during ontogeny[13] gives suppression the jump in maintaining unresponsiveness to self tissues.

We would suggest that although suppressor T cells might have the potential to react with many determinants on the sheep red blood cell, there is something about the mechanism of presentation of the antigen which results in selection of those suppressor clones that react with glycophorin. We would thus predict that a mutation that prevents the expression of glycophorin on the surface of red blood cells would produce a much higher risk for development of autoimmune hemolytic anemia.

The concept of the antigen bridge remains a useful construct in the description of immune processes. Several mechanisms have been proposed to explain how cells which directly bind their antigenic determinants (B cells, suppressor T cells) can interact with cells which associate with antigen less directly (helper cells, contrasuppressor cells)[31]. These mechanisms will not be further discussed here.

The antigen bridge provides a mechanism by which both suppressor cells and contrasuppressor cells can interact with the helper T cell via preferred determinants allowing regulation of this target. We have proposed, in addition, that these preferred determinants and the mechanism by which they are selected by regulatory subpopulations is indicative of potentially extremely important processes of self-nonself discrimination. The view from the bridge, then, is of an immunoregulatory panorama which features antigen subsets as one of the key points of interest.

REFERENCES

1. K. Landsteiner, "The Specificity of Serological Reactions," Springfield, Ill., Thomas, (1936).
2. Z. Ovary and B. Benacerraf, Immunological specificity of the secondary response with dinitrophenylated proteins. Proc. Soc.Exp.Biol., N.Y. 114:72 (1963).
3. H. N. Claman, E. A. Chaperon and R. F. Triplett, Thymus-marrow cell combination: Synergism in antibody production. Proc. Soc.Exp.Biol., N.Y. 122:1167 (1966).
4. A. J. S. Davies, E. Leuchars, V. Wallis, R. Marchant and E. V. Elliot, The failure of thymus derived cells to produce antibody. Transplantation 5:222 (1967).
5. N. A. Mitchison, Antigen recognition responsible for the induction in vitro of the secondary response. Cold Sring Harbor Symp. Quant. Biol., 32:431 (1967).
6. R. Rajewsky, Tolerance specificity and the immune response to lactic dehydrogenase isoenzymes, Cold Spring Harbor Symp. Quant.Biol. 32:547 (1967).
7. G. F. Mitchell and J. F. A. P. Miller, Cell to cell interaction in the immune response. II. The source of hemolysin-forming

cells in irradiated mice given bone marrow and thymus or thoracic duct lymphocytes. J.Exp.Med. 128:821 (1968).

8. N. A. Mitchison, Mechanism of action of antilymphocyte serum. Fed.Proc. 29:222 (1970)

9. G. M. Iverson, The ability of CBA mice to produce anti-idiotype to 5563 myeloma protein. Nature 227:273 (1970).

10. G. Moller, (ed) Regulatory role of antigenic determinants. Immunol.Rev. Vol. 39 (1978).

11. R. K. Gershon, D. D. Eardley, S. Durum, D. R. Green, F. W. Shen, K. Yamauchi, H. Cantor and D. B. Myrphy, Contrasuppression: A novel immunoregulatory activity. J.Exp.Med. 153:1533 (1981)

12. D. R. Green, R. K. Gershon D. D. Eardley, Functional deletion of different Ly-1 T cell inducer subset activities by Ly-2 suppressor T lymphocytes. Proc.Natl.Acad.Sci.USA 78:3819 (1981)

13. D. R. Green, D. D. Eardley, A. Kimura, D. B. Murphy, K. Yamauchi and R. K. Gershon, Immunoregulatory circuits which modulate responsiveness to suppressor cell signals: Characterization of an effector cell in the contrasuppressor circuit. Eur.J. Immunol. (in press).

14. M. Fresno, G. Nabel, L. McVay-Boudreau, H. Furthmayr and H. Cantor, Antigen specific T lymphocyte clones. I. Characterization of a T lymphocyte clone expressing antigen-specific suppressive activity. J.Exp.Med. 153:1246 (1981).

15. G. M. Iverson, D. D. Eardley, C. A. Janeway and R. K. Gershon, Isolation of circulating IgT with the use of anti-idiotype immunoabsorbants. Proc.Natl.Acad.Sci.USA in press (1981).

16. C. Schiff, Regulation of the murine immune response to sheep red blood cells: The effects of educating lymphocytes in vitro with glycoproteins isolated from the sheep erythrocyte membrane, M.D. Thesis, Yale University School of Medicine, (1980).

17. L. Adorini, M. A. Harvey, A. Miller and E. E. Sercarz, The fine specificity of regulatory T cells. II. Suppressor and helper T cells are induced by different regions of hen egg white lysozyme (HEL) in a genetically non-responder mouse strain. J.Exp.Med. 150:293 (1979).

18. L. S. Wicker, C. D. Benjamin, M. E. Katz, E. E. Sercarz and A. Miller, The similar and highly restrictive epitope specificity of Ts and primary B cells, (ABSTR) Fed.Proc. 41:302 (1980).

19. D. R. Green and R. K. Gershon, Regulatory T cells in hyperimmune mice, manuscript in preparation.

20. C. R. Parish, Appearance of non-specific suppressor T cells during in vitro culture, Immunology 33:597 (1977).

21. B. Benacerraf and R. N. Germain, The immune response genes of the major histocompatibility complex, Immunol.Rev. 38:70 (1978).

22. E. E. Sercarz, R. L. Yowell and L. Adorini, Immune response genes control the helper-suppressor balance, in "Immune System: Genetics and Regulation," (E.E. Sercarz, L.A.

Herzenberg and C.F. Fox, eds), Academic Press, New York (1977).

23. K. Yamauchi, D. R. Green, D. D. Eardley, D. B. Murphy and R. K. Gershon, Immunoregulatory circuits that modulate responsiveness to suppressor cell signals: The failure of B10 mice to respond to suppressor factors can be overcome by quenching the contrasuppressor circuit, J.Exp.Med. 153:1547.

24. J. S. Britz, P. W. Askenase, W. Ptak, R. M. Steinman and R. K. Gershon, Specialized antigen-presenting cells: Splenic dendritic cells, and peritoneal exudate cells induced by mycobacteria active effector T cells that are resistant to suppression. J.Exp.Med, 155:1344.

25. W. O. Weigle, Termination of acquired immunological tolerance to protein antigens following immunization with altered protein antigens, J.Exp.Med. 116:913 (1962).

26. R. K. Gershon and K. Kondo, Tolerance to sheep red cells: Breakage with thymocytes and horse red cells, Science 175:996 (1972)

27. C. A. Janeway, Jr., B. Broughton, E. Dzierzak, B. Jones, D.D. Eardley, S. Durum, K. Yamauchi, D. R. Green and R. K. Gershon, Studies of T lymphocyte function in B cell deprived mice, in "Immunoglobulin Idiotypes and their Expression. ICN-UCLA Symposia on Molecular and Cellular Biology," Vol. XX (C.A. Janeway, Jr., E.E. Sercarz, H. Wigzell and C.F. Fox, eds), Academic Press, NY. 661 (1981).

28. U. Hurtenbach, D. H. Sachs and G. M. Shearer, Protection against graft versus host associated immunosuppression in F_1 mice. I. Activation of F_1 regulatory cells by host specific anti-MHC antibodies, J.Exp.Med. 155:1344.

29. L. A. Herzenberg, Allotype suppression and epitope specific regulation, Immunol.Today, in press.

30. C. Henry and N. K. Jerne, Competition of 19S and 7S antigen receptors in the regulation of the primary immune response, J.Exp.Med. 128:133 (1968).

31. M. Feldmann, P. Beverley, P. Erb, S. Howie, S. Kontiainen, A. Maoz, M. Mathies, I. McKenzie and J. Woody, Current concepts of the antibody response - heterogeneity of lymphoid cells, interactions, and factors, Cold Spring Harbor Symp.Quant.Biol. 41:113 (1976).

ANTIGEN STRUCTURES USED BY REGULATORY T CELLS IN THE INTERACTION AMONG T SUPPRESSOR, T HELPER AND B CELLS

Urszula Krzych, Audree Fowler
and Eli E. Sercarz

Departments of Microbiology and Biological Chemistry
University of California, Los Angeles, CA 90024

INTRODUCTION

Subpopulations of T cells displaying helper and suppressor activities profoundly regulate antibody responses to protein antigens. In the mouse system, these two functional cell types can easily be manipulated with the use of cytotoxic antibodies directed against their phenotypic markers, i.e., Lyt-1 appearing on T helper cells and Lyt-23 expressed by the suppressor and cytotoxic T cell lineages[1]. In systems permitting the analysis of modes of cellular communication, these two cell types seem to interact in their regulation of antibody responses via idiotypic complementarity or antigen-bridging. It is also well documented that the induction of T cells, at least helper T cells, is controlled and restricted by the MHC. Evidence supporting MHC restriction in suppressor cell induction is also available[2,3]. Despite all this accumulated information, we still have a limited understanding about the exact manner in which Th and Ts encounter and react with the epitopes on a protein antigen or how these cell types interact with each other. Even at the level of T-B collaboration it is still not known if a precise spatial relationship must exist between the epitope seen by the Th and that seen by the B cell. Would a single Th cell suffice to present all epitopes on the molecule to ambient B cells or must there be an exact steric relationship between the two epitopes? The same question can be asked about the relationship between epitopes recognized by Ts and those recognized by Th cells. It is clear in 3 different systems - the Ir gene control of the anti-lysozyme response in H-2^b mice[4], the sheep glycophorin-specific suppression of SRBC responses[5] and the suppression of the entire anti-beta-galactosidase-fluorescein (GZ-FITC) response by certain GZ peptides[6] that Ts of a single specificity can obliterate the entire response to the antigen. This might

easily have been explained by postulating non-specific effects, pre-
sumable mediated by soluble factor(s), but in the GZ case it has
been demonstrated that even in the presence of GZ-specific Ts and
soluble GZ, the anti-KLH-FITC response proceeds normally. Thus at
least in some cases, a physical bridge is needed to connect the Ts
receptor and the Th receptor. Similar specificity controls have
been performed in the Th-B cell collaboration experiments with results indicating the necessity for antigen-bridging between the interacting cells[7].

     In most studies, functional T cells are induced by immunizing
with the whole antigen, and then depending on the dose, time and
mode of immunization, either T helper (Th), or T suppressor (Ts)
cell activity is subsequently measured. Few studies exist in which
the fine specificity of both Th and Ts to the same antigen have been
explored and even fewer where peptides bearing defined determinants
of a protein antigen have been used to induce a specific T cell
function. In the lysozyme system, the N-terminal-C-terminal peptide,
termed N-C, induces T suppressor cells in nonresponder B6 mice, and
another peptide, L2 (=aa 13-105), prepared by cyanogen-bromide cleavage, is capable of turning on T helper cells in the same strain[8].
Since the predominant antibody is of N-C specificity, it appears
that L2-specific Th can use an antigen bridge in communicating with
N-C specific B cells[9]. Glucagon also displays a T helper epitope
which is distinct from the B cell determinants on the same antigen.
However, no suppressor determinant has been defined[10]. Identification
of the epitope specificity of each cell type allows construction of
possible cellular networks with an explanation of how they become
engaged in the antibody response to a multideterminant protein antigen.

     In an attempt to further elucidate the issue of T helper-T
suppressor cell interaction, we have been investigating the functional
Th-Ts specificity repertoire directed against beta-galactosidase
(GZ) of E.coli in the CBA/J mouse. GZ is a tetrameric protein whose
monomers are also quite large (1021 a.a.=116,250 daltons)[11]. Our
hope was to generalize the information regarding limited functional
T cell recognition repertoires and the nature of cell cooperation,
which previously had been gathered from studies on a representative
of the small 100-150 amino acid-containing proteins (i.e., lysozyme).
Little is known about the physical state in which a large protein
and its epitopes exist when confronted by the variety of cells in
the immune system. A large protein might be transformed into a
smaller entity after in vivo macrophage treatment, or it might actually be utilized in situ in its native state. GZ appeared to be
suitable for exploring these topics because it is one of the few
large proteins whose amino acid sequence is known, and because of
the ready availability of its cyanogen bromide and trypsin-cleaved
peptides (70% of the whole GZ molecule was available in the form of
purified peptides from the Zabin-Fowler laboratory at UCLA).

Several questions regarding T cell repertoires to GZ first had to be clarified. Was the proliferative T cell repertoire similar to the helper T cell repertoire? How extensive were the Th and Ts repertoires? Did they overlap? Afterwards, questions could be asked about the interactions among cells of given specificities. Could each Th cell serve as a target for each specificity of Ts?

By assessing the ability of GZ peptides to induce Th or Ts for the in vitro GZ-FITC response, certain surprising conclusions were reached. Thus, despite the wide recognition of many potential determinants on this huge antigen by the proliferating T cells, the functional Th and Ts cells were restricted to very few regions on the GZ monomer.

A restricted Th repertoire despite extensive T cell recognition

The assessment of the proliferative T cell repertoire in response to GZ was achieved by using a panel of available cyanogen bromide (CB) peptides for T cell priming in vivo and subsequently by restimulating the peptide-primed T cells with either native GZ or RCM-GZ in vitro. Previous observations that there exists equality at the T cell level between GZ and RCM-GZ allowed for an interchangeable use of both antigens. The overall results indicated that the peptide recognition repertoire was rather extensive along the GZ molecule: of the CB peptides tested (see Figure 1 for a list of these) only 2 failed to prime for a T cell proliferative response to GZ. Particularly noticeable was the CB-20 (a.a.) peptide which regardless of its large size, repeatedly failed to imprint the immune machinery, using the criterion of recall with either GZ or RCM-GZ.

Parallel experiments were conducted to test if these proliferating T cells induced by the majority of the CB peptides, would collaborate with B cells in the presence of GZ-FITC to yield anti-FITC PFC. The results, shown in the upper panel of Figure 1 were unexpected, in that only 2 peptides, CB-2(a.a. 3-92) and CB-10 (a.a. 378-418) generated GZ-specific Th cells useful in promoting a specific anti FITC-PFC response. The other peptides showed no, or marginal helper activity in this context. The possibility exists that the inactive peptides may equally well stimulate induction of Th cells which, however, can not express their function in the FITC hapten system used in this study. Alternatively, other functional T cell subpopulations may show antigen-induced proliferation, but not be able to help B cells.

The T suppressor repertoire is as limited as the T-helper repertoire

In previous experiments assessing the T suppressor cell repertoire to GZ it was found[6] that CB-2 primed spleen cells could suppress

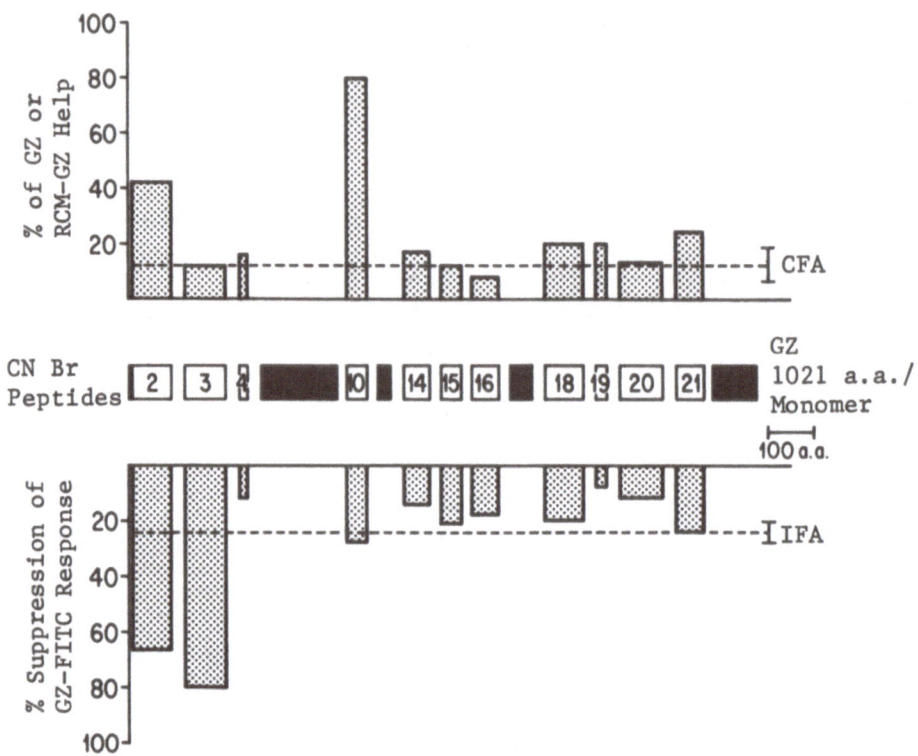

Fig. 1. Help or suppression is induced by CB fragments of GZ. For
        Th measurement, purified T cells from lymph nodes primed
        with either 50μg GZ or RCM-GZ, or equivalent molar doses
        of the indicated CB peptides in CFA, were titrated into
        FITC primed splenic B cells and cultured for 4 days in the
        presence of 10μg/ml GZ-FITC. When Ts were being evaluated,
        spleen cells were titrated into GZ primed lymph nodes (as
        a source of Th targets). The width of each bar in the
        figure reflects the number of a.a. residues in the CB pep-
        tide; filled rectangles indicate peptides not available
        for testing. Induction of helper activity by the native or
        reduced GZ or CB-derived peptides was quantitated as anti-
        FITC PFC on GRBC-FITC. The results represent composite data
        from several experiments. The responses from each experi-
        ment have been normalized to the percent of the response
        achieved after GZ or RCM-GZ priming. (These responses
        range from 150 to 300 PFC/culture). The maximal T cell
        activity after peptide priming is plotted; this usually
        occurred at the point when 10% of the input cells were pep-
        tide primed T cells.

the anti-FITC response induced by GZ-FITC. Our recent experiments, using the same panel of CNBr peptides as had been used for the Th induction, revealed an equally restricted number of determinants utilizable for triggering Ts. Results presented in the lower half of Figure 1 show that indeed CB-2 and only one other peptide, CB-3 (93-187), were capable of evoking GZ-specific Ts able to suppress the anti-FITC response to GZ-FITC. Other peptides had suppressive activity no greater than cells primed with an irrelevant protein antigen, KLH, or IFA alone.

Restricted and non-overlapping Th and Ts epitopes

Since CB-2 contained both Th and Ts inducing determinants it became important to investigate whether they overlapped or were distinct epitopes. The CB-2 peptide is quite large, approximately the size of lysozyme, and so it is reasonable to assume that functionally unique epitopes would be specified by distinct regions. More extensive submapping of the functional epitopes within the CB-2 peptide was accomplished by using trypsin cleaved peptides which ranged from a pentapeptide to a tridecapeptide. The nonapeptide T6 was the only peptide capable of inducing Th cells which would respond to GZ-FITC (Figure 2); on the other hand, T3 and T4-priming induced Ts cells which turned off the anti-FITC PFC response helped by GZ-primed Th cells. Thus the major conclusion reached from these experiments is that the Th and Ts epitopes on the GZ molecule are nonoverlapping. The large tryptic peptide, T8, extending from a.a. residue 60-140 and therefore including a CB-2 and CB-3 moiety, seemed to induce Ts only, using the same in vitro system as readout.

The rarity of Th and Ts determinants has both a recognitive aspect as well as an interactive one. In considering recognition, we would simply point out that there may be a limited recognition of amino-acid sequence by MHC structures which are involved in associating with protein antigens. The suggestion put forth by Benacerraf[12] seems quite reasonable: certain stretches of 3-4 amino acids have affinity for particular MHC molecules, whether the eventual association is covalent and enzymatically fostered, or whether there are strong ionic or hydrophobic attractive forces. We can predict that whatever the nature and specificity of the recognition between MHC molecules and antigen, it will be different for restricting elements in the Ia vs. K/D regions which are utilized for helper and cytotoxic T cell induction, respectively. Although the Ts restriction element is unknown, we feel that the data showing the non-overlapping between Ts and Th epitopes suppests that the "chemistry" of the MHC attachment in the presentation of antigen to Ts and Th will be disparate.

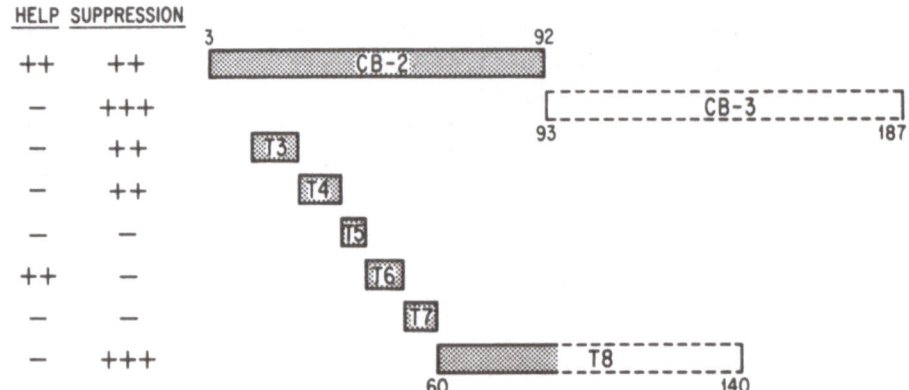

Fig. 2.   Functional epitopes within CB-2 and CB-3.   Tryptic peptides
          T3 through T8 were used for priming mice in either the Th
          (10-day peptide primed lymph nodes) or the Ts (28-day pep-
          tide primed spleen).   GZ primed Th populations were used as
          targets for the suppressor T cells.   GZ-FITC was the anti-
          gen in all cases and the anti-FITC response was assessed.

The targets of suppression

     With regard to the interactive aspect of Ts-Th-B relations, we
now turned our attention to the structure of the antigenic entities
involved.   It remained to clarify the actual cellular targets of
these Ts since the majority of the anti-GZ-FITC response was suppressible by CB-2, CB-3 or T8 generated Ts.   Initially, experiments
were performed to ascertain that the observed suppression involved
Th rather than B cell targets.   KLH generated Th activity was measured in the presence of KLH-FITC, with the result that these responses
remained unchanged regardless of the presence of GZ-peptide induced
Ts, even in the presence of GZ.

     In order to explore the specificity relationships of particular
Ts with Th, the 2 available peptide-primed Th types were each used
as targets of Ts.   Furthermore, it was hoped that we could indirectly
learn which were the true Th engaged after GZ or RCM-GZ priming.
Thus the experiments similar to those in Figures 1 and 2 were performed with each of the target Th mentioned above, and each of the
three distinct Ts populations.

     Contrary to our expectations, we observed that CB-10 induced
Th seemed to be totally unaffected by the suppression exerted by any
of the Ts cells (see Figure 3).   Equally striking was the observation
that CB-2 induced Th activity was resistant to Ts generated by CB-3,
although sensitive to T8-induced Ts.   Clearly, T8-specific Ts are
different from CB-3-specific Ts.   It should be stressed again that
both GZ and RCM-GZ induced Th can serve as targets for the Ts cells
from each of these 3 sources.

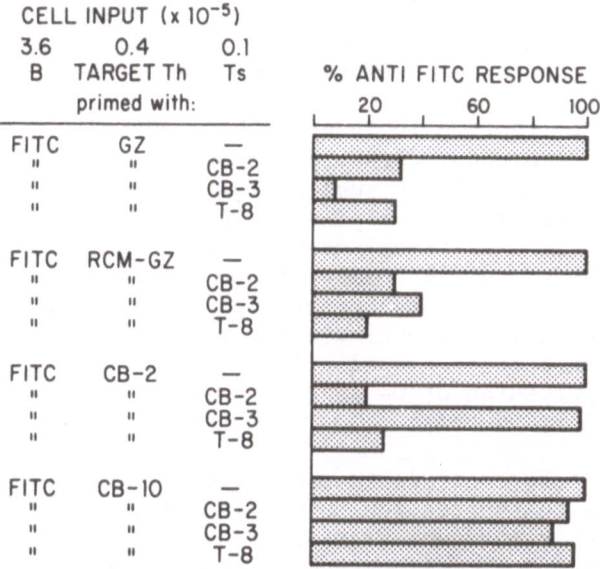

Fig. 3. Limited interaction between Th and Ts. The system is
         similar to that described in the legends to Figures 1 and
         2. Three different inducers of Ts were tested – CB-2,
         CB-3, and T8 while four targets Th were compared – GZ',
         RCM-GZ', CB-2', and CB-10' populations.

These data strongly suggest that in the context of a large
antigen such as GZ, and probably also for small protein antigens,
the interactions between Th and Ts are highly ordered, involving T
cells with carefully matched specificities. Although we are not
privy to the secret rules of matching, the fact remains that Ts and
Th epitopes are non-overlapping and distinct.

Which T cells are triggered by GZ?

In fact, the predominant epitope for Th in GZ may not even
exist on the cyanogen bromide peptides. The evidence presented above
in Figure 3 shows that CB-3 induced Ts cells effectively suppress
GZ and RCM-GZ specific Th responses, yet are inefficient in turning
off either CB-2 or CB-10 induced Th activities in the anti-FITC PFC
response. This strongly suggests the conclusion that a Th cell
whose specificity has not been previously detected is induced after
GZ priming and is suppressible by CB-3-specific Ts cells. This may

be the major and dominant Th following GZ stimulation! It is conceivable that the cleavage of the methionine separation CB-2 and
CB-3 by cyanogen bromide destroys this dominant Th-inducing epitope.

The argument that CB-10 induced Th cells may <u>not</u> be represented
in the GZ primed Th population could provide an explanation of the
insensitivity of CB-10 Th to suppression by any of the populations
of Ts. It is possible that a private Ts cell close to CB-10 is
raised after GZ immunization, and interferes with CB-10-specific
help. On the contrary, it could also be argued that the CB-10
specific Th cells represent a minor population among the GZ induced
Th cells and do in fact contribute minimally to the overall helper
response. Perhaps the residual helper activity seen after CB-2,
CB-3 and T8 suppression of GZ Th cells is that cohort furnished by
the CB-10 specific Th cells.

We are able to conclude that whatever Th are induced by GZ are
all susceptible to suppression induced by CB-2, CB-3 and T8, suggesting a highly restricted utilization of N-terminal regions of beta-
galactosidase by the T cell system. Conceivably, there is an N-
terminal to C-terminal polarity in the processing of a protein by
the immune machinery, giving little opportunity to internal regions
of a large protein to influence the specificity of response.

The collaboration between Th and B cells

In considering the nature of the interaction between Th and B,
our approach is to consider this as a distinct and late act in the
Drama of Antibody Formation quite apart from the initial activation
of the Th precursor. Act I, Scene I would portray the presentation
of an antigen fragment to precursor Th cells. Only much later,
somewhere in Act II or III would a scene occur in which the activated,
mature Th is ready to collaborate with an ambient B cell. At this
stage it has been postulated[13] that the 3-dimensional epitope contained within the B cell receptor is a different one from that reactive with the T cell.

With a protein the size of lysozyme, it has been shown that Th
directed against one epitope on the L2 fragment can collaborate with
B cells directed toward a distant epitope involving both the N and
C-termini[4]. In wondering about the interactive complex used in cooperation around GZ as antigen, the simplest prediction would be
that some version of the monomer or tetramer could be employed. In
Figure 4 two possibilities are suggested. In the first, I, it is
assumed that the only T cells which can help are those which can
recognize an epitope on the surface of the antigen which bears some
resemblance to the small peptide that originally activated the Th
precursor. An alternative, II, pictures a "fractured molecule" whose
structurally native epitope is protected within a B cell receptor

I. SURFACE DISPLAY

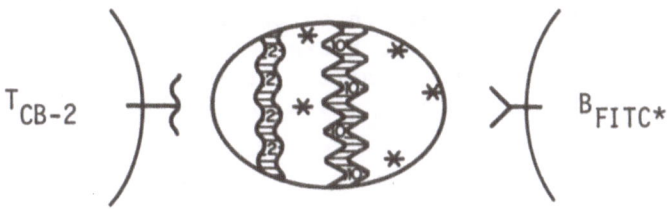

II. FRACTURED ANTIGEN

Fig. 4.   Easter Egg portrayal of T and B Cell Interaction. I. Restriction to CB-2 and CB-10 specificities can be attributed
to the failure of Th cell recognition of any other peptides
on the surface of the native molecule, which is supposedly
used in T-B interaction. II. A portion of the entity used
in T-B collaboration is assumed to be native (right-side)
and in place within a B-cell receptor while the antigen is
continually "eaten away" to a stable point, viz., at CB-10.

leaving exposed a peptide fragment, cleaved recently by proteolytic
enzymes, and mimicking the initial activating immunogen. In each
of these models, it is assumed that the direct attachment to amino
acid residues of MHC molecules is not a sine qua non for T cell reactivity at the late, interactive phase of collaboration.

However, our results in the suppression target experiments point
to the notion that a much smaller fragment is involved than the
fractured egg, seen in the "Easter Egg" models of Figure 4, which
represent about one-half of the GZ monomer. In fact, our current
view is that the fragments made from GZ which are used for T-T interactions may be as small as 40 a.a. residues and probably no bigger
than 100. However, the chemical nature of the entity involved in
B cell triggering may be very similar to II, where the 3-dimensional
moiety of the "fractured molecule" is protected within the B cell
receptor while the susceptible, fragmented end lies exposed beyond
the edges of the receptor.

We are aware of another model of T-B collaboration set forth by
Grey and Chesnut (this volume). According to this view, native antigen makes contact with the B cell receptor, the antigen is endocytosed,
processed, and re-expressed on the surface of the B cell in a complex
with Ia molecules. In this formulation, it is difficult to see how
a steric relationship among T cell and B cell epitopes would affect
and limit the response.

The nature of the antigenic fragment bridge used in Th-Ts cell interaction

If a cryptic immunodominant epitope (CIE) is truly the major Th
specificity activated by GZ immunization, then the conclusion we can
reach about the suppressibility of this CIE-specific Th is that the
antigen fragment used in its regulation must contain the CIE as well
as a closely-associated Ts-inducing epitope.

The lack of sensitivity of the CB-2 induced Th to CB-3-specific
Ts strongly implies a "regional" effect in the interaction of Th and
Ts. Thus, we can make the assumption that an optimal suppressor
determinant-helper determinant (SD-HD) distance exists on the antigen molecule. This distance is probably smaller than either
CB-2 or CB-3 and may be as small as 20 amino acid residues. It may
be instructive to compare this SD-HD distance to that on lysozyme,
where it may be able to be defined more exactly. The SD in H-2^b
mice is known to include phe-3 and the helper cell reactivity is
directed against an epitope within tryptic peptide 11 (a.a. 74-96).
Because of the nature of the lysozyme molecule in which the N and
C-termini are joined by disulfide bonds, the inter SD-HD distance
may not be much larger than 20-30 a.a. Linear distances surely are
not as relevant as the topological distance along the relevant fragment.

Actually, if we consider the overlapping fragments carved out
of GZ by the antigen-processing machinery, we probably must consider
several overlapping fragments rather than a single relevant entity.
As a minimum, the "initiating fragment" for Ts must have a suppressor
epitope and an anchorage site (=AS) for the MHC restriction element,
while the initiating fragment for Th must in parallel have a helper
epitope HE and a different AS for its MHC restriction element. We
can postulate that the activation of different T cells occurs simultaneously and independently as long as these minimal-size antigenic
fragments exist, although activation can no doubt take place within
the context of a much larger peptide. For effective Th-B collaboration, the decisive factor may very well be the determinant (BD) in
contact with the B cell receptor. For example, if FITC residues
are limiting, and one is in place at the N-terminal α-amino group,
no other FITC derivatizable lysine residues are available until

CB-5. Certain potential Th (e.g. within CB-3 and 4) may be excluded from using this N-terminal FITC for bridging because they are too distant. Apparently, the CIE and T6 are an acceptable distance from this BD and therefore are capable of inducing help for FITC.

In addition to such a fragment containing an HD and a BD, another chunk of antigen might coexist which connects HD with SD and conceivably, there will be even larger fragments containing all three: HD, SD and BD. Activated Ts and Th can then interact using one of these latter two fragments. As is evident in this presentation, the HD is the one element in common on the antigenic fragments used for communication between Ts and Th or Th and B. In Figure 5, "6" and "X": refer to two such HD; e.g., tryptic peptide 6 can be considered the common fulcrum in the 4-star-6-FITC SD-HD-BD system. The fragments used for T-T interaction are shown within a dark shadow while the minimal T-B interacting fragments are speckled. Presumably a structure containing all 3 determinants is unnecessary. It is evident from this work that the detection of a suppressor determinant depends directly on a closely associated HD and BD, and an appropriate molecular fragment including both determinants (which becomes available after GZ processing).

An interesting and possibly useful distinction between the terms "determinant" and "epitope" in T cell immunology arises in this context. Let us consider a helper "determinant". This can be considered an area on the antigen whose existence is created by the presence of several amino acids comprising an anchorage site for a particular Ia restriction element. From this single AS, a variety of "epitopes" can be presented to ambient T cells. Manca et al. in our laboratory[14] have shown that from a single AS, T cells recognizing different fine specificities can be triggered.

At the B cell level, the concept has been put forth[15] of "favored determinants" which may be special in that they may be particularly stable in their 3 dimensional conformation. A large variety of epitopes may be defined within this determinant region. When a non-haptenated protein is the antigen, intrinsically favored determinants may be just as rare as are the FITC haptens in the GZ system. In this case, the limitation in B cell determinants may reduce the number of potentially usable Th-inducing determinants. Alternatively, proteins may present a wealth of potential B cell determinants and epitopes and thereby not limit the revelation of the full complement of HD. A direct test of this question is now underway to see whether the spectrum of CB peptides which can serve as Th inducers changes when the experiments focus on other B cell determinants, e.g. those on native GZ, or on the azobenzene arsonate hapten in ABA-GZ, where tyrosine is preferentially derivatized rather than lysine.

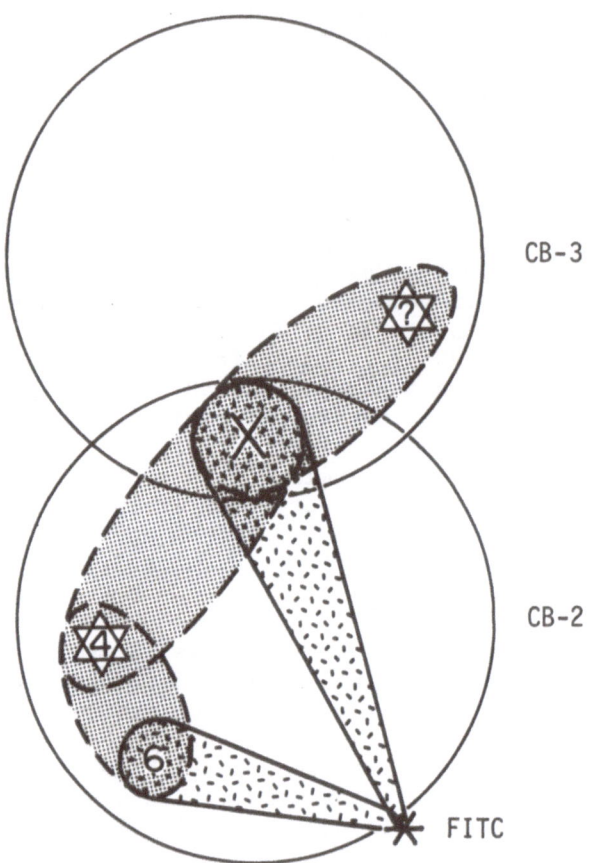

Fig. 5. Relationship of Ts, Th and B-cell determinants. CB-2 and
        CB-3 are shown as overlapping only insofar as the CIE, the
        cryptic immunodominant epitope, contains elements of each
        peptide. X indicates the CIE which connects directly to
        the FITC hapten. Suppressor determinants (Stars of David)
        on CB-3 or on tryptic peptide 4 are found on fragments in
        association with the X epitope, as indicated by the shadowed
        ovals.

Concluding summary statements

        Even on a large-protein antigen such as beta-galactosidase, Ts
and Th inducing determinants are not widespread.

        Suppressor determinants and helper determinants are non-over-
lapping, suggesting that the anchorage sites on the GZ molecule and
the associated MHC restriction elements are different for Ts and Th.

Ts and Th cell interaction seems to involve small fragments of antigen, produced during antigen processing. The stringency concerning the targets which can be used by particular Ts probably reflects the requirement for the suppressor determinant to adjoin the helper determinant on a relevant interaction fragment.

ACKNOWLEDGMENTS

This work has been supported by NIH Grants AI-11183 and AI 04181, and a Grant from the University of California Cancer Research Coordinating Committee. Urszula Krzych was partially supported by an NIH postdoctoral award. We thank Vicky Godoy for preparation of this manuscript.

REFERENCES

1. H. Cantor and E.A. Boyse, Functional subclasses of T lymphocytes bearing different Ly antigens. I. The generation of functionally distinct T-cell subclasses is a differentiative process independent of antigen, J.Exp.Med. 141:1376-1389 (1975).
2. B.A. Araneo and E.E. Sercarz, MHC restriction and positive selection of suppressor cells revealed by mixture with inducer T cells, in: "T Cell Recognition," B. Pernis and H. Vogel, eds., Academic Press, New York, pp.329-342 (1980).
3. B. Araneo, D.W. Metzger, R. Yowell, E.E. Sercarz, Positive selection of major histocompatibility complex-restricted suppressor T cells bearing the predominant idiotype in the immune response to lysozyme, PNAS. 78:499-503 (1981).
4. E.E. Sercarz, R.L. Yowell, D. Turkin, A. Miller, B. Araneo, and L. Adorini, Different functional specificity repertoires for suppressor and helper T cells, Immunol.Rev. 39:108-136 (1978).
5. D.R. Green and R.K. Gershon, A view from the bridge: antigenic determinants in immunoregulation. This volume.
6. D. Turkin and E.E. Sercarz, Key antigenic determinants in the regulation of the immune response, PNAS. 74:3894-3987 (1977).
7. J.W. Goodman, D.E. Nitecki, S. Fong, and Z. Kaymakcalan, Antigen bridging in T cell-B cell interaction: fact or fiction? This volume.
8. L. Adorini, M.A. Harvey, A. Miller and E.E. Sercarz, The fine specificity of regulatory T cells. II. Suppressor and helper T cells are induced by different regions of hen egg-white lysozyme (HEL) in a genetically nonresponder mouse strain, J.Exp. Med. 150:293-306 (1979).
9. M.A. Harvey, L. Adorini, A. Miller, and E.E. Sercarz, Lysozyme induced T suppressor cells and antibodies bear a predominant idiotype, Nature, 281:594-596 (1979).
10. G. Senyk, E.B. Williams, D.E. Nitecki, and J.W. Goodman, The functional dissection of an antigen molecule: specificity of

humoral and cellular immune response to glucagon, J.Exp.Med.
133:1294 (1971).

11. A.V. Fowler and I. Zabin, Amino acid sequence of β-galactosidase,
    XI. Peptide ordering procedures and the complete sequence,
    J.Biol.Chem. 253:5521-5525 (1978).

12. B. Benacerraf, A hypothesis to relate the specificity of T ly-
    mphocytes and the activity of I region-specific Ir genes in
    macrophages and B lymphocytes, J.Immunol. 120:1809-1812 (1978).

13. N.A. Mitchison, K. Rajewsky, and R.B. Taylor, Cooperation of
    antigenic determinants and of cells in the induction of anti-
    bodies, in: "Developmental Aspects of Antibody Formation and
    Structure," J. Sterzl and I. Riha, eds., Academia Publishing
    House of Czechoslovak Academy of Sciences, p.547-564 (1970).

14. F.M. Manca, J.A. Clarke, A. Miller, and E.E. Sercarz, Differing
    T cell specificities exist for a single determinant region on
    the lysozyme molecule, Submitted for publication.

15. M.Z. Atassi, Precise determination of protein antigenic sites
    has unravelled the molecular immune recognition of proteins
    and provided a prototype of synthetic mimicking of other pro-
    tein binding sites, Molec.Cell Biochem. 32:21 (1980).

ACTIVATION OF HUMAN B LYMPHOCYTES BY

ANTIGEN SPECIFIC T CELL LINES

Antonio Lanzavecchia

Cattedra di Immunologia
Università di Genova
Genova, Italia

INTRODUCTION

A major breakthrough in cellular immunology in the last years is the possibility of expanding in vitro lines or clones of normal T cells with antigenic specificity and certain functional properties[1,2]. This has been made possible by two important technological advances: i) the possibility of propagating and enriching the cultures of antigen-specific cells by repeated exposures to a given antigen in vitro[3,4] and ii) the discovery of the T cell growth factor (IL-2) [5]. We have used this approach to establish lines of human T cells with specificity for soluble antigens and with helper activity on B cell proliferation and differentiation[6]. We report here some of the properties of these long term cultured T blasts (Tb) and some aspects of their interaction with antigen presenting cells and B cells.

RESULTS AND DISCUSSION

Establishment of lines of antigen specific Tb

Peripheral blood mononuclear cells (PBM) from donors responsive to Tetanus Toxoid (TT) or Diphtheria Toxoid (DT) were stimulated with the appropriate antigen in vitro. After 7 days the blast cells

---

Abbreviations: IL-1=Interleukin-1; TT=Tetanus Toxoid; DT=Diphtheria Toxoid; PBM=Peripheral Blood Mononuclear Cells; PHA=Phytohaemagglutinin; APC=Antigen Presenting Cells; Tb=T cell blasts.

were isolated by centrifugation on a gradient of Percoll and cultured in media supplemented with a 30% supernatant of phytohaemagglutinin (PHA) activated tonsil cells as a source of IL-2. When cultured in IL-2, the T cells grew with a doubling period of 2-3 days. Every 3-5 weeks, Tb were restimulated with antigen in the presence of mitomycin-C treated autologous PBM for 7 days. By using alternate periods of growth in IL-2 and of restimulation with antigen, 8 lines were maintained in culture for several months. All of them consistently displayed the original antigenic specificity and helper function.

All the lines were OKT3+, OKT4+, OKT8-[7], formed rosettes with sheep erythrocytes, and lacked surface Ig or Fc receptors. Like all of the activated T cells, they were Ia+[8].

Antigen specific, Ia-restricted activation of Tb: requirement for an antigen presenting cell

The requirements for the activation of Tb by antigen were investigated using a short term proliferation assay as shown in a representative experiment reported in Table 1. Tb responded by proliferation solely to the antigen initially used to raise the line from PBM. This response was obtained only in the presence of autologous but not of allogeneic mitomycin-C treated PBM, which had the likely function of antigen presentation[9,10].

The role of Ia antigens in the process of antigen presentation to Tb is supported by the inhibition of the antigen-induced T cell proliferation by an anti-Ia heteroantiserum[11] used at dilutions that do not inhibit the PHA induced proliferation.

In analogy with the previous data on peripheral blood human T cells[12,13] these findings demonstrate that there is a requirement for an antigen-presenting cell (APC) in order to activate Tb with antigen and that Ia antigens on the APC may play a role as restriction structures for the activation.

Tb upon activation by antigen can induce polyclonal proliferation and differentiation of autologous B cells

Tb were added to T cell depleted PBM (containing B cells and macrophages) and stimulated with the specific antigen. From a representative experiment in Table 2 it is evident that high numbers of plasma cells were generated in a number which exceed the initial input of B cells 5 to 10 fold. This shows that both proliferation and differentiation of B cells to plasma cells were occurring. The B cell response was likely to be polyclonal as the Ig produced lacked specificity for the antigen used for stimulation (data not shown).

Table 1. Antigen specific activation of Tb: requirement for an autologous antigen presenting cell

Responding cells* (10⁴ Tb anti TT)	Antigen Presenting Cells (10⁵ mitocycin-C treated PBM)	Thymidine incorporation*** (CPM) in cultures stimulated by:			
		TT	DT	PHA	–
+	none	220	180	300	200
+	autologous	35000	250	41500	300
+	allogeneic	450	300	42200	250
+	autologous + anti Ia**	1600	280	40500	220

* Tb raised against TT and maintained in culture for more than 3 months were cocultured in flat bottom microplates with autologous or allogeneic mitomycin-C treated PBM (as a source of APC) and stimulated with TT,DT (both at 20μg/ml or PHA. PBM were responsive to both DT and TT.

** Anti Ia antiserum[11] was added at the beginning of cultures at a 1/500 final dilution.

*** Thymidine incorporation was measured after 48 hours.

Table 2.   Antigen activated Tb induce a polyclonal
           proliferation and differentiation of B cells

	No. of plasmacells/culture generated in the presence of:		
Cells in culture*	TT	DT	–
5x10⁴ non-T cells	<200	<200	<200
5x10⁴ non-T cells + 10⁴ Tb anti TT	150000	400	<200
5x10⁴ non-T cells + 10⁴ Tb anti DT	200	120000	<200

*5x10⁴ T cell-depleted PBM (non-T cells) containing B
cells and macrophages were cultured alone or together
with 10⁴ Tb specific for either TT or DT and stimulated
with TT or DT (20μg/ml). The number of plasma cells
generated after 7 days was determined by immunofluorescence.

Supernatants of antigen activated Tb fail to induce B cell activation

The polyclonal B cell activation observed above may be related
to the production of a soluble mediator with the property of a polyclonal B cell activator. This would not be dissimilar from that observed in some experimental systems with peripheral blood T cells[14].
However, we constantly failed to detect such an activity in the
supernatant of Tb stimulated with antigen in the presence of the
appropriate APC. Although incapable of stimulating peripheral
blood resting B cells, these supernatants could support the proliferation and differentiation of lectin-activated or even "background" activated tonsil B cells.

The polyclonal activation of B cells by antigen specific Tb requires high antigen doses

In some murine systems, high concentrations of antigen are required to elicit a proliferative response of T cells, while approximately 1000 times lower doses generally induce specific antibody
production in vitro. These different antigen requirements may be
interpreted in terms of cooperation of antigen specific T cells with
either the macrophage or the antigen specific B cell. In fact, while
the macrophage absorbs antigen non specifically and therefore is
likely to need high antigen concentrations, antigen specific B cells
may concentrate soluble antigens on their surface through their

specific membrane Ig receptors. It is possible that the polyclonal response observed using antigen specific Tb is due to the fact that at high antigen concentrations all the B cells bind to the antigen in a non-specific manner[15] (i.e., not through Ig receptors) and are thus activated by interaction with antigen specific T cells. In these conditions they would not differ from macrophages.

Consistent with this hypothesis, the proliferative response of Tb and the polyclonal B cell activation were found to have super-imposable dose-response curves (with a maximum at high doses of antigen (20-100μg/ml of TT) (data not shown).

Tb have to recognize antigen on B cells in order to induce them to polyclonal antibody synthesis

The problem of whether or not B cells have to interact directly with Tb in order to be triggered was investigated using antigen-pulsed PBM. As shown in Table 3, antigen pulsed PBM were capable of supporting Tb proliferation and of being in turn triggered to Ig production.

In another series of experiments Tb were mixed with an excess of antigen pulsed, mitomycin-C treated PBM. Small numbers of PBM (pulsed or not with antigen) were added to the above mixture as a source of responding B cells, in the presence or absence of soluble antigen. The results shown in Table 4 indicate that responding B cells when pulsed with antigen produced Ig molecules whereas the same cells that had not been pulsed failed to do so unless soluble antigen was added.

These data suggest that Tb activate B cells directly, possibly by recognizing antigen on their surface.

CONCLUDING REMARKS AND SPECULATIONS

The experiments reported show that lines of human T cells with specificity for soluble antigens and helper activity can be raised and maintained in vitro by repeated stimulations with antigen and addition of IL-2. These cells recognize specific antigens together with the Ia determinants of the APC. The latter are likely to be a subset of Ia$^+$ adherent macrophages endowed with a special stimulatory capacity for T cells[9,10]. The antigen induced interaction of specific T cells with APC leads to T cell proliferation and to the production of soluble mediators which promote growth and maturation of activated but not of resting B cells. B cell activation in turn may be brought about in antigen bearing B cells by direct interaction with Tb. The requirement for antigen to be present on the B cells in order to obtain activation suggests a classical T-B antigen bridge[16] as the

Table 3. Antigen pulsed PBM can support T cell activation and Ig production*

10^4Tb anti TT	10^5 PBM	TT in culture	Thymidine incorporation (CPM)	IgM produced** (ng/ml)
+	TT-pulsed***	-	45000	6900
+	TT-pulsed	+	46000	6600
+	not pulsed	-	150	<20
+	not pulsed	+	44000	4500

* Culture conditions as in Tables 1 and 2.
** Determined by an enzyme immunoassay in the 7 day culture supernatant.
***PBM were pulsed with 100μg/ml TT for 3 hours at 37°C and washed extensively.

Table 4. Tb have to recognize antigen on B cells in order to induce them to a polyclonal antibody synthesis*

10^4 Tb anti TT	Source of APC (10^5 TT-pulsed, mitomycin-C treated PBM)	Source of responding B cells	Free TT in culture	Thymidine incorporation (CPM)	IgM produced (ng/ml)
+	+	–	+	45000	<20
+	+	10^4 PBM	–	44000	250
+	+	10^4 PBM	+	40000	1950
+	+	10^4 TT-pulsed PBM	–	45500	2200
+	+	10^4 TT-pulsed PBM	+	41000	2100

*Culture conditions as in Figure 3.

prevailing mechanism under these experimental conditions. It is possible that the antigen concentrations necessary to pulse APC may be sufficient to provide enough antigen to pulse also B cells irrespective of their antigen specificity. This would cause their polyclonal B cell activation when the antigen absorbed on their surface is recognized by Tb, possibly in conjunction with the relevant Ia structures.

The specificity of the system may be revealed at low antigen concentrations. In such a case one might predict that only B cells with antigen specific Ig receptors will be able to concentrate sufficient antigen to be activated by antigen specific T cells.

Taken as a whole the results obtained so far provide a rationale for the polyclonal Ig production during the response to antigen and suggest a suitable model of T-B cell interaction in man.

This work was supported in part by grants No CT80.0044/04 and No CT80.00408.04 from Consiglio Nazionale delle Ricerche, Roma, Italy.

REFERENCES

1. G. Möller, (Ed.), T cell stimulating growth factors, Immunological Rev. Munksgaard, Copenhagen, 51: (1980).
2. G. Möller, (Ed.), T cell clones, Immunological Rev. Vol.54 (1981).
3. H.R. MacDonald, H.D. Engers, J.C. Cerottini, and K.T. Brunner, Generation of cytotoxic T lymphocytes in vitro II. Effect of repeated exposures to alloantigens on the cytotoxic activity of long-term mixed leukocyte cultures, J.Exp.Med. 140:718 (1974).
4. S.Z. Ben Sasson, W.E. Paul, E.M. Shevach, and I. Green, In vitro selection and extended culture of antigen specific T lymphocytes. II. Mechanisms of selection, J.Immunol. 115:1723 (1975).
5. F.W. Ruscetti, D.A. Morgan, and R.C. Gallo, Functional and morphological characterization of human T cells continuously grown in vitro, J.Immunol. 119:131 (1977).
6. A. Lanzavecchia, M. Ferrarini, and F. Celada, Human T cell lines with antigen specificity and helper activity. (Submitted).
7. E.L. Reinherz and S.F. Schlossman, Regulation of the immune response. Inducer and suppressor T lymphocyte subsets in human beings, New Engl.J.Med. 303:370 (1980).
8. S.M. Fu, N. Chiorazzi, C.Y. Wang, G. Montazeri, H.G. Kunkel, H.S. Ko, and A.B. Gottlieb, Ia bearing T lymphocytes in man: their identification and role in generation of allogeneic helper activity, J.Exp.Med. 148:1423 (1978).
9. A.S. Rosenthal, Determinant selection and macrophage function in genetic control of the immune response, Immunological Rev. 40:135 (1978).

10. R.H. Schwartz, A. Yano, and W.E. Paul, Interaction between antigen presenting cells and primed T lymphocytes, Immunological Rev. 40:153 (1978).

11. R.J. Winchester, C.Y. Wang, J. Halper, and T. Hoffman, Studies with B cell allo and heteroantisera: parallel reactivity and special properties, Scand.J.Immunol. 5:795 (1976).

12. B.O. Bergholtz and E. Thorsby, Macrophage-dependent response of immune T lymphocytes to PPD in vitro. Influence of HLA-D compatibility, Scand.J.Immunol. 6:679 (1977).

13. H. Hansen-Sønderstrup, B. Rubin, S.F. Sørensen, and A. Svejgaard, Importance of HLA-D antigens for cooperation between human monocytes and T lymphocytes, Eur.J.Immunol. 8:520 (1978).

14. R.S. Geha, Regulation of human B cell activation, Immunological Rev. 45:275 (1979).

15. J.L. Greenstein, J. Leary, P. Horan, J.W. Kappler, and P. Marrack, Flow sorting of antigen binding B cell subsets, J. Immunol. 124:1472 (1980).

16. N.A. Mitchison, The carrier effect in the secondary response to hapten-protein conjugates. II. Cellular cooperation, Eur.J. Immunol. 1:18 (1971).

T HELPER CELL RECOGNITION OF TWO TYPES

OF MOUSE LAMBDA LIGHT CHAINS

Trond Jørgensen, Bjarne Bogen
and Kristian Hannestad

Institute of Medical Biology
University of Tromsø School of Medicine
9001 Tromsø, Norway

SUMMARY

We have studied the immune responses of mouse memory T helper
cells (Th) against the λ_2-chain of IgA myeloma protein 315 and the
λ_1-chain of IgA myeloma protein J558. The Mitchison type adoptive
transfer system was used to assay Th responses. The immune response
to $V\lambda_2$-315 was influenced by H-2 linked genes as revealed by H-2
congenic mice. The recognition of $V\lambda_2$ was independent of protein
conformation since both the folded and unfolded domain primed Th that
responded to a boost with the native complete myeloma protein 315.
Despite the high degree of amino acid sequence homology (88%) between
$V\lambda_2$-315 and $V\lambda_1$-J558, only the free form of λ_1-chain was recognized
in this assay system. In contrast, $V\lambda_2$ was recognized in both the
assembled and free form. Possible explanations for this difference
were discussed. Th primed with λ_2 cross-reacted with the free form
of λ_1 but not vice versa, i.e. the cross-reaction was unidirectional.
This allowed certain conclusions to be drawn concerning the antigenic
complexity of the two types of λ-chains.

One of the postulates of the Jerne's network hypothesis is stated
as follows: "since V-domains present the only means for discrimination
between two lymphocytes that are otherwise identical (with respect
to heterogeneity I), we can state a priori that V-domains (i.e. paratopes and idiotopes) must be the primary targets of any specific
regulatory mechanism in the immune system"[1]. A second assumption is
that "-- every paratope in the immune system can recognize a number
of different idiotopes present in the same immune system, and vice
versa, i.e. that the immune system is a network of V-domains"[1]. To
understand the potential impact of T cells in the network it is

important to obtain information about how V-domains are recognized by T cells.

For several reasons we chose to study the T helper (Th) cell recognition of myeloma protein 315 (M315), a IgA(λ_2) myeloma protein which binds DNP- and TNP-lysine with moderately high affinity[2]. First, it is immunogenic in BALB/c mice which develop an antibody response specific for idiotopes that require assembled (H315 + L315) chains[3]. The idiotope recognized by the bulk of the BALB/c antibodies is closely associated with the antibody combining site since the hapten DNP-lysine blocked the binding of M315 to the anti-idiotypic antibodies[3]. Second, BALB/c mice producing antibodies to M315 were resistant to challenge with a dose of MOPC315 tumor cells that killed non-immune mice, demonstrating that variable regions (V) can function as tumor specific transplantation antigens[4]. Third, the Fv fragment is readily produced, from which the two V-domains can be obtained for studies of the immune response to the assembled and free domains[5].

The organization of the mouse λ light chain genes has recently been analyzed[6]. Four constant region (C) genes and only two V genes were found. The genes are arranged in two clusters, $V\lambda_1$, J_3, C_3, J_1, C_1 and $V\lambda_2$, J_2, C_2, J_4, C_4. The λ gene family is thus the simplest of the three Ig multigene families in the mouse. This fact together with the known amino acid sequences of λ_2-315 [7] and eighteen λ_1 chains[8] make the λ chain system uniquely suitable for studies of the immunogenicity and antigenic structure of V-domains.

Our studies to date have revealed that Fv-315, Fv-315 affinity labelled with bromoacetyl-DNP-L-lysine and $V\lambda_2$ elicit Th that respond to a boost with the complete M315[9],[10]. In contrast, V_H-315 appears to be non-immunogenic for BALB/c mice, at least at the level of detection of the Mitchison type adoptive transfer system[10]. These observations indicate that M315 possesses a major immunogenic site for Th that is spatially distinct from the site recognized by the bulk of anti-idiotypic antibodies; this site is expressed on the free V_L-domain as well as on the assembled V_L-domain of the complete M315 protein. We also found that when various strains of mice were immunized with either V_H or $V\lambda_2$-315, Th from mice of the H-2^d haplotype were high responders to $V\lambda_2$ and low responders to V_H, while mice of the H-2^k haplotype were high responders to V_H and low responders to $V\lambda_2$ [11]. Furthermore, the congenic strain BALB.K (H-2^k) responded like mice of the k haplotype, supporting the conclusion that immune response (IR) genes linked to the major histocompatibility complex (MHC) regulate these Th responses[11].

In this report we describe further studies on the IR-gene control of Th response to $V\lambda_2$-315, employing another congenic pair of mouse strains (C3H and C3H-H-2°). Second, we present evidence that Th primed with the unfolded $V\lambda_2$-315 recognize the native folded

$V\lambda_2$ domain of the complete M315. Third, we compare the recognition of the λ_1 chain of the IgA myeloma protein J558 with that of the λ_2-315.

MATERIALS AND METHODS

Antigens. The following procedures have been described[9,10]: purification of M315 and its V_L and V_H domains, conjugation of 4-hydroxy-3-iodo-5-nitrophenylacetyl (NIP)-azide to proteins, and iodination of NIP-caproic acid. Myeloma protein J558 (IgAλ_1) was purified from ascites by a modification of the method of Hiramoto et al[12]. Unfolded $V\lambda_2$-315 was produced by cleaving the intrachain desulfide bridge with 0.005M dithiothreitol in the presence of 5M guanidine HCl followed by alkylation with 0.01M iodoacetamide, pH 8.3.

Immunization. Hapten-priming of B cells with NIP$_7$-BSA or NIP$_6$ /10^5m.w. KLH and priming of T helper cells with carrier (λ_2-315, λ_1-J558, $V\lambda_2$-315, M315 or J558) was done with a single intraperitoneal injection of immunogen in Freund's complete adjuvant 6-8 weeks before cell transfer as described[9,10].

Adoptive cell transfer. Spleen cells ($\frac{1}{2}$ spleen equivalent) from hapten- and carrier-primed donors were transferred i.v. to 500 rad irradiated syngeneic recipients as described[9,10]. The recipients were boosted the day after transfer with 100-200 μg of NIP-substituted myeloma protein, Fab fragment or free L-chain. Anti-NIP antibody responses were measured 10-11 days after boost by a modified Farr assay[9].

RESULTS

Th response to λ_2-315 is influenced by H-2 linked genes. In these experiments Th of donor mice were primed with $V\lambda_2$-315 and the recipients were boosted with NIP$_3$-Fab-315. A strong anti-NIP response of the recipients indicated that the donors of Th were high responders; in contrast, those strains in which the adoptive anti-NIP response was very low we concluded were low responders to $V\lambda_2$-315.

Previous observations revealed that mouse strains of the H-2^d haplotype (BALB/c and DBA/2) were high responders to $V\lambda_2$, while mice of the k haplotype (CBA, C3H and AKR) were low responders[11]. The Th of (C3H x BALB/c)F1 mice were high responders, showing that the responder phenotype was inherited in a dominant fashion[11].

The importance of MHC genes in these responses was proven with H-2 congenic strains (Table 1). Thus, BALB.K (H-2^k) which only differs from BALB/c at the H-2 complex was a low responder to $V\lambda_2$,

Table 1. Th recognition of Vλ_2-315 is IR-gene regulated

Strain immunized	I-region haplotype	Anti-NIP response of recipients after boost with NIP$_3$-Fab-315
BALB/c	d	4+*
BALB.K	k	0
C3H	k	0
C3H.H-2^O	d	4+

Th priming: Vλ_2-315 in FCA.
*Farr assay. (See ref. 9 for details)

while mice of the C3H-H-2^O strain, which is H-2^d (except for the D-region which is of k haplotype origin) were high responders.

Th primed with unfolded Vλ_2-315 recognize native Vλ_2 in Fab-315. Previous studies have shown that Th of BALB/c mice primed with the folded (intrachain disulfide bridge intact) Vλ_2-315 responded when challenged with the complete M315 or its Fab fragment, indicating that the antigenic site recognized by Vλ_2-primed Th was expressed on both the free and assembled form of Vλ_2.

Th primed with the unfolded (intrachain disulfide bridge cleaved and carboxymethylated) Vλ_2-315 also responded when boosted with the complete M315 (Table 2). This observation, which has been reported[13], indicates that Th recognizes an antigenic site on Vλ_2 which is expressed on the unfolded domain and therefore is independent of conformation.

Comparison of the immunogenicity of λ_1 of J558 and λ_2 of M315. The V- and C-domains of these λ-chains differ by 14 amino acids[14] and 29 amino acids[7], respectively. Thus the V-domains of these chains exhibit a high degree (88%) of homology, which invited an exploration of the immunogenicity and cross-reactivity of these chains at the level of Th.

Table 2. Th primed with the unfolded Vλ_2-315 domain recognize the native myeloma protein 315

Priming of helper cell donors	Anti-NIP response of recipients after boost with NIP$_4$
MEM	0
Native Vλ_2-315	4+
Unfolded Vλ_2-315	4+

We found that in spite of their homology the antigenic properties of these chains displayed several striking differences (Table 3). λ_1 of J558 was immunogenic in its free, but not in its assembled form (group 10 vs. group 8 and 9); the antigenic site of λ_1 has not yet been localized with respect to the V- or C-domain. By contrast, λ_2 of M315 was immunogenic both in its free (group 3) and assembled forms (group 1 and 2) as shown before[9]. The helper effect of T cells primed with the complete M315 was not stronger than that of Th primed with λ_2-315 (group 1 and 2) suggesting that the bulk of the Th population recognized only λ_2 and that the H chain did not contribute significantly to the antigenic specificity of M315. The antigenic site of λ_2 could be assigned to its V-domain (group 6) as shown before[10]. However, this does not rule out that there also exists an antigenic determinant in the C-region; the lack of an isolated intact C-domain has precluded an examination of this possibility. The free form of λ_2-315 primed Th for cross-reactivity with free λ_1 (group 4), but not with assembled λ_1 (group 5). Surprisingly, $V\lambda_2$ did not cross-prime (group 7); the simplest explanation is that the cross-reactive site is localized in the C-domain, but the lesser degree of homology between the C- than V- domains argues against it. Alternatively, this site could be situated in the V-region, and shortening the λ_2 chain by amputation of its C-domain could alter the presentation of the V-domain in such a way that the cross-reactive site is no longer recognized. Finally, it is notable that the priming for cross-reactivity was unidirectional as λ_1, in contrast to λ_2, lacked this property (group 11 vs. group 4). We have not yet completed the analysis of whether the anti-λ_1 response is also regulated by IR-genes.

Table 3. Recognition of λ L-chains by T helper cells from BALB/c mice

Antigen used for:		Secondary anti-NIP response of recipients
Th priming	Boost	
NIP-conjugated:		Farr assay:
1. M315(α λ_2)	Fab-315	3+
2. λ_2-315	Fab-315	4+
3. λ_2-315	λ_2-315	4+
4. λ_2-315	λ_1-J558	4+
5. λ_2-315	J558(α λ_1)	0
6. $V\lambda_2$-315	λ_2-315	4+
7. $V\lambda_2$-315	λ_1-J558	0
8. J558(α λ_1)	Fab-J558	0
9. λ_1-J558	Fab-J558	0
10. λ_1-J558	λ_1-J558	3+
11. λ_1-J558	λ_2-315	0

DISCUSSION

Evidently $V\lambda_2$-315 is recognized by isologous Th in the folded
or unfolded and in the assembled or free form. The specificity of
the recognition is shown by the fact that λ_2-315-primed Th did not
respond to a boost with the complete myeloma protein J558. These
observations demonstrate that Th exist which focus on a conformation-
independent antigenic site expressed by an individual Ig V-domain[13].
In contrast, most serologically defined idiotypes require the quaternary structure of assembled ($V_H + V_L$) domains for their expression;
the serologically defined idiotype of M315 which is ultimately associated with the DNP-lysine binding site[3] is an example of this
rule. This striking example from the idiotypic universe of the
difference between T and B cell recognition recalls analogous observations with conventional non-Ig protein antigens[15,16]. The effect
of MHC-linked genes on the Th response to $V\lambda_2$-315 provides another
similarity between the recognition of M315 idiotype and some conventional protein antigens; evidently, at least some V-domains are subject to the same IR-gene control mechanisms as are several conventional antigens. Several groups have, however, observed a new kind
of Th cell specific for determinants that seem to resemble serologically defined complex idiotypes. These T-cells act in combination
with conventional carrier specific helper cells but are distinct
from the latter by their idiotype specificity. These helper cells
are detected by their ability to promote certain secreting cells[17-21].
In some instances these idiotype-specific helpers can be depleted
by incubation on plastic dishes coated with the immunoglobulin bearing the idiotype[17,20] and in one system they are present before
immunization (natural helper cells)[19] and the interaction with B
cells bearing the complementary idiotype is not MHC-restricted[22].
The fine specificity of this class of Th cells has, however, not yet
been defined.

It is interesting to speculate upon the potential effect of
$V\lambda_2$ specific regulatory T cells. If it is assumed that the Th of
the present report, detected by their ability to deliver carrier
specific help, are also able to regulate the responses of B cells
that bear $V\lambda_2$-like domains, it follows that these Th will not only
interact with clones producing the M315 serological idiotype.
According to this reasoning $V\lambda_2$-like domains have the potential to
associate with a large number of V_H domains other than V_H-315, and
all pairs with a $V\lambda_2$-like member can be envisaged to serve as targets
for $V\lambda_2$-specific regulatory T cells. If a B cell-T helper cell
interaction alone is sufficient to trigger and drive B cell differentiation and proliferation, these $V\lambda_2$-specific Th could induce a
polyclonal B cell activation dominated by λ_2-chains. More likely,
however, two signals are required for B cell activation in which
case the targets for the $V\lambda_2$-specific Th would be limited to $V\lambda_2$-
bearing clones that have been stimulated with a given antigen.

Also in this situation, however, the $V\lambda_2$-domain would be expected to be associated with several different V_H-domains. The $V\lambda_2$-specific helper may thus promote an increase in the proportion of specific antibodies bearing λ_2-chains. Such promotion should be detectable with all antigens for which $V\lambda_2$-domains can contribute (together with appropriate V_H-domains) complementary combining sites.

The experience with the $V\lambda_2$-domain of myeloma protein 315 led us to explore whether the λ_1 chain of the IgA myeloma protein J558 was perceived similarly. We found that despite the extensive (88%) homology between $V\lambda_1$ and $V\lambda_2$, only the free form of the λ_1 chain was recognized by Th in this system. The simplest interpretation is that we are observing a response to a cryptic antigenic site that is exposed upon dissociation of the H and L chains. In that case one cannot readily assign a role in the network for these helpers unless B cells display free L chains on the surface in addition to complete Ig, for which there is little evidence. Antigenic determinants specific for free L chains and hidden in the complete Ig molecule are well known from heterologous serological systems[23] but have not, to our knowledge, been described for T cell recognition of isologous Ig.

Three alternatives present themselves as explanations for the apparent failure of Th to recognize the assembled λ_1 of J558. The basis for the first is that λ_1-J558 is encoded by the germ-line $V\lambda_1$ gene[24] and that λ_1 accounts for 80-90% of all λ chains in serum [25]. It is thus quite possible that immunological tolerance due to clonal deletion has developed for the assembled λ_1 but not for its free form which is presumably present in minimal concentration due to clearance in the kidney. In contrast, λ_2-315 has diversified somatically in five positions from the germ-line encoded $V\lambda_2$ sequence[14,26] and could therefore be a rare variant; in addition, $C\lambda_2$ only comprises 10% of all serum λ chains[25]. The serum soncentration of the $V\lambda_2$-315 may therefore be too low for tolerance to develop. Alternatively, the poor immunogenicity of the assembled λ_1 chain could be inherent in its structure. Third, a regulatory effect could be operating. If, for example, the H-chain subunit of the boost antigen (see group 9, Table 3) contained a suppressor-cell activating determinant, removal of the H-chain could allow a response to develop against the free L-chain. In that case, the antigenic site on λ_1-J558 could still be available on the complete myeloma protein, and it is therefore conceivable that the germ-line gene encoded $V\lambda_1$-region is immunogenic in association with another V-domain than V_H-J558. It may be recalled that in the hen egg white lysozyme system the genetic unresponsiveness of the B10 strain seems to be a consequence of the predominant activation of suppressor cells by a small region of lysozyme[27], and that natural suppressor cells exist against the idiotype of myeloma protein 460 [28]. The present experiments do not allow a choice between these three alternatives.

The cross-reaction between λ_2–315 and free λ_1–J558 elucidates some of the antigenic complexity of these chains. The λ_2–315 chain must bear at least two antigenic sites for Th cells, one which is expressed on both the free and assembled V-domain and another which seems to demand the free form of the entire λ_2 chain. Only the latter determinant primed Th for cross-reactivity with λ_1–J558. The localization of the cross-reactive determinant is unknown; the inability of Vλ_2 to cross-prime Th for responses to λ_1 points to the C-domain, while the much higher degree of amino acid sequence homology between the V-domains speaks for the latter. Conceivably, local patches of high homology between the two Cλ-domains could exist in the regions that form contacts with the CH1 domain due to the higher conservation of these regions and could provide the structural basis for a cross-reactive site in the C-domain that would be hidden in the complete Ig molecule. It is notable that the cross-reaction between λ_1 and λ_2 at the Th level was unidirectional: λ_2–315 primed Th cross-reacted with λ_1 but not vice versa. Since Th primed with λ_1 responded to a boost with λ_1, but not to λ_2, λ_1 presumably also possesses two antigenic sites, one private which is immunogenic, i.e. able to prime uneducated T cells, and a second cross-reactive site that is only recognized by memory Th cells primed by a similar determinant on λ_2–315.

The inability of λ_2–315 and λ_1–J558 primed Th cells to respond when boosted with the assembled, in contrast to the free, form of λ_1–J558 imposes certain limitations on antigen processing. The result indicates that the assembled λ_1 of the complete myeloma protein J558 is not processed in such a way that it is rendered immunogenically equivalent to the free form of λ_1.

ACKNOWLEDGMENTS

This work was supported by The Norwegian Research Council for Science and the Humanities, The Norwegian Association for Fighting Cancer and The Norwegian Cancer Association. We thank Anne Støre for excellent technical assistance and Henny Johansen for typing the manuscript.

REFERENCES

1. N. K. Jerne, The immune system; a web of V-domains, Harvey Lectures 70, 93 (1976).

2. H. Eisen, E. S. Simms and M. Potter, Mouse myeloma protein with antihapten activity; the protein produced by plasma cell tumor MOPC 315, Biochemistry 7, 4126 (1968).

3. S. Sirisinha and H. N. Eisen, Autoimmune-like antibodies to the ligand binding sites of myeloma proteins, Proc. Natl. Acad. Sci., USA 68, 3130 (1971).

4. R. G. Lynch, R. J. Graff, S. Sirisinha, E. S. Simms and H. N. Eisen, Myeloma proteins as tumor specific transplantation antigens, Proc. Natl. Acad. Sci., USA 69, 1540 (1972).

5. J. Hochman, D. Inbar and D. Givol, An active antibody fragment (Fv) composed of the variable portions of heavy and light chains, Biochemistry 12, 1130 (1973).

6. B. Blomberg, A. Traunecker, H. Eisen and S. Tonegawa, Organization of four mouse λ light chain immunoglobulin genes, Proc. Natl. Acad. Sci., USA 78, 3765 (1981).

7. E. S. Dugan, R. A. Bradshaw, E. S. Simms and H. N. Eisen, Amino acid sequence of the light chain of mouse myeloma protein (MOPC-315), Biochemistry 12, 5400 (1973).

8. M. G. Weigert, I. M. Cesari, S. J. Yonkovich and M. Cohn, Variability in the lambda light chain sequences of mouse antibody, Nature, (Lond.) 228, 1045 (1970).

9. T. Jørgensen and K. Hannestad, Specificity of T- and B-lymphocytes for myeloma protein 315, Eur. J. Immunol., 7, 426 (1977).

10. T. Jørgensen and K. Hannestad, T helper lymphocytes recognize the V_L domain of the isologous mouse myeloma protein 315, Scand. J. Immunol., 10, 317 (1979).

11. T. Jørgensen and K. Hannestad, H-2 linked genes control immune responses to V-domains of myeloma protein 315, Nature, (Lond.) 288, 396 (1980).

12. R. Hiramoto, V. H. Ghanta, J. R. McGhee, R. Schrohenloher and N. M. Hamlin, Use of dextran conjugated columns for the isolation of large quantities of MOPC 104E IgM, Immunochemistry, 9, 1251 (1972).

13. T. Jørgensen, B. Bogen and K. Hannestad, Recognition of variable (V) domains of myeloma protein 315 by B- and T-lymphocytes, In Immunoglobulin Idiotypes, ICN-UCLA Symposia on Molecular and Cellular Biology, eds. C. A. Janeway, E. E. Sercarz, H. Wigzell and C. F. Fox, Academic Press, in press (1981).

14. S. Tonegawa, A. M. Maxam, R. Tizard, O. Bernard and W. Gilbert, Sequence of a mouse germ-line gene for a variable region of an immunoglobulin light chain, Proc. Natl. Acad. Sci., USA 75, 1485 (1978).

15. K. Thompson, M. Harris, E. Benjamini, G. Mitchell and M. Noble, Cellular and humoral immunity: a distinction in antigenic recognition, Nature New Biology, 238, 20 (1972).

16. V. Schirrmacher and H. Wigzell, Immune responses against native and chemically modified albumins in mice, J. Exp. Med., 136, 1616 (1972).

17. R. Woodland and H. Cantor, Idiotype-specific T-helper cells are required to induce idiotype + B-memory cells to secrete antibody, Eur. J. Immunol., 8, 600 (1978).

18. D. Hetzelberger and K. Eichmann, Recognition of idiotypes in lymphocyte interactions; I. Idiotypic selectivity in cooperation between T and B lymphocytes, Eur. J. Immunol., 8, 846 (1978).

19. K. Eichmann, I. Falk and K. Rajewsky, Recognition of idiotypes
    in lymphocyte interactions; II. Antigen-independent co-
    operation between T and B lymphocytes, Eur. J. Immunol., 8,
    853 (1978).
20. K. Bottomly and D. E. Mosier, Mice whose B cells cannot produce
    the T15 idiotype also lack an antigen-specific helper T cell
    required for T15 expression, J. Exp. Med., 150, 1399 (1979).
21. L. Adorini, M. Harvey and E. E. Sercarz, The fine specificity
    of regulatory T cells; IV. Idiotypic complementarity and
    antigen-bridging interactions in the anti-lysozyme response,
    Eur. J. Immunol., 9, 906 (1979).
22. K. Bottomly and D. E. Mosier, Antigen-specific helper T cells
    required for dominant idiotype expression are not H-2 re-
    stricted, J. Exp. Med., 154, 411 (1981).
23. Y. Yagi, U. S. Rutishauser and D. Pressman, Release of anti-L-
    chain antibody from antigen-antibody complexes by Fd-fragment,
    Immunochemistry 5, 67 (1968).
24. O. Bernard, N. Hozumi and S. Tonegawa, Sequences of mouse immuno-
    lobulin light chain genes before and after somatic changes,
    Cell, 15, 1133 (1978).
25. T. Cotner and H. N. Eisen, The natural abundance of λ_2-light
    chains in inbred mice, J. Exp. Med., 148, 1388 (1978).
26. S. Tonegawa, A. M. Maxam, R. Tizard, O. Bernard and W. Gilbert,
    Sequence of a mouse germ-line gene for a variable region of
    an immunoglobulin light chain, Proc. Natl. Acad. Sci., USA 75,
    1485 (1978).
27. L. Adorini, M. Harvey, A. Miller and E. Sercarz, Fine specificity
    of regulatory T cells; II. Suppressor and helper T cells are
    induced by different regions of hen egg white lysozyme in a
    genetically nonresponder mouse strain, J. Exp. Med., 150, 293
    (1979).
28. C. Bona and W. E. Paul, Cellular basis of regulation of expression
    of idiotype; I. T-suppressor cells specific for MOPC-460
    idiotype regulate the expression of cells secreting anti-
    trinitrophenyl antibodies bearing 460 idiotype, J. Exp. Med.,
    149, 592 (1979).

EFFECTS OF ANTI-IDIOTYPIC SERA (Ab-2) AND MONOCLONAL IDIOTYPIC
ANTIBODY (Ab-1) ON THE IMMUNE RESPONSE TO A SIMPLE POLYPEPTIDE
ANTIGEN WITH ONLY TWO IMMUNOLOGICALLY ACTIVE EPITOPES. ANALYSIS
OF THE RESPONSE AT THE UNIDETERMINANT LEVEL

Michael Weaver, Lydia K. Sikora and Julia G. Levy

Department of Microbiology
University of British Columbia
Vancouver, British Columbia, Canada V6T 1W5

It is now well recognized that the specific selection of B cell
subclones is mediated by way of cell surface immunoglobulins exhibiting the V domains which that cell is programmed to produce as
secretory Ig when it undergoes division and differentiation. Thus,
there is no ambiguity in designating Ig as the B cell receptor which
as such will exhibit dependence on the molecular configuration of the
epitope to which it binds. Although an enormous amount of work has,
in the past decade, gone into the elucidation of the molecular properties of the equally specific receptor on the T cell, many questions
still remain unanswered. It is not yet clear whether the T cell
receptor and the antigen specific factors secreted by subpopulations
of T cells constitute the same or distinct molecular species.
Although it is generally accepted that V_L markers can be found on T
cells, no V_H markers have yet been detected and the role of V_L in
antigen recognition has not been clarified. The most definitive
work in this area has utilized small haptenic molecules (azobenzene
arsonate, NIP, etc.) as antigens rather than the more natural polypeptides. With use of these haptens it has been shown quite con-
vincingly that specific T cells bear idiotypic determinants which
react with anti-idiotypic antisera raised against the major idiotypes
expressed in the serum of immunized animals. It would therefore
appear that when such sterically stable structures as these haptens
are used as antigens, both reactive B and T cells share common idiotypes, presumably by way of commonly expressed V_L gene products[1,2].
Added to this already complex picture is the possibility that idio-
type-anti-idiotype reactions between B and T cells play a role in
immune responsiveness without the basic requirement for the presence
of antigen.

429

When the question regarding B and T cell recognition of protein antigens is addressed, the picture may not be as clear cut. Ig receptors on B cells are dependent for their interaction not only on the primary structure of an antigen epitope but also on its secondary and tertiary structure. At the T cell level of recognition, there are a number of examples in the literature indicating that secondary and tertiary structures are not of great significance. It has been found that antibodies raised to native lysozyme do not cross-react with its reduced and carboxymethylated derivative, while cell-mediated immunity was cross-reactive regardless of whether animals were immunized with native or denatured material[3]. Similar observations were made by Parish using native and chemically modified flagellin[4,5]. More recently, studies on native and denatured tetanus toxin have shown essentially the same results[6]. There are two explanations for these data: either, B cells (or Ig) recognize determinants which are configuration-dependent while T cells recognize different determinants which are not configuration-dependent and are distinct from those recognized by B cells, or, the same determinants are recognized by both cell types but the T cell receptor is not as configuration-dependent as is the B cell receptor. It has recently been suggested by Erb and Feldmann that a macrophage population which may be involved in antigen trapping and subsequent presentation to T cells digests antigen so extensively before presentation in combination with I-A, that most of its original configuration-dependent epitopes are lost. If differential antigen processing is involved, then other accessory cells must exist which trap and present antigen preferentially to B cells in a less degraded manner. Where network mechanisms are to be studied with complex antigens, such considerations only add to the complexity and difficulty of experimental design.

In our laboratory, the molecular bases of the immune response to the ferredoxin molecule have been studied. This molecule, of bacterial origin and comprising 55 amino acid residues was found to have only two antigenic determinants at the level of B cell responses in the rabbit[7]. These determinants were located within the NH_2-terminal heptapeptide and the COOH-terminal pentapeptide of the molecule. These two epitopes were subsequently shown to have antigenic activity at the T cell level in guinea pigs[8]. Thus, in this molecule, it would appear that even across species barriers, specific antigenic determinants are recognized at both the B and T cell level. More recently, it has been shown that the immune response to ferredoxin (Fd) in mice is under strict MHC control[9] and that these animals also respond essentially to either the NH_2- or COOH-terminal determinants, as ascertained by the isolation and characterization of anti-Fd secreting hybridomas[10]. MHC control of the anti-Fd response has also been shown to extend to the magnitude of the response to either determinant. In high responder mice, which are of the $H-2^k$ haplotype, 80% of the antibody produced is directed to the C determinant, while the remaining antibody is reactive with the N determinant.

Because this molecule is so limited in terms of its antigenic determinants, it was considered by us to be a relatively simple model for dissection of cellular function at the uni-determinant level. Since both B and T cells respond to the same epitopes on Fd, it was also considered a possible tool for examining idiotype expression and control in lymphocyte subsets.

The experimental approach used in this study has involved the production in mice of hybridomas secreting antibody specific for either the NH_2- or COOH-terminal epitopes (these constitute the Ab-1 population). By raising anti-idiotype antiserum (Ab-2) to the monoclonal antibodies, it was assumed that it would be possible to establish the prevalence of individual idiotype expression in various strains of mice at the B cell level, to establish the properties of the anti-idiotypic antiserum in controlling the immune response to Fd, and to determine whether the idiotype was expressed on subpopulations of T cells.

In Figure 1, the sequence of the Fd molecule is shown; also the immunologically active sequences are depicted as the N- or C-determinant. It is also demonstrated that the molecule is susceptible to trypsin cleavage at only one peptide bond, between lys-3 and ala-4. Thus, trypsin cleavage effectively destroys one of the antigenic determinants, while leaving the remaining 52 residues of

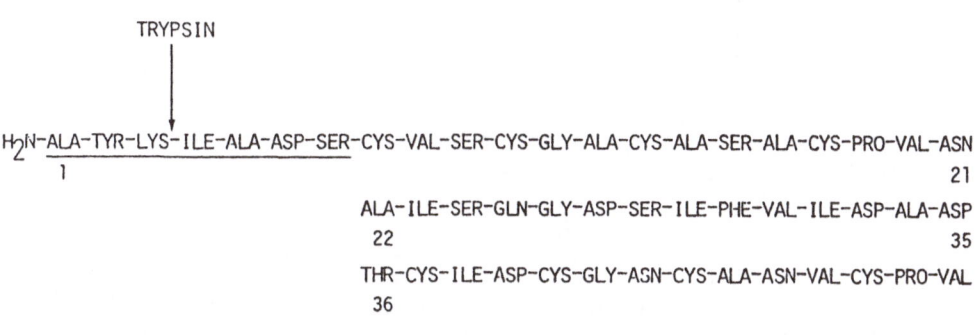

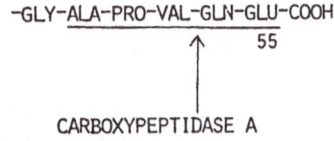

Fig. 1. Amino acid sequence of Fd. The underlined sequences indicate the two major antigenic determinants. The arrows show the point of trypsin cleavage in the NH_2 terminal determinant and the point at which carboxypeptidase A activity is terminated in the COOH-terminal determinant.

the molecule intact. Similarly, carboxypeptidase A can remove the two COOH-terminal amino acids, but will not degrade the molecule past val-53 because of pro at position 52. This digestion process then effectively removes the COOH-terminal antigenic determinant. Using these enzymatically degraded fragments in an inhibition assay in the ELISA, we were able to characterize hybridomas as specific for the N- or C-determinant (Figure 2).

This manuscript reports on preliminary results found using anti-idiotypic antisera raised against two hybridoma monoclonal antibodies, one directed to the N-determinant (anti-Fd-1) and one to the C-determinant (anti-Fd-2). It is realized that these two hybridomas represent only a small part of the immunologic mosaic of the immune response to a relatively weak antigen, but this approach is essential if definitive data regarding idiotype expression and its control in the immune response are to be obtained.

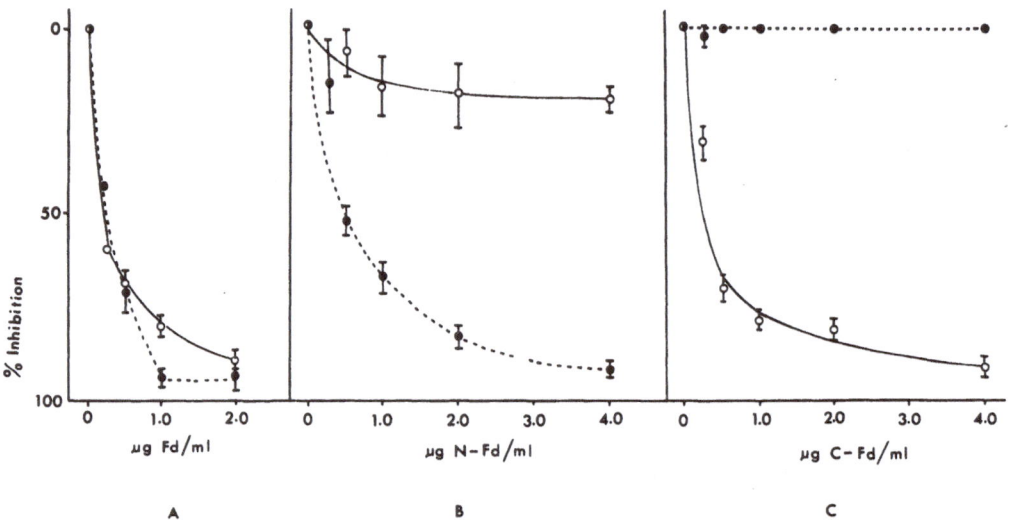

Fig. 2.   Inhibition of antibodies to ferredoxin by enzyme derived fragments. Optimal dilutions of a representative B10.BR antiserum (constituted almost entirely of C-determinant specific antibody) or Fd-1 ascites were incubated overnight with whole Fd or the enzyme degraded fragments ranging from 0 to 4 µg per ml. In the second phase, unblocked antibody was detected by reactions with solid phase Fd in the ELISA. The reactivity expressed on the abscissa represents the value of test samples normalized to the activity of antibody controls incubated overnight with no antigen. O-O, B10.BR serum 1/250; ●-●, Fd-1 ascites 1/10,000.

Both hybridomas were derived by fusion of B10.BR Fd-immune splenocytes with SP2/0 cells according to standard procedures[11]. B10.BR mice (H-2^k) are high responders to Fd as are all H-2^k strains, and this responsiveness maps to the I-A region of the MHC[9].

The first hybridoma (Fd-1) has specificity for the N-determinant. Inhibition studies using Fd or its selectivity degraded derivatives showed that Fd-1 binding to Fd on ELISA plates could be effectively blocked with either Fd or carboxypeptidase A degraded Fd, but was not inhibited by trypsin degraded Fd, indicating specificity for the N-determinant (Figure 2). The second fusion product reacted with the C-determinant since it bound to either Fd or trypsin digested Fd on ELISA plates, but not with carboxypeptidase digested material (Figure 3). Anti-idiotypic antisera to Fd-1 and Fd-2 were raised in rabbits, which, after appropriate absorptions were used in a RIA to assess the expression of these idiotypes in Fd-immune serum from a variety of mouse strains. Some of the early results for Fd-1 are shown in Table 1. It can be seen that the Fd-1 idiotype is expressed in a significant percentage of individual B10.BR (H-2^k) immune sera, and is present in SJL (H-2^s) immune serum. It is also found in other immune sera from mice with B10 backgrounds. It would thus appear that this idiotype is a commonly expressed one in Fd-immune serum and shows linkage to the Ig heavy chain allotype, which is IgCH^b. The anti-Fd-2 serum on the other hand did not react in any significant way with anti-Fd serum with which it was tested, indicating that this idiotype is neither common nor conserved at the B cell level and does not constitute even a frequently occurring minor idiotype in Fd-immune serum.

The effects of anti-Fd-1 and complement on Fd-immune splenocyte populations used in syngeneic adoptive transfer experiments was tested. Results are shown in Figure 4. Immune splenocytes were passed over nylon wool or treated with anti-Thy-1 plus C' to yield T and B cell enriched populations prior to treatment with anti-Fd-1 plus C'. The separated cell populations were reconstituted and injected into irradiated recipients which were boosted with Fd and bled a week later. It can be seen that when T cells or whole spleen populations were treated with anti-Fd-1 and transferred, the anti-Fd response of recipients was uniformly higher than that of control animals. These results imply that this idiotype is expressed on subpopulations of T cells which possibly play a regulatory role in the immune response. When serum from these animals, or those in which whole spleen cells had been treated with anti-Fd-1 plus C' prior to adoptive transfer, was examined for idiotype expression it was noted that while the relative amount of anti-N antibody had not been altered, the absolute amount of Fd-1 idiotype in the serum was increased about 4 fold. This did not account for the marked increase in the overall anti-Fd response, so it was clear that the anti-idiotype treatment of the adoptively transferred cells was having an effect at two levels. The actual quantitation of the test antisera

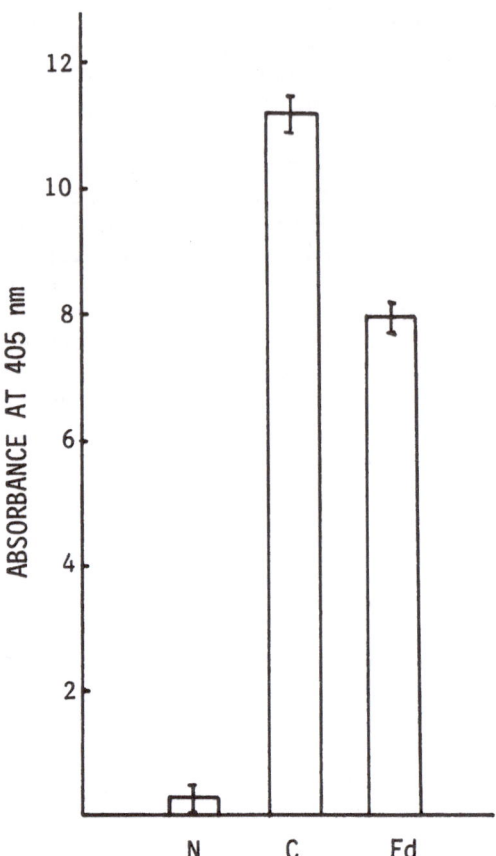

Fig. 3.   The binding of hybridoma Fd-2 to either the N or C fragment
          or to Fd on ELISA plates.   Antigen or fragments were
          affixed to plates at 4 µg per ml and monoclonal antibodies
          were reacted for 30 min at a dilution of 1:50,000.

and idiotype expression are shown in Table 2.   Several facts are
apparent from these serological analyses: the anti-idiotype antisera
plus C' do not apparently kill idiotype-bearing B cells since idiotype is increased in the immune serum; the mechanism by which anti-
Fd-1 brings about increased Fd-1 expression is mediated by T cell
subpopulations but the actual process cannot yet be explained although the observation is consistent with several network models;
the response to the C-determinant is also markedly increased by
treatment with anti-Fd-1 plus C' and this again is presumably
mediated by T cell subsets which are antigen binding (id+) and generate help for the anti-C response.   A possible model by which both
of these observations can be explained by network interactions is
shown in Figure 5.   In this model the B_N^+ cell represents the idiotype

Table 1. Anti-idiotype (Fd-1) expression in the serum of various strains of mice after immunization with Fd

	Number of animals	Haplotype	Allotype	Anti-Fd µg/ml	Anti-N µg/ml	Fd-1 µg/ml	% Fd-1
B10.BR	11	k	b	9.03±6.44 (0.9-21.3)	1.81±1.29 (0.18-4.26)	0.31±0.37 (0 -1.36)	22.9±33.3 (0 -100)
B10.S	pool	s	b	1.20	0.66	0.088	13.3
SJL	4	s	b	1.35±0.72 (0.96-2.40)	0.74±0.39 (0.46±1.32)	0.25±0.10 (0.18-0.40)	36.8±12.9 (27.3-55.7)
CBA	6	k	j	4.39±2.45 (1.54-7.81)	0.88±0.49 (0.31-1.56)	0	0

Table 2. Analysis of sera taken from irradiated B10.BR mice adoptively transferred with Fd-immune splenocytes which had been treated with anti-Fd-1 + C or RIg + C. Results are the average of six mice in each group. Animals were injected with Fd and bled 3 weeks later.

Treatment of Fd-immune spleen cells	Anti-Fd (ng/ml)	% N-antibody	Fd-1 expression (ng/ml)	Fd-1 (% of total N antibody)
NRS + C	1367 ± 585	21.70 ± 2.07	87.2	29.1
Anti-Fd-1 + C	3550 ± 1290	21.60 ± 3.44	358.6	47.5

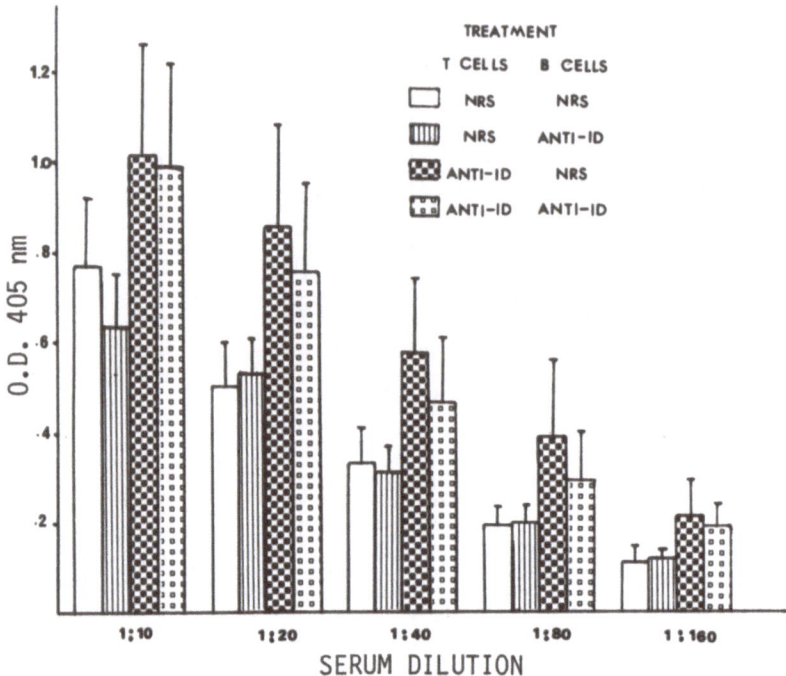

Fig. 4. Effect of anti-Fd-1 on the adoptive secondary response to
Fd in B10.BR mice. Fd-primed T and/or B cells were treated
in vitro with anti-Fd-1 prior to transfer to irradiated
B10.BR recipients. Animals were immunized with Fd and bled
10 days later. Serum was tested in the ELISA.

bearing/antigen binding B cell which may be stimulated by either
anti-idiotype T cells (T⁻) or by traditional T helper cells with
specificity for the C-determinant and acting via the hapten carrier
effect (these are not shown). Because anti-Fd-1 has a stimulatory
effect on B_N^+ expression indirectly through T cells, it is necessary
to introduce a second T cell population (T⁺) which bears the Fd-1
idiotype and is influenced by the presence of anti-Fd-1 (Ab-2).
These cells could increase B_N^+ expression by operating through the
T⁻ population. It is this population also (the T⁺) which could
function in generating the help required to increase the production
of C-determinant specific antibody. Since T⁺ will bind the N-determinant it could effectively generate the help required for presentation of the C-determinant to the appropriate B cells. This model
has been kept simple, and we do not ignore the possibility that there
may be a number of functionally distinct T⁺ cells (helpers, sup-
pressors) which may be preferentially affected by the treatments
(anti-Fd-1 and or complement) used here.

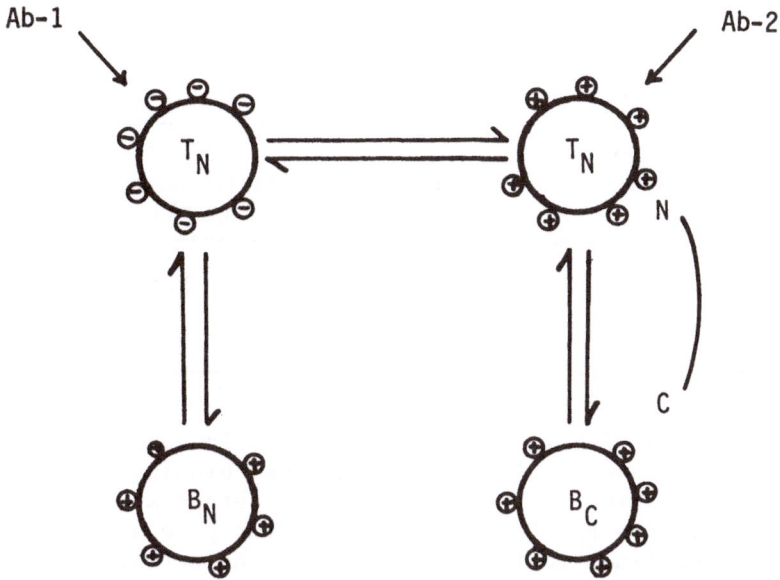

Fig. 5. A possible network model. B_N^+ constitutes the idiotype
positive B cell which is regulated via T^- (anti-idiotypic)
cells. T^- populations are subject to regulation by way of
T^+ cells which regulate the network by way of T^- but also
influence immune responsiveness by being capable of binding
the antigen (epitope N) and generating some kind of amplification or help to B_C^+ cells which bear unrelated idiotypes which bind the second epitope (epitope C). Treatment
with anti-Fd-1 (Ab-2) brings about an increase in expression
of antibody from both B_N^+ and B_C^+ cells. Treatment with
Fd-1 (Ab-1) results only in increased expression by B_C^+
cells.

A further experiment on this idiotypic model has recently been
carried out to observe the effects of Fd-1 (Ab-1) on the primary
response in B10.BR mice to Fd. Animals were given 10 µg of Fd-1 7
days prior to priming with Fd. They were bled 14 days later and
their sera analyzed for titre, idiotype expression and ratio of N:C
antibody. Results are shown in Table 3. Animals treated with Fd-1
produced considerably higher titres to Fd than did their controls.
When the serum was analyzed for N:C ratio or idiotype expression it
was found that there was a significant decrease in idiotype expression, but the relative concentration of N-specific antibody was
unchanged in comparison to control animals. The major shift in antibody titre therefore was attributable to a significant increase in
C-specific antibody. In order to include these data in the model
shown in Figure 5, one can suggest that the Ab-1 reacts with the T^-
population and that this interaction will influence B_N^+ expression

Table 3.   Analysis of sera taken from B10.BR mice treated 7 days
           prior to immunization with Fd with 10µg of Fd-1. Mice
           were bled at 14 days.

Treatment	Anti-Fd (µg/ml)	% N-antibody	Fd-1 expression (ng/ml)	Fd-1 (% total N antibody)
Fd	4.3 ± 6.3	18.93 ± 10.26	208 ± 1.22	48.4%
Fd-1 + Fd	18.9 ± 18.6	19.12 ± 6.91	224 ± 106	6.2%

by inhibiting the Fd-1 idiotype. Also, T^- stimulation with Ab-1
could cause increased activity of T^+ antigen binding T helper cells
which could then bind the N-determinant of ferredoxin and produce
greater help to the B_C^+ C-determinant binding population. Thus, in
this system one finds that perturbation of the network with either
Ab-1 or Ab-2 can result in increased help at the level of antigen
presentation to unrelated populations of B cells binding the second
antigenic determinant. Because two distinct effects are seen with
Ab-2, it may be possible to use this system further in dissecting
subpopulations of idiotype bearing T cells.

The second anti-idiotype tested here (anti-Fd-2) is directed to
a monoclonal antibody with specificity to the C-determinant. This
idiotype is only infrequently expressed in the serum of Fd-immune
B10.BR mice and is never seen in other strains. Preliminary experiments were conducted in which either anti-idiotypic serum (anti-Fd-1
or anti-Fd-2) was injected into Fd-primed or unprimed mice 7 days
prior to subsequent antigen challenge. In these experiments, anti-
Fd-1 had no significant effect on the Fd responses of the treated
mice. This could indicate that the regulatory cells described above
and defined by in vitro treatment and adoptive transfer do not exert
a dominant enough effect in vivo to yield a measurable effect. By
contrast, and somewhat surprisingly, anti-Fd-2 when administered in
vivo had a profound effect on anti-Fd responses in both primed and
naive B10.BR mice. Typical results are shown in Figure 6. When
resulting antisera were examined for idiotype expression, it was
found that no changes had occurred and Fd-2 was not expressed at
increased frequency. One would predict then that increased antibody
levels must be functioning through the second mechanism proposed
here, i.e., increased help. Perhaps the most surprising observation
made in these preliminary studies with anti-Fd-2 was obtained when
DBA/2 mice (non-responder haplotype) received anti-Fd-2 prior to
antigen challenge. In these experiments, the apparent non-responder
status of these animals was ablated and one sees titres comparable
to that seen in B10.BR mice (Figure 6). Thus, it appears that while

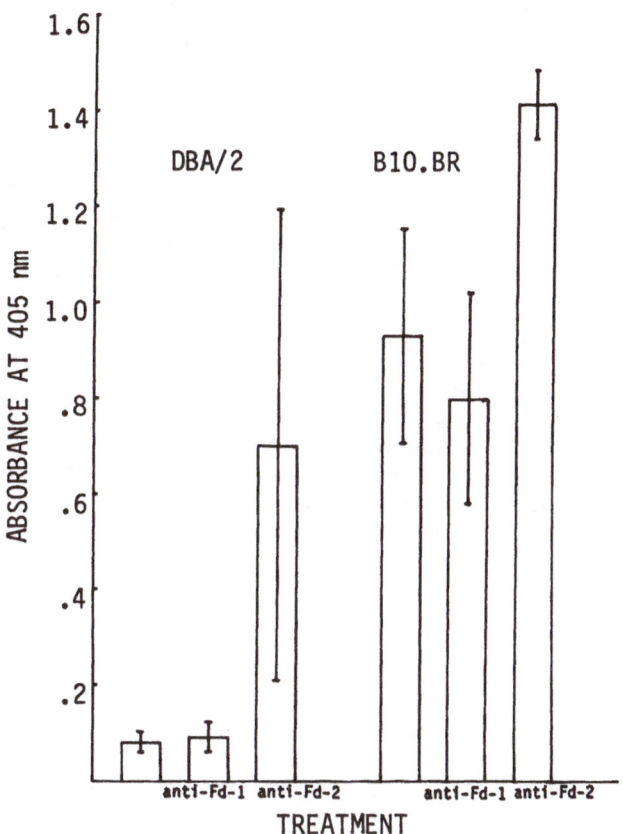

Fig. 6. The effects of anti-Fd-1 and anti-Fd-2 in vivo on the Fd
response in primed DAB/2 (non-responder) and B10.BR
(responder) mice. While anti-Fd-1 has no measurable effect,
anti-Fd-2 enhances the B10.BR response and ablates the
non-responder status in DBA/2 mice. Antigen was administered 7 days after i.p. adminstration of 10 μg of anti-
idiotype and animals were bled 10 days later. Anti-Fd
titres were measured by ELISA. Sera shown here were all
at a dilution of 1:160.

this idiotype (Fd-2) is not expressed with any frequency at the B
cell level, it may well represent a major, dominant regulatory idiotype on T cell subpopulations. Network perturbations with anti-
idiotype do not appear to function at any level through stimulation
of idiotype positive B cells. Because we have insufficient data at
this time, it is impossible to determine how these observations may
be fitted into the model presented here other than to predict that
anti-idiotypic sera may stimulate help. These predictions can be
tested since, if this is the case, higher levels of N-specific antibody would be detected in anti-idiotype treated mice.

Most of the work which has been done using anti-idiotypic anti-sera directed to monoclonal antibodies specific for a given epitope have shown that this kind of perturbation results in either an enhanced or suppressed response to that epitope depending on the way in which the experiments have been carried out. In the experiments discussed here, we have shown that the anti-idiotypic sera act indirectly via T cells to enhance the expression of idiotype positive B cells. What perhaps makes these experiments of some interest in network studies is that we present clear evidence that at least some of those idiotype positive T cells affected bring about amplification of responses to the other, unrelated epitope on the Fd molecule. This gives evidence not only for the existence of a closed network operating via idiotype anti-idiotype interactions, but also for the participation of idiotype positive antigen binding T cells causing amplification of idiotypically unrelated B cells via classical hapten-carrier bridging. Although this aspect of network has been predicted, to our knowledge, these observations constitute the first direct evidence for such an occurrence.

IX: EFFECTOR MECHANISMS FOR CELL SIGNALLING RESULTING FROM ANTIGEN-ANTIBODY INTERACTIONS

INTRODUCTION

How does the binding of antigen to antibody trigger the immune response in cells? The complete answer to this question will not be simple, for different types of cells participate in the immune defense responding to different types of immune complexes composed of antigen and IgG or IgM or IgA or IgE. And the responses are different; for example, macrophages process immune complexes, B and T lymphocytes are triggered through various stages of differentiation by them, and mast cells and basophils are induced to secrete products. Nevertheless, it is possible that certain basic features of the interaction with immune complexes will be found to be common to all or most cell types. We may ask, for example, whether activation takes place through an associative model, a distortive model or an allosteric model, both with respect to the changes which take place within the antibody molecule upon binding antigen at the combining site, and also with respect to the receptors for the antibody molecules located on the surfaces of the responding cells. That is, how are the immunological signals generated on the antibody and on the receptor, and how are these signals transmitted across the plasma membrane?

Of the cellular systems being studied, one of the simplest and, at this time best understood appears to be the activation of mast cells by immune complexes formed of antigen and IgE. The surface receptors on mast cells can bind monomeric IgE very tightly prior to exposure to antigen. Then, as may be common for many cell types, the addition of a multivalent antigen causes aggregation of the IgE and attached receptor. Thus, receptor molecules are pulled together, and this appears to be sufficient for the generation of the immunological signal at the exterior surface of the mast cell; however, the molecular mechanism by which the signal is transmitted across the membrane to trigger the processes resulting in the secretion of the granules filled with vasoactive products is still not understood and remains a key question to be answered.

In the first paper in Topic IX, Henry Metzger reviews the role of IgE and its receptor in mediating secretion. He discusses the generation of the immunological signal, the structural properties of the IgE receptor, and the early biochemical events following IgE-mediated stimulation of mast cells. His short review contains much pertinent information and conclusions, some of which are likely to be generally applicable to the activation of cells by immune complexes.

One of the early biochemical events following the interaction between immune complexes and cells is thought to be changes in the levels of intracellular calcium. It is essential to develop new probes which are sensitive to alterations in calcium levels, and can be introduced into small cells, such as lymphocytes. In the second paper of Topic IX, Tullio Pozzan and colleagues report the development of a fluorescent Ca^{++} probe which, in its esterified form, will cross the plasma membrane, and then be hydrolysed to the parental molecule. With this probe, the authors then study Ca^{++} uptake by thymocytes stimulated by Con A, and by splenocytes stimulated by anti-Ig_m.

In marked contrast to the apparent simplicity of the activation mechanism of mast cells by immune complexes, the activations of B and T lymphocytes require additional mitogenic signals for complete responses. In the third paper of this Topic by Drs. Berman and Ascher, the molecular requirements for the mitogenic signal are discussed in detail, and appear to involve the exposure of a site which is masked in the intact IgG, but can appear in immune complexes, with heat aggregation or with papain digestion. Further processing of these altered IgG or Fc by macrophages then produces a 14000 dalton mitogenic subfragment. This mitogen causes B cell proliferation without the presence of T cells; however, T cells are required for polyclonal antibody response. In the fourth paper, Dr. Weigle and colleagues extend this discussion of the modulation of immune reactivity by the Fc piece, showing a requirement for the Lyt$^+$ 1 population of T cells which may act on both effector T cells and proliferating B cells by secreting an active factor, probably (Fc) TRF, which can act synergistically with IL-2.

Finally, in the last paper of this Topic, Tadamitsu Kishimoto and his coworkers investigate the response to the T cell replacing factor (TRF) produced by PHA stimulation of T cells. Using a human B lymphoblastoid cell line capable of differentiating into IgG-producing cells in the presence of TRF, they show that a cytoplasmic

substance which can induce differentiation, is generated by limited
proteolysis of a precursor protein by a serine esterase which is
activated by the TRF.

THE ROLE OF IgE AND ITS RECEPTOR IN MEDIATING SECRETION

Henry Metzger

National Institutes of Health
National Institute of Arthritis, Metabolism and
Digestive Diseases, Arthritis and Rheumatism Branch
Section on Chemical Immunology
Bethesda, Maryland 20205

In this review I shall describe our current view about antigen-induced, antibody-mediated stimulation of mast cells and basophils, and shall also give some comparative data on similar effector systems. I will focus particularly on those molecular events which occur early. These are of special interest because interference with these early steps may offer an especially fruitful opportunity for therapeutic intervention with a minimum of harmful side reactions.

ROLE OF ANTIBODY

Clearly the role of the immunoglobulin is critical. It provides the necessary link between the stimulating antigen and the responding cell. The antigen reacts with the 'Fab' portion of the antibody; the responding cell with the 'Fc' region. These two regions of the immunoglobulin are covalently bound to each other, but how are they functionally connected? Considerable attention has been directed to this problem; in particular to the question of whether conformational changes in the immunoglobulin permit transmission of some signal from the sites which bind antigen to those in the Fc region which bind the effector molecules. No final answer is available nor are we yet certain that there is one answer which applies to all systems. However, most of the evidence indicates that such antigen-induced conformational changes do not play a significant role in antibody-mediated stimulation[1,2].

NATURE OF INTERACTION BETWEEN ANTIBODY AND THE EFFECTOR SYSTEM

Whereas detailed information on the antigen-combining sites of antibodies is available[3], no such precise data exist with regard to the areas on the Fc region which interact with effector systems. The greatest information about the latter is available with respect to the region which interacts with the C1 component of complement, ie. with C1q[4]; the regeions which interact with so-called 'Fc-receptors' are much less well defined. There are several studies which indicate that the integrity of more than one Fc domain is required[5] although the actual site of binding may be more limited.

There is no evidence that formation of covalent bonds between the immunoglobulin (ligand) and the effector system (receptor) occurs. The interaction involves non-covalent bonds between the two components directly; in no instance has a co-factor such as an alkali metal or other small molecule been implicated. In one instance - the IgE-mast cell system - the affinity constant of the receptor for the mono-meric ligand is extremely high ($\geqslant 10^{10} M^{-1}$)[6,7]. In other cases it is sufficiently low that significant binding only occurs when aggregation of the ligand permits multiple cooperative interactions[8]. It is interesting that in all cases where an effector system which is stimulated by antigen-antibody complexes has been defined, it is capable of such multiple interactions. This is necessarily the case with cell-surface receptors and is also true of the non-cell associated classical pathway of complement activation.

NATURE OF RECEPTORS

A. The mast cell receptor for IgE. Since the principal object of this review is to consider this system, I will describe it in some detail. Certain salient features of these receptors can be dis-covered by examination of intact cells. The principal findings are that the receptors are numerous[6], mobile, unclustered, and uni-valent[9,10]. Their extremely high binding constant has already been mentioned. There are several other characteristics of the receptor in situ which can be discovered after disruption of the cells. These properties as well as many others which can be determined after iso-lation and purification of the receptor are collected in Table 1. Some fundamental questions still remain to be answered, such as the orientation of the component parts of the receptor in the plasma membrane, and almost nothing is known about the fine details of the structure. Nevertheless, we have been able to develop a plausible model (Figure 1) which at least can be used as a basis for developing experiments aimed at elucidating further details.

B. Other receptors. Considerably more is known about the structure of the Fc receptor for monomeric IgE than for any other Fc receptors; in the latter cases little is certain as yet. They

Table 1. Structural Properties of Cell Surface Receptor for IgE

Overall structure		Reference
Molecular formular	α β	11
Size	80,000	11, 12
Structure of α-chain		
Size	50,000	13 - 15
Composition:polypeptide	34,000	15
carbohydrate	16,000	15
Domains	α1 + α2	16
size	α1 > α2	16
carbohydrate	α1 > α2	16
surface labeling	α2 >> α1	16
Structure of β-chain		
Size	30,000	11, 17, 18
Composition	No carbohydrate	18
Domains	β1 + β2	18
size	β1 ≈ 20,000	18
	β2 ≈ 10,000	18
labeling with		
hydrophobic probe	β1	17, 18
interaction with	β1	18
phosphorylated	β2	(unpublished observations)*

*C. Fewtrell, A. Goetze and H. Metzger (1981).

appear to be glycoproteins but estimates of their size by elecro-
phoresis in polyacrylamide gels using ionic detergents, have varied
widely[19]. With regard to the other major effector system mentioned,
the classical complement pathway, substantial information on the
primary receptor -Cl- is known[4].

THE INITIAL MOLECULAR EVENT

    A. The IgE-mast cell system. It has been determined from a
variety of studies that aggregation of the cell-bound IgE is a
necessary and sufficient stimulus to produce IgE-mediated degranulation. The only other extrinsic components required are those
necessary to support an actively metabolizing cell and calcium[20].
More recently, several aspects of the aggregation event have been
clarified further:

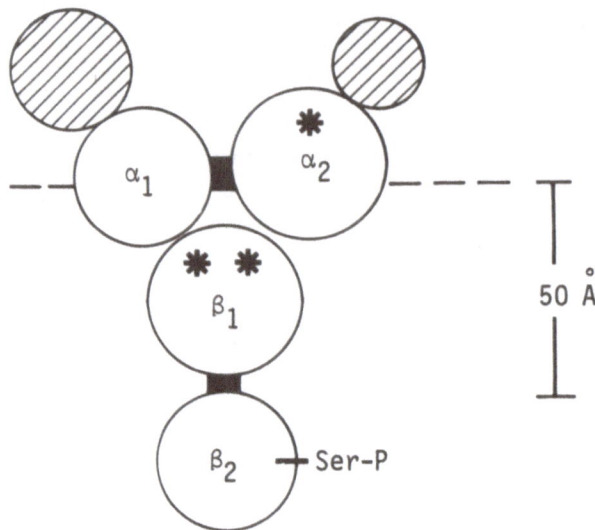

Fig. 1.   Schematic representation of the receptor for IgE. This
          figure is modified from the one in Ref.16 to show the ap-
          parent site of phosphorylation as described in
          C. Fewtrell, A. Goetze and H. Metzger (ms. submitted).
          As in that figure the horizontal line represents the sur-
          face of the outer leaflet of the plasma membrane bilayer;
          the hatched areas represent carbohydrate; the black areas,
          sites of proteolytic attack; the single asterisk, the
          principal site of surface labeling; the double asterisk,
          the site of labelling by iodonaphthyl azide. The circles
          represent spheres whose volumes are proportional to their
          masses.

(1) The minimal aggregation required is dimer formation, and such
    dimers can provide the "unit signals" which initiate the degranu-
    lation process[21]. However, our recent studies using a tumor
    analog of rodent basophils, suggest that larger clusters may
    induce more effective signals[22]. This effect of larger aggre-
    gates cannot be observed when normal rat mast cells are
    studied[21,22] and only hints of a similar phenomenon are observed
    with normal human basophils[23]. For reasons discussed else-
    where[22], we believe that the tumor cells are permitting us to
    visualize an aspect of the triggering mechanism which is diffi-
    cult to demonstrate but is nevertheless applicable to normal
    cells also.
(2) The second aspect which has recently been clarified is that it
    is the aggregation of the receptor and not the IgE bound to it
    which is critical. This was demonstrated by showing that recep-
    tor aggregation per se[24,25] even in the total absence of IgE[25]
    leads to stimulation of the cells.

B. Other systems. With other cellular Fc receptors, aggregation of the antibodies which bind to them have similarly been shown to be critical[1]. In some of these it has also been shown that whereas dimers are sufficient to initiate a signal, larger oligomers may be more effective[26]. Similarly in the classical pathway of complement activation studies have shown that dimers work but that higher antibody aggregates are more effective[27].

THE SECOND STEP

A. The IgE-mast cell system. There is still no information about the immediate molecular consequence of receptor aggregation in this system. A variety of biochemical perturbations have been described which may be important during the early stages of degranulation triggered by IgE (Table 2) but in none of these has the receptor itself been implicated directly. Clarification of these events and how the receptor for IgE relates to them, is the major challenge for those of us engaged in exploring this system.

B. Other Systems. A provocative report has appeared recently which assigns phospholipase A_2 activity to the receptor on lymphocytes which binds the Fc regions of human IgG[37]. Materials with indistinguishable properties were isolated by affinity chromatography from columns conjugated with IgG or lecithin alternatively. The characterization of both the structure of this material and its relationship to the enzymatic activity remain as yet preliminary, but these findings may represent a significant advance.

Considerable progress has been made in defining how aggregates of IgG combine with the multivalent C1q and thereby activate in turn the C1r and C1s proteases in the complement pathway[4]. Because it is not cell-bound and is relatively abundant, this system should be capable of being substantially elucidated over the next several years.

Table 2. Early Biochemical Events Following IgE-Mediated Stimulation of Mast Cells

Event	Reference
Transient increase in cAMP	28, 29
Transient increase in phospholipid methylation	30, 31
Increased phosphatidyl inositol metabolism	32, 33
Increased $^{45}Ca^{2+}$ uptake	34, 35
Increased $^{45}Ca^{2+}$ efflux	36

CONCLUDING REMARKS

Significant advances have been made in clarifying the structure of the receptor for IgE on mast cells and basophils which is importantly involved in immediate hypersensitivity reactions. While knowledge about the receptors' structure has developed more rapidly than about the functional mechanism, there are reasons for optimism that the latter also will be clarified. Other systems which are similarly responsive to antigen-induced antibody aggregation may provide important clues as to what to look for.

REFERENCES

1. H. Metzger, The effect of antigen on antibodies: Recent studies, Contemp. Topics Molec. Immuno. 7:191 (1978).
2. C. Fewtrell, M. Geier, A. Goetze, D. Holowka, D. E. Isenman, J. F. Jones, H. Metzger, M. Novia, D. Siekmann, E. Silverton and K. Stein, Mediation of effector functions by antibodies: Report of a workshop, Molecular Immunol. 16:741 (1979).
3. L. M. Amzel and R. J. Poljak, Three-dimensional structure of immunoglobulins, Ann. Rev. Biochem. 48:961 (1979).
4. R. R. Porter and K. B. Reid, The biochemistry of complement, Nature 175:699 (1978).
5. K. J. Dorrington, The structural basis for the functional versatility of immunoglobulin G, Can. J. Biochem. 56:1087 (1978).
6. A. Kulczycki, Jr., and H. Metzger, The interaction of IgE with rat basophilic leukemia cells. II. Quantitative aspects of the binding reaction, J. Exp. Med. 140:1676 (1974).
7. G. Rossi, S. A. Newman and H. Metzger, Assay and partial characterization of the solubilized cell surface receptor for immunoglobulin E, J. Biol. Chem. 252:704 (1977).
8. D. M. Segal and E. Hurwitz, Binding of affinity cross-linked oligomers of IgG to cells bearing Fc receptors, J. Immunol. 118:1338 (1977).
9. J. Schlessinger, W. W. Webb, E. L. Elson and H. Metzger, Lateral motion and valence of Fc receptors on rat peritoneal mast cells, Nature 264:550 (1976).
10. G. R. Mendoza and H. Metzger, Distribution and valency of receptor for IgE on rodent mast cells and related tumor cells, Nature 264:548 (1976).
11. D. Holowka, H. Hartmann, J. Kanellopoulos and H. Metzger, Association of the receptor for immunoglobulin E with an endogenous polypeptide on rat basophilic leukemia cells, J. Receptors Res. 1:41 (1980).
12. S. A. Newman, G. Rossi and H. Metzger, The molecular weight and valency of the cell receptor for IgE. Proc. Natl. Acad. Sci. 74:869 (1977).
13. D. H. Conrad and A. Froese, Characterization of the target cell receptor for IgE II Polyacrylamide gel analysis of the sur-

face IgE receptor from normal rat mast cells and from rat
basophilic leukemia cells, J. Immunol. 116:319 (1976).

14. A. Kulczycki, Jr., T. A. McNearney and C. W. Parker, The rat
basophilic leukemia cell receptor for IgE. Characterization
as a glycoprotein, J. Immunol. 117:661 (1976).

15. J. M. Kanellopoulos, T. Y. Liu, G. Poy and H. Metzger, Composition and subunit structure of the cell receptor for immunoglobulin E, J. Biol. Chem. 255:9060 (1980).

16. A. Goetze, J. Kanellopoulos, D. Rice and H. Metzger, Enzymatic
cleavage products of the α-subunit of the receptor for IgE,
Biochemistry (in press).

17. D. Holowka, C. Gitler, T. Bercovici and H. Metzger, Reaction of
5-iodonaphthyl-1-nitrene with the receptor for IgE on normal
and tumor mast cells, Nature 289:806 (1981).

18. D. Holowka and H. Metzger, Further characterization of the β-
component of the receptor for immunoglobulin E, Molec.
Immunol. (in press).

19. I. S. Mellman and J. C. Unkeless, Purification of a functional
mouse Fc receptor through the use of a monoclonal antibody,
J. Exp. Med. 152:1048 (1980).

20. B. Uvnas, Histamine storage and release, Fed. Proc. 33:2172
(1974).

21. D. M. Segal, J. D. Taurog and H. Metzger, Dimeric immunoglobulin
E serves as a unit signal for mast cell degranulation, Proc.
Natl. Acad. Sci. USA 74:2993 (1977).

22. C. Fewtrell and H. Metzger, Larger oligomers of IgE are more
effective than dimers in stimulating rat basophilic leukemia
cells, J. Immunol. 125:701 (1980).

23. A. Kagey-Sobotka, M. Dembo, B. Goldstein, H. Metzger and
L. M. Lichtenstein, Qualitative characteristics of histamine
release from human basophils by covalently crosslinked IgE,
J. Immunol. (in press).

24. T. Ishizaka and K. Ishizaka, Triggering of histamine release
from rat mast cells by divalent antibodies against IgE-
receptors, J. Immuno. 120:800 (1978).

25. C. Isersky, J. D. Taurog, G. Poy and H. Metzger, Triggering of
culture mastocytoma cells by antibodies to the receptor for
IgE, J. Immunol. 121:549 (1978).

26. P. H. Plotz, R. P. Kimberly, R. L. Guyer and D. M. Segal, Stable
model immune complexes produced by bivalent affinity labeling
haptens: Invivo survival, Mol. Immunol. 16:721 (1979).

27. J. Tschopp, T. Schulthess, J. Engel and J. C. Jaton, Antigen-
independent activation of the 1st component of complement C1
by chemically crosslinked rabbit IgG oligomers FEBS Letters
112:152 (1980).

28. T. J. Sullivan, K. L. Parker, A. Kulczycki, Jr., and C. W. Parker,
Modulation of cyclic AMP in purified rat mast cells. III.
Studies on the effects of concanavalin A and anti IgE on
cyclic concentrations during histamine release, J. Immunol.
117:713 (1976).

29. R. A. Lewis, S. T. Holgate, L. J. Roberts II, J. E. Maquire,
    J. A. Oates and K. F. Austen, Effects of indomethacin on
    cyclic nucleo-tide levels and histamine release from rat
    serosal mast cells, J. Immunol. 123:1663 (1979).

30. F. Hirata, J. Axelrod and F. T. Crews, Concanavalin A stimulates
    phospholipid methylation and phosphatidyl serine decarboxy-
    lation in rat mast cells, Proc. Natl. Acad. Sci. 76:4813
    (1979).

31. T. Ishizaka, F. Hirata, K. Ishizaka and J. Axelrod, Stimulation
    of phospholipid methylation, Ca^{2+} influx and histamine release
    by bridging of IgE receptors on rat mast cells, Proc. Natl.
    Acad. Sci. 76:4813 (1980).

32. S. Cockcroft and B. D. Gomperts, Evidence for a role of phos-
    phatidylipsitol turnover in stimulus-secretion coupling:
    studies with rat peritoneal mast cells, Biochem. J. 178:681
    (1979).

33. D. A. Kennerly, T. J. Sullivan and C. W. Parker, Activation of
    phospholipid metabolism during mediator release from stimu-
    lated rat mast cells, J. Immunol. 122:152 (1979).

34. J. C. Foreman, M. B. Hallet and J. L. Mongar, The relationship
    between histamine secretion and ^{45}Ca uptake by mast cells,
    J. Physiol. 271:193 (1977).

35. T. Ishizaka, J. C. Foreman, A. R. Sterk and K. Ishizaka, Induc-
    tion of calcium flux across the rat mast cell membrane by
    bridging IgE receptors, Proc. Natl. Acad. Sci. USA 76:5858
    (1979).

36. C. Fewtrell and H. Metzger, Stimulus-secretion coupling in rat
    basophilic leukaemia cells, in"Biochem. of Acute Allergic
    Reactions," K. F. Austen and E. Becker, eds., (in press).

37. T. Suzuki, R. Sadasivan, T. Taki and G. M. Helmkamp, Jr., Studies
    of Fc receptors of human β lymphocytes: Phospholipase A₂
    activity of Fc γ receptors, Biochemistry 19:6037 (1980).

A NEW METHOD FOR MONITORING INTRACELLULAR FREE Ca^{2+} IN LYMPHOCYTES.

CONCANAVALIN A AND ANTI-Ig_m INDUCE AN EARLY RISE OF CYTOPLASMIC

FREE Ca^{2+}

T. Pozzan, T. R. Rink[1] and R. Y. Tsien[1]

C.N.R. Unit for the Study of Physiology of Mitochondria
and Institute of General Pathology
University of Padova, Italy
[1]Physiological Laboratory
University of Cambridge, England

INTRODUCTION

A wide variety of extracellular agents stimulate resting lymphocytes to enter the cell cycle and/or to redistribute their surface receptors in an energy-dependent mechanism. The ultimate mechanism by which the binding of a ligand to its surface receptor switches on the necessary intracellular processes remains unknown. Nearly every "second messenger" found in other cell system, pH, cyclic nucleotides, membrane potential, Ca^{2+}, has been discussed as a possible carrier of the signal in lymphocytes. Among these candidates $[Ca^{2+}]_{in}$ has received the most attention.

Many attempts have been made to find changes in transmembrane ^{45}Ca fluxes in response to mitogens[1,2] or/and agents known to induce cross-linking and capping of membrane receptors[3], but the results have been often contradictory. Moreover the key variable $[Ca^{2+}]_{in}$, which eventually controls the Ca^{2+} dependent intracellular events can not be monitored by ^{45}Ca while a direct injection of Ca^{2+} indicators such as aequorin and arsenazo III, or impalement with microelectrodes is not applicable to small cells as lymphocytes. We have recently described a method to trap a new fluorescent Ca^{2+} indicator in intact lymphocytes[4] by using its non polar ester derivative which crosses the plasma membrane and is then hydrolyzed intracellularly back to the parental molecule. In this paper we will briefly summarize the method for measuring resting $[Ca^{2+}]_{in}$ level in lymphocytes

and we will describe the effects of T cell mitogens and anti surface
Ig antibodies (anti-Ig) on $[Ca^{2+}]_{in}$.

MATERIALS AND METHODS

Quin 2 acetoxymethylester (Quin 2 AM) was synthesized according
to the general procedure described by Tsien[5]. Thymocytes and splenocytes from Balb/C mice were isolated as previously described[5]. All
experiments were performed at 37° and all chemicals were analytical
grade.

RESULTS

Tsien[5] has recently synthesized a series of new Ca^{2+} chelators
and indicators which exhibit a high affinity for Ca^{2+} ($K_d = 10^{-7}M$),
and a high selectively versus Mg^{2+} and H^+. Loading of cells by
means of intracellularly hydrolyzable acetoxymethyl derivatives of
a Ca^{2+} chelator (BAPTA) has been recently described by Tsien[7]. In
this paper we have used the fluorescent derivative of BAPTA, Quin 2,
corresponding to structure 2 of Tsien[5]. Quin 2-acetoxy-methylesther
(Quin $_2$-AM) hydrolysis by cellular esterases is easily detected by
the gradual shift in the emission spectrum from that of the ester
peaking at 430nm to that of the final indicator peaking at 492nm.
The hydrolyzed Quin 2-AM is trapped intracellularly as demonstrated
by centrifuging and resuspending the cells in fresh medium. The
indicator is presumably free in the cytoplasm of lymphocytes. This
last conclusion is based on the following lines of evidence:

(a) All Quin 2 is released by digitonin at concentrations which do
    not affect lymphocytes mitochondrial and lysosomal integrity.
(b) High voltage electric discharge, which makes small holes in the
    plasma membrane, releases most of Quin 2 while the majority of
    the intracellular proteins are still retained in the lymphocytes.
(c) The excitation and emission spectra of Quin 2 inside the cells
    are identical to the spectra obtained in simplified medium.
(d) Quin 2 inside the lymphocytes is neither bound nor excluded by
    the nucleus.

Quin 2 responds to changes of $[Ca^{2+}]$ with an increased quantum
efficiency. The difference between 0 Ca^{2+} and 100µM Ca^{2+} is about
6 fold. The general procedure for calibrating Quin 2 fluorescence
in the intact cells in terms of $[Ca^{2+}]_{in}$ is the following:

The fluorescence signal of Quin 2 trapped in cells is recorded and
then the dye released from the cells with digitonin or Triton X 100.
The K_d of Quin 2 is 115nM in the presence of 1mM Mg^{2+} and if the
cells are lysed in the usual medium which contains 1mM Ca^{2+} the dye
should be completely saturated and F_{max} (maximum fluorescence)

therefore obtained. The Ca^{2+} level of the lysate can then be adjusted to known levels and compared to the fluorescence of Quin 2 before lysis. In practice it is sufficient to know, in each experiment, the fluorescence at saturating Ca^{2+}, (F_{max}) and the fluorescence at 0 Ca^{2+} (F_{min}) after cell lysis and the $[Ca^{2+}]_{in}$ before lysis can be calculated according to the following formula:

$$[Ca^{2+}]_{in} = K_d (F - F_{min})/(F_{max} - F)$$

where K_d is the apparent dissociation constant, F is the fluorescence of Quin 2 inside the cells and F_{min} and F_{max} are the fluorescence of Quin 2 at very low and very high Ca^{2+} respectively. It must be noted that K_d is affected by the intracellular $[Mg^{2+}]$ and $[H^+]$. We have determined both $[Mg^{2+}]_{in}$ and pH_{in} of lymphocytes[8] which are $\sim$ 1mM and 7.05 (at pH_{out} 7.4) respectively. The apparent K_d of Quin 2 at these concentrations of Mg^{2+} and H^+ is 115nM.

The mean value for $[Ca^{2+}]_{in}$ of mouse lymphocytes from spleen and thymus is 119±9nM. It can be demonstrated that the $[Ca^{2+}]_{in}$ is not affected by the amount of Quin 2 trapped and that Quin 2 loading up to 2-3mM have no relevant short-term toxic effects.

Figure 1 shows the effect on $[Ca^{2+}]_{in}$ of mitogenic doses of Concanavalin A (Con A) in mouse thymocytes. After a lag phase of about 30-60 secs the fluorescence rises, indicating a rise of $[Ca^{2+}]_{in}$ to a new steady-state level which is maintained for several tens of minutes. Assuming that all the cells in the suspension respond to Con A with a rise in $[Ca^{2+}]_{in}$, the fluorescence increase should correspond to a 2 fold increase in $[Ca^{2+}]_{in}$. This rise in cytoplasmic $[Ca^{2+}]$ is strictly dependent on extracellular Ca^{2+}, it

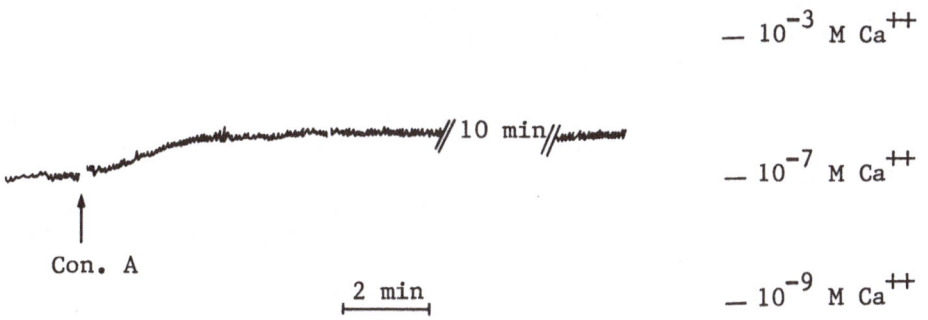

$— 10^{-3}$ M Ca⁺⁺

10 min

$— 10^{-7}$ M Ca⁺⁺

Con. A

2 min

$— 10^{-9}$ M Ca⁺⁺

Fig. 1.   Effect of Con A on $[Ca^{2+}]_{in}$ in thymocytes. The medium contained 140mM NaCl, 5mM KCl, 1mM Na_3PO_4, 5.5mM glucose, 0.5mM $MgSO_4$, 1mM $CaCl_2$, 20mM Hepes (pH 7.4), 37°C. Mouse thymocytes: 5 x 10⁶ cells x ml⁻¹. Quin 2 inside the cells: 1.5nmol x 10⁷ cells⁻¹. Con A when added 1µg x ml⁻¹.

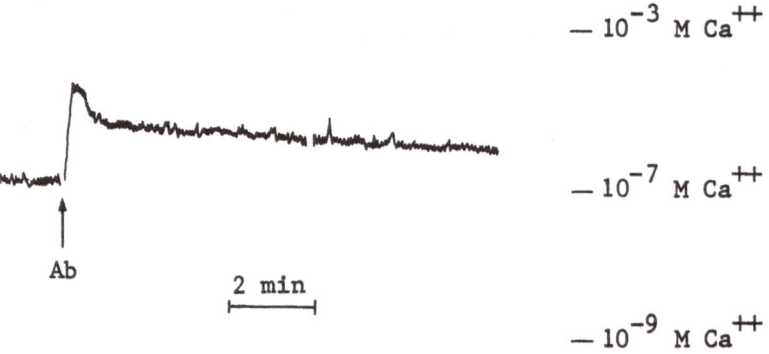

$- 10^{-3}$ M Ca^{++}

$- 10^{-7}$ M Ca^{++}

Ab

2 min

$- 10^{-9}$ M Ca^{++}

Fig. 2. Effect of anti-Ig$_m$ on $[Ca^{2+}]_{in}$ in spleen lymphocytes.
Medium as in Figure 1. Mouse splenocytes 5 x 10^6 cells x
ml^{-1}, 60% Ig$_m$ positive. Quin 2 inside the cells: 0.75nmol
x 10^7 cells. Ab(anti-Ig$_m$) when added 40µg x ml^{-1}.

is prevented by metabolic inhibitors, it is blocked or reversed by
agents known to raise intracellular cyclic AMP such as theophylline
dibutyryl-cyclic AMP and cholera toxin. The rise in $[Ca^{2+}]_{in}$ is
higher but rapidly reversed at supraoptimal Con A concentrations,
and it is prevented or reversed by anti Con A antibody concentrations
which effectively block mitogenic stimulation.

Figure 2 shows the effect of anti-Ig on Quin 2 fluorescence in
spleen cells. There is a rapid rise of fluorescence immediately
after anti-Ig addition which then decreases to the original level
paralleling capping of the receptors. The anti-Ig effect is specific
for B cells, and therefore it can be calculated that the $[Ca^{2+}]_{in}$,
shortly after anti-Ig addition, can rise above 1µM. There are several important differences between the $[Ca^{2+}]_{in}$ rise induced by Con A
in the T cells and that induced by anti-Ig in the B cells.

(a) Con A effect is strictly dependent on external Ca^{2+}, while anti-
    Ig effect can be observed, although reduced in amplitude and
    duration, also in the absence of external Ca^{2+}.
(b) Con A induced $[Ca^{2+}]_{in}$ rise is maintained for more than 30
    minutes while anti-Ig $[Ca^{2+}]_{in}$ rise is completely reversed in
    less than 15 minutes.
(c) There is no effect of drugs affecting cyclic AMP on the anti-Ig
    induced $[Ca^{2+}]_{in}$ rise.

The rise in $[Ca^{2+}]_{in}$ induced by anti-Ig, although it is an
important consequence of the crosslinking of the surface Ig of the
B cells, is not necessary for capping to occur, since it can be prevented without affecting either the rate nor the extent of capping.

DISCUSSION

The method for monitoring intracellular $[Ca^{2+}]$ described in this report has allowed the direct measurement of $[Ca^{2+}]_{in}$ in lymphocytes, cells for which the usual techniques, impalement with microelectrodes and microinjection of dyes, have been so far unsuccessful. The method is easy and reproducible and in theory applicable to a variety of cells, in particular small cells in suspension. We have demonstrated that two ligands which crosslink surface receptors induce a rise of $[Ca^{2+}]_{in}$ in lymphocytes. It is interesting that both ligands, Con A and anti-Ig, are polyclonal mitogens for T and B cells respectively, although anti-Ig must be chemically modified (Fc portion removed or the molecule attached to insoluble beads), before becoming effectively mitogenic. We have shown elsewhere[4] that other polyclonal mitogens, PHA, A 23187, ionomicin, at mitogenic concentrations, cause a sustained rise of $[Ca^{2+}]_{in}$ in lymphocytes. These data are therefore in agreement with the proposed hypothesis that a rise of intracellular $[Ca^{2+}]$ is an early obligatory event of mitogenic stimulation[8,9]. The hypothesis is also supported by the observation that agents known to prevent or abort mitogenic stimulation in lymphocytes, like drugs which increase cyclic AMP or antilectins antibodies, prevent or abort the $[Ca^{2+}]_{in}$ rise.

ACKNOWLEDGEMENTS

Dr. Tullio Pozzan held an EMBO long term fellowship. This work was partially supported by SPC Grant to the "Cell Biology Group for the Control of Eucaryotic Cell Growth".

REFERENCES

1. M. H. Freedman, M. C. Raff and B. Gomperts, Induction of increased calcium uptake in mouse T lymphocytes by Concanavalin A and its modulation by cyclic nucleotide, Nature 255:378 (1975).
2. T. R. Hesketh, G. A. Smith, M. D. Houslay, G. B. Warren and J. C. Metcalfe, Is an early calcium flux necessary to stimulate lymphocytes? Nature 267:490 (1977).
3. J. Braun, R. I. Sha'afi and E. R. Unanue, Crosslinking by ligands to surface immunoglobulin triggers mobilization of intracellular ⁴⁵Ca²⁺ in B lymphocytes, J. Cell. Biol. 82:755 (1979).
4. T. Pozzan, T. J. Rink and R. Y. Tsien, Intracellular free Ca²⁺ in intact lymphocytes, J. Physiol. 318:12 (1981).
5. R. Y. Tsien, New calcium indicators and buffers with high selectivity against magnesium and protons: design, synthesis and prototype structures, Biochemistry 19:2396 (1980).
6. T. Pozzan, A. N. Corps, C. Montecucco, T. R. Hesketh and J. C. Metcalfe, Cap formation by various ligands on lympho-

cytes shows the same dependence on high cellular ATP levels,
Biochim. Biophys. Acta 602:558 (1980).

7. R. Y. Tsien, A non disruptive technique for loading calcium
    buffers and indicators into cells, Nature 290:527 (1981).

8. T. Pozzan, T. J. Rink and R. Y. Tsien, Cytoplasmic pH and free
    Mg^{2+} in pig mesenteric lymphocytes, J. Physiol. (in press).

9. R. Y. Tsien, T. Pozzan and T. J. Rink, T cell mitogens cause
    early changes in cytoplasmic free Ca^{2+} and membrane potential
    in lymphocytes, Nature (in press).

B LYMPHOCYTE STIMULATION AND SUPPRESSION BY THE Fc PORTION OF IMMUNOGLOBULIN

Monique A. Berman and Michael S. Ascher

Department of Medicine
University of California
Irvine, California 92717

MOLECULAR REQUIREMENTS FOR THE MITOGENIC SIGNAL

The Fc region of immunoglobulin (Ig), whether in the form of an Fc fragment[1], aggregated Ig[1,2], or an immune complex[3] can induce normal murine splenic B cells and human peripheral blood B lymphocytes to proliferate and differentiate to polyclonal antibody-secreting cells. Only the Fc portion of Ig is able to induce this response. Fab or F(ab')$_2$ fragments, either soluble or heat aggregated, are inactive. Some change has to occur in the Fc part of the molecule since unaggregated intact IgG is inactive. It appears that the alterations of IgG that occur with heat aggregation and after interaction of IgG antibody with antigen, as well as by splitting Fc from the parent molecule by papain digestion, all expose a site in the Fc portion that is masked in the intact molecule and which is critical in mitogenesis. The mitogenic signal of Fc fragments does not appear to depend on approximation of Fc regions by aggregation. Mitogenically active human Fc fragments obtained by cleavage with papain have a sedimentation rate of Fc monomers and do not appear to aggregate in the culture medium[1]. Further, heat aggregation or chemical crosslinking of purified Fc fragments does not increase their mitogenicity (unpublished observation).

For _in vitro_ stimulation with IgG, only heating at 63°C produced mitogenically active aggregates. Aggregation with DMSO, ethanol, or acetone which has been shown to produce IgG aggregates with high anti-complementary activity, did not produce mitogenically active aggregates[4]. Heating appears to induce a critical change in the IgG molecule not seen with other methods of aggregation. A conformational change may occur during heating of IgG. Alternatively, the heating of IgG preparations may provide a brief period of proteolysis by

459

the trace amounts of contaminating proteases. Several reports of
proteolytic activity in IgG preparations have appeared. This activity
has been attributed to contamination with plasmin and cathepsin
B[5,6], but could even be due to intrinsic proteolytic activity in
immunoglobulins[7,8].

After heat aggtegation, papain cleavage, or antigen-antibody
complex formation, further processing of the modified Fc protions
seems to be required for their stimulation of B lymphocytes. This
step can proceed in the presence of macrophages or macrophage supernatants during a 1 hour incubation at 37°C, as demonstrated by
Morgan and Weigle[9]. Native IgG is resistant to this step but Fc
fragments[9], IgG aggregates[4], and immune complexes[3] produce a mitogenic Fc subfragment. The mitogenic subfragment of 14,000mol/wt
can be recovered from macrophage supernatants and can bind to an
anti-Fc affinity column[9]. The subfragment alone is sufficient to
stimulate B lymphocyte cultures in the absence of macrophages,
whereas untreated Fc requires macrophages. The macrophage enzyme
producing this subfragment and the exact origin of the Fc subfragment
in the IgG molecule remain to be determined. It may reside in the
$C\gamma 3$ domain since the pepsin fragment of IgG (pFc'), as well as a
fragment obtained by trypsin cleavage of Fc fragments, also representing the $C\gamma 3$ domain, are both able to induce DNA synthesis[10].

Recently, we observed that papain Fc fragments which elute as
a single protein peak by Sephadex gel filtration in phosphate buffer,
can be resolved into 28,000 to 40,000mol/wt (component I) and 14,000
mol/wt (component II) regions by gel filtration in 3M guanidine.
Both components are mitogenic, but component II (14,000mol/wt) is
active at 10-100 fold lower concentrations. This suggests that
this active component is already cleaved off during papain digestion,
but that it remains non-covalently bound and can be dissociated.
It has not been determined yet if this 14,000mol/wt component found
after papain digestion is identical to the 14,000mol/wt component
described by Morgan and Weigle[9] which can be isolated from macrophage
supernatants after incubation of macrophages with Fc fragments.
Different papain Fc preparations contained different ratios of component I [Fc(I)] to component II [Fc(II)]. Batches which were very
active in mitogenesis, contained up to twice the amount of component
II compared to component I on a weight basis. Over the past two
years several batches of Fc fragments were prepared which were found
to be nearly inactive, yet they appeared identical to active batches
on SDS-polyacrylamide gel electrophoresis and Ouchterlony analysis.
Gel filtration in 3M guanidine showed that they contained large
amounts of Fc(I). Attempts to produce more Fc(II) through prolonged
digestion with papain were not successful. However, by exposure to
pH4 for 1 hour at 37°C followed by neutralization, previously inactive papain Fc preparation were found to acquire significant activity (Table 1). The increase in activity was found to correlate
with an increase of Fc(II) in the pH4 treated preparations, but the

Table 1. Altered Mitogenicity of Fc After Exposure to pH4*

| | CPM ^{3}H-Thymidine Uptake (± S.E.) on Day 5† | |
	Control	Fc (50µg/ml)
untreated papain Fc	1,856 (± 125)	2,922 (± 17)
acid treated Fc	-	115,586 (± 1,178)

*Fc fragments were incubated in phosphate buffer acidified
with acetic acid for 1 hour at 37°C and subsequently neu-
tralized by dialysis against phosphate-buffered saline
(pH 7.0). Balb/c spleen cells (5 x 10^5/culture) were cultured in RPMI-1640 containing 0.5% autologous mouse serum
and 2-mercaptoethanol[1].
†Mean CPM from triplicate cultures.

direct causal relationship between the two phenomena has yet to be
clarified. It appears that incubation at pH4 allows further digestion, perhaps by a contaminating serum enzyme with an acidic pH
optimum (ie. cathepsin B, cathepsin D), or by small amounts of en-
zyme(s) remaining from the papain digestion. Alternatively, acid
treatment may produce a conformational change. Such changes induced
by exposure to acid pH have been reported for Fc fragments of human
IgD[11] for Fc of IgG[12] and for intact rabbit IgG[13,14]. Interestingly,
pH4 treatmetnt of gammaglobulin has also been used to reduce side
effects of gammaglobulin preparations used clinically for intravenous
administration. Altering or destroying the Fc portion of IgG by
enzyme treatment also reduced the reactogenicity of such preparations. However, the efficacy of such preparations is greatly
reduced, while acid treatment at pH4 does not reduce Fc-associated
effector functions[15,16].

     Acid treatment results in the appearance of a second precipitin
line in double diffusion against anti-Fc serum, while untreated Fc
fragments show a single line. Purified Fc(I) and Fc(II) subfragments
produce single precipitin lines. Fc(I) shows identity with untreated
papain Fc and with one of the lines in acid treated Fc, Fc(II) gives
a reaction of identity with the second (new) line in acid treated
Fc and with one of two lines in pFc', but it lacks a determinant
expressed on Fc(I). The Fc(II) subfragment also fails to bind to
protein A, while Fc(I) does. These findings are consistent with a
Cγ3 origin of Fc(II).

Murine spleen cells were found to respond to Fc fragments from mouse, human, goat and bovine IgG[1]. In contrast, Fc from rabbit IgG were not stimulatory and isolation of pepsin fragments (pFc') and pH4 exposure did not produce mitogenic subfragments. Rabbit Fc preparations all contained crystals and perhaps such crystalline Fc competes for a receptor but it is unable to provide the proliferative signal. Nevertheless, the finding that these pure Fc preparations are not mitogenic, emphasizes that some additional undefined factor plays a role in mitogenic stimulation. Papain or pepsin digestion of rabbit IgG may not expose the active site(s) of the Fc region. Alternatively, rabbit Fc fragments may lack the association of an additional, unknown component which bears the mitogenic activity, or they may contain a suppressive factor. Suggestive evidence for the latter possibility was obtained during acid treatment of certain nonactive human Fc preparations where a small portion of the material was found to precipitate. This aggregated material is highly suppressive in normal and mitogenic-stimulated spleen cell cultures, but has not been characterized further.

TARGET CELLS FOR THE MITOGENIC SIGNAL

In both man[17] and mouse[1] only B lymphocytes were found to respond to Fc fragments with proliferation. Macrophages are required for the stimulation with intact Fc fragments, but not with pFc' fragments or with the 14,000mol/wt Fc(II) component. The presence of T cells is required for the stimulation of polyclonal antibody synthesis but not for proliferation[1,18]. T lymphocytes do not themselves proliferate, but are stimulated to produce a helper factor[19,20] which increased the proliferative[21] and cytotoxic responses[22] to stimulation by allogenic cells.

In mouse spleen, Fc fragment-responsive cells are found mainly in the Fc receptor bearing (FcR+) B cell population. The FcR⁻ population (depleted of FcR+ cells by rosetting) showed low but varying proliferative responses to Fc fragments. The enriched FcR+ population failed to show a stronger proliferative response than unseparated spleen cells[23]. In spleens from antigen-stimulated animals, the response to Fc fragments was nearly always higher in the FcR⁻ population. These observations suggest that the mitogenic signal does not involve binding to an Fc receptor which is detectable by rosetting and binding techniques. An enhanced proliferative response to Fc is thus far unique to spleen cells enriched for complement receptor bearing (CR+) cells[23]. Likewise, responsiveness to Fc stimulation correlates with the appearance of splenic CR+ lymphocytes, not with the presence of FcR+ B cells. Studies on the ontogenic appearance of the complement receptor have shown the absence of CR+ B lymphocytes in newborn mouse spleen cells and their gradual appearance 1 to 3 weeks after birth[28,29]. At this time, partial Fc fragment stimulation (30-50%) can be observed[23]. FcR+ B cells are already present in the 17-day fetal liver and in newborn spleen[26], yet no Fc fragment response can be observed with these tissues. B cells can be separated into two subpopulations based on their expression of the differentiation antigen Lyb-5[27]. The Lyb-5 determinant is found on a late-appearing subpopulation of B-cells in normal mice but is virtually absent in mutant CBA/N mice[27,28]. Adult CBA/N mice (at 3 months of age) could not be triggered to proliferate with Fc fragments while they responded well to the B-cell mitogen LPS[23]. This finding also suggests that Fc fragments can only trigger late maturing B cells. It has been reported that Lyb-5[+], but not Lyn-5[-] B cells can respond to soluble T helper factors present in supernatants from concanavalin-A stimulated spleen cells[29]. This observation parallels the finding with Fc-induced stimulation, and may indicate that the Fc signal acts through a receptor for a growth or T helper factor rather than through an Fc receptor.

Fc FRAGMENT-MEDIATED ENHANCEMENT AND SUPPRESSION OF ANTIBODY

RESPONSES

In addition to being potent mitogens and polyclonal activators, Fc fragments also have the capacity to amplify antibody responses in vivo and in vitro[30,31]. Our experiments show that the time of addition of Fc fragments to sheep red blood cell (SRBC)-stimulated mouse spleen cell cultures is crucial. Enhancement of the plaque forming cell (PFC) response to SRBC (measured on day 5 of culture) is seen only when Fc fragments are added at the initiation of culture. With delayed addition (day 2 to 5), up to 95% suppression of the primary SRBC response has been consistently observed (Table I). Optimal suppression results from addition of Fc fragments on day 3 to primary cultures and on day 1 to 2 to secondary cultures. Proliferation of spleen cells in these SRBC-stimulated cultures was unaffected by Fc-mediated suppression. This suggests that Fc may interfere with antibody secretion or late differentiation of plasma cell precursors. In this respect Fc is clearly different from SIRS (soluble immune response suppressor) which inhibits B cell proliferation[32]. The kinetics of Fc-mediated suppression closely ressemble the suppression of the anti-SRBC response by soluble FcR described by Fridman et al.[33]. It is not clear, however, how the two phenomena of Fc and FcR mediated suppression are related.

The suppression by Fc fragments can be observed in T cell depleted, interleukin-2 (IL-2) and SRBC stimulated cultures (Table 2). It appears, therefore, that the suppression does not require the induction of suppressor T cells, although these may be induced when T cells are present. The Fc fragment may act directly on plasma cell precursors or on macrophages, or it may interfere with the IL-2 signal.

Table 2. Modulation of the Antibody Response by Fc Fragments

Spleen cells	SRBC	IL-2*	Fc (10μg/ml) Added (Day)	Direct Anti-SRBC PFC/culture (day 5)
Unseparated	−	−	−	13
Unseparated	+	−	−	350
Unseparated	+	−	+ (0)	1896
Unseparated	+	−	+ (3)	52
B cells	+	−	−	12
B cells	+	+	−	110
B cells	+	−	+ (0)	8
B cells	+	+	+ (0)	300
B cells	+	+	+ (3)	14

Balb/c spleen cells (10^6/well) were cultured in microtiter plates using RPMI containing 7.5% newborn calf serum. 1μl of a 1% SRBC suspension was added to each culture.
*IL-2 was provided by Dr. J. Watson[40].

The suppression of antibody responses by Fc fragments appears independent of the mitogenic signal associated with Fc fragments. This conclusion is based on the following observations:

(1) Preparations with low or no mitogenic activity were as capable of suppressing as mitogenic preparations,
(2) submitogenic doses of Fc fragments (1μg/ml or less) were sufficient to produce suppression, and
(3) no change in proliferation was observed in suppressed cultures.

The finding that nonmitogenic Fc fragments such as rabbit Fc and certain nonmitogenic batches of human Fc were suppressive, supports our hypothesis that mitogenesis and suppression depend on different molecular structures of the Fc region.

ROLE OF Fc FRAGMENTS IN VIVO

Numerous studies have described immunoregulatory effects of antigen-antibody complexes which depend on the Fc portion of IgG[31]. The stimulatory and suppressive effects of purified Fc fragments described here may represent effects that are normally produced by antigen-antibody complexes in vivo. Fc fragments may mimic a conformational signal, by presenting the same determinant(s), exposed by enzymatic digestion, which surface or secreted antibodies express as a consequence of antigen binding. Although we are studying fragments that are obtained by enzymatic cleavage in vitro, we would like to speculate that a similar proteolytic activation occurs in vivo. There is evidence that similar Fc fragments exist in vivo. IgG fragments which appear identical to Fc and Fc' fragments produced by papain have been found in normal human plasma and urine[34,35], suggesting that they are natural products of IgG catabolism. Fc fragments can also be produced with purified human plasmin[36] or with proteolytic enzymes from lysosomal fractions of bovine spleen[37]. In vivo, fragments could be produced in lysosomes and subsequently released, or they may be produced extracellularly by secreted proteases. Analogous to the mechanism described for mitogenesis by epidermal growth factor and for serine proteases[38,39], various Fc fragment and subfragment-mediated effects may then depend on the interaction with carrier proteins or factors, or on the interaction with different receptors produced by the target cells.

ACKNOWLEDGEMENTS

We are grateful to Dr. L. H. Perrin, Hospital Cantonal Geneva, and to Dr. G. A. Gutman, University of California, Irvine, for support during part of these studies.

REFERENCES

1. M. A. Berman and W. O. Weigle, B-lymphocyte activation by the Fc region of IgG, J. Exp. Med. 146:241 (1977).
2. E. L. Morgan and W. O. Weigle, Aggregated human gammaglobulin-induced proliferation and polyclonal activation of murine B lymphocytes, J. Immunol. 125:226 (1980).
3. E. L. Morgan and W. O. Weigle, Regulation of B lymphocyte activation by the Fc portion of immunoglobulin, J. Supramolecular Struct. 14:201 (1981).
4. E. L. Morgan and W. O. Weigle, Aggregated human gammaglobulin-induced proliferation and polyclonal activation of murine B lymphocytes, J. Immunol. 125:226 (1980).
5. B. Robert and R. S. Bockman, Studies on the proteolytic activity of gammaglobulin preparations, Biochem. J. 102:554 (1967).
6. J. S. Finlayson, Immune globulins, Sem. Thromb. Hemostasis 6:44 (1979).
7. S. Erhan and L. D. Greller, Do immunoglobulins have proteolytic activity?, Nature 251:353 (1974).
8. E. J. Victoria and L. C. Mahan, Proteolysis of red cell membrane proteins by immunoglobulin G preparations, Mol. Immunol. 18:699 (1981).
9. E. L. Morgan and W. O. Weigle, Regulation of Fc fragment induced murine spleen cell proliferation, J. Exp. Med. 151:1 (1980).

10. M. A. Berman, H. L. Spiegelberg and W. O. Weigle, lymphocyte stimulation with Fc fragments. I. Class, subclass, and domain of active fragments, J. Immunol. 122:89 (1979).

11. R. Jefferies, J. B. Mathews and P. M. Bayley, Studies of human IgG myeloma proteins. Conformational changes induced in the Fc delta fragment on heating or exposure to acid pH, Immunochemistry 15:19 (1978).

12. A. C. Ghose and B. Jirgensons, Conformational studies on the tryptic digestion fragments of human immunoglobulin G, Archs. Biochem. Biophys. 144:384 (1971).

13. D. A. Charwood and S. Utsumi, Conformation changes and dissociation of Fc fragments of rabbit immunoglobulin as a function of pH, Biochem. J. 112:357 (1969).

14. M. Vandenbranden, J. L. De Coen, R. Jeener, L. Kanarek and J. M. Ruyschaert, Interactions of gamma-immunoglobulins with lipid mono- or bilayers and liposomes. Existence of two conformations of gamma-immunoglobulins of different hydrophobicities, Mol. Immunol. 18:621 (1981).

15. S. Barandun, A. Morell and F. Skvaril, Clinical use of intravenous gamma-globulin, Biblthca haemat. 46:170 (1980).

16. S. Barandun, F. Skvaril and A. Morell, Prophylaxe and Therapie mit Gamma Globulin, Schweiz. med. Wschr. 106:533 (1976).

17. E. L. Morgan and W. O. Weigle, Polyclonal activation of human B lymphocytes by Fc fragments. I. Characterization of the cellular requirements for Fc fragment-mediated polyclonal antibody secretion by human peripheral blood B lymphocytes, J. Exp. Med. 154:778 (1981).

18. E. L. Morgan and W. O. Weigle, Polyclonal activation of human murine B-lymphocytes by Fc fragments. I. The requirement for two signals in the generation of the polyclonal antibody response induced by Fc fragments, J. Immunol. 124:1330 (1981).

19. M. L. Thoman, E. L. Morgan and W. O. Weigle, Polyclonal activation of murine B lymphocytes by Fc fragments. II. Replacement of T cells by a soluble helper T cell-replacing factor (TRF), J. Immunol. 125:1630 (1980).

20. M. L. Thoman, E. L. Morgan and W. O. Weigle, Fc fragment activation of T lymphocytes. I. Fc fragments trigger Lyt-1$^+$23$^-$ T lymphocytes to release a helper T cell-replacing factor, J. Immunol. 126:632 (1981).

21. E. L. Morgan, M. L. Thoman and W. O. Weigle, Enhancement of T lymphocyte functions by Fc fragments of immunoglobulins. I. Augmentations of allogeneic mixed lymphocyte culture reactions requires I-A or I-B subregion differences between effector and stimulator cell populations, J. Exp. Med. 153:1161 (1981).

22. E. L. Morgan, M. L. Thoman and W. O. Weigle, Enhancement of T lymphocyte functions by Fc fragments of immunoglobulin. II. Augmentation of the cell-mediated lympholysis response occurs through an Lyt-1$^+$2$^-$ helper T cell, J. Immunol. 127:2526 (1981).

23. M. A. Berman, E. L. Morgan and W. O. Weigle, Lymphocyte stimu-
    lation with Fc fragments. II. Requirements for mature B
    lymphocytes, Cell. Immunol. 52:341 (1980).
24. M. C. Gelfand, G. J. Elfenbein, M. M. Frank and W. E. Paul,
    Ontogeny of B lymphocytes. II. Relative rates of appearance
    of lymphocytes bearing surface immunoglobulin and complement
    receptors, J. Exp. Med. 139:1125 (1974).
25. M. C. Gelfand, D. H. Sachs, R. Lieberman and W. E. Paul, Ontogeny
    of B lymphocytes. III. H-2 linkage of a gene controlling
    the rate of appearance of complement receptor lymphocytes,
    J. Exp. Med. 139:1142 (1974).
26. I. Scher, A. Ahmed and S. O. Sharrow, Murine B lymphocyte hetero-
    geneity: distribution of complement receptor-bearing and
    minor lymphocyte-stimulating B lymphocytes among cells with
    different densities of total surface Ig and IgM, J. Immunol.
    119:1938 (1977).
27. A. Ahmed, I. Scher, S. O. Sharrow, A. H. Smith, W. E. Paul,
    D. H. Sachs and K. W. Sell, B lymphocyte heterogeneity:
    development of an alloantiserum which distinguishes B lympho-
    cyte differentiation alloantigens, J. Exp. Med. 145:101
    (1977).
28. A. Ahmed and I. Scher, Murine B cell heterogeneity defined by
    anti-Lyb5, an alloantiserum specific for a late-appearing B
    lymphocyte subpopulation, in:"B lymphocytes in the immune
    response," M. Cooper, D. E. Mosier, E. S. Vitetta and I. Scher
    eds., Elsevier-North Holland, Inc., New York, p 117 (1979).
29. A. Singer, J. Morrissey, S. Hathcock, A. Ahmed, I. Scher and
    R. J. Hodes, Role of the major histocompatibility complex
    in T cell activation of B cell subpopulations. Lyb-5+ and
    Lyb-5- B cell subpopulations differ in their requirement for
    major histocompatibility complex-restricted T cell recog-
    nition, J. Exp. Med. 154:501 (1981).
30. E. L. Morgan, S. M. Walker, M. L. Thoman and W. O. Weigle,
    Regulation of the immune response. I. The potentiation of
    in vivo and in vitro immune responses by Fc fragments, J. Exp.
    Med. 152:113 (1980).
31. W. O. Weigle and M. A. Berman, Role of the Fc portion of antibody
    in immune regulation, in:"Cells of immunoglobulin synthesis,"
    B. Pernis, H. J. Vogel, eds., Acad. Press, New York, 1:627
    (1980).
32. T. Tadakuma and C. W. Pierce, Mode of action of soluble immune
    response suppressor (SIRS) produced by Con A-activated spleen
    cells, J. Immunol. 120:481 (1978).
33. W. H. Fridman, C. Rabourdin-Combe, C. Neauport-Sautes and
    R. H. Gisler, Characterization and function of T cell Fc gamma
    receptor, Immunol. Rev. 56:51 (1981).
34. M. W. Turner and D. S. Rowe, A naturally occurring fragment re-
    lated to the heavy chains of immunoglobulin G in normal urine,
    Nature 210:130 (1966).

35. I. Berggard and H. Nennich, Fc fragment of immunoglobulin G in normal human plasma and urine, Nature 214:69 (1976).

36. F. Skvaril, L. Theilkas, M. Probst, A. Morell and S. Barandun, IgG subclasses IgG subclass composition and immunochemical characteristics of plasmin-treated human gammaglobulin, Vox Sang. 30:334 (1976).

37. J. Fehr, LoSpalluto and M. Ziff, Digestion of immunoglobulin G by lysosomal enzymes, Fedn. Proc. 28:496 (1969).

38. D. A. Low, J. B. Baker, W. C. Koonce and D. Cunningham, Released protease-nexin regulates cellular binding, internalization, and degradation of serine proteases, Proc. Natl. Acad. Sci. 78:2340 (1981).

39. D. J. Knauer and D. D. Cunningham, Epidermal growth factor carrier protein binds to cells via a complex formed with released carrier protein nexin, Proc. Natl. Acad. Sci. (in press).

40. J. Watson, S. Gillis, J. Marbrook, D. Mochizuki and K. A. Smith, Biochemical and biological characterization of lymphocyte regulatory molecules. I. Purification of a class of murine lymphokines, J. Exp. Med. 150:849 (1979).

MODULATION OF IMMUNOLOGICAL REACTIVITY BY THE Fc PIECE OF

IMMUNOGLOBULIN

William O. Weigle, Edward L. Morgan
Marilyn L. Thoman

Department of Immunopathology
Scripps Clinic and Research Foundation
La Jolla
California 92037

INTRODUCTION

Although the specific reactivities of antibodies are restricted to the variable regions located in the Fab portions of the molecule, the biological functions of these proteins are governed by the Fc region of the molecule. In addition to opsonization, complement fixation, placental transfer, metabolism of the intact molecule and anaphylaxis, the Fc region is responsible for immune regulation by antibody and antigen-antibody complexes (reviewed in Ref. 1). It is well documented that passive antibody can modulate the immune response to a subsequent injection of specific antigen[2-11]. This modulation may be in the form of either enhancement or suppression, depending on the antibody-antigen ratios. Pre-formed complexes injected directly into animals have also been reported to modulate the specific response to the antigen. Although extremely large amounts of antibody given either in the form of complexes or free antibody appears to suppress the immune response by masking antigenic determinants and sequestering them from the immune system[5,8], more physiological doses of antibody modulate the immune response through an active process in which the Fc fragment of the antibody is an absolute requirement[3,6].

ACTIVATION OF LYPHOCYTES BY Fc FRAGMENTS

Proliferation

The finding that isolated Fc fragments obtained by papain degradation of human IgG causes a proliferative response in murine B
lymphocytes offers an approach to further define the role of the Fc
region in antibody regulation of the immune response. The proliferative response could be induced by the Fc fragments of most classes
and subclasses of human immunoglobulins as well as by that of mouse
IgG[1,12]. More recently it has been shown that the proliferative
response of mouse B cells to the Fc fragment of human IgG requires
macrophages[13]. Depletion of macrophages from the spleen cell cultures abrogated the ability of the Fc fragment to stimulate these
cells to proliferate (Figure 1) and the reactivity could be constituted by adding back plastic adherent cells. Of special interest
is that the addition of Fc fragments to irradiated adherent cells
treated with anti-T cell sera and complement resulted in an active
supernatant which can stimulate macrophage-depleted B cells to proliferate[14]. The active supernatants are generated very shortly after
the addition of Fc fragments. Although this supernatant activity
obviously is not the result of carryover of Fc fragments, the reac-

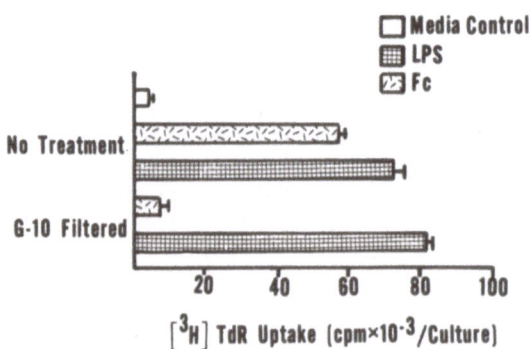

Fig. 1. The proliferative response of Sephadex G-10 filtered mouse
         spleen cells. Lipopolysaccharide (LPS) was used at a con-
         centration of 50 g and Fc at 250 g/ml. Reprinted from
         E. L. Morgan and W. O. Weigle, J. Exp. Med. 150:256 (1979).

Table 1.   The Ability of the Anti-Fc Affinity Column to Remove the
           Mitogenic Activity from the Adherent Cell Supernate

Supernatant Source	Treatment	cpm ± SE	% Reduction*
Fc adherent†	None	39,813 ± 822	–
Fc adherent	Anti-Fc§ (1x)	12,393 ± 548	69
	Anti-Fc (2x)	3,229 ± 781	92
Fc adherent	Anti-BSA‖ (1x)	37,166 ± 1,089	7
	Anti-BSA (2x)	44,119 ± 3,334	0

*(1-[affinity column filtered/nonfiltered]) x 100.
†The 14,000-mol wt fraction from the Sephadex G-50 chromatographic
 separation of Fc adherent cell supernate.
§1 x - filtered once, 2 x - filtered twice, 100µl were added to
 each culture well. The material was filtered theough an anti-Fc
 affinity column before use.
‖The material was filtered through an anti-BSA affinity column before use.
(Reprinted from E. L. Morgan and W. O. Weigle, J. Exp. Med. 151:1
 1980).

tivity can be removed by passing the supernatant over anti-Fc
columns, but not colums containing anti-bovine serum albumin (Table
1). Analysis of these supernatants by Sephadex G-50 chromatography
with appropriate markers indicates that the activity resides in a
14,000 molecular weight (m.w.) peptide (Figure 2) apparently cleaved
from the Fc fragments by macrophage enzymes. Other fragments, which
constitute most of the CH_3 domain, obtained from intact IgG by digestion with either plasmin or pepsin are able to stimulate mouse B
cells to proliferate in the absence of macrophages. Thus, it appears
that the Fc fragments may react with their Fc receptors on macrophages, where they are cleaved by macrophage enzymes to yield active
peptides which are capable of stimulating B cells to proliferate.
The Fc subfragments and/or the CH_3 domain of human IgG, apparently
no longer have reactivity to the macrophage Fc receptor since this
reactivity apparently requires both the CH_2 and CH_3 domains[15].

Polyclonal Activation

     In addition to B cell proliferation, the Fc fragments of human
IgG stimulate the B cells to polyclonally secrete immunoglobulin[16].
The addition of Fc fragments to mouse spleen cells results in a
polyclonal antibody response as evidenced by an increase in the

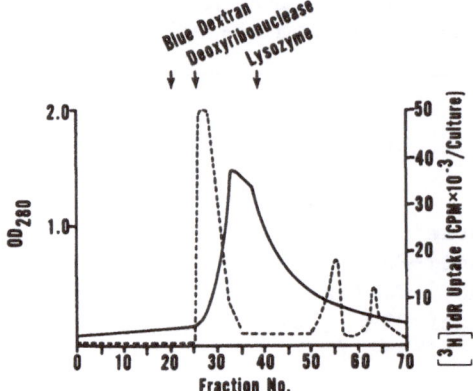

Fig. 2. Sephadex G-50 superfine chromatographic separation of Fc-
adherent cell supernate. Each column fraction was assayed
for protein content by O.D. and for mitogenic activity.
The molecular weight of the standards were: blue dextran
(50,000), deoxyribonuclease (30,000) and lysozyme (14,300).
O.D.280 (----), mitogenic activity(——). Reprinted from
E. L. Morgan and W. O. Weigle, J. Exp. Med. 151:1 (1980).

Table 2. The Ability of Fc Fragments to Induce a Polyclonal Antibody Response

Stimulator	Anti-TNP PCF/10⁶ Cultured Cells ± S.E.[a]	
	Expt. 1	Expt. 2
None	10 ± 2	22 ± 5
Fc fragment[b]	190 ± 15	182 ± 25
LPS[c]	479 ± 48	381 ± 78

[a]Direct PFC response.
[b]One hundred micrograms of Fc/culture.
[c]Twenty micrograms of LPS/culture.
(Reprinted from E. L. Morgan and W. O. Weigle, J. Immunol. 124:1330
1980).

number of plaque-forming cells (PFC) to trinitrophenyl hapten
(Table 2). In addition to macrophages, this polyclonal response
induced by Fc requires T cells. The T cell requirement for the response, however, can be replaced with a thymus-replacing factor
generated by activation of mouse T cells with Concanavalin A[17].
Furthermore, the addition of Fc fragments to mouse splenic T cells
results in the generation of a thymus-replacing factor (TRF) which
can also replace the T cells in the Fc-generated polyclonal response[18]. Thus, Fc-generated TRF [(Fc)TRF] when added to nude spleen
cells with Fc fragments results in a polyclonal response. However,
if macrophages are removed from the nude spleen cells, the addition
of Fc and (Fc)TRF does not cause a polyclonal response (Table 3).
On the other hand, the addition of the Fc subfragment (generated
by the addition of Fc fragments to macrophages) and (Fc)TRF causes
polyclonal responses in nude spleen cells depleted of macrophages.
This latter TRF is different than Interleukin-2 in that it has no
T cell growth factor or co-stimulator activity[19].

It is also of interest that Fc fragments can generate a polyclonal response in human peripheral blood lymphocytes (PBL) and, as
with the mouse system, this response requires both T cells and macrophages[20]. The dose of Fc fragment required for the polyclonal response in human PBL is approximately 1-2 logs less than that needed
for the polyclonal response in mouse spleen cells (Figure 3). The
ability to generate polyclonal responses in this latter system varies

Table 3. Effect of Adherent Cell Depletion on (Fc)TRF Production[a]

Spleen cells	Direct Anti-TNP PFC/10^6 ± SE
Untreated	110 ± 10
Single Sephadex G-10 filtration[b]	74 ± 17
Multiple Sephadex G-10 filtrations[b]	22 ± 10
Anti-Ia and C treated[c]	<10

[a]Various spleen cell populations were cultured 24 hr with 50 g/ml
Fc fragments. Culture supernates (10 g) were tested for the
ability to replace T cell function in Fc-induced polyclonal antibody responses.
[b]Spleen cells (50 x 100) were applied to 9ml columns of Sephadex
G-10. Approximately 50% were recovered and directly cultured
with Fc fragments, or refiltered.
[c]Spleen cells were treated with anti-Ia antiserum and C as described in Materials and Methods.
(Reprinted from M. L. Thoman, E. L. Morgan and W. O. Weigle, J.
Immunol. 126:632 (1981)).

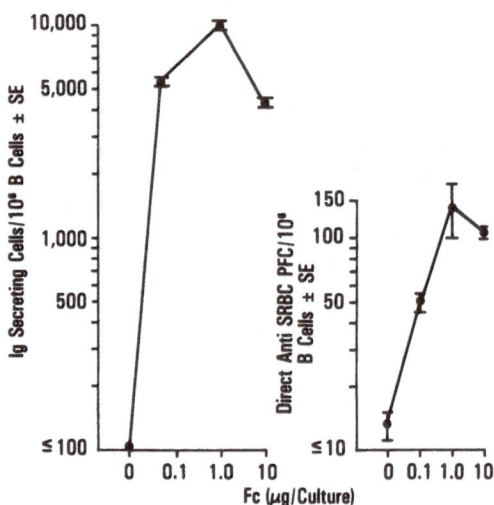

Fig. 3.   Dose response of Fc-fragment-mediated polyclonal antibody
          response by human PBL. Total Ig secretion was measured
          on day 6 from cultures that contain 1×10^5 B cells and
          2×10^5 T cells. The anti-SRBC response was measured on
          day 5 from cultures that contain 1×10^5 B cells and $1 \times$
          10^5 T cells. (Reprinted from E. L. Morgan and
          W. O. Weigle, J. Exp. Med. 154:778 (1981)).

considerably from individual to individual. Again as with the mouse
spleen cells, the active supernatant can be generated by adding
the Fc fragment to human peripheral blood monocytes (Figure 4).
The addition of Fc to adherent monocytes results in an active supernatant which, when added to PBL depleted of monocytes, results in
a polyclonal response. Analysis of the supernatant for polyclonal
activity by Sephadex G-75 chromatography with appropriate markers
demonstrated that the activity of the Fc fragment-monocyte supernatants resided in a subfragment with a molecular weight of 19,000,
a molecular weight similar to that found with mouse macrophages
and Fc fragments (Figure 5).

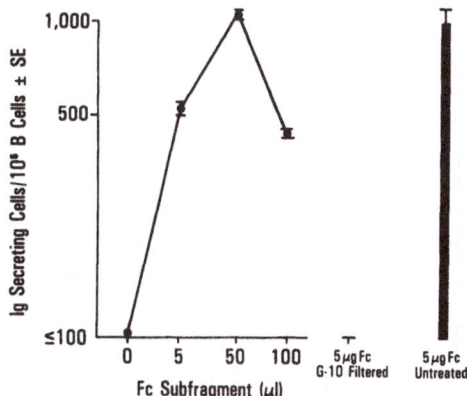

Fig. 4. Ability of human adherent monocytes to digest Fc fragments
         into polyclonal active Fc subfragments. Increasing amounts
         of Fc subfragments were added to monocyte-depleted (G-10
         filtered) cell populations, and the response was measured
         on day 6 of cultures that contained 1 x 10^5 B cells and
         1 x 10^5 T cells. (Reprinted from E. L. Morgan and
         W. O. Weigle, J. Exp. Med. 154:778 (1981)).

ACTIVATION OF LYMPHOCYTES BY COMPLEXED IMMUNOGLOBULIN

Aggregated Immunoglobulin

     It is of special importance that the generation of proliferative
and polyclonal responses is not limited to the Fc fragments of immunoglobulin. Aggregated preparations of human IgG are also capable
of causing proliferative and polyclonal responses in mouse spleen
cells[21]. Monoclonal IgG and grossly aggregated material (obtained
by centrifugation of IgG-heated at 63°C for 25 minutes) have neither
proliferative nor polyclonal activity. However, soluble heat-
aggregated IgG (obtained by precipitation of heat-aggregated IgG at
0.62M sodium sulfate Na_2SO_4) gives very good proliferative response
in mouse spleen cells (Table 4). The proliferative response with
this material also requires macrophages but not T cells, whereas
the polyclonal response requires both T cells and macrophages. Addition of the aggregated preparations to macrophages results in an
active supernatant and the activity can be removed by passage over

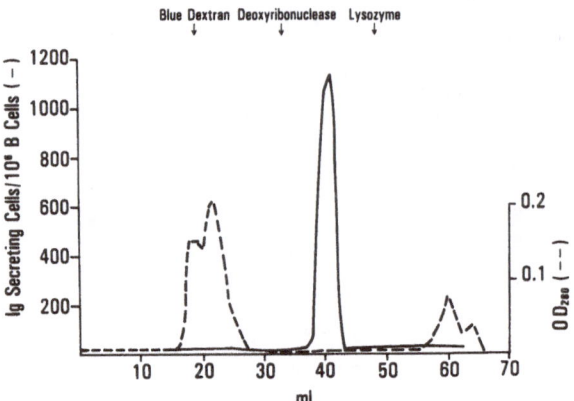

Fig. 5.   Sephadex G-75 chromatographic separation of Fc-adherent
          cell supernate. Responder cell populations were Sephadex
          G-10 filtered before culture, and the response was
          measured on day 6 of culture. (Reprinted from
          E. L. Morgan and W. O. Weigle, J. Exp. Med. 154:778
          (1981)).

Table 4.   The Ability of AHGG to Induce Proliferation and Poly-
           clonal Antibody Production

Stimulator	cpm ± S.E.[a]	Direct Anti-TNP PFC/ 10^6 Cultured Cells ± S.E.[a]
	2,792 ± 385	6 ± 2
AHGG[b]	31,002 ± 1,939	116 ± 8
Fc fragments[c]	45,555 ± 1,401	143 ± 6

[a]Proliferation and polyclonal antibody production were measured on
 day 3 of culture.
[b]100µg/culture heat AHGG.
[c]100µg/culture Fc fragments from HGG.
(Reprinted from E. L. Morgan and W. O. Weigle, J. Immunol. 125:226
 (1980).

an anti-Fc column. Furthermore, analysis of the supernatant by Sephadex chromatography reveals an active subunit of approximately 14,000m.w.

Antigen-Antibody Complexes

It was previously shown that the addition of antibody-antigen complexes, formed at certain ratios, cause a proliferative response in mouse spleen cells[1]. More recently, it was shown that the formation of complexes by adding a constant level of mouse anti-ovalbumin with increasing amounts of ovalbumin (OVA) to cultures of mouse spleen cells results in polyclonal responses at critical antigen-antibody ratios[22]. The polyclonal response by these complexes requires both T cells and macrophages. Thus, it appears that as with the Fc fragments, complexes formed with immunoglobulin either by heat aggregation or addition of antigen to specific antibody result in activation of both T and B cells. The mechanism involved in all three systems appear to be the same. Similar activation of either mouse or human lymphocytes does not occur with intact molecules of immunoglobulin or their Fab fragments. Thus, it appears that the Fc fragments and immunoglobulin in the form of either heat-aggregated IgG or immune complexes contain an enzyme attack site that is not present on free intact immunoglobulin. This enzyme attack site apparently allows degradation of the Fc region into subfragments that activate B cells to proliferate and T cells to generate a (Fc)TRF that is capable of causing the proliferating B cells to differentiate into antibody-secreting cells.

REGULATION OF IMMUNE REACTIVITIES BY Fc FRAGMENTS

Antibody

In addition to the nonspecific activation of T and B cells in the above fashion, Fc fragments of IgG have been shown to modulate a variety of immune reactivities that are antigen-specific. This modulation is usually seen as an enhancement of immune reactivity, especially in situations where suboptimal concentrations of the antigen are employed. The addition of Fc fragments of human IgG to the primary _in vivo_ and secondary _in vitro_ anti-sheep red blood cell (SRBC) response of mouse spleen cells is markedly enhanced by addition of Fc fragments of human IgG when suboptimal concentrations of SRBC are added (Figure 6)[23]. A similar enhancement has been observed in the _in vivo_ response with suboptimal concentrations of SRBC. It appears that enhancement occurs through an Lyt 1[+] helper T cell since the T cell-independent responses are not enhanced by Fc fragments. Furthermore, when T cell help is replaced by TRF, no enhancement is observed.

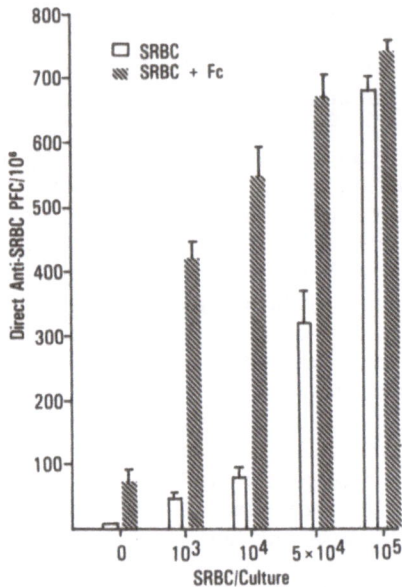

Fig. 6.   Enhancement of the secondary in vitro anti-SRBC response
          with Fc fragments. Increasing numbers of SRBC were added
          with 100μg (▨) or alone (□) to in vitro cultures.
          The response was measured on day 4. (Reprinted from
          E. L. Morgan et al., J. Exp. Med. 152:113 (1980).

Antigen-Specific T Cell Proliferation

     Further evidence that Fc fragments enhance immune reactivity
through Lyt 1[+] cell is shown by the enhancement of the antigen-
specific T cell proliferative response by Fc fragments[24]. The proliferating cell in this response has been shown to be an Lyt 1[+]
cell[25]. In the present experiments, mice were immunized at the base
of the tail with OVA in complete Freund's adjuvant. Cells obtained
from the draining lymph node were stimulated in vitro with OVA or
OVA and Fc fragments and the subsequent proliferative response
measured. As with the PFC response, marked enhancement of the T

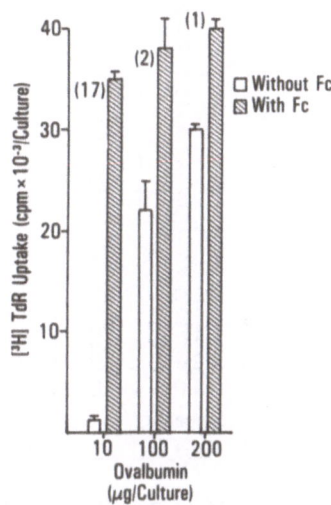

Fig. 7.   Nylon wool-purified lymph node-derived T cells were cul-
          tured with increasing amounts of OVA with or without 100μ
          Fc fragments/culture.   The response was measured on day 5
          of culture.   Numbers in parentheses indicate enhancement
          index.   (Reprinted from E. L. Morgan, M. L. Thoman and
          W. O. Weigle, J. Exp. Med. 153:1161 (1981)).

cell proliferative response was seen by adding Fc fragments with
suboptimal concentrations of OVA and only minimal enhancement was
seen with maximal concentrations of the antigen (Figure 7).   Removal
of Lyt 2[+] cells and Lyt 123[+] cells by anti-Lyt 2 antisera plus complement affected neither the proliferative response nor the ability
of Fc fragments to enhance that response, thus, reinforcing the assumption that the enhancement is occurring through an Lyt 1[+] cell.

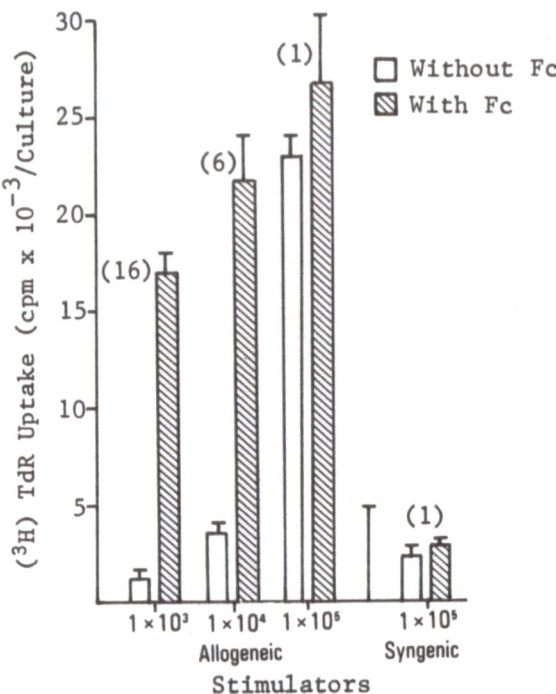

Fig. 8. Nylon-purified C5BL/6 splenic T cells were cultured with
         increasing numbers of irradiated CBA/CaJ spleen cells
         with or without 100μg Fc fragments/culture. The response
         was measured on day 4 of culture. (Reprinted from
         E. L. Morgan, M. L. Thoman and W. O. Weigle, J. Exp. Med.
         153:1161 (1981)).

Mixed Lymphocyte Reaction
─────────────────────────

    In another study, the effect of Fc fragments on the mixed
lymphocyte reaction (MLR) was determined using a one-way stimulation
assay where differences between the stimulator and responder cells
were throughout the entire H-2 region[25]. Again, marked enhancement
of the MLR was observed when Fc fragments were added with suboptimal
numbers of stimulator cells, but only minimal enhancement was observed with optimal numbers of stimulator cells (Figure 8). It
was of particular interest that the Fc fragment-mediated enhancement
of the MLR is restricted to those responses where the allogeneic
differences between stimulator and responder cells are in the I-A
and/or I-B subregions of the major histocompatibility locus. By
employing B10 congenic mice, it was observed that differences in

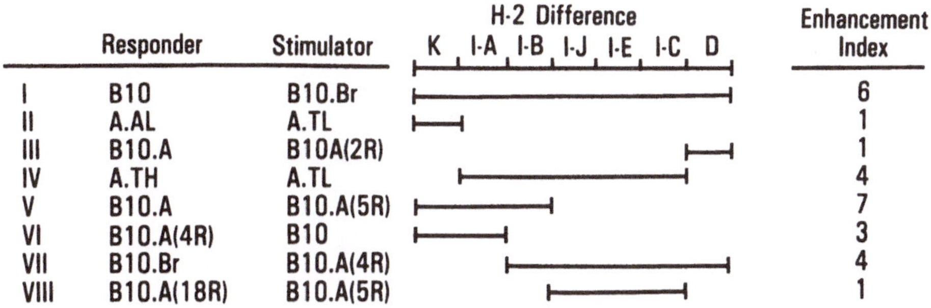

	Responder	Stimulator	H-2 Difference K I-A I-B I-J I-E I-C D	Enhancement Index
I	B10	B10.Br		6
II	A.AL	A.TL		1
III	B10.A	B10A(2R)		1
IV	A.TH	A.TL		4
V	B10.A	B10.A(5R)		7
VI	B10.A(4R)	B10		3
VII	B10.Br	B10.A(4R)		4
VIII	B10.A(18R)	B10.A(5R)		1

Fig. 9. Allogeneic MLR between mouse strains with limited H-2 differences. Suboptimal numbers (1 x 10^4) of stimulator cells were employed. When optimal E:S ratios were used, all strain combinations produced an MLR 20,000cpm. The response was measured on day 4 of culture. (Reprinted from E. L. Morgan, M. L. Thoman and W. O. Weigle, J. Exp. Med. 153:1161 (1981)).

I-A and/or I-B subregions were necessary for obtaining an enhanced MLR. It can be seen in Figure 9 that H-2K or H-2D region differences played no role in this Fc-mediated enhancement. Of special importance was the finding that the Lyt phenotype of the T cells responding to I-A and/or I-B region differences is Lyt 1$^+$23$^-$, whereas the non-enhanceable H-2D region MLR is mediated by Lyt 123$^+$ cells. The Lyt phenotype of the I-A and/or I-B region MLF effector cells was deduced from experiments where anti-Lyt 1 but not anti-Lyt 2 treatment eliminated the anti-I region MLR, whereas the phenotype of the H-2D region MLR effector cells was established from experiments where both anti-Lyt-1 and anti-Lyt-2 treatment eliminated the anti-H-2D MLR.

Cell-Mediated Lympholysis

It has also been shown that the Fc fragment of human IgG augments specific T cell-mediated cytolytic responses[26]. This Fc fragment-mediated enhancement is restricted to responses where the allogeneic differences between effector and stimulator cells encompass the H-2I region. Significant enhancement of cell-mediated lympholysis (CML) response occurs when differences in the effector and target cells were at H-2K+I+D. Fc-mediated enhancement of CML also occurred when differences between effector and target cells were limited to the H-2I region (Figure 10). In contrast, when differences in the strain combinations were only at the H-2D region, no enhancement was seen. As in the case of enhancement of other immune reactivities, the cellular site for the Fc enhancement of the CML response is the Lyt 1$^+$ helper T cell. Although Lyt 1$^+$

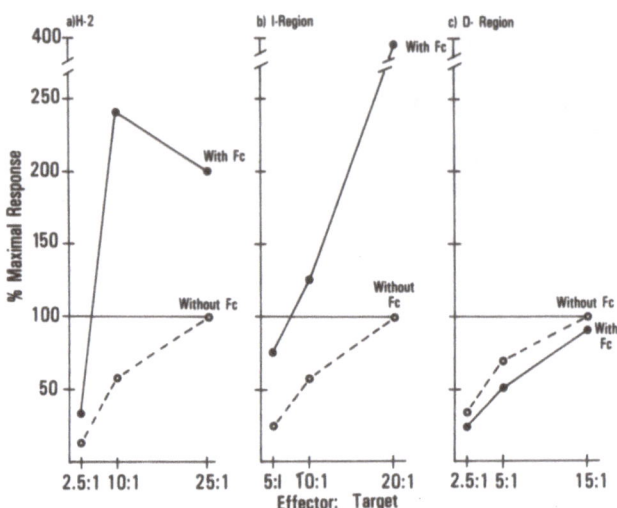

Fig. 10.   Comparison of the ability of Fc fragments to augment CML
           responses across (a) H-2 minor; (b) I-region; and (c) D-
           region differences.   100% maximal response is defined as
           the maximal response obtained in the absence of Fc frag-
           ments.   50μg/culture Fc was added to culture during the
           generation of cytotoxic response.   (Reprinted from
           E. L. Morgan, M. L. Thoman and W. O. Weigle, J. Immunol.
           (in press)).

helper T cells had to be present to obtain Fc-augmented CML responses, the substitution of Interleukin-2 for the Lyt 1[+] helper T
cells permitted a normal cytotoxic response.   However, this response
could not be enhanced by Fc fragments.   Furthermore, reaction between Fc fragments and the helper cell population alone was all
that was required for an enhanced cytotoxic response.

      Thus, it appears that in certain situations, at least Fc frag-
ments can substantially potentiate the immune reactivities involved.
They can markedly enhance both in vitro and in vivo humoral immune
responses when suboptimal concentrations of the antigens are employed.   Only T cell-dependent antibody responses are enhanced and

the enhancement appears to be via a T cell helper population. Similarly, Fc fragments augment T cell-mediated responses when suboptimal
concentrations of stimulating antigens are employed. With the MLR,
CML and the antigen-induced T cell proliferative responses, the enhancement appears to be through a T cell that bears the Lyt 1^+23^-
phenotype. Enhancement of the MLR and CML responses requires I-A
and/or I-B MHC sub-region differences between stimulator and responder
populations.

SUMMARY

Regions in the Fc piece of the immunoglobulin molecule appear
to be capable of modulating a variety of immune reactivities either
in the form of Fc fragments, heat-aggregated immunoglobulin or
antibody-antigen complexes. The ability of these agents to induce
proliferation requires macrophages and T cells. The role of the
macrophage appears to be to cleave active subfragments from the Fc
regions which are capable of activating B cells to proliferate and
T cells to secrete a (Fc)TRF that stimulates the proliferating B
cells to differentiate into antibody-secreting cells. The enhancement of the various immune reactivities by the Fc fragment apparently
results from activating the Lyt 1^+ population of T cells to release
an active factor, probably (Fc)TRF, which may act on both effector
T cells and proliferating B cells. The (Fc)TRF acts synergistically
with IL-2 in the enhancement of the in vitro antibody response.
This TRF may also act synergistically with IL-2 in the potentiation
of in vitro T cell reactivities.

It is far easier to understand how these interactions occur
in vitro, than to visualize the in vivo events involved in such putative regulation by antibody in the form of immune complexes. It
may follow from this model that antibody-antigen complexes, when
aggregated in critical ratios, readily fix to Fc receptors on macrophages and as the result of such aggregation, enzyme attack sites
are generated which allow macrophage enzymes to cleave active peptides from the Fc region of the antibody molecule. The active peptides, being in the microenvironment of the antigen-specific reactive
T and B cells, may stimulate the B cell to proliferate and the T
cell to release factors which synergize with IL-2 to potentiate both
effector T cell function and B cell differentiation. The specificity
of this enhancement would reside in the fact that the peptides and
resulting lymphokines would be diluted beyond effective concentrations when they diffuse only a short distance from this microenvironment. Furthermore, complement components and their active
peptides would also be released in this microenvironment and may
well participate in a similar immune regulation. Complement components including C3[27-29], C3b[30,31], C3a[32,33], C3c[34], C3d[34], and
C5a[35,36] have all been implicated in either enhancing or inhibiting
immune reactivity.

ACKNOWLEDGEMENTS

This is publication number 2626 from the Department of
Immunopathology, Scripps Clinic and Research Foundation, La Jolla,
California 92037. Supported in part by US Public Health Service
grants A107007, AG00783 and A115761 to William O. Weigle; CA30654
to Edward L. Morgan; and Biomedical Research Support Grant RPO-5514.
Edward L. Morgan is recipient of U.S.P.H.S. Research Career Develop-
ment Award CA00765. Marily L. Thoman is recipient of U.S.P.H.S.
National Research Service Award A106085.

The authors wish to thank Nancy Kantor for technical excellence,
and Alice Bruce for secretarial expertise.

REFERENCES

1. W. O. Weigle and M. A. Berman, Role of the Fc portion of antibody
    in immune regulation, in:"Cells of Immunoglobulin Synthesis,"
    H. Vogel and B. Pernis, eds., p 223, Academic Press, New York
    (1979).
2. J. W. Uhr and G. Möller , Regulatory effect of antibody on the
    immune response, Adv. Immunol. 8:81 (1968).
3. M. Hoffman and J. W. Kappler, Two distinct mechanisms of immune
    suppression by antibody, Nature 272:64 (1978).
4. N. R. St.C. Sinclair, R. K. Lees, G. Fagan and A. Birnbaum,
    Regulation of the immune response. VIII. Characterization
    of antibody-mediated suppression of an in vitro cell-mediated
    suppression of an in vitro cell-mediated immune response,
    Cell. Immunol. 16:330 (1975).
5. J.-C. Cerottini, P. J. McConahey and F. J. Dixon, Specificity
    of the immunosuppression caused by passive administration of
    antibody, J. Immunol. 103:268 (1969).
6. W. M. Watson and F. W. Fitch, Suppression of the antibody re-
    sponse to SRBC with F(ab)2 and IgG in vitro, J. Immunol.
    110:1427 (1973).
7. G. Terres, S. L. Morrison, G. S. Habicht and R. D. Stoner,
    Appearance of early "primed state" in mice following the con-
    comitant injections of antigen and specific antiserum, J.
    Immunol. 108:1473 (1972).
8. M. Feldman and E. Diener, Antibody-mediated suppression of the
    immune response in vitro, J. Immunol. 108:93 (1972).
9. C. S. Pincus, M. E. Lamm and V. Nussenzweig, Regulation of the
    immune response: Suppression and enhancing effects of pass-
    ively administered antibody, J. Exp. Med. 133:987 (1971).
10. J. G. Walker and G. W. Siskind, Studies on the control of anti-
    body synthesis. Effect of antibody affinity upon its ability
    to suppress antibody formation, Immunology 14:21 (1968).
11. G. G. B. Klaus, The generation of memory cells. II. Generation
    of B memory cells with preformed antigen-antibody complexes,
    Immunology 34:643 (1978).

12. M. A. Berman and W. O. Weigle, B-lymphocyte activation by the
    Fc region of IgG, J. Exp. Med. 146:241 (1977).
13. E. L. Morgan and W. O. Weigle, The requirement for adherent cells
    in the Fc fragment-induced proliferative response of murine
    spleen cells, J. Exp. Med. 150:256 (1979).
14. E. L. Morgan and W. O. Weigle, Regulation of Fc fragment-induced
    murine spleen cell proliferation, J. Exp. Med. 151:1 (1980).
15. N. Haeffner-Cavillon, K. J. Dorrington and M. J. Klein, Studies
    on the Fcγ receptor of the murine macrophage-like cell line
    P388D₁. II. Binding of human IgG subclass proteins and
    their proteolytic fragments, J. Immunol. 123:1914 (1979).
16. E. L. Morgan and W. O. Weigle, Polyclonal activation of murine
    B lymphocytes by Fc fragments. III. The requirements for
    two signals in the generation of the polyclonal antibody re-
    sponse induce by Fc fragments, J. Immunol. 124:1330 (1980).
17. M. L. Thoman, E. L. Morgan and W. O. Weigle, Polyclonal acti-
    vation of murine B lymphocytes by Fc fragments. II. Replace-
    ment of T cells by a soluble helper. T cell replacing factor
    (TRF), J. Immunol. 125:1630 (1980).
18. M. L. Thoman, E. L. Morgan and W. O. Weigly, Fc fragment acti-
    vation of T lymphocytes. I. Fc fragments trigger Lyt 1⁺,2,
    3⁻ T lymphocytes to release a helper T cell-replacing factor,
    J. Immunol. 126:632 (1981).
19. M. L. Thoman and W. O. Weigle, Preliminary chemical and biologi-
    cal characterization of (Fc)TRF: An Fc fragment-induced T
    cell-replacing factor, J. Immunol. 128:590 (1982).
20. E. L. Morgan and W. O. Weigle, Polyclonal activation of human
    lymphocytes by Fc fragments, J. Exp. Med. 154:778 (1981).
21. E. L. Morgan and W. O. Weigle, Aggregated human α globulin-
    induced proliferation and polyclonal activation of murine B
    lymphocytes, J. Immunology 125:226 (1980).
22. E. L. Morgan and W. O. Weigle, Regulation of B lymphocyte acti-
    vation by the Fc portion of immunoglobulin, J. Supramol.
    Struct. 14:201 (1980).
23. E. L. Morgan, M. L. Thoman, A. M. Walker and W. O. Weigle,
    Regulation of the immune response. I. Potentiation of in
    vivo and in vitro immune responses by Fc fragments, J. Exp.
    Med. 152:113 (1980).
24. E. L. Morgan, M. L. Thoman and W. O. Weigle, Enhancement of T
    lymphocyte functions by Fc fragments of immunoglobulins. I.
    Augmentation of allogeneic mixed lymphocyte culture reactions
    requires I-A or I-B subregion differences between effector
    and stimulator cell populations, J. Exp. Med. 153:1161 (1981).
25. G. Corradin, H. M. Etlinger and J. M. Chiller, Lymphocyte speci-
    ficity to protein antigens. I. Characterization of the
    antigen-induced in vitro cell-dependent proliferative response
    with lymph node cells from primed mice, J. Immunol. 119:1048
    (1977).
26. E. L. Morgan, M. L. Thoman and W. O. Weigle, Enhancement of T
    lymphocyte functions by Fc fragments of immunoglobulin. II.

Augmentation of the cell-mediated lympholysis response occurs through an Lyt $1^+,2^-$ helper T cell, J. Immunol. 127:2526 (1982).

27. M. B. Pepys, Role of complement in induction of antibody production in vivo. Effect of cobra factor and other <3-reactive agents on thymus-dependent and thymus-independent antibody response, J. Exp. Med. 140:126 (1974).

28. G. B. Klaus and J. H. Humphrey, The generation of memory cells. I. The role of C3 in the generation of B memory cells, J. Immunology 33:31 (1977).

29. C. G. Romball, R. J. Ulevitch and W. O. Weigle, Role of C3 in the regulation of a splenic PFC response in rabbits, J. Immunol. 124:151 (1980).

30. P. Dukor, F. M. Dietrich, R. H. Gisler, G. Schumann and D. Bitler-Suerman, Possible targets of complement action in B cell-triggering, Prog. Immunol. 3:99 (1974).

31. K.-U. Hartmann and V. A. Bakisch, Stimulation of murine B lymphocytes by isolated C3b, J. Exp. Med. 142:600 (1975).

32. B. W. Needlemen, J. M. Weiler and T. L. Feldbush, The third component of complement inhibits human lymphocyte blastogenesis, J. Immunol. 126:1586 (1981).

33. E. L. Morgan, W. O. Weigle and T. E. Hugli, Anaphylatoxin-mediated regulation of the immune response. I. C3a-mediated suppression of human and murine humoral immune responses, J. Exp. Med. (in press).

34. H. A. Schenkein and R. J. Ganco, Inhibition of lymphocyte blastogenesis by C3c and C3d, J. Immunol. 122:1126 (1979).

35. D. E. Chenoweth, M.G. Goodman and W. O. Weigle, Demonstration of a specific receptor for human C5a anaphylatoxin on murine macrophages, J. Exp. Med. (in press).

36. M. G. Goodman, D. E. Chenoweth and W. O. Weigle, Potentiation of the primary humoral immune response in vitro, J. Immunol. (in press).

T CELL REPLACING FACTOR (TRF)-INDUCED IgG PRODUCTION IN A HUMAN B

CELL LINE AND THE MECHANISM OF TRANSMEMBRANE SIGNALING THROUGH

TRF-ACCEPTORS

Tadamitsu Kishimoto, Yoshitsugu Miki
Hiroyuki Kishi, Atsushi Muraguchi, Kazuyuki Yoshizaki
and Yichi Yamamura

Division of Cellular Immunology
Institute for Molecular and Cellular Biology and
   Department of Medicine
Osaka University, Osaka Japan

INTRODUCTION

Binding of a given antigen with surface immunoglobulins (sIg) initiates a complex series of activation processes of B lymphocytes into immunoglobulin (Ig)-producing cells under the influence of helper T cells. In spite of intensive investigations of the triggering and regulation of immunocompetent cells, the subcellular biochemical mechanisms operative in those phenomena remain poorly understood. One of the central problems in analyzing such events is an extensive diversity of the cell types in the immune system as well as a complexity of interactions between them. The diversity is reflected not only in the large repertoire of antigenic specificities expressed by lymphocytes, but also in the many subsets of lymphocytes with distinct functions. In such situations, tumors derived from cells of the immune system of established cell lines have been used as models of clones of immune cells which can be subjected to biochemical and molecular analysis[1-5]. If those neoplastic cells or cell lines can be affected by external signals and activated into different states of differentiation, they especially prove useful for the molecular analysis of signaling and activation of lymphocytes.

In this study we have found a human B lymphoblastoid cell line which is capable of differentiation into IgG producing cells in the

presence of B cell specific differentiation factor (TRF, T cell replacing factor). By utilizing this cell line, we analysed the
mechanisms of the transduction of TRF-mediated signals into inside
the cells. The study shows the generation of the cytoplasmic
substance, which appears to be involved in the signal transmission
from membrane to nuclei, by the limited proteolysis of the precursor
protein with TRF-activated serine esterase.

MATERIALS AND METHODS

Cells: Ap Epstein-Barr virus-transformed human B blastoid cell
line, CESS, was a gift from Dr. P. Ralph (Sloan-Kettering Institute
for Cancer Research, Rye, N.Y.). Approximately 10% of cells expressed surface IgG and 0.5 to 1% of total cells secreted IgG when
measured by reverse plaque assay[6]. They did not bear any IgM or
IgD on their surface and none of them secreted any IgM. CESS cells
were fractionated into surface IgG-bearing (CESSγ[+]) and IgG-negative
(CESSγ[−]) populations by adherence to anti-γ-coated dishes according
to the method of Wysocki and Sato[7] and CESSγ[+] cells were employed
for the experiments.

Reagents: PHA-P (lot 3110-56) was obtained from Difco (Detroit,
MI). Diisopropylfluorophosphate (DFP),phenylmethylsulfonylfluoride
(PMSF) were purchased from Sigma Chemical Co. (St. Louis, MO.).
Diisopropylmethylphosphate (DMP), p-nitrophenylethylpenthylphospho-
nate (p-NPEPP), and phenylethylpenthylphosphonate (PEPP) were gifts
from Dr. E. L. Becker (University of Connecticut, Farmington,
Connecticut).

TRF: T cell-derived soluble factor(s) were prepared by the
method of Gillis et al.[8]. Human palatine tonsils were obtained at
tonsillectomy from patients with chronic tonsillitis and were separated into lymphocytes as described previously[9]. Cells (1×10^6/ml)
were stimulated with 0.1% PHA. After 48 hr incubation, cell-free
supernatants were recovered, concentrated 10 times with Amicon (YM-5)
and fractionated on Sephadex G-100. Fractions with the molecular
weight between 15,000 to 20,000 were employed as TRF. In some
experiments, further purification by chromatofocusing was carried
out and TRF activity was isolated from IL-2 or B cell growth factor
(BCGF) activity as described[10].

Measurement of IgG producing cells: A reverse plaque assay[11]
was employed by using protein A-coated SRBC and affinity-purified
anti-γ antibody.

Red cell-mediated microinjection: For the introduction of the
cytoplasm into human red blood cells (HRBC), 1ml cytoplasm, which
had been obtained from 1×10^7 TRF-stimulated cells and dialysed
against reverse phosphate buffered saline, rPBS (137mM KCl, 2.7mM
NaCl, 8.1mM Na_2HPO_4, 1.5mM KH_2PO_4, 4mM $MgCl_2$, pH 7.2), was mixed

with 0.1ml packed volume of HRBC and dialysed against a sufficient volume of 10 fold diluted rPBS with stirring for 30 minutes at 4°C. HRBC were resealed by dialysing against isotonic rPBS for 30 minutes and then, the HRBC ghosts containing the cytoplasm were washed with PBS. Fusion of HRBC ghosts and CESS cells was performed according to the method of Furusawa et al.[12] by using HVJ. Efficiency of the injection was examined by the introduction of FITC-conjugated avidin and 80% of cells could be injected by this method. The injected CESS cells were cultured for 40 h and IgG producing cells were enumerated by reverse plaque assay.

RESULTS AND DISCUSSION

    IgG induction in CESS cells with TRF: Ten thousand cells in 0.2ml medium were incubated with varying concentrations of TRF and the numbers of IgG producing cells were measured at 48 hr by reverse plaque assay. As shown in Figure 1, an increase of IgG producing cells was proportional to the concentration of TRF and 10 units of

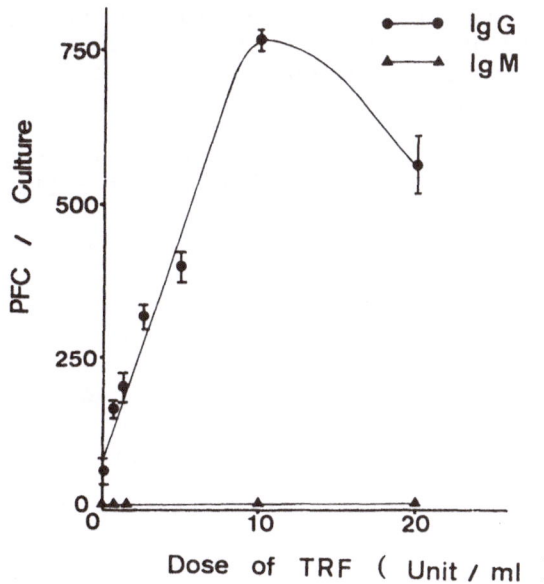

Fig. 1. Dose response of TRF in the induction of IgG in CESS cells. Cells (1 x 10^4/0.2ml) were cultured in the presence of varying concentrations of TRF for 48 hr and IgG (●——●) and IgM (▲——▲) producing cells were assessed by reverse plaque assay.

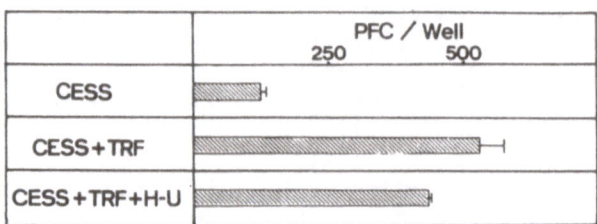

Fig. 2. Effect of hydroxyurea (HU, 0.1mM) on TRF-induced IgG
         production in CESS cells.

TRF induced the maximum increase of IgG producing cells. An increase of IgG producing cells was observed at 24 hr and reached to
its maximum level at 48 hr. The addition of 0.1mM hydroxyurea (HU),
which completely blocked the proliferation of CESS cells, did not
show any inhibitory effect on TRF-induced IgG production in the
cells (Figure 2). Northern blotting analysis with the genomic DNA
containing human γ_1 coding sequence as a probe showed a TRF-induced
increase of mRNA for γ_1 chain (data not shown). These results showed
that CESS cells were capable of differentiation into IgG producing
cells in the presence of TRF without any requirement of cell division. Thus, the cells may represent the final differentiation stage
of B cells.

   Expression of TRF-acceptors on CESS cells: In order to study
whether acceptors for TRF were expressed on CESS cells, absorption

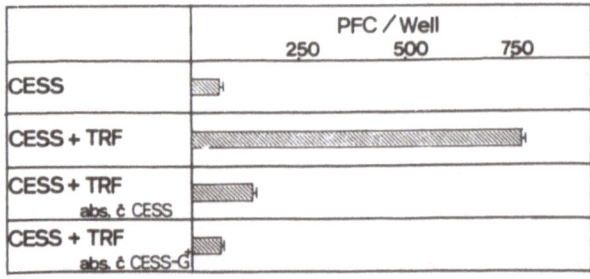

Fig. 3. Absorption of TRF activity with CESS cells. 10 units of
         TRF were absorbed with 1×10^7 cells for 2 hr at $4°C$ and
         absorptions were repeated three times. After absorption,
         TRF activity to induce IgG in CESS cells was examined.

experiments were carried out. Ten units of TRF were incubated with 1×10^7 cells for 2 hr at 4°C and absorption was repeated 2 times. After absorption, the activity of TRF to induce IgG in CESS cells was tested. As shown in Figure 3, TRF which had been adsorbed with CESS cells, did not induce IgG in CESS cells. As a control experiment, absorption of TRF with an IL-2-dependent human cytotoxic T cell line (CTL) was carried out and CTL cells did not absorb any TRF activity. In a marked contrast, CESS cells did not absorb any IL-2 activity, showing that TRF and IL-2 were distinct molecules and CESS cells expressed acceptors for TRF but not for IL-2. Furthermore, CESS cells did not absorb B cell growth factor (BCGF) activity when tested by employing anti-μ-stimulated human B cells. The result showed that CESS cells did not express BCGF acceptors and suggested that the expression of BCGF- and TRF-acceptors was dependent on the activation stage of B cells.

Involvement of stimulus-activatable serine esterase in the processes of the transmission of TRF-mediated signals: Addition of 1×10^{-3} M DFP with TRF inhibited the TRF-induced IgG production in CESS cells and inhibitory effect was proportional to the concentration of DFP between 10^{-5} to 10^{-3} M as shown in Table 1. Nonphosphorylating analogue of DFP, DMP, did not show any inhibitory effect, whereas the other organophosphorous inhibitors, such as p-NPEPP inhibited TRF-induced IgG production in CESS cells (Table 2). These results indicated the involvement of serine esterase in the transmission of TRF-mediated signals. Pre-incubation of CESS cells with DFP in the absence of TRF did not affect TRF-induced IgG production in CESS cells, showing the involvement of stimulus-activatable serine esterase in the activation processes of CESS cells.

Involvement of stimulus-activatable serine esterase was further confirmed by employing synchronized CESS cells. CESS cells were synchronized by thymidine block and synchronized cells were pulse-stimulated with TRF for the study when TRF signals could be delivered into inside the cells. IgG production in CESS cells was observed only when TRF signals were provided to the cells at the G_1 phase (Table 3). The addition of DFP with TRF to the cells at the G_1 phase blocked TRF-induced IgG production but DFP did not show any inhibitory effect when it was added to the cells which were at S-G_2-M phase. These results clearly showed the involvement of stimulus-activatable serine esterase in the processes of TRF-mediated activation of B cells.

Not only serine esterase activation but also methylation of membrane phospholipids was induced by stimulation of CESS cells with TRF. Two dimensional thin-layer chromatography showed an increase of ^{3}H-methyl incorporation into phosphatidylcholine. Several inhibitors of transmethylation, such as 5'-S-isobutyl-5'-deoxyadenosine (SIBA) or 2-hydroxyethylhydrazine (HEH), could block TRF-induced IgG production as well as serine esterase activation. On the other hand,

DFP, serine esterase inhibitor, did not block TRF-induced methylation of phospholipids. As schematically summarized in Figure 4, the result shows that TRF induces the methylation of phospholipids which sets the stage for the activation of serine esterase. Then, TRF-activated serine esterase may be involved in limited proteolysis of precursor proteins into the active substance which is responsible for the signal transmission from membrane to nuclei.

Table 1. Effect of a Serine Esterase Inhibitor, DFP, on TRF-Induced IgG Production in CESS Cells

TRF-stimulation[a]	Inhibitor (DFP)	IgG PFC/culture
−	−	60 ± 5
+	−	204 ± 2
+	1×10^{-3} M	38 ± 1
+	1×10^{-4} M	75 ± 7
+	1×10^{-5} M	107 ± 4

[a] 1×10^{4} cells/0.2ml were cultured for 48 hr in the presence or absence of 10 units/ml TRF.

Table 2. Effect of Several Organophosphorous Inhibitors or DFP Analogues on TRF-Induced IgG Production in CESS Cells

TRF-stimulation[a]	Inhibitors	IgG PFC/culture
−	−	127 ± 40
+	−	512 ± 40
+	DFP 1×10^{-3} M	110 ± 2
+	DMP 1×10^{-3} M	469 ± 13
+	p-NPEPP 1×10^{-4} M	116 ± 14
+	PEPP 1×10^{-4} M	228 ± 15

[a] 1×10 cells/0.2ml were cultured for 48 hr in the presence or absence of 10 units/ml TRF. Organophosphorous inhibitors were present through the whole culture period.

Table 3.   Effect of DFP on TRF-induced IgG Pro-
           duction in the Synchronized CESS Cells

Cell cycle		IgG PFC[b]/culture
$S-G_2-M$	G_1	
–	–	60 ± 10
TRF	–	50 ± 10
–	TRF	142 ± 12
–	TRF + DFP[a]	44 ± 2
TRF	TRF	156 ± 36
TRF + DFP	TRF	174 ± 14

[a]The concentration of DPF and TRF were 1×10^{-3} M
and 10 units/ml.

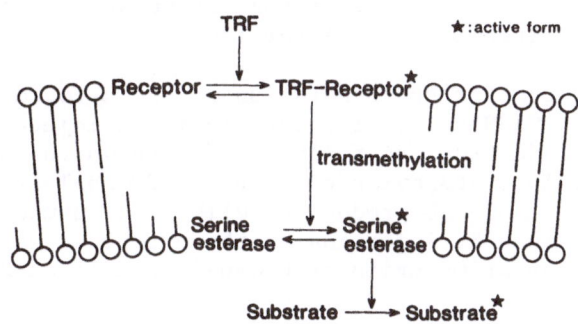

Fig. 4.   Schematic models of transmembrane signaling through TRF-
          acceptors.

Demonstration of the cytoplasmic factor responsible for the
signal transmission: In order to demonstrate that the substance,
which may be involved in the signal transmission, is generated by
TRF-stimulation as depicted in Figure 4, red cell-mediated microinjection method has been employed. CESS cells were incubated
with 10 units/ml of TRF for 8 hrs. After activation with TRF, cells
were washed, disrupted and cytoplasmic fractions free from membranes,
microsomal fractions and nuclei were obtained by ultracentrifugation.
The cytoplasmic fraction thus obtained was introduced into ghosts
of horse red blood cells (HRBC), which were then fused with non-
stimulated CESS cells and an increase of IgG-producing cells was
followed.

As described in Table 4, injection of the cytoplasmic factor from TRF-stimulated cells into non-stimulated cells induced a significant increase of IgG producing cells with the comparison of the cells injected with cytoplasmic factor from non-stimulated cells. Injection of TRF itself did not induce any increase of IgG producing cells. These results suggest that the factor(s), which are capable of inducing IgG production in CESS cells, are generated in the cells stimulated with TRF. Dose response relationship between the concentrations of the cytoplasmic factor(s) and increase of IgG producing cells in the cells injected with the cytoplasmic factors was observed. The result also supported the notion that TRF-stimulation induced the cytoplasmic factor(s) responsible for the signal transmission from membrane to nuclei. In order to exclude the possibility that the cytoplasmic factor(s) from TRF-stimulated cells included IgG or m-RNA specific for IgG and they were responsible for IgG induction in non-stimulated cells, absorption of the cytoplasmic factor(s) with anti-IgG conjugated column or treatment of the factor(s) with insoluble RNase was carried out. As shown in Table 5, neither absorption with anti-IgG nor treatment with RNase abrogated the activity of the factor(s) to induce an increase of IgG producing cells in non-stimulated cells, showing that the active substance in the cytoplasm was not IgG itself or m-RNA for IgG.

In order to confirm the generation of the cytoplasmic factor(s) in TRF-stimulated CESS cells, the cytoplasm was injected into normal B cells and Igs-induction was followed. As shown in Table 6, injection of the cytoplasm into Staphylococcus aureus Cowan-I-stimulated B cells induced an increase of IgG producing cells. The number of IgG producing cells obtained by injection of the cytoplasm was comparable to that observed by stimulation of Cowan-I-stimulated B cells with TRF.

Table 4. Induction of IgG in CESS Cells by Injection of the Cytoplasmic Factor(s) Derived from TRF-Stimulated Cells

Injected cytoplasm	IgG PFC/10^4 cells[a]
-	62 ± 2
from non-stimulated cells	76 ± 8
from TRF-stimulated cells[b]	180 ± 13
TRF	60 ± 8

[a]IgG producing cells were enumerated 40 hr after injection.
[b]Cytoplasm was obtained 8 hr after TRF stimulation.

Table 5. Exclusion of the Possibility that the Active Cytoplasmic Substances were IgG or m-RNA for IgG.

Treatment of the cytoplasmic factor(s)[a]	IgG PFC/10^4 cells
Without factor(s)	61 ± 24
Non-treated factor(s)	245 ± 18
Treatment with RNase	225 ± 18
Absorption with anti-IgG	223 ± 42

[a]After treatment, the factor(s) were injected into non-stimulated cells.

Table 6. IgG Induction in Cowan-I-stimulated normal B cells by injection of cytoplasm from TRF-stimulated CESS cells

Treatment of Cowan-I-stimulated B cells	IgG PFC/culture
-	56 ± 13
Injection of the cytoplasm	
from TRF-stimulated CESS cells	647 ± 23
from NON-stimulated CESS cells	159 ± 16
from a T cell line (CEM)	45 ± 6
Stimulation with TRF	706 ± 52

As previously described, involvement of stimulus-activatable serine esterase in the transmission of TRF-mediated signals has been suggested. Thus, we studied whether TRF-activated serine esterase was responsible for the generation of the cytoplasmic factor(s). Incubation of the cells with TRF in the presence of DFP did not generate the cytoplasmic factor(s). In contrast, the presence of actinomycin D did not inhibit the generation of the cytoplasmic factor(s). Taken collectively, these results suggest that the active substance(s) involved in the transmission of TRF-mediated signals are generated by limited proteolysis of the precursor proteins in resting cells with TRF-activated serine esterase.

Similar result was observed in our previous experiments with anti-Ig-stimulated rabbit lymphocytes. In a series of experiments[13-16], we have demonstrated the genration of the cytoplasmic factor(s) which transmitted the signals given through Ig-receptors in rabbit B cells, ie. (i) anti-Ig-stimulation induced an increase of the phosphorylation of non-histone nuclear proteins (NHP) in nuclei[14], (ii) incubation of non-stimulated nuclei with the cytoplasm from anti-Ig-stimulated cells induced an increase of NHP-specific protein kinase activity in non-stimulated nuclei[15], (iii) incubation of the cytoplasm from non-stimulated cells with the membrane fraction from anti-Ig-stimulated cells induced active cytoplasmic factor(s) responsible for the activation of NHP-specific protein kinase in nuclei[16]. All of those experiments observed in anti-Ig-stimulated rabbit B cells and TRF-stimulated human B cell line suggested the generation of the cytoplasmic factor(s) involved in the epigenetic regulation of the activation of gene expression in B cells.

Several studies have suggested the presence of the cytoplasmic factors which are responsible for the activation or extinction of gene expression of eukaryotic cells, such as, (i) induction of liver specific enzymes in erythroleukemic cells[17] or in fibroblast[18] by fusion with cytoplasts of rat hepatoma cells, (ii) fusion of cytoplasts derived from growth factor-stimulated 3T3 cells with non-stimulated 3T3 cells induced replication of those cells[19], (iii) extinction of hemoglobin synthesis in erythroleukemia cells[20] or albumin synthesis in liver cells[21] by fusion with enucleated cytoplasts of fibroblast cells. In those experiments, however, enucleated cytoplasts were employed and the presence of the active substances in cytoplasm was not directly demostrated. Thus, the present experiment is the first successful demonstration of the soluble cytoplasmic factor(s) involved in the activation of gene expression. As red cell-mediated microinjection is a simple and reproducible method, chemical characterizations of the factors involved in the signal transmission will be attained by employing this experimental system.

SUMMARY

IgG secretion was induced in a human B blastoid cell line, CESS, by the addition of partially purified T cell-derived helper factor(s) (TRF). Cell proliferation was not required for TRF-induced IgG induction in CESS cells. Absorption experiments demonstrated the presence of acceptors for TRF but not for IL-2 or BCGF on the surface of CESS cells. Involvement of stimulus-activatable serine esterase in the transmission of TRF-mediated signals into inside cells was demonstrated by employing several organophosphorous inhibitors and synchronized cells. The involvement of TRF-induced methylation of phospholipids was also demonstrated and the result suggested that phospholipid methylation set the stage for the activation of serine esterase. Red cell-mediated microinjection was employed for the demonstration of the cytoplasmic substance(s), which were generated in TRF-stimulated cells and might be involved in the transmission of membrane-mediated signals into inside the cells. Thus, injection of the cytoplasm obtained from TRF-stimulated cells induced IgG induction in non-stimulated cells. The generation of the active substance was blocked by DFP but not by actinomycin D, suggesting the involvement of the limited proteolysis of the precursor molecules into the active substances by TRF-activated serine esterase.

REFERENCES

1. T. Kishimoto, T. Hirano, T. Kuritani, Y. Yamamura, P. Ralph and R. A. Good, Induction of IgG production in human B lympho-blastoid cell lines with normal human T cells, Nature 271:756 (1978).
2. S. M. Fu, N. Chiorazzi, H. G. Kunkel, J. P. Halper and S. R. Harris, Induction of in vitro differentiation and im-munoglobulin synthesis of human leukemic B lymphocytes, J. Exp. Med. 148:1570 (1978).
3. S. Strober, E. S. Gronovicz, M. R. Knapp, S. Slarin, E. S. Vitetta, R. A. Warnke, B. Kotzin and J. Schröder, Im-munobiology of a spontaneous murine B cell leukemia (BCL₁), Immunol. Rev. 48:169 (1979).
4. O. Saiki, T. Kishimoto, T. Kuritani, A. Muraguchi and Y. Yamamura, In vitro induction of IgM-secretion and switching of IgG-production in human B leukemic cells with the help of T. cells, J. Immunol. 124:2609 (1980).
5. N. Sakaguchi, T. Kishimoto, H. Kikutani, T. Watanabe, N. Yoshida, A. Shimizu, Y. Yamawaki-Kataoka, T. Honjo and Y. Yamamura, Induction and regulation of immunoglobulin expression in a murine pre-B cell line, 70Z/3. I. Cell cycle-associated induction of sIgM expression and κ-chain synthesis in 70Z/3 cells by LPS stimulation, J. Immunol. 125:2654 (1980).
6. A. Maraguchi, T. Kishimoto, Y. Miki, T. Kuritani, T. Kaieda, K. Yoshizaki and Yuichi Yamamura, T cell-replacing factor-(TRF) induced IgG secretion in a human B blastoid cell line and demonstration of acceptors for TRF, J. Immunol. 127:412 (1981).
7. L. J. Wysocki and V. L. Sato, "Panning" for lymphocytes: a method for cell selection, Proc. Natl. Acad. Sci. USA 75:2844 (1978).
8. S. Gillis, K. A. Smith and J. Watson, Biochemical characterization of lymphocyte regulatory molecules. II. Purification of a class of rat and human lymphokines, J. Immunol. 124:1954 (1980).
9. A. Muraguchi, T. Kishimoto, T. Kuritani, T. Watanabe and Y. Yamamura, In vitro immune response of human peripheral lymphocytes. V. PHA- and protein A-induced human B colony formation and analysis of subpopulations of B cells, J. Immunol. 125:564 (1980).
10. K. Yoshizaki, T. Nakagawa, T. Kaieda, A. Muraguchi, Y. Yamamura and T. Kishimoto, Induction of proliferation and Igs-production in human B leukemic cells by anti-immunoglobulins and T cell factors, J. Immunol. 128:1296 (1982).

11. E. Gronovicz, A. Coutinho and F. Melchers, A plaque assay for all cells secreting Ig and a given type or class, Eur. J. Immunol. 6:588 (1976).

12. M. Furusawa, T. Nishimura, M. Yamaizumi and Y. Okada, Injection of foreign substances into single cells by cell fusion, Nature 249:449 (1974).

13. T. Kishimoto, T. Miyake, Y. Nishizawa, T. Watanabe and Y. Yamamura, Triggering mechanism of B-lymphocytes. I. Effect of anti-immunoglobulin and enhancing soluble factor on differentiation and proliferation of B-cells, J. Immunol. 115: 1179 (1975).

14. T. Kishimoto, Y. Nishizawa, H. Kikutani and Y. Yamamura, Biphasic effect of cyclic AMP on IgG production and on the changes of non-histone nuclear proteins induced with anti-immunoglobulin and enhancing soluble factor, J. Immunol. 118:2027 (1977).

15. Y. Nishizawa, T. Kishimoto, H. Kikutani and Y. Yamamura, Induction and properties of cytoplasmic factor(s) which enhance nuclear nonhistone protein phosphorylation in lymphocytes stimulated by anti-Ig, J. Exp. Med. 146:653 (1977).

16. T. Kishimoto, H. Kikutani, Y. Nishizawa, N. Sakaguchi and Y. Yamamura, Involvement of anti-Ig-activated serine protease in the generation of cytoplasmic factor(s) that are responsible for the transmission of Ig-receptor-mediated signals, J. Immunol. 123:1504 (1979).

17. T. V. Gopalakrishnan and W. F. Anderson, Epigenetic activation of phenylalanine hydroxylase in mouse erythroleukemia cells by the cytoplast of rat hepatoma cells, Proc. Natl. Acad. Sci. USA 76:3932 (1979).

18. L. A. Lipsich, J. R. Kates and J. J. Lucas, Expression of a liver-specific function by mouse fibroblast nuclei transplanted into rat hepatoma cytoplasts, Nature 281:74 (1979).

19. J. C. Smith and C. D. Stiles, Cytoplasmic transfer of the mitogenic response to platelet-derived growth factor, Proc. Natl. Acad. Sci. USA 78:4363 (1981).

20. T. V. Gopalakrishnan, E. B. Thompson and W. F. Anderson, Extinction of hemoglobin inducibility in Friend erythroleukemia cells by fusion with cytoplasm of enucleated mouse neuroblastoma or fibroblast cells, Proc. Natl. Acad. Sci. USA 74:1642 (1977).

21. C. R. Kahn, R. Bertolotti, M. Ninio and M. C. Weiss, Short-lived cytoplasmic regulators of gene expression in cell cybrids, Nature 290:717 (1981).